SPRINGER
LABORATORY

C. Kessler (Ed.)

Nonradioactive Labeling and Detection of Biomolecules

With 58 Figures and 22 Tables

Springer-Verlag
Berlin Heidelberg New York
London Paris Tokyo
Hong Kong Barcelona
Budapest

Priv.-Doz. Dr. Christoph Kessler
Boehringer Mannheim GmbH, Forschungszentrum Penzberg,
Nonnenwald 2, W-8122 Penzberg, FRG

Ludwig-Maximilians-Universität München,
Institut für Biochemie, Karlstr. 21,
W-8000 München 2, FRG

ISBN 3-540-55482-3 Springer Verlag Berlin Heidelberg New York
ISBN 0-387-55482-3 Springer-Verlag New York Berlin Heidelberg

Library of Congress Cataloging-in-Publication Data

Nonradioactive labeling and detection of biomolecules / C. Kessler, ed.
Includes bibliographical references and index. ISBN 3-540-55482-3 (alk. paper):
− ISBN 0-387-55482-3 (alk. paper : U.S.) 1. Biomolecules−Labeling.
I. Kessler, Christoph. QP519.9.A37N65 1992 574.19'285−dc20

© Springer-Verlag Berlin Heidelberg 1992
   Printed in Germany

The use of general descriptive names, registered names, trademarks, etc. in this
publication does not imply, even in the absence of a specific statement, that such names
are exempt from the relevant protective laws and regulations and therefore free for
general use.

Product liability: The publishers cannot guarantee the accuracy of any information about
dosage and application contained in this book. In every individual case the user must
check such information by consulting the relevant literature.

Typesetting: Storch GmbH, Wiesentheid

66/3145 − 5 4 3 2 1 0 − Printed on acid-free paper

For My Wife Susanne

Preface

During the last two decades the use of nonradioactive markers instead of radioisotopes has gained increasing importance in bioanalysis. Major disadvantages of isotope use, such as rapid decay, radiation exposure, the need of specific isotope facilities, and the elimination of radioactive waste by trained personnel, have been overcome by the development of nonradioactive labeling and detection techniques. In contrast to radioisotopes, the nonradioactive modification groups are characterized by long-term stability, general, rapid, but also easy-to-handle modification procedures, by the existence of highly sensitive detection systems, and by the avoidance of hazardous materials.

In addition to specialized labeling techniques, the majority of recent advancements have been in the development of universal nonradioactive indicator systems which permit analysis of different biomolecules by using molecule-specific formats but uniform detection reactions.

The main interest in the use of nonradioactive bioanalytical approaches focuses on analytes like nucleic acids, proteins and glycans. These molecules are of great importance in basic research regarding gene expression and downstream processing of the respective proteins. However, other more medically oriented applications of nonisotopic analysis techniques in immunology, virology, microbiology, and human genetics, or in paternity testing and forensic applications as well as in various areas of biotechnology such as food analysis, plant and animal breeding, environmental analysis, or safety analyses of recombinant laboratories are of increasing interest.

Because of the large variety of different approaches used for nonradioactive labeling of nucleic acids, proteins, and glycans by enzymatic, chemical, and photochemical reactions and because of the different detection protocols including optical, chemi-, bio- and electrochemiluminescent and fluorescent signal generation, this book was designed to cover most of the relevant labeling and detection protocols in a single publication. In addition, it also covers recently developed amplification procedures and application formats for blot and in situ analyses as well as quantitative formats for the sensitive detection of biomolecules in solution.

The book is organized into 22 chapters that — in addition to an introduction focusing on general aspects of nonradioactive bio-analytics — are grouped into four major sections: standard nonradioactive labeling and detection systems, specialized nonradioactive detection systems, enhanced systems, and application formats. Each chapter is accompanied by a list of the necessary standard reagents, buffers, and nonradioactive compounds. Two appendices compile the addresses of suppliers for the required reagents and equipment.

The central aim of this book is to introduce the reader to the principles of the various methods and also to provide a collection of the respective protocols. The book thus covers the wide spectrum of general and specialized technology mainly on the basis of lab protocols. It should become an up-to-date and useful practical tool for nonradioactive bioanalysis, stimulating the direct and convenient use of nonradioactive labeling and detection methods in experiments in basic and applied fields of genetic engineering, biotechnology, and medicine.

The editor is very grateful to the more than fifty authors who have supported the realization of this book by providing excellent contributions, and also to many collegues for helpful discussions, and critical reading of parts of the manuscript.

Icking-Dorfen, August 1992   Christoph Kessler

Contents

List of Contributors

Arnold, N.
Institute of Anthropology
and Human Genetics,
University of Munich,
Richard-Wagner-Str. 10/I,
8000 München 2, FRG

Baldino, F.
Cephalon Inc.,
145 Brandywine Parkway,
West Chester, PA 19380, USA

Bayer, E. A.
The Weizmann Institute,
Department of Biophysics,
76100 Rehovot, Israel

Bronstein, I.
Tropix Inc., 47, Wiggins Avenue,
Bedford, MA 01730, USA

Cardullo, R. A.
Dept. of Biology,
The University of California,
Riverside, CA 92521, USA

Carr, R. I.
Dept. of Medicine,
Division of Rheumatology,
Dalhousie University,
Halifax Clivic Building,
5938 University Avenue,
Halifax, N.S., B3H 1V9, Canada

Cohen, B.
Howard Hughes Medical Institute,
Dept. of Cell Biology,
Baylor College of Medicine,
Houston, TX 77030, USA

Collins, M.
Genetics Institute,
87 Cambridge Park Drive,
Cambridge, MA 02140-2387, USA

Coutlée, F.
Départment de Microbiologie
et Maladies infectieuses
et Centre de Recherche,
Hopital Notre-Dame,
1560 Sherbrooke Est,
Montréal pQ,
H2L 4M1, Canada

Diamandis, E. P.
Department of Clinical
Biochemistry,
Toronto Western Hospital,
399 Bathurst Street,
Toronto,
Ontario M5T 2S8, Canada

Dooley, S.
Institut für Humangenetik,
Universität des Saarlandes,
6650 Homburg/Saar, FRG

Durrant, I.
Amersham Laboratories,
White Lion Road,
Amersham,
Buckinghamshire HP7 9LL, UK

Epplen, J. T.
Institut für Genetik,
Abteilung für Molekulare Genetik,
Ruhr Universität Bochum,
Universitätsstr. 150,
4630 Bochum, FRG

Geiger, R. E.
Dillizer Str. 31,
8036 Herrsching, FRG

Genersch, E.
Bruckenfischerstr. 17,
8000 München 90, FRG

Guder, H.-J.
Boehringer Mannheim GmbH,
Bahnhofstr. 9−15,
8132 Tutzing, FRG

Guesdon, J. L.
Laboratoire de Prédéveloppement
des Sondes, Institut Pasteur,
28 Rue du Dr. Roux,
75724 Paris Cedex, France

Haselbeck, A.
Boehringer Mannheim GmbH,
Nonnenwald 2,
8122 Penzberg, FRG

Heino, P.
Dept. of Virology,
Karolinska Institute, SBL,
105 21 Stockholm, Sweden

Herrington, C. S.
University of Oxford,
Nuffield Dept. of Pathology and
Bacterology,
John Radcliffe Hospital,
Headington,
Oxford OX3 9DU, UK

Hillan, K. J.
University Dept. of Pathology,
Western Infirmary,
Glasgow G11 6NT, UK

Höltke, H.-J.
Boehringer Mannheim GmbH,
Nonnenwald 2,
8122 Penzberg, FRG

James, E.
Cangene Corp.,
3403 American Drive,
Missisauga,
Ontario L4V 1T4, Canada

Josel, H.-P.
Boehringer Mannheim GmbH,
Bahnhofstr. 9−15,
8132 Tutzing, FRG

Kenten, J. H.
Igen Inc.,
1530 East Jefferson St.,
Rockville MD 20852, USA

Kessler, C.
Boehringer Mannheim GmbH,
Nonnenwald 2,
8122 Penzberg, FRG

Koch, J.
Inst. of Human Genetics,
The Bartholin Building,
University of Aarhus,
8000 Aarhus C., Denmark

Kunz, W.
Institut für Genetik,
Universität Düsseldorf,
Universitätsstraße 1,
4000 Düsseldorf, FRG

Lazar, J. G.
Digene Diagnostics Inc.,
2301-B Broadbirch Drive,
Silver Spring, MD 20904, USA

Little, M. C.
Becton Dickinson Research Center,
21 Davis Drive, P.O. Box 12016,
Research Triangle Park,
NC 27709-2016, USA

Marich, J. E.
Syngene Inc.,
10030 Barnes Canyon Road,
San Diego, CA 92121, USA

Martin, C. S.
Tropix Inc.,
47, Wiggins Avenue,
Bedford, MA 01730, USA

Nur, I.
Organics Ltd.,
P.O. Box 360,
Yavne 70650, Israel

Patel, N. H.
Carnegie Institution,
115 West University Parkway,
Baltimore, MD 21210, USA

Plas, P. F. E. M., van de,
Aurion, ImmunoGold Reagents
and Accessories,
Costerweg 5,
6702 AA Wageningen,
The Netherlands

Pohl, T. M.
GATC GmbH,
Fritz-Arnold-Str. 23,
7750 Konstanz, FRG

Raap, A. K.
Dept. of Cytochemistry and
Cytometry,
Leiden University,
Wassenaarseweg 72,
2333 AL Leiden,
The Netherlands

Rashtchian, A.
Molecular Biology Research &
Development,
Life Technologies, Inc.,
8717 Grovemont Circle,
Gaithersburg, MD 20877, USA

Rüger, B.
Boehringer Mannheim GmbH,
Nonnenwald 2,
8122 Penzberg, FRG

Rüger, R.
Boehringer Mannheim GmbH,
Nonnenwald 2,
8122 Penzberg, FRG

Sagner, G.
Boehringer Mannheim GmbH,
Nonnenwald 2,
8122 Penzberg, FRG

Schmidt, E. R.
Institut für Genetik,
Universität Mainz,
Saarstr. 21,
6500 Mainz 1, FRG

Segev, D.
ImClone Systems Inc.,
180 Varick Street,
New York, NY 10014, USA

Seibl, R.
Boehringer Mannheim GmbH,
Nonnenwald 2,
8122 Penzberg, FRG

Shimer, G. H.
OmniGene Inc.,
85 Bolton Street,
Cambridge, MA 02140, USA

Söderlund, H.
Technical Research Center
of Finland, VIT Biotechnical
Laboratory,
Tietotie 2,
02151 Espoo, Finland

Stackebrandt, E.
Dept. of Microbiology,
The University of Queensland,
4072 Queensland, Australia

Starzinski-Powitz, A.
J. W. Goethe-Universität,
Institut der Anthropologie
und Humangenetik für Biologen,
Siesmayerstraße 70,
6000 Frankfurt/M. 11, FRG

Tautz, D.
Zoologisches Institut,
Universität München,
Luisenstr. 14,
8000 München 90, FRG

Walter, T.
Boehringer Mannheim GmbH,
Nonnenwald 2,
8122 Penzberg, FRG

Zuber, U.
Lehrstuhl für Genetik,
Universität Bayreuth,
Universitätsstr. 30,
8580 Bayreuth, FRG

1 General Aspects of Nonradioactive Labeling and Detection

CHRISTOPH KESSLER

1.1 Introduction

In the course of the past decade, increasing attempts to detect basic biological substances such as nucleic acids, proteins, and glycans by nonradioactive bioanalytical indicator systems have been made. In addition to special techniques such as nucleic acid detection via ethidium bromide intercalation (Bauer and Vinograd, 1968; Nathans and Smith, 1975; Ausubel et al., 1987), protein visualization by Coomassie staining (Bennett and Scott, 1971; Zehr et al., 1989), or the detection of glycans by periodate-Schiff staining (Kapitany and Zebrowski, 1973), an increasing number of indicator systems are being developed which are characterized by a higher specificity and sensitivity. Analytical systems which permit the analysis of different biomolecules by uniform detection principles are of particular interest.

The aims of this book are to summarize the properties of the most commonly applied systems used in nonradioactive labeling and detection of biomolecules such as nucleic acids, proteins, and glycans and to give detailed protocols for a variety of standard labeling and detection methods. Besides focusing on standard protocols, both additional protocols for more spezialized detection systems and various applications to basic and applied molecular biology are given. Furthermore, the protocols of the most important nucleic acid amplification systems are also provided. These may be used in combination with nonradioactive nucleic acid detection systems for identification of DNA or RNA on a single molecule level.

References

Ausubel FM, Brent R, Kingston RE, Moore DD, Seidman JG, Smith JA, Struhl A (1987) Current protocols in molecular biology. Greene Publishing Associates and Wiley-Interscience, New York

Bauer W, Vinograd J (1968) Interaction of closed circular DNA with intercalative dyes. I. Superhelix density of SV40 DNA in the presence and absence of dye. J Mol Biol 33:141−171

Bayer EA, Wilchek M (1980) The use of the avidin-biotin complex as a tool in molecular biology. Meth Biochem Anal 26:1−45

Bennett J, Scott KJ (1971) Quantitative staining of fraction I protein in polyacrylamide gels using Coomassie brilliant blue. Anal Biochem 43:173−182

Kapitany RA, Zebrowski EJ (1973) A high resolution PAS stain for polyacrylamide gel electrophoresis. Anal Biochem 56:361−369

Nathans D, Smith HO (1975) Restriction endonucleases in the analysis and restructuring of DNA molecules. Ann Rev Biochem 44:273–293
Zehr BD, Savin TJ, Hall RE (1989) A one-step low background Coomassie staining procedure for polyacrylamide gels. Anal Biochem 182:157–159

1.2 The Concept of Nonradioactive Bioanalytics

The concept of nonradioactive bioanalytical indicator systems is based on the detection of the various biological target molecules (analytes) by selective interaction with specific binding partners (probes). To these probes, appropriate detector systems are coupled, either directly by covalent binding or indirectly by additional, specific, high-affinity interaction. The accompanying table lists the currently possible binding pairs which permit the detection of analytes via the corresponding probes. Reviews describing the detection of nucleic acids, proteins/haptens, and glycans are given by Wilchek and Bayer (1987), Matthews and Kricka (1988), Linke and Küppers (1988), Keller and Manak (1989), and Kricka (1992).

Bioanalytic binding pairs

Target molecules (analytes)	Binding partners (probes)
Nucleic acids	Nucleic acid probes
Nucleic acids	Nucleic acid binding proteins
Proteins	Antibodies
Transport proteins	Components of metabolism
Glycoproteins	Lectins
Enzymes	Cofactors/effectors/inhibitors
Receptors	Secondary metabolites
Secondary metabolites	Antibodies
Membranes	Liposomes
Lymphoid cells	Mitogens/antigens
Metal ions	Complex-forming agents

The nonradioactive indicator systems developed in the last few years have primarily been adapted to the detection of nucleic acids since corresponding nonradioactive detection systems for proteins and haptens have already been established. The nonradioactive systems increasingly substitute analogous radioactive procedures for those which were based on the incorporation of radioactive isotopes such as ^3H, ^{14}C, ^{32}P, ^{35}S, or ^{125}I (Maitland et al., 1987). In contrast to isotopic labels nonradioactive modification groups are stable; in addition, there is neither an accumulation of radioactive waste nor the need of an isotope laboratory guided by trained personnel.

Most of the recently developed nonradioactive systems are based on the enzymatic, photochemical, or chemical incorporation of a reporter group

(Kessler, 1991; Kricka, 1992) which can be detected with high sensitivity by optical, luminescence, fluorescence, or metal-precipitating detection systems (Urdea et al., 1988; Coutlee et al., 1989; Kricka, 1992). Furthermore, electrochemical detection systems using pH electrodes or sensor technology have also been described (Briggs, 1987; Downs et al., 1987; Hafemann et al., 1988; McKnabb and Tedesco, 1989; Ikariyama et al., 1989). Attempts are being made to use specific labeling and detection pairs not only for the detection of different target molecule specificities, but also for the detection of a large variety of different kinds of biomolecules (universal detection systems).

In many systems, the specific detection on target molecules is accomplished by specific reaction with a single binding partner conjugated with a reporter group and by removing the excess conjugate by washing (separation formats). The use of two supplementary binding partners characterized by distinguishable properties in the formed complex allows for homogeneous reaction procedures without the need of washing steps (separation-free formats).

Separation formats include sandwich and competition/replacement assays (Linke and Küppers, 1988; Jungell-Nortamo et al., 1988; Vary, 1987; Collins et al., 1989); examples of separation-free formats are enzyme complementation and energy transfer systems (Hendersen et al., 1986; Cardullo et al., 1988) and fluorescence depolarization assays (Hicks, 1984; Schray et al., 1988) (for application formats see also Part IV).

References

Briggs J (1987) Biosensors emerge from the laboratory. Nature 329:565–566
Cardullo RA, Agrawal S, Flores C, Zamecnik PC, Wolf DE (1988) Detection of nucleic acid hybridization by nonradioactive fluorescence resonance energy transfer. Proc Natl Acad Sci USA 85:8790–8794
Collins M, Fritsch EF, Ellwood MS, Diamond SE, Williams JI, Brewen JG (1988) A novel diagnostic method based on strand displacement. Mol Cell Probes 2:15–30
Coutlee F, Viscidi P, Yolken H (1989) Comparison of colorimetric, fluorescent, and enzymatic amplification substrate systems in an enzyme immunoassay for detection of DNA-RNA hybrids. J Clin Microbiol 27:1002–1007
Downs MEA, Kobayashi S, Karube I (1987) Review. New DNA technology and the DNA biosensor. Anal Lett 20:1897–1927
Hafemann DG, Parce JW, McConnell HM (1988) Light-addressable potentiometric sensor for biochemical systems. Science 240:1182–1185
Henderson DR, Friedmann SB, Harris JD, Manning WB, Zoccoli MA (1986) CEDIA, a new homogeneous immunoassay system. Clin Chem 32:1637–1641
Hicks JM (1984) Fluorescence immunoassay. Hum Pathol 15:112–116
Ikariyama Y, Shimada N, Yukiashi T, Aizawa M, Yamauchi S (1989) Microbiosensing device for real time determination. Bull Chem Soc Japan 62:1864–1868
Jungell-Nortamo A, Syvänen AC, Luoma P, Söderlund H (1988) Nucleic acid sandwich hybridization: enhanced reaction rate with magnetic microparticles as carriers. Mol Cell Probes 2:281–288
Kessler C (1991) The digoxigenin:anti-digoxigenin (DIG) technology – a survey on the concept and realization of a novel bioanalytical indicator system. Mol Cell Probes 5:161–205

Keller GH, Manak MM (1989) DNA probes. Stockton Press, New York

Kricka LJ (1992) Nonisotopic DNA probe techniques. Academic Press, San Diego

Krieg PA, Melton DA (1987) In vitro RNA synthesis with SP6 RNA polymerase. Meth Enzymol 155:397−415

Linke R, Küppers R (1988) Nicht-isotopische Immunoassays − ein Überblick. In: Borsdorf R, Fresenius W, Günzler H, Huber W, Kelker H, Lüderwald I, Tölg G, Wisser H (eds) Analytiker Taschenbuch, Springer Verlag, Berlin/Heidelberg, pp 127−177

Maitland NJ, Cox MF, Lynas C, Prime S, Crane I, Scully C (1987) Nucleic acid probes in the study of latent viral disease. J Oral Pathol 16:199−211

McKnabb RR, Tedesco JL (1989) Measuring contaminating DNA in bioreactor derived monoclonals. BioTechnol 7:343−347

Schray KJ, Artz PG, Hevey RC (1988) Determination of avidin and biotin by fluorescence polarization. Anal Chem 60:853−855

Urdea MS, Warner BD, Running JA, Stempien M, Clyne J, Horn T (1988) A comparison of nonradioisotopic hybridization assay methods using fluorescent, chemiluminescent and enzyme-labeled synthetic oligodeoxyribonucleotide probes. Nucleic Acids Res 16:4937−4956

Vary CPH (1987) A homogeneous nucleic acid hybridization assay based on strand displacement. Nucleic Acids Res 15:6883−6897

Wilchek M, Bayer EA (1987) Labeling glycoconjugates with hydrazide reagents. Meth Enzymol 138:429−442

1.3 Direct and Indirect Labeling and Detection Systems

The various nonradioactive bioanalytical indicator systems can be classified as direct or indirect (see figure). Both kinds of assays differ in the number of components and thus the number of reaction steps used for the detection reaction. Whereas direct systems are mostly used for detection of standardized target biomolecules, the more flexible indirect systems are

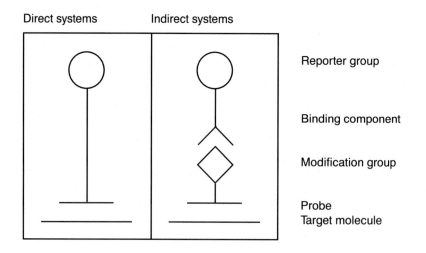

Direct systems   Indirect systems

Reporter group

Binding component

Modification group

Probe
Target molecule

often applied to the rapid detection of different target biomolecules of variable specificity. Sensitivity of detection can further be enhanced by coupled signal or target-specific amplification reactions (see Parts II and III).

1.3.1 Direct Systems

In direct systems, the analyte-specific probes are directly and covalently linked with the signal-generating reporter group; thus, the detection of nucleic acids in direct systems consists of:

1. Hybrid formation between analyte and labeled probe
2. Signal generation via the reporter group directly bound to the probe

Frequently used direct reporter groups are fluorescent dyes such as fluorescein and rhodamine (Lichter et al., 1990) and marker enzymes such as alkaline phosphatase (AP) (Jablonski et al., 1986) or horseradish peroxidase (HRP) coupled to chemiluminescence (Pollard-Knight et al., 1990) or silver enhancement detection systems (Taub, 1986).

The advantage of direct systems is that they allow for the detection of target molecules through only a single interaction; however, the disadvantage is that for each type of target molecule an individual conjugate consisting of both the specific binding partner and the coupled detector system must be available.

1.3.2 Indirect Systems

In contrast to direct systems, the reporter group in indirect systems is not linked directly but indirectly through an additional noncovalent interaction between a modification group of the probe and a universal indicator molecule which binds to the probe. Therefore, indirect systems first require alteration of the analyte-specific probe by introduction of a particular modification group. This modification group binds through an additional noncovalent interaction to a universal reporter group. The detection of nucleic acids in indirect systems is thus divided into three reaction steps:

1. Hybrid formation between analyte and modified probe
2. Specific noncovalent interaction between the modified probe and the binding partner coupled with the reporter group
3. Signal generation via the reporter group indirectly bound to the probe

Indirect detection systems therefore differ from direct detection systems in that the signal-generating detection system is not directly coupled with the molecule-specific probe but rather the binding is mediated by an additional interaction between the modification group and the indicator molecule.

Binding partners of modified biomolecules

Modification of analyte: probe	Example(s)	Reference(s)
Vitamins		
Vitamin: binding protein	Biotin: Avidin or Streptavidin	Bayer and Wilchek (1980); Langer et al. (1981); Wilchek and Bayer (1988); Bayer and Wilchek (1990)
Haptens		
Hapten: hapten-specific antibody	Digoxigenin (DIG): >DIG	Kessler et al. (1990); Höltke et al. (1990); Seibl et al. (1990); Mühlegger et al. (1990); Höltke and Kessler (1990); Schmitz et al. (1991)
	Dinitrophenyl (DNP): >DNP	Keller et al. (1988; 1989)
	Fluorescein isothiocyanate (FITC): >FITC	Serke and Pachmann (1988); Parsons (1988)
	Biotin (bio): >bio	Langer-Safer et al. (1982); Agrawal et al. (1986); Binder (1987)
	N-2-Acetylaminofluoren (AAF): >AAF	Tchen et al. (1984); Landegent et al. (1984; 1985); Cremers et al. (1987)
	N-2-Acetylamino-7-iodofluoren (AAIF): >AAIF	Tchen et al. (1984); Syvänen et al. (1986)
	5C-Bromodesoxyuridine (5-BrdU): >5-BrdU	Traincard et al. (1983); Porstman et al. (1985); Sakamoto et al. (1987)
	5C-Sulfite-desoxycytidine (SO_3-dC): >SO_3-dC	Herzberg (1984); Pezella et al. (1987); Hyman et al. (1987)
	Ethidium (Et): >Et	Albarella and Anderson (1985); Dattagupta et al. (1985)
Conformation of nucleic acid		
Nucleic acid hybrid: conformation-specific antibody	RNA/DNA: >RNA/DNA	Van Prooijen-Knegt et al. (1982); Stollar and Rashtchian (1987); Rashtchian et al. (1987); Coutlee et al. (1989a, 1989b, 1989c)
	RNA/RNA: >RNA/RNA	Coutlee et al. (1989b)
	DNA: >DNA	McKnabb et al. (1989)
Sequence of nucleic acid		
Nucleic acid sequence: binding protein	ssDNA: *E. coli* ssb protein	Syvänen et al. (1985); McKnabb et al. (1989)
	dsDNA: histone	Renz (1983); Bulow and Link (1986)
	T7 promoters A1/A2/A3: *E. coli* RNA polymerase	Paau et al. (1983)
	5-Aza-dC: DNA methyltransferase	Reckmann and Rieke (1987)
	lac operon: lac repressor	Dattagupta et al. (1988)

Modification of analyte: probe	Example(s)	Reference(s)
	Protein A-NA: IgF-Fc	Dattagupta et al. (1984); Czichos et al. (1989)
	S peptide-NA: S protein	Rabin et al. (1985)
Modification of heavy metal ions		
Heavy metal: mercaptane	Hg^{2+}: glutathione-TNP: >TNP	Baumann et al. (1983); Hopman et al. (1986a; 1986b)
Polyadenylation		
Polyadenylation: polythymidine ends	$(dA)_x$: $(dT)_x$	Woodhead and Malcolm (1984); Kumar et al. (1988); Parsons (1988)
Polyadenylation: polynucleotide phosphorylase	$(dA)_x$: PNP/pyruvate kinase/ATP-coupled luciferase reaction	Vary et al. (1986); Gillam (1987)

PNP: polynucleotide phosphorylase   5-Aza-dC: 5-azadeoxycytidine
ssb: single-strand binding protein   NA: nucleic acid
TNP: trinitrophenyl

A variety of interaction pairs consisting of modification group and binding partner are already in use. The table shows a list of important interaction pairs described for the nonradioactive detection of nucleic acids. The different kinds of interactions with the respective indicator group are accomplished by the selective binding of a specific modification group or of an altered conformation of the modified binding partner. For a review on labeling of nucleic acids see Wilchek and Bayer (1990) and Kessler (1992).

Aside from the well known systems using a vitamin, e.g., biotin (Langer et al., 1981), or a hapten, e.g., digoxigenin (Kessler, 1991), bromodeoxyuridine (Sakamoto et al., 1987), sulfone (Hyman, 1987), or immunogold (Tomlinson et al., 1988), alternative kinds of interactions with the respective modification groups have been established by either the selective binding of heavy metal ions, e.g., mercury (Hopman et al., 1986), or nucleic acid conformations, e.g., DNA:RNA hybrids (Coutlee et al., 1989a; 1989b).

The disadvantage of indirect systems is that interaction between another binding pair is required for detection of the target molecules. However, the advantage of these systems is that the detection components are universal, i.e., they can be used for the detection of nucleic acids, proteins, haptens, and glycans. Furthermore, differentiation within individual kinds of biomolecules is also possible with identical detection systems.

Indirect systems are hence more versatile for biomolecule detection. For this reason, indirect systems are not only used in basic research but also in the more applied fields of genetic engineering, biotechnology, or medicine, where the detection of different target molecules is essential for solving specific problems. In the case of genetic engineering, for example, this holds true for cloning and overexpression of eukaryotic genes; in addition, the proteins expressed are often glycosylated. Thus, possible target molecules during gene expression are recombinant DNA, mRNA transcripts, expressed proteins, and sugar modifications. It is also of interest to characterize the purified proteins with respect to homogeneity, glycan specificity, and the absence of residual nucleic acids.

References

Agrawal S, Christodoulou C, Gait MJ (1986) Efficient methods for attaching nonradioactive labels to the 5′ ends of synthetic oligodeoxyribonucleotides. Nucleic Acids Res 14:6227−6245

Albarella JP, Anderson LH (1985) Detection of polynucleotide sequence in medium and when single stranded nucleic acids are present by using probe, intercalator and antibody. Eur Pat Appl 0146815

Albarella JP, Anderson LH, Carrico RJ (1985) Detection of polynucleotide sequence in sample of nucleic acids by using nucleic acid probe and contact of duplexes with immobilized antibody. Eur Pat Appl 0146039

Baumann JG, Wiegant J, van Duijin P (1983) The development, using poly(Hg-U) in a model system, of a new method to visualize cytochemical hybridization in fluorescence microscopy. J Histochem Cytochem 31:571−578

Bayer EA, Wilchek M (1980) The use of the avidin-biotin complex as a tool in molecular biology. Meth Biochem Anal 26:1−45

Bayer EA, Wilchek M (1990) Avidin-biotin technology. Methods in Enzymology, Vol 184. Academic Press, San Diego

Binder M (1987) In situ hybridization at the electron microscope level. Scanning Microsc 1:331−338

Bulow S, Link G (1986) A general and sensitive method for staining DNA and RNA blots. Nucleic Acids Res 14:3973

Coutlee F, Bobo L, Mayur K, Yolken RH, Viscidi RP (1989a) Immunodetection of DNA with biotinylated RNA probes: a study of reactivity of a monoclonal antibody to DNA-RNA hybrids. Anal Biochem 181:96−105

Coutlee F, Viscidi P, Yolken RH (1989b) Comparison of colorimetric, fluorescent, and enzymatic amplification substrate systems in an enzyme immunoassay for detection of DNA-RNA hybrids. J Clin Microbiol 27:1002−1007

Coutlee F, Yolken RH, Viscidi RP (1989c) Nonisotopic detection of RNA in an enzyme immunoassay using a monoclonal antibody against DNA-RNA hybrids. Anal Biochem 181:153−162

Cremers AF, Jansen in de Wal N, Wiegant J, Dirks RW, Weisbeek P, Van der Ploeg M, Landegent JE (1987) Nonradioactive in situ hybridization. A comparison of several immunocytochemical detection systems using reflection-contrast and electron microscopy. Histochem 86:609−615

Czichos J, Koehler M, Reckmann B, Renz M (1989) Protein-DNA conjugates produced by UV irradiation and their use as probes for hybridization. Nucleic Acids Res 17:1563−1572

Dattagupta N, Knowles W, Marchesi VT, Crothers DM (1984) Nucleic acid-protein conjugate. Eur Pat Appl 0154884

Dattagupta N, Rae PMM, Knowles WJ, Crothers DM (1985) Nucleic acid detection probe comprises hybridisable single stranded part of nucleic acid connected to nonhybridisable nucleic acid with specific recognition site. Eur Pat Appl 0147665

Dattagupta N, Rae PMM, Knowles WJ, Crothers DM (1988) Use of nonhybridizable nucleic acids for the detection of nucleic acid hybridization. US 4724202

Gillam IC (1987) Nonradioactive probes for specific DNA sequences. Trends Biotech 5:332−334

Herzberg M (1984) Molecular genetic probe, assay technique, and a kit using this molecular genetic probe. Eur Pat Appl 0128018

Höltke H-J, Kessler C (1990) Nonradioactive labeling of RNA transcripts in vitro with the hapten digoxigenin (DIG); hybridization and ELISA-based detection. Nucleic Acids Res 18:5843−5851

Höltke H-J, Seibl R, Burg J, Mühlegger K, Kessler C (1990) Nonradioactive labeling and detection of nucleic acids: II. Optimization of the digoxigenin system. Mol Gen Hoppe-Seyler 371:929−938

Hopmann AHN, Wiegant J, Tesser GI, Van Duijn P (1986a) A nonradioactive in situ hybridization method based on mercurated nucleic acid probes and sulfhydryl-hapten ligands. Nucleic Acids Res 14:6471−6488

Hopman AHN, Wiegant J, van Duijn P (1986b) A new hybridocytochemical method based on mercurated nucleic acide probes and sulhydryl-hapten ligands. I. Stability of the mercurysulfhydryl bond and influence of the ligand structure on immunochemical detection of the hapten. Histochem 84:169−178

Hyman HC, Yogev D, Razin S (1987) DNA probes for detection and identification of Mycoplasma pneumoniaea and Mycoplasma genitalium. J Clin Microbiol 25:726−728

Jablonski E, Moomaw EW, Tullis RH, Ruth JL (1986) Preparation of oligodeoxynucleotide-alkaline phosphatase conjugates and their use as hybridization probes. Nucleic Acids Res 14:6115−6128

Keller GH, Cumming CU, Huang DP, Manak MM, Ting R (1988) A chemical method for introducing haptens onto DNA probes. Anal Biochem 170:441−450

Keller GH, Huang DP, Manak MM (1989) Labeling of DNA probes with a photoactivatable hapten. Anal Biochem 177:392−395

Kessler C (1991) The digoxigenin:anti-digoxigenin (DIG) technology − a survey on the concept and realization of a novel bioanalytical indicator system. Mol Cell Probes 5:161−205

Kessler C (1992) Nonradioactive nucleic acid labeling methods. In: Kricka LJ (ed) Nonisotopic DNA probe techniques, Academic Press, pp 29−92

Kessler C, Höltke H-J, Seibl R, Burg J, Mühlegger K (1990) Nonradioactive labeling and detection of nucleic acids: I. A novel DNA labeling and detection system based on digoxigenin:antidigoxigenin ELISA principle (digoxigenin system). Mol Gen Hoppe-Seyler 371:917−927

Kumar A, Tchen P, Roullet F, Cohen J (1988) Nonradioactive labeling of synthetic oligonucleotide probes with terminal deoxynucleotidyl transferase. Anal Biochem 169:376−382

Landegent JE, Jansen in de Wal N, Baan RA, Hoeijmakers JH, Van der Ploeg M (1984) 2-Acetylaminofluorene-modified probes for the indirect hybridocytochemical detection of specific nucleic acid sequences. Exp Cell Res 153:61−72

Landegent JE, Jansen in de Wal N, Ploem JS, Van der Ploeg M (1985) Sensitive detection of hybridocytochemical results by means of reflection-contrast microscopy. J Histochem Cytochem 33:1241−1246

Langer PR, Waldrop AA, Ward DC (1981) Enzymatic synthesis of biotin-labeled polynucleotides: novel nucleic acid affinity probes. Proc Natl Acad Sci USA 78:6633−6637

Langer-Safer PR, Levine M, Ward DC (1982) Immunological method for mapping genes on Drosophila polytene chromosomes. Proc Natl Acad Sci USA 79:4381−4385

Lichter P, Tang CJC, Call K, Hermanson G, Evans GA, Housman D, Ward DC (1990) High-resolution mapping of human chromosome 11 by in situ hybridization with cosmid clones. Science 247:64−69

McKnabb S, Rupp R, Tedesco JL (1989) Measuring contamination DNA in bioreactor-derived monoclonals. Bio/Technology 7:343−347

Mühlegger K, Huber E, von der Eltz H, Rüger R, Kessler C (1990) Nonradioactive labeling and detection of nucleic acids: IV. Synthesis and properties of the nucleotide compounds of the digoxigenin system and of photodigoxigenin. Mol Gen Hoppe-Seyler 371:939−951

Paau A, Platt SG, Sequeiro L (1983) Assay method and probe for polynucleotide sequence. UK Pat Appl 2125964

Parsons G (1988) Development of DNA probe-based commercial assay. J Clin Immunoassay 11:152−160

Pezzella M, Pezzella F, Galli C, Macchi B, Verani P, Sorice F, Baroni CD (1987) In situ
hybridization of human immunodeficiency virus (HTLV-III) in cryostat sections of
lymph nodes of lymphadenopathy syndrome patients. J Med Virol 22:135–142

Pollard-Knight D, Read CA, Downes MJ, Howard LA, Leadbetter MR, Pheby SA,
McNaughton E, Syms A, Brady MAW (1990) Nonradioactive nucleic acid detection
by enhanced chemiluminescence using probes directly labeled with horseradish
peroxidase. Anal Biochem 185:84–89

Porstmann T, Ternynck T, Avrameas S (1985) Quantitation of 5-bromo-2-deoxyuridine
incorporation into DNA: an enzyme immunoassay for the assessment of the lymphoid
cell proliferative response. J Immunol Methods 82:169–179

Rabin BR, Taylorson CJ, Hollaway MR (1985) Assay method using enzyme fragments
as labels and new enzyme substrates producing coenzymes or prosthetic groups. Eur
Pat Appl 0156641

Rashtchian A, Elredge J, Ottaviani M, Abbott M, Mock G, Lovern D, Klinger J, Parsons G (1987) Immunological capture of nucleic acid hybrids and application to nonradioactive DNA probe assay. Clin Chem 33:1526–1530

Reckmann B, Rieke E (1987) Verfahren und Mittel zur Bestimmung von Nucleinsäuren.
Eur Pat Appl 0286958

Renz M (1983) Polynucleotide-histone H1 complexes as probes for blot hybridization.
EMBO J 2:817–822

Sakamoto H, Traincard F, Vo-Quang T, Ternynck T, Guesdon JL, Avrameas S (1987)
5-Bromodeoxyuridine in vivo labeling of M13 DNA, and its use as a nonradioactive
probe for hybridization experiments. Mol Cell Probes 1:109–120

Schmitz GG, Walter T, Kessler C (1991) Nonradioactive labeling of oligonucleotides in
vitro with the hapten digoxigenin (DIG) by tailing with terminal transferase. Anal
Biochem 192:222–231

Seibl R, Höltke H-J, Rüger R, Meindl A, Zachau H-G, Rasshofer G, Roggendorf M,
Wolf H, Arnold N, Wienberg J, Kessler C (1990) Nonradioactive labeling and detec-
tion of nucleic acids: III. Applications of the digoxigenin system. Mol Gen Hoppe-
Seyler 371:939–951

Serke S, Pachmann K (1988) An immunocytochemical method for the detection of
fluorochrome-labeled DNA probes hybridized in situ with cellular RNA. J Immunol
Meth 112:207–211

Stollar BD, Rashtchian A (1987) Immunochemical approaches to gene probe assays.
Anal Biochem 161:387–394

Syvänen AC, Alanen M, Söderlund H (1985) A complex of single-strand binding protein
and M13 DNA as hybridization probe. Nucleic Acids Res 13:2789–2802

Syvänen AC, Tchen P, Ranki M, Söderlund H (1986) Time-resolved fluorometry: a sensitive method to quantify DNA-hybrids. Nucleic Acids Res 14:1017–1028

Taub F (1986) An assay for nucleic acid sequences, particularly genetic lesions. PCT Int
Appl WO 86/03227

Tchen P, Fuchs RPP, Sage E, Leng M (1984) Chemically modified nucleic acids as
immunodetectable probes in hybridization experiments. Proc Natl Acad Sci USA
81:3466–3470

Tomlinson S, Lyga A, Huguenel E, Dattagupta N (1988) Detection of biotinylated
nucleic acid hybrids by antibody-coated gold colloid. Anal Biochem 171:217–222

Traincard F, Ternynck T, Danchin A, Avrameas S (1983) An immunoenzymic procedure for the demonstration of nucleic acid molecular hybridization. Ann Immunol
134:399–405

Van Prooijen-Knegt AC, Van Hoek JF, Bauman JG, Van Duijin P, Wool IG, Van der
Ploeg M (1982) In situ hybridization of DNA sequences in human metaphase chromo-
somes visualized by an indirect fluorescent immunocytochemical procedure. Exp Cell
Res 141:397–407

Vary CPH, McMahon FJ, Barbone FP, Diamond SE (1986) Nonisotopic detection
methods for strand displacement assays of nucleic acids. Clin Chem 32:1696–1701

Wilchek M, Bayer EA (1988) The avidin-biotin complex in bioanalytical applications.
Anal Biochem 171:1–32

Woodhead JL, Malcolm ADB (1984) Nonradioactive gene-specific probes. Biochem
Soc Trans 12:279–280

1.4 Labeling Methods

Besides the various methods for linking fluorescent dyes or marker enzymes directly to nucleic acid probes, a broad repertoire of enzymatic, photochemical, and chemical labeling methods can be used for introduction of modification groups (see table).

The most frequently used sites of base modification are the C-5 position of uracil and cytosine, the C-6 of thymine, and the C-8 of guanine and adenine; these positions are not involved in hydrogen bonding. In addition, the N^4 position of cytosine and the N^6 position of adenine have also been used for base modification; however, these sites are involved in

Possible modes of modification of nucleic acid probes

Reactive groups of binding partner: modifying agent	Reference(s)
Enzymatic modification	
dsDNA: M-dNTP/E. coli DNA polymerase I	Rigby et al. (1977); Höltke et al. (1990)
ssDNA: M-dNTP/primer/ Klenow polymerase	Langer et al. (1981); Gregersen et al. (1987); Höltke et al. (1990)
3'-OH-ssDNA/RNA: M-dNTP, M-ddNTP/terminal transferase	Riley et al. (1986); Pitcher et al. (1987); Schmitz et al. (1991)
ssRNA: M-dNTP/primer/ reverse transcriptase	Vary et al. (1986)
dsNA transcription unit: M-NTP/ SP6, T7, T3 RNA polymerase	McCracken (1989); Theissen et al. (1989); Höltke and Kessler (1990)
Photochemical modification	
ss,dsDNA/ss,dsRNA: M-azidobenzoyl/v	Ben-Hur and Song (1984); Forster et al. (1985); Cimino (1985); Mühlegger et al. (1990)
dsDNS/dsRNA: M-NS-intercalator/v	Brown et al. (1982); Dattagupta and Crothers (1984); Sheldon et al. (1985, 1987); Dattagupta et al. (1989)
Chemical modification	
DNA · CHO: M-hydrazide	Reisfeld et al. (1987)
DNA · Hg^{2+}: M-mercaptane	Dale et al. (1975); Bergstrom and Ruth (1977); Langer et al. (1981); Ward et al. (1982); Baumann et al. (1983); Hopman et al. (1986)
DNA · SH: M-amine	Renz and Kurz (1984); Landes (1985); Al-Hakim and Hull (1986)
DNA · NH_2: M-amine/SO_3^{2-}	Draper and Gold (1980); Viscidi et al. (1986); Gillam and Tener (1986)
Allylamine-oligonucleotide: M-N-hydroxysuccinimide ester	Langer et al. (1981); Cook et al. (1988); Urdea et al. (1988)
5'-p-Oligonucleotide: M-N-hydroxysuccinimide ester/ diaminoethyl, -hexyl	Agrawal et al. (1986)
3'-MF-CPG-oligonucleotide: M-N-hydroxysuccinimide ester	Kempe et al. (1985); Nelson et al. (1989a; 1989b)

Abbreviations: ds, double-stranded; NA, nucleic acid

hydrogen bonding. Oligonucleotides may also be modified at either the 5′ or 3′ terminus; internal labeling at the 2′-position of the deoxyribose and at the phosphodiester bridge is also possible.

1.4.1 Enzymatic Labeling

Enzymatic labeling reactions are catalyzed by a number of DNA-dependent DNA (DNAP) or RNA (RNAP) polymerases, RNA-dependent DNA polymerases (RT) or terminal transferases (Langer et al., 1981; Ausubel et al., 1987; Takahashi et al., 1989; Kessler et al., 1990; Höltke and Kessler, 1990; Schmitz et al., 1991). In these procedures, nucleotide analogues modified with particular haptens, e.g., biotin, digoxigenin (DIG) or fluorescein, are used instead of or in combination with their non-modified counterparts. In the case of DNA fragments and oligodeoxynucleotides, hapten-dUTP, hapten-dCTP, or hapten-dATP is applied as the modified substrate (Langer et al., 1981; Gebeyehu et al., 1987); for labeling RNA, hapten-UTP is used as the nucleotide analogue (Höltke and Kessler, 1990).

For DNA molecules, homogeneous labeling can be obtained by random priming with Klenow polymerase, nick translation with *Escherichia coli* DNA polymerase I or by the polymerase chain reaction (PCR) with *Taq* DNA polymerase. In the latter reaction, vector-free double- or single-stranded probes may be synthesized. DNA end labeling can be achieved by a tailing reaction with terminal transferase preferentially at 3′ protruding ends, by a fill-in reaction with Klenow polymerase at 5′ protruding ends, or a T4 DNA polymerase replacement reaction by sequential action of 3′ → 5′ exonuclease and 5′ → 3′ polymerase at 3′ protruding fragment termini guided by the absence or presence of the respective nucleotides.

Starting from RNA as template, labeled DNA probes can be synthesized with viral reverse transcriptases, e.g., from AMV (avian myoblastosis virus) and MoMLV (Moloney murine leucaemia virus) by oligodeoxynucleotide-primed RNA-dependent DNA synthesis (reverse transcription) using hapten-modified deoxynucleotides (e.g., bio-dUTP or DIG-dUTP) in addition to nonmodified deoxynucleotides as substrates.

RNA can be homogeneously labeled by synthesizing "run-off" transcripts catalyzed by the phage-coded, DNA-dependent SP6, T7, or T3 RNA polymerases. The promoter-directed RNA probe synthesis results in vector-free RNA hybridization probes (as opposed to PCR-generated vector-free DNA hybridization probes).

Oligodeoxynucleotides can be enzymatically 3′ end labeled by a tailing reaction with terminal transferase; internal enzymatic labeling is obtained with Klenow polymerase starting from short primers that bind to complementary sequences within the 3′ region of the template oligodeoxynucleotide strand.

1.4.2 Photochemical Labeling

The photochemical derivatization of nucleic acids can be accomplished using azide-containing compounds. This approach results in the photochemical dissociation of nitrogen and the subsequent reaction of the resulting nitrene intermediate with the molecule to be labeled (Forster et al., 1985; Mühlegger et al., 1990).

As regards nucleic acids, however, a number of photoreactive substances are also known which intercalate into nucleic acids and are then covalently linked with flanking bases through a photoreaction (Dattagupta and Crothers, 1984). Examples of these types of photoreactive compounds are furocoumarin compounds such as psoralen or angelic acid, acridine dyes such as acridine orange, phenanthrolines such as ethidium bromide, and phenazines, phenothiazines, and quinones.

1.4.3 Chemical Labeling

In the case of nucleic acids, chemical derivatization can be performed by a number of alternative reactions (for review see Matthews and Kricka, 1988; Kricka, 1992). For example, sulfite-catalyzed transamination of the N^6 amino group of cysteine; mercury derivatization of the C-5 position of pyrimidines and of the C-8 position of purines with $Hg(Ac)_2$ and subsequent reaction with corresponding mercaptanes; bromide derivatization at the same position with N-bromosuccinimide and subsequent reaction with corresponding amines; substitution of amino groups with bifunctional reagents such as 3-(4-bromo-3-oxabutan-1-sulphonyl)-propionic acid-hydroxysuccinimide-ester or glutaraldehyde (Sodja and Davidson, 1978); as or mercaptan derivatization of amine residues of the nucleotide base by 3-(2-pyridinedithionic)-propionic acid-hydroxysuccinimide-ester or N-acetyl-N(p-glyoxybenzoyl)cysteamine (Ehrat et al., 1986).

Chemical labeling of oligonucleotides can be performed internally or at both termini during chemical oligonucleotide synthesis by the incorporation of allylamine-modified protected synthesis components (Haralambidis et al., 1987; Cook et al., 1988). These allylamine-derivatized oligonucleotides react with N-hydroxysuccinimide (NHS) esters coupled with the haptens of interest. End labeling can be obtained at the 5′ end by reaction with activated ethyl or hexyl derivatives (Agrawal et al., 1986) or by direct labeling of 5′ phosphorylated oligonucleotides with functionalized hapten derivates (Kempe et al., 1985); end labeling at the 3′ end is achieved by use of multifunctional (MF) carriers, e.g., controlled pore glass (MF-CPG) (Nelson et al., 1989a; 1989b). Finally, 5′ end labeling via NH_2 or SH groups has also been reported. The introduction of suitable marker groups into oligonucleotides via modified phosphodiester linkages is also possible (for references see Kricka, 1992).

Chemical labeling of proteins and glycans is accomplished by reaction with group-specific reagents which react specifically with free amino groups, e.g., *N*-hydroxysuccinimide (NHS) ester/*p*-nitrophenyl (PNP) ester; mercaptan groups, e.g., *p*-diazobenzene (DAB) ester; aldehyde groups, e.g., hydrazides (HZ); or aromatic ring structures such as phenols or imidazoles, e.g., DAB ester.

References

Agrawal S, Christodoulou C, Gait MJ (1986) Efficient methods for attaching non-radioactive labels to the 5′ ends of synthetic oligodeoxyribonucleotides. Nucleic Acids Res 14:6227–6245

Al-Hakim AH, Hull R (1986) Studies towards the development of chemically synthesized nonradioactive biotinylated nucleic acid hybridization probes. Nucleic Acids Res 14:9965–9976

Ausubel FM, Brent R, Kingston RE, Moore DD, Seidman JG, Smith JA, Struhl A (1987) Current protocols in molecular biology. Greene Publishing Associates and Wiley-Interscience, New York

Baumann JG, Wiegant J, van Duijin P (1983) The development, using poly(Hg-U) in a model system, of a new method to visualize cytochemical hybridization in fluorescence microscopy. J Histochem Cytochem 31:571–578

Ben-Hur E, Song PS (1984) The photochemistry and photobiology of furocoumarins (psoralens). Adv Radiat Biol 11:131–171

Bergstrom DE, Ruth JL (1977) Preparation of carbon-5 mercurated pyrimidine nucleosides. J Carbohydr (Nucleos Nucleot) 4:257–269

Brown DM, Frampton J, Goelet P, Karn J (1982) Sensitive detection of RNA using strand-specific M13 probes. Gene 20:139–144

Cimino GD, Gamper HB, Isaacs ST, Hearst JE (1985) Psoralens as photoactive probes of nucleic acid structure and function: organic chemistry, photochemistry and biochemistry. Annu Rev Biochem 54:1151–1193

Cook AF, Vuocolo E, Brakel CL (1988) Synthesis and hybridization of a series of biotinylated oligonucleotides. Nucleic Acids Res 16:4077–4095

Dale RMK, Martin E, Livingston DC, Ward DC (1975) Direct covalent mercuration of nucleotides and polynucleotides. Biochemistry 14:2447–2457

Dattagupta N, Crothers DM (1984) Labeled nucleic acid probes and adducts for their preparation. Eur Pat Appl 0131830

Draper DE, Gold L (1980) A method for linking fluorescent labels to polynucleotides: application to studies or ribosome-ribonucleic acid interactions. Biochemistry 19:1774–1781

Ehrat M, Cecchini DJ, Giese RW (1986) Substrate-leash amplification with ribonuclease S-peptide and S-protein. Clin Chem 32:1622–1630

Forster AC, McInnes JL, Skingle DC, Symons RH (1985) Nonradioactive hybridization probes prepared by the chemical labeling of DNA and RNA with a novel reagent, photobiotin. Nucleic Acids Res 13:745–761

Gebeyehu G, Rao PY, SooChan P, Simms DA, Klevan L (1987) Novel biotinylated nucleotide analogs for labeling and colorimetric detection of DNA. Nucleic Acids Res 15:4513–4534

Gillam IC, Tener GM (1986) N^4-(6-aminohexyl)cytidine and -deoxycytidine nucleotides can be used to label DNA. Anal Biochem 157:199–207

Gregersen N, Koch J, Koelvraa S, Petersen KB, Bolund L (1987) Improved methods for the detection of unique sequences in Southern blots of mammalian DNA by non-radioactive biotinylated DNA hybridization probes. Clin Chim Acta 169:267–280

Haralambidis J, Chai M, Tregear GW (1987) Preparation of base-modified nucleosides suitable for nonradioactive label attachment and their incorporation into synthetic oligodeoxyribonucleotides. Nucleic Acids Res 15:4857–4876

Höltke H-J, Kessler C (1990) Nonradioactive labeling of RNA transcripts in vitro with the hapten digoxigenin (DIG); hybridization and ELISA-based detection. Nucleic Acids Res 18:5843–5851

Höltke H-J, Sagner G, Kessler C, Schmitz G (1991) Sensitive chemiluminescent detection of digoxigenin-labeled nucleic acids: a fast and simple protocol and its application. BioTechniques 12:104−113

Höltke H-J, Seibl R, Burg J, Mühlegger K, Kessler C (1990) Nonradioactive labeling and detection of nucleic acids: II. Optimization of the digoxigenin system. Mol Gen Hoppe-Seyler 371:929−938

Hopman AHN, Wiegant J, Tesser GI, Van Duijn P (1986) A nonradioactive in situ hybridization method based on mercurated nucleic acid probes and sulfhydryl-hapten ligands. Nucleic Acids Res 14:6471−6488

Hopman AHN, Wiegant J, van Duijn P (1986) A new hybridocytochemical method based on mercurated nucleic acide probes and sulfhydryl-hapten ligands. I. Stability of the mercurysulfhydryl bond and influence of the ligand structure on immunochemical detection of the hapten. Histochemistry 84:169−178

Kempe T, Sundquist WI, Chow F, Hu SL (1985) Chemical and enzymatic biotin-labeling of oligonucleotides. Nucleic Acids Res 13:45−57

Kessler C, Höltke H-J, Seibl R, Burg J, Mühlegger K (1990) Nonradioactive labeling and detection of nucleic acids: I. A novel DNA labeling and detection system based on digoxigenin:anti-digoxigenin ELISA principle (digoxigenin system). Mol Gen Hoppe-Seyler 371:917−927

Kricka LJ (1992) Nonisotopic DNA probe techniques. Academic Press, San Diego

Landes GM (1985) Labeled DNA. Eur Pat Appl 0138357

Langer PR, Waldrop AA, Ward DC (1981) Enzymatic synthesis of biotin-labeled polynucleotides: novel nucleic acid affinity probes. Proc Natl Acad Sci USA 78:633−637

Matthews JA, Kricka LJ (1988) Analytical strategies for the use of DNA probes. Anal Biochem 169:1−25

McCracken S (1989) Preparation of RNA transcripts using SP6 RNA polymerase. In: Keller GH, Manak MM (eds) DNA Probes, Stockton Press, New York, pp 119−120

Mühlegger K, Huber E, von der Eltz H, Rüger R, Kessler C (1990) Nonradioactive labeling and detection of nucleic acids: IV. Synthesis and properties of the nucleotide compounds of the digoxigenin system and of photodigoxigenin. Mol Gen Hoppe-Seyler 371:939−951

Nelson PS, Frye RA, Liu E (1989a) Bifunctional oligonucleotide probes synthesized using a novel CPG support are able to detect single base pair mutations. Nucleic Acids Res 17:7187−7194

Nelson PS, Sherman-Gold R, Leon R (1989b) A new and versatile reagent for incorporating multiple primary aliphatic amines into synthetic oligonucleotides. Nucleic Acids Res 17:7179−7186

Pitcher DG, Owen RJ, Dyal P, Beck A (1987) Synthesis of a biotinylated DNA probe to detect ribosomal RNA cistrons in Providencia stuartii. FEMS Microbiol Lett 48:283−287

Reisfeld A, Rothenberg JM, Bayer EA, Wilchek M (1987) Nonradioactive hybridization probes prepared by the reaction of biotin hydrazide with DNA. Biochem Biophys Res Commun 142:519−526

Renz M, Kurz C (1984) A colorimetric method for DNA hybridization. Nucleic Acids Res 12:3435−3444

Rigby PWJ, Dieckmann M, Rhodes C, Berg P (1977) Labeling deoxyribonucleic acid to high specific activity in vitro by nick translation with DNA polymerase I. J Mol Biol 113:237−251

Riley LK, Marshall ME, Coleman MS (1986) A method for biotinylating oligonucleotide probes for use in molecular hybridization. DNA 5:333−337

Schmitz GG, Walter T, Kessler C (1991) Nonradioactive labeling of oligonucleotides in vitro with the hapten digoxigenin (DIG) by tailing with terminal transferase. Anal Biochem 192:222−231

Sheldon EL, Kellogg DE, Watson RE, Levinson CH, Erlich HA (1986) Use of nonisotopic M13 probes for genetic analysis: application to class II loci. Proc Natl Acad Sci USA 83:9085−9089

Sodja A, Davidson N (1978) Gene mapping and gene enrichment by the avidin-biotin interaction: use of cytochrome-c as a polyamine bridge. Nucleic Acids Res 5:385–401

Takahashi T, Mitsuda T, Okuda K (1989) An alternative nonradioactive method for labeling DNA using biotin. Anal Biochem 179:77–85

Theissen G, Richter A, Lukacs N (1989) Degree of biotinylation in nucleic acids estimated by a gel retardation assay. Anal Biochem 179:98–105

Urdea MS, Warner BD, Running JA, Stempien M, Clyne J, Horn T (1988) A comparison of nonradioisotopic hybridization assay methods using fluorescent, chemiluminescent and enzyme labeled synthetic oligonucleotide probes. Nucleic Acids Res 16:4937–4956

Vary CPH, McMahon FJ, Barbone FP, Diamond SE (1986) Nonisotopic detection methods for strand displacement assays of nucleic acids. Clin Chem 32:1696–1701

Viscidi RP, Connelly CJ, Yolken RH (1986) Novel chemical method for the preparation of nucleic acids for nonisotopic hybridization. J Clin Microbiol 23:311–317

Ward DC, Waldrop AA, Langer PR (1982) Modified nucleotides and their use. Eur Pat Appl 0063879

1.5 Detection Methods

In nonradioactive detection systems, along with marker enzymes which catalyze the conversion of color substrates, luminescence and fluorescence approaches are also often applied. An overview of the various detection systems used in direct or indirect systems is given in the accompanying tables. The generated signal can further be amplified by coupled enzymatic signal enhancement cascade reactions.

1.5.1 Direct Detection Systems

In direct systems the most popular labels are marker enzymes such as AP (Jablonski et al., 1986) or HRP (Renz and Kurz, 1984; Taub, 1986; Pollard-Knight, 1990) and fluorescent tags such as fluorescein ($\lambda_{max\ emission}$ = 527 nm) or rhodamine ($\lambda_{max\ emission}$ = 622 nm) (Serke and Pachmann, 1988). Whereas the coupled enzymes are often used in blot or soluble formats, the fluorescent dyes are especially useful in tissue sections for in situ detection in multiplex mapping of metaphase chromosomes (Lichter et al., 1990). In the hybridization protection assay (HPA), a chemiluminescent acridinium dye is linked to the probe via an alkaline-labile linker (Arnold et al., 1989). Hydrolysis of this linker takes place with single-stranded acridine-modified probes, whereas with the double-stranded hybrids the rate of hydrolysis is strongly reduced. This difference in hydrolysis is due to intercalation of the acridinium dye specifically into the double-stranded helix only; this intercalation stabilizes the linker against alkaline hydrolysis.

The difference in the rates of hydrolysis is used in quantitative homogeneous assay formats. Signal generation is obtained with unbound probe under alkaline conditions; by hydrolysis of the labile linker arm a

Nonradioactive detection systems

Format	Optical detection systems	Reference(s)	Luminescence detection systems	Reference(s)
Blot	AP/BCIP, NBT	Lojda et al. (1973); Anderson and Deinard (1974); Franci and Vidal (1988)	AP/AMPPD; AP/Lumigen™ PPD, Lumiphos™ 530; AP/CSPD™	Schaap et al. (1987); Bronstein et al. (1989a, 1989b); Bronstein et al. (1989a)
	AP/FAST dyes	West et al. (1990); Höltke et al. (1992)	β-Gal/AMPDG	Bronstein et al. (1989a)
	β-Gal/BCIG(X-gal),	Lojda et al. (1973); Anderson and Deinard (1974); Franci and Vidal (1988)		
	HRP/TMB	Bos et al. (1981)		
	Immunogold	Tomlinson et al. (1988); Guérin-Reverchon et al. (1989)		
Solution	AP/p-NPP	Garen and Levinthal (1960)	AP/AMPPD; AP/Lumigen™ PPD, Lumiphos™ 530; AP/CSPD™	Schaap et al. (1987); Bronstein et al. (1989a; 1989b)
	β-Gal/CPRG	Wallenfels et al. (1960)	β-Gal/AMPGD	Bronstein et al. (1989a)
	POD/ABTS®	Galatti (1979); Porstmann et al. (1981)	POD/luminol	Pollard-Knight et al. (1990)
			Xanthine oxidase/ cyclic dihydrazides	Baret and Fert (1989)
	GOD: POD-pair/ ABTS	Taub (1986)		
	Hexokinase: G-6- PDH-pair/ABTS®	Albarella et al. (1985a; 1985b)	G-6-PDH/phena- zinium salts	Tsuji et al. (1987)
			Hydrolysis of acridinium esters	Arnold et al. (1989)
			AP/D-luciferin-O- phosphate: firefly luciferase/ ATP/O$_2$	Hauber and Geiger (1987); Kricka (1988); Geiger et al. (1989)
In situ	AP/BCIP, NBT	Heiles et al. (1988); Tautz and Pfeifle (1989); Seibl et al. (1990);	AP/AMPPD	Bronstein and Voyta (1989)
	POD/TMB	Lichter et al. (1990)		
	Immunogold	Lichter et al. (1990); Tomlinson et al. (1988); Guérin-Reverchon et al. (1989)		

Nonradioactive, indirect, fluorescnce detection systems

Format	Fluorescence detection systems	Reference(s)	Electrochemical detection systems	Reference(s)
Blot	Fluorescein/ Rhodamine/ Hydroxy-coumarin	Agrawal et al. (1986)		
	AP/AMPPD, fluorescein/rhodamine/ hydroxy-coumarin	Voyta et al. (1988); Beck and Köster (1990)		
Solution	AP/4-MUF-P	Fernley and Walker (1965)	Urease/urea	McKnabb and Tedesco (1989)
	Attophos™	Donahue et al. (1991)	Ruthenium-II complexes	Blackburn et al. (1991)
	β-Gal/4-MUF-β-Gal	Buonocore et al. (1980)		
	POD/homovanillinic acid-o-dianisidine/ H_2O_2	Twai et al. (1983); Arkawa et al. (1982)		
	Aromatic peroxalate compounds/H_2O_2	Hemmilä et al. (1984); Lovgren et al. (1985)		
	Eu^{3+} micelles and Eu^{3+} chelates	Diamandis (1988); Diamandis et al. (1988; 1989)		
In situ	Fluorescein/ Rhodamine/ Hydroxy-coumarin	Agrawal et al. (1986)		

cyclooxetane ring intermediate is formed followed by rapid conversion to an excited N-methylacridone, which emits light upon relaxation to the ground state.

1.5.2 Indirect Detection Systems

In indirect systems, the whole range of detection reactions can be utilized. The binding groups recognizing the modified probes are covalently bound to molecules which, in turn, generate the signal to be measured. In most cases, these conjugated molecules are enzymes that generate an optical, luminescent, or fluorescent reaction product through a coupled catalytic substrate reaction (for review see Kricka, 1992).

1.5.2.1 Optical Detection

The best known marker enzymes are AP from calf intestine (Ishikawa et al., 1983), HRP (Wilson and Nakane, 1978), and β-galactosidase from E. coli (β-Gal) (Inoue et al., 1985). Urease (McKnabb et al., 1989), glucose oxidase (GOD), and microperoxidase are also used but to a lesser extent.

Depending on the kind of substrate used, the enzymatically catalyzed substrate reactions yield either insoluble precipitates on blots and in in situ approaches or soluble color products for quantitative measurements, e.g., in microtiter plates. Sensitive substrates resulting in insoluble precipitates include 5-bromo-4-chloro-3-indolyl phosphate (BCIP or X phosphate)/ nitroblue tetrazolium salt (NBT) for AP (Franci and Vidal, 1988), 3.3',5.5'-tetramethylbenzidine (TMB) for peroxidase (Bos et al., 1981), and 5-bromo-4-chloro-3-indolyl β-galactoside (BCIG or X-gal)/NBT for β-Gal (Lojda et al., 1973); the respective soluble substrates are p-nitro-phenyl phosphate (p-NPP), 2,2-azino-di-[3-ethyl-benzthiazoline sulfate] (ABTS®) (Gallati, 1979; Porstmann et al., 1981), and chlorophenol red-β-D-galactopyranoside (CPRG) (Wallenfels et al., 1960).

Increased sensitivity and speed of detection can be obtained in optical systems by coupling of an enzymatic cascade reaction using the following components (Self, 1985; Johannson et al., 1985; Stanley et al., 1985): alcohol dehydrogenase (ADH); diaphorase (DP); nicotinamide-dinucleo-tide, oxidated form (NAD); and nicotinamide-dinucleotide, reduced form (NADH) with p-iodonitro-tetrazolium purple (INT violet) as the color substrate.

In the first reaction step, a primary NADP substrate is dephosphory-lated to NAD by hybrid-bound AP. The NAD cofactor activates in a cyclic reaction a coupled secondary redox system that contains the two enzymes ADH and DP. In each cycle the generated NAD is reduced to NADPH+ H^+ by linked oxidation of ethanol to acetaldehyde by ADH; this reaction is coupled to a second redox reaction in which NADPH + H^+ is oxidized to NAD and again linked with the reduction of INT violet to formazan. The intensively stained formazan can be photometrically quantified: ($\lambda_{max\ emission}$ = 465 nm/ethanol,dimethylformamide).

1.5.2.2 Luminescence Detection

Aside from optical systems, a number of alternative detection systems have been developed based on luminescent substrates.

In *bioluminescence systems,* a bioluminescent substrate is released by an initial enzymatic reaction mediated by a hybrid-bound marker enzyme (Miska and Geiger, 1987; Hauber and Geiger, 1987; Gould and Subra-mani, 1988; Kricka, 1988; Geiger et al., 1989). Using D-luciferin-O-phos-phate and AP, D-luciferin is released by phosphate hydrolysis and sub-sequently converted into oxyluciferin, AMP, and PP, in a coupled reaction catalyzed by the enzyme luciferase from *Photinus pyralis* in the presence of O_2 and ATP. This luciferase-catalyzed oxidation of D-luciferin into oxyluciferin is accompanied by the emission of light ($\lambda_{max\ emission}$ = 600 nm). In *chemiluminescence systems,* soluble chemiluminescent substrates are used with AP or β-Gal by exchanging the optical BCIP/NBT or BCIG/ NBT substrates for the corresponding 1,2-dioxetane luminescence substrates 3-(2'-spirodamantane)-4-methoxy-(3''-phosphoryloxy)-phenyl-1,2-dioxetane (Lumigen™PPD/Lumiphos™530/CSPD™) as well as 3-

(4-methoxyspiro[1,2-dioxetane-3,2'-tricyclo[3.3.1.1^{3,7}]decan]-4-yl)phenyl phosphate or 3-(4-methoxyspiro [1,2-dioxetane-3,2'-tricyclo [3.3.1^{3,7}] decan]-4-yl) phenyl β-galatoside (AMPPD or AMPGD) (Schaap et al., 1987; Voyta et al., 1988; Bronstein and Kricka, 1989; Bronstein and Voyta, 1989; Thorpe and Kricka, 1989; Bronstein et al., 1989a, 1989b; Beck and Köster, 1990; Höltke et al., 1991). In the AMPPD and AMPGD chemiluminescent systems, enzymatic cleavage of a phosphate (AMPPD) or β-galactoside (AMPGD) residue forms the unstable AMPD$^-$ anion, which decomposes to produce light. The chemiluminescence intensity can be increased by inclusion of the 1,2-dioxetane derivatives in micelles or in covalently linked polymers which, in turn, contain fluorophores such as fluorescein that can be activated by the appropriate emitted light. By selecting suitable fluorophores, a secondary fluorescence signal can be generated as follows: hydroxy-coumarin ($\lambda_{max\ emission}$ = 467 nm), fluorescein ($\lambda_{max\ emission}$ = 527 nm), rhodamine ($\lambda_{max\ emission}$ = 622 nm).

Electrochemiluminescence detection is possible with ruthenium-II-tris(bipyridyl) complexes ([Ru(bpy)$_3$]$^{2+}$ salts) at the surface of an electrode (Blackburn et al., 1991).

1.5.2.3 Fluorescence Detection

Analogous micelles or chelates as with enhanced chemiluminescence are used with time-resolved fluorescence (TRF), in which the probes are coupled with lanthanide ions (e.g., Eu^{3+} or Tb^{3+}) complexed by micelles (Hemmilä et al., 1984; Lovgren et al., 1985) or chelating agents (Soini and Kojola, 1983; Diamandis, 1988; Evangelista et al., 1988; Oser and Valet, 1988; Diamandis et al., 1988; 1989). However, in contrast to chemiluminescence, the fluorescent signal is not generated by enzymatic activation of the substrate but by activation via illumination with photons, forcing the lanthanide ions into excited states. The micelles or chelates stabilize these activated states; light emission is therefore retarded. This retardation has the advantage that there is no or only little overlap between exciting and emitting light, resulting in background-free detection of fluorescent signals. Besides TRF-based fluorescence detection, AP-catalyzed fluorescence detection using Attophos^{TM} is possible (Donahue et al., 1991).

References

Agrawal S, Christodoulou C, Gait MJ (1986) Efficient methods for attacking non-radioactive labels to the 5' ends of synthetic oligodeoxyribonucleotides. Nucleic Acids Res 14:6227–6245

Albarella JP, Anderson LH, Carricio RJ (1985a) Detection of polynucleotide sequence in sample of nucleic acids by using nucleic acid probe and contact of duplexes with immobilized antibody. Eur Pat Appl 0146039

Albarella JP, DeRiemer LHA, Carrico RJ (1985b) Hybridization assay employing labeled pairs of hybrid binding reagents. Eur Pat Appl 0144914

Anderson GL, Deinard AS (1974) Nitroblue tetrazolium (NBT) test. Review Am J Med Technol 40:345–353

Arakawa H, Maeda M, Tsuji A (1982) Chemiluminescence enzyme immunoassay of 17-hydroxyprogesterone using glucose oxidase and *bis*(2,4,6-trichlorophenyl)oxalate-fluorescent dye system. Chem Pharm Bull 30:3036−3039

Arnold LJ, Hammond PW, Wiese WA, Nelson NC (1989) Assay formats involving acridinium ester-labeled DNA probes. Clin Chem 35:1588−1594

Baret A, Fert V (1989) T_4 and ultrasensitive TSH immunoassays using luminescent enhanced xanthine oxidase assay. J Biolumin Chemilumin 4:149−153

Beck S, Köster H (1990) Applications of dioxetane chemiluminescent probes to molecular biology. Anal Chem 62:2258−2270

Blackburn GF, Shah HP, Kenten JH, Leland J, Kamin RA, Link J, Peterman J, Powell MJ, Shah A, Talley DB, Tyagi SK, Wilkins E, Wu T-G, Massey RJ (1991) Electrochemiluminescence detection for development of immunoassays and DNA probe assays for clinical diagnosis. Clin Chem 37:1534−1539

Bos ES, van der Doelen AA, van Rooy N, Schuurs AH (1981) 3,3′,5,5′-Tetramethyl-benzidine as an Ames test negative chromogen for horse-radish peroxidase in enzyme-immunoassay. J Immunoassay 2:187−204

Bronstein I, Edwards B, Voyta JC (1989a) 1,2-Dioxetanes; novel chemiluminescent enzyme substrates. Applications to immunoassay. J Biolumin Chemilumin 4:99−111

Bronstein I, Kricka LJ (1989) Clinical applications of luminescent assay for enzymes and enzyme labels. J Clin Lab Anal 3:316−322

Bronstein I, Voyta JC (1989) Chemiluminescent detection of herpes simplex virus I DNA in blot and in situ hybridization assay. Clin Chem 35:1856−1857

Bronstein I, Voyta JC, Edwards B (1989b) A comparison of chemiluminescent and colorimetric substrates in a hepatitis B virus DNA hybridization assay. Anal Biochem 180:95−98

Buonocore V, Sgambati O, De Rosa M, Esposito E, Gambacorta A (1980) A constitutive β-galactosidase from the extreme thermoacidophile archaebacterium Caldariella acidophila: properties of the enzyme in the free state and in immobilized whole cells. J Appl Biochem 2:390−397

Diamandis EP (1988) Immunoassay with time-resolved fluorescence spectroscopy: principles and applications (Review). Clin Biochem 21:139−150

Diamandis EP, Bhayana V, Conway K, Reichstein E, Papanastasiou-Diamandis A (1988) Time-resolved fluoroimmunoassay of cortisol in serum with a europium chelate as label. Clin Biochem 21:291−296

Diamandis EP, Morton RC, Reichstein E, Khoasravi MJ (1989) Multiple fluorescence labeling with europium chelators. Application to time-resolved fluoroimmunoassays. Anal Chem 61:48−53

Donahue C, Neece V, Nycz C, Weng JMH, Walker GT, Vonk GP, Jurgensen S (1991) The San Diego Conference on Nucleic Acids: The leading edge. San Diego, CA, Abstract 23

Evangelista RA, Pollak A, Allore B, Templeton EF, Morton RC, Diamandis EP (1988) A new europium chelate for protein labeling and time-resolved fluorometric applications. Clin Biochem 21:173−178

Fernley HN, Walker PG (1965) Kinetic behaviour of calf-intestinal alkaline phosphatase with 4-methylumbelliferyl phosphate. Biochem J 97:95−103

Franci C, Vidal J (1988) Coupling redox and enzymic reactions improves the sensitivity of the ELISA-spot assay. J Immunol Methods 107:239−244

Gallati H (1979) Horseradish peroxidase: a study of the kinetics and the determination of optimal reaction conditions using hydrogen peroxide and 2,2′-azinobis(3-ethyl-benzthiazoline-6-sulfonic acid) (ABTS) as substrate. J Clin Chem Clin Biochem 17:1−7

Garen A, Levinthal C (1960) A fine-structure genetic and chemical study of the enzyme alkaline phosphatase of E. coli. I. Purification and characterization of alkaline phosphatase. Biochim Biophys Acta 38:470−483

Geiger R, Hauber R, Miska N (1989) New, bioluminescence-enhanced detection system for use in enzyme activity tests, enzyme immunoassays, protein blotting and nucleic acid hybridization. Mol Cell Probes 3:309−328

Gould SJ, Subramani S (1988) Review. Firefly luciferase as a tool in molecular and cell biology. Anal Biochem 175:5−13

Guérin-Reverchon I, Chardonnet Y, Chignol MC, Thivolet J (1989) A comparison of methods for the detection of human papillomavirus DNA by in situ hybridization with biotinylated probes on human carcinoma cell lines: application to wart sections. J Immunol Meth 123:167−176

Hauber R, Geiger R (1987) A new, very sensitive, bioluminescence-enhanced detection system for protein blotting. I. Ultrasensitive detection systems for protein blotting and DNA hybridization. J Clin Chem Clin Biochem 25:511−514

Heiles HBJ, Genersch E, Kessler C, Neumann R, Eggers HJ (1988) In situ hybridization with digoxigenin-labeled DNA of human papillomavirus (HPV 16/18) in HeLa and SiHa cells. BioTechniques 6:978−981

Hemmilä I, Dakubu S, Mukka V-M, Siitari H, Lövgren T (1984) Europium as a label in time-resolved immunofluorometric assays. Anal Biochem 137:335−343

Höltke HJ, Sagner G, Kessler C, Schmitz G (1991) Sensitive chemiluminescent detection of digoxigenin-labeled nucleic acids: a fast and simple protocol and its application. BioTechniques 12:104−113

Höltke HJ, Ettl I, Finken M, West S, Kunz W (1992) Multiple nucleic acid labeling and rainbow detection. Anal Biochem, in press

Inoue S, Hashida S, Tanaka K, Imagawa M, Ishikawa E (1985) Preparation of monomeric affinity-purified Fab'-β-D-galactosidase conjugate for immunoenzymometric assay. Anal Lett 18:1331−1344

Ishikawa E, Imagawa M, Hashida S, Yoshitake S, Hamaguchi Y, Ueno T (1983). Enzyme labeling of antibodies and their fragments for enzyme immunoassay and immunohistochemical staining. J Immunoassay 4:209−327

Iwai H, Ishihara F, Akihama S (1983) A fluorometric rate assay for peroxidase using the homovanillic acid-o-dianisidine-hydrogen peroxide system. Chem Pharm Bull 31:3579−3582

Jablonski E, Moomaw EW, Tullis RH, Ruth JL (1986) Preparation of oligodeoxynucleotide-alkaline phosphatase conjugates and their use as hybridization probes. Nucleic Acids Res 14:6115−6128

Johannsson A, Stanley CJ, Self CH (1985) A fast highly sensitive colorimetric enzyme immunoassay system demonstrating benefits of enzyme amplification in clinical chemistry. Clin Chim Acta 148:119−124

Kricka LJ (1988) Review. Clinical and biochemical applications of luciferase and luciferins. Anal Biochem 175:14−21

Kricka LJ (1992) Nonisotopic DNA probe techniques. Academic Press, San Diego

Lichter P, Tang CJC, Call K, Hermanson G, Evans GA, Housman D, Ward DC (1990) High-resolution mapping of human chromosome 11 by in situ hybridization with cosmid clones. Science 247:64−69

Lojda Z, Slaby J, Kraml J, Kolinska J (1973) Synthetic substrates in the histochemical demonstration of intestinal disaccharidases. Histochemie 34:361−369

Lovgren T, Hemmilä I, Pettersson K, Halonen P (1985) Time-resolved fluorometry in immunoassay. In: Collins WP (ed) Alternative immunoassays, John Wiley and Sons, Chichester, England

McKnabb S, Rupp R, Tedesco JL (1989) Measuring contamination DNA in bioreactor derived monoclonals. Bio/Technology 7:343−347

Miska W, Geiger R (1987) Synthesis and characterization of luciferin derivatives for use in bioluminescence enhanced enzyme immunoassay. I. New ultrasensitive detection systems for enzyme immunoassay. J Clin Chem Clin Biochem 25:23−30

Oser A, Roth WK, Valet G (1988) Sensitive nonradioactive dot-blot hybridization using DNA probes labeled with chelate group substituted psoralen and quantitative detection by europium ion fluorescence. Nucleic Acids Res 16:1181−1196

Pollard-Knight D, Read CA, Downes MJ, Howard LA, Leadbetter MR, Pheby SA, McNaughton E, Syms A, Brady MAW (1990) Nonradioactive nucleic acid detection by enhanced chemiluminescence using probes directly labeled with horseradish peroxidase. Anal Biochem 185:84−89

Porstmann B, Porstmann T, Nugel E (1981) Comparison of chromogens for the determination of horseradish peroxidase as a marker in enzyme immunoassay. J Clin Chem Clin Biochem 19:435−439

Renz M, Kurz C (1984) A colorimetric method for DNA hybridization. Nucleic Acids Res 12:3435−3444

Schaap AP, Sandison MD, Handley RS (1987) Chemical and enzymatic triggering of 1,2-dioxetanes. Alkaline phosphatase-catalyzed chemiluminescence from an aryl phosphate-substituted dioxetane. Tetrahedron Lett 28:1159−1162

Seibl R, Höltke H-J, Rüger R, Meindl A, Zachau H-G, Rasshofer G, Roggendorf M, Wolf H, Arnold N, Wienberg J, Kessler C (1990) Nonradioactive labeling and detection or nucleic acids: III. Applications of the digoxigenin system. Mol Gen Hoppe-Seyler 371:939−951

Self CH (1985) Enzyme amplification − a general method applied to provide an immunoassisted assay for placental alkaline phosphatase. J Immunol Methods 76:389−393

Soini E, Kojola H (1983) Time-resolved fluorometer for lanthianide chelates − a new generation of nonisotopic immunoassays. Clin Chem 29:65−68

Stanley CJ, Johannsson A, Self CH (1985) Enzyme amplification can enhance both the speed and the sensitivity of immunoassays. J Immunol Methods 83:89−95

Taub F (1986) An assay for nucleic acid sequences, particularly genetic lesions. PCT Int Appl WO 86/03227

Tautz D, Pfeifle C (1989) A nonradioactive in situ hybridization method for the localization of specific RNAs in Drosophila embryos reveals translational control of the segmentation gene hunchback. Chromosoma 98:81−85

Tomlinson S, Lyga A, Huguenel E, Dattagupta N (1988) Detection of biotinylated nucleic acid hybrids by antibody-coated gold colloid. Anal Biochem 171:217−222

Tsuji A, Maeda M, Arakawa H, Shimizu S, Tanabe K, Sudo Y (1987) Chemiluminescence enzyme immunoassay using invertase, glucose-6-phosphate dehydrogenase and β-D-galactosidase as label. In: Scholmerich J, Anderson R, Kapp A, Ernst M, Woods WG (eds) Bioluminescence and chemiluminescence, Wiley, Interscience, Chichester, England, pp 233−235

Voyta JC, Edwards B, Bronstein I (1988) Ultrasensitive chemiluminescent detection of alkaline phosphatase activity. Clin Chem 34:1157

Wallenfels K, Lehmann J, Malhotra OP (1960) Untersuchungen über milchzuckerspaltende Enzyme − Die Spezifität der β-Galactosidase von E. coli ML309. Biochem Z 333:209−225

West S, Schröder J, Kunz W (1990) A multiple-staining procedure for the detection of different DNA fragments on a single blot. Anal Biochem 190:254−258

Wilson MB, Nakane PK (1978) Recent development in the periodate method of conjugating horseradish peroxidase (HRPO) to antibodies. In: Knapp W, Holubar K, Wick G (eds) Immunofluorescence and related staining techniques, Elsevier/North Holland Biomedical Press, New York, Amsterdam, pp 215−224

1.6 Guide to the Use of Information in this Book

Following this overview of nonradioactive labeling and detection of biomolecules, the remaining parts of this book provide a short description, accompanied by detailed protocols, of: (a) standard labeling and detection systems (Part I), (b) specialized nonradioactive detection systems (Part II) and (c) enhanced amplification systems (Part III).

In Part IV detailed protocols of selected applications are given using blot, in situ or soluble formats.

Accompanying each chapter are lists of the necessary standard reagents, buffers, and commercially available nonradioactive compounds and the respective kits.

The central aim of this book is to give the reader both the principles of the various methods and the respective protocols. Therefore each chapter is comprised of:

1. A short description of the method including its characteristics, limitations, and application(s)
2. A reaction scheme
3. A detailed standard protocol including necessary reagents, buffers, and equipment
4. Special hints which have been compiled to simplify application and allow for troubleshooting.

For each of the various applications, a representative example is also described. Cross-reference is given to the respective labeling and detection methods used for the particular applications. For direct use of the protocols in the lab the relevant technical information regarding the availability of the needed equipment and material is collected in the appendix.

I Standard Nonradioactive Labeling and Detection Systems

2 Overview of Nonradioactive Labeling Systems

CHRISTOPH KESSLER

A variety of labeling systems for nucleic acids, proteins, and glycans have been developed in the past decade. The table shows an overview of the important labeling and detection systems, including cross-references to the descriptions in the other chapters and sections in this book. Among the indirect approaches the most sensitive systems are: (a) the biotin:(strept-) avidin (bio) and (b) the digoxigenin:anti-digoxigenin (DIG) system which realize the detection of less than picogram levels of DNA or RNA. However, other indirect systems are also of interest especially in particular applications such as the study of cell proliferation (5-BrdU), in situ studies (sulfone), or staining of tissue sections in addition to histological staining (immunogold). The described direct alkaline phosphatase (AP)- or horse-radish peroxidase (HRP)-based systems (SNAP/ECL) are useful, for example, for hybridization with standard probes in fingerprinting or in membrane-based sequencing approaches.

In the biotin system, the modification of the binding component is mediated by the vitamin biotin (Langer et al., 1981; Wilchek and Bayer, 1988). The incorporated biotin is detected by the binding of the indicator protein avidin, isolated from egg white, or streptavidin, isolated from *Streptomyces avidinii* bacteria. Each protein has four high-affinity binding sites for biotin; the binding constant is $K = 10^{15} \text{mol}^{-1}$ (Chaiet and Wolf, 1964; Greene, 1975). The biotin label has been widely used in a variety of different assays, including both direct and indirect formats, for detection of nucleic acids, proteins, and glycans on blots, in solution, or in situ (for references see Bayer and Wilchek, 1990).

An inherent property of the biotin system is that an endogenous vitamin, vitamin H, is used as a modification group. Therefore, background reactions may occur with endogenous biotin especially during in situ analysis of material of natural origin. Another property of the biotin system is the tendency of the two binding proteins towards elevated nonspecific interaction with blotting membranes even if the membrane surfaces are treated with membrane-blocking substances.

Background reactions have been reduced by deglycosylating avidin (Jones et al., 1987) or by preincubating the blotting membranes with buffers of high ionic strength, by blocking the membranes with lactoproteins, or by forming complexes between avidin and the acidic lysozyme protein (Duhamel and Johnson, 1985; Hiller et al., 1987).

Nonradioactive labeling systems

Mode of labeling	Mode of detection	Cross reference	Development/ availability	Reference(s)
Nucleic acid systems on polynucleotide basis				
Enzymatic modification				
Biotin-dUTP/ nick translation	Streptavidin-AP	Sect. 4.1	Boehringer Mannheim/ Enzo	Langer et al. (1981)
Biotin-dUTP/tailing	Streptavidin-AP	Sect. 4.1	Boehringer Mannheim/ Enzo	Brakel and Engelhardt (1985)
Biotin-dATP/nick translation	Streptavidin-AP	Sect. 4.1	LTI	Gebeyehu et al. (1987)
Digoxigenin dUTP/ random priming	Anti-digoxigenin-AP	Sect. 3.2	Boehringer Mannheim	Kessler et al. (1990)
Digoxigenin-UTP/ transcription	Anti-digoxigenin-AP	Sect. 3.2	Boehringer Mannheim	Höltke and Kessler (1990)
Chemical modification				
AAIF	Secondary AP ab	–	INSERM	Tchen et al. (1984)
	Secondary Eu^{2+} ab	–	Orion	Syvänen et al. (1986)
Sulfone	Secondary ab	Chap. 6	Orgenics	Proverenny et al. (1979)
POD	Direct	–	EMBO	Renz and Kurz (1984)
	Direct/luminol	Chap. 8	Amersham	Pollard-Knight et al. (1990)
	Direct	Chap. 9	Digene	Taub (1986)
Biotin-angelicin	Secondary ab	–	Miles	Albarella et al. (1989)
Biotin-psoralen	Streptavidin-POD TRF	–	Cetus/MPI	Sheldon et al. (1986); Oser et al. (1988)
Photobiotin	Streptavin-AP	Sect. 4.1	Boehringer Mannheim/ BRESA/LTI	Forster et al. (1985)
	Streptavidin-Eu^{2+}	–	Orion	Dahlen et al. (1987)
	Gold ab	Chap. 7	Miles	Tomlinson et al. (1988)
Photo-DNP	Secondary ab	–	Biotec Research	Keller et al. (1989)
Photodigoxigenin	Antidigoxigenin-AP	Sect. 3.2	Boehringer Mannheim	Mühlegger et al. (1990)
Biotin transamination	Streptavidin-AP	–	Johns Hopkins University	Viscidi et al. (1986)
Digoxigenin transamination	Anti-digoxigenin- β-gal	Sect. 3.2	Boehringer Mannheim	Graf and Lenz (1984)
Eu^{2+} transamination	TRF	–	Wallac	Dahlen et al. (1988)
Hg^{2+} derivatization	HS hapten/secondary ab	–	University of Leiden	Hopman et al. (1986)
5-BrdU derivatization	Secondary ab	Chap. 5	Biotech Research	Keller et al. (1988)
Biotin hydrazide	Streptavidin-AP	–	Weizman Institute/ Showa University	Reisfeld et al. (1987); Takahashi et al. (1989)
Diazobiotin	Streptavidin-AP	–	Weizmann Institute	Rothenberg and Wilchek (1988)
Biotin-DNA binding protein	Streptavidin acid phosphatase	–	Orion	Syvänen et al. (1985)
Nucleic acid systems on oligonucleotide basis				
Enzymatic modification				
Biotin-dUTP/tailing	Streptavidin-AP	Sect. 4.1	Boehringer Mannheim/ INSERM/LTI	Kumar et al. (1988)
Digoxigenin-dUTP/ tailing	Anti-digoxigenin-AP	Sect. 3.2	Boehringer Mannheim	Schmitz et al. (1991)

Nonradioactive labeling systems

Mode of labeling	Mode of detection	Cross reference	Development/ availability	Reference(s)
Chemical modification				
Aminocytosine-oligo-marker enzyme	Direct	Chap. 10	Molecular Biosystems/ Chiron	Jablonski et al. (1986); Urdea et al. (1988)
5'-Amino-oligo-marker enzyme	Direct	–	University Adelaide	Li et al. (1987)
			EMBO	Sproat et al. (1987)
HS-oligo-marker enzyme	Direct	–	Salk Institute	Chu and Orgel (1988)
Protein systems				
Chemical modification				
Biotin-NHS/amino groups	Streptavidin-AP	Sect. 4.2	Amersham/Boehringer Mannheim/Pierce/Sigma	Wilchek and Bayer (1988)
Biotin-MH/ mercaptane groups	Streptavidin-AP	Sect. 4.2	Amersham/Boehringer Mannheim/Pierce/Sigma	Strauss (1984); Bayer et al. (1985)
Biotin-DAB/tyr,his hydroxyl groups	Streptavidin-AP	Sect. 4.2	Weizman Institute	Wilchek et al. (1986)
Biotin-PNP/ aromatic groups	Streptavidin-AP	Sect. 4.2	Weizman Institute	Wilchek et al. (1986)
Photobiotin	Streptavidin-AP	Sect. 4.2	Boehringer Mannheim/ LTI/Pierce	Lacey and Grant (1987)
Digoxigenin-NHS/ amino groups	Anti-digoxigenin-AP	Sect. 3.3	Boehringer Mannheim	Haselbeck and Hösel, personal communication
Digoxigenin-PM/ mercaptane groups	Anti-digoxigenin-AP	Sect. 3.3	Boehringer Mannheim	Haselbeck and Hösel, personal communication
Photodigoxigenin	Anti-digoxigenin-AP	Sect. 3.3	Boehringer Mannheim	Haselbeck and Hösel, personal communication; Mühlegger et al. (1990)
Glycan systems				
Chemical modification				
Biotin-HZ	Streptavidin-AP	Sect. 4.2	Weizman Institute	Wilchek and Bayer (1987)
Digoxigenin-HZ	Anti-digoxigenin-AP	Sect. 3.3	Boehringer Mannheim	Haselbeck and Hösel (1990)
Digoxigenin-labeled lectins GNA, SNA, MAA, PNA, DSA	Anti-digoxigenin-AP	Sect. 3.3	Boehringer Mannheim	Haselbeck et al. (1990)

Due to the above-described properties of the biotin system, an alternative, the DIG system, of equal sensitivity but with reduced nonspecific background reactions was developed (Kessler et al., 1990). The hapten: anti-hapten DIG system is based on the specific interaction between the cardenolide steroid DIG and a high-affinity, DIG-specific antibody (Höltke et al., 1990). Since the cardenolide DIG occurs exclusively in *Digitalis* plants (Hegnauer, 1971), nonspecific interactions with endogenous cellular substances are strongly reduced in other biological materials using DIG as a modification group. Only in human sera have anti-DIG cross-reacting binding activities been described. However, these binding activities can be specifically counteracted by pretreating the serum (Armbruster and Greene, 1989).

The DIG system has been widely applied in the direct and indirect detection of nucleic acids, proteins, and glycans on blots, in solution, or in situ formats (for reviews see Kessler, 1990; Kessler, 1991).

With the 5-BrdU system the bases of both DNA and RNA are readily brominated under mild conditions using dilute aqueous bromine or N-bromosuccinimide. Bromination occurs at the C-8 position of purines and at the C-5 position of cytosine and possibly at the C-6 position of thymine. The brominated intermediates can be detected directly via 5-BrdU-specific antibodies coupled with reporter molecules (Sakamoto et al., 1987) or by reaction with amines such as 1,6-diaminohexane prebound with detectable haptens such as disophenol (2,6-diiodo-4-nitrophenol, DNP) via one of the two amino functions (Keller et al., 1988).

With the sulfone system a sulfone group is introduced as a hapten into the nucleic acid by sulfonation (Verdlov et al., 1974), which is obtained with a high concentration of sodium bisulfite, at position C-6 of cytidine residues. The resulting sulfone derivative is relatively unstable but can be stabilized by substituting the amino group at C-4 of the cytosine base with the nucleophilic reagent methylhydroxylamine. Cytosines are transformed by this reaction into N^4-methoxy-5,6-dihydrocytosine-6-sulfonate derivatives. These cytosine derivatives of nucleic acid probes can directly be detected by sulfone-specific antibodies coupled with reporter groups (Herzberg, 1984; Nur et al., 1989).

With the immunogold system hapten-specific antibodies are coupled directly with gold particles of standardized size (Tomlison et al., 1988). After hybrid formation, the hybrid-fixed gold particles act as starters for subsequent enhancing silver precipitation; enhancement is obtained by enlarging the primary gold particles by forming layers of silver coating. Direct detection of these silver-enhanced particles is possible with a light microscope. In addition, the tissue section can be histologicaly stained (Saman, 1986).

Direct coupling of the reporter enzyme alkaline phosphatase (AP) is possible with short synthetic oligodeoxynucleotides; in this case the probes are covalently cross-linked to the marker enzyme AP using the homobifunctional reagent disuccinimidyl suberate (Jablonski et al., 1986). Oligodeoxynucleotides in the range of 21–26 bases can easily be labeled by this method. The oligodeoxynucleotides are modified in a first reaction step with a linker arm, which contains a terminal reactive primary amine function. Enzyme coupling is obtained by acylation of enzyme amino groups with N-hydroxysuccinimidyl ester. The linking reaction results in defined enzyme:oligodeoxynucleotide complexes consisting of one enzyme label per oligodeoxynucleotide, with a mass ratio of protein to DNA of about 20. AP is detected either with dye substrates, e.g., 5-bromo-4-chloro-3-indolyl phosphate/nitroblue tetrazolium salt (BCIP/NBT) [small nuclear AP (SNAP) probes], or via phosphorylated phenyl dioxetane mediated chemiluminescence using e.g., 3-(4-methoxyspiro [1,2-dioxetane-3,2'-tricyclo[3.3.1.1^{3,7}]decan]-4-yl) phenyl phosphate (AMPPD) [nonisotopic chemiluminescent enhanced (NICE) probes].

Direct cross-linking of the reporter enzyme horseradish peroxidase (HRP) to probe DNA is achieved by coupling a performed enzyme-polyethyleneimine complex (Renz and Kurz, 1984) and the bifunctional cross-linking reagent glutaraldehyde (Pollard-Knight et al., 1990). The labeled probe can be used for hybridization without any further high-performance liquid purification.

With this labeling method, DNA probes ranging from 50 to several thousand base pairs (bp) can be cross-linked with the marker enzyme. Approximately one active peroxidase molecule in linked every 50−100 bp. Therefore, probes smaller than 50 bp do not react with the labeling components. For chemiluminescent detection of HRP hybrid complexes, HRP oxidizes the chemiluminescent substrate luminol in the presence of light-intensifying enhancer compounds such as p-iodophenol. For detection during in situ analyses a silver enhancement reaction may also be applied.

References

Albarella JP, Minegar RL, Patterson WL, Dattagupta N, Carlson E (1989) Monoadduct forming photochemical reagents for labeling nucleic acids for hybridization. Nucleic Acids Res 17:4293−4308

Armbruster DA, Greene DT (1989) Indole compounds do not cause false positives with the TDx cannabinoid assay. Clin Chem 35:323

Bayer EA, Wilchek M (1990) Avidin-biotin technology. Methods in Enzymology, Vol 184. Academic Press, San Diego

Bayer EA, Zalis MG, Wilchek M (1985) 3-(N'-Malaimido-propionyl)biocytin: a versatile thio-specific biotinylating reagent. Anal Biochem 149:529−536

Brakel DL, Engelhardt DL (1985) DNA hybridization method using biotin. In: Kingsbury DT, Falcow S (eds) Rapid detection and identification of infectious agents, Academic Press, New York, pp 235−245

Chaiet L, Wolf FJ (1964) The properties of streptavidin, a biotin-binding protein produced by Streptomycetes. Arch Biochem Biophys 106:1−5

Chu ECF, Orgel LE (1988) Ligation of oligonucleotides to nucleic acids or proteins via disulfide bonds. Nucleic Acids Res 16:3671−3691

Dahlen P, Hurskainen P, Lovgren T, Hyypia T (1988) Time-resolved fluorometry for the identification of viral DNA in clinical specimens. J Clin Microbiol 26:2434−2436

Dahlen R, Syvänen AC, Hurskainen P, Kwiatkowski M, Sund C, Ylikoski J, Söderlund H, Lovgren T (1987) Sensitive detection of genes by sandwich hybridization and time-resolved fluorometry. Mol Cell Probes 1:159−168

Duhamel RC, Johnson DA (1985) Use of nonfat dry milk to block nonspecific nuclear and membrane staining by avidin conjugates. Histochem Cytochem 33:711−714

Forster AC, McInnes JL, Skingle DC, Symons RH (1985) Nonradioactive hybridization probes prepared by the chemical labeling of DNA and RNA with a novel reagent, photobiotin. Nucleic Acids Res 13:745−761

Gebeyehu G, Rao PY, SooChan P, Simms DA, Klevan L (1987) Novel biotinylated nucleotide analogs for labeling and colorimetric detection of DNA. Nucleic Acids Res 15:4513−4534

Graf H, Lenz H (1984) Derivatized nucleic acid sequence and its use in detection of nucleic acids. Deutsche Offenlegungsschrift 3431536

Greene NM (1975) Avidin. In: Anfinsen CB, Edsall JT (eds) Advances in protein chemistry. Academic Press, New York, pp 85−133

Haselbeck A, Hösel W (1990) Description and application of an immunological detection system for analyzing glycoproteins in blots. Gycoconj J 7:63−74

Haselbeck A, Schickaneder E, von der Eltz H, Hösel W (1990) Structural characterization of glycoprotein carbohydrate chains by using digoxigenin-labeled lectins on blots. Anal Biochem 191:25−30

Hegnauer R (1971) Pflanzenstoffe und Pflanzensystematik. Naturwissenschaften 58:585−598

Herzberg M (1984) Molecular genetic probe, assay technique, a kit using this molecular genetic probe. Eur Pat Appl 0128018

Hiller Y, Gershoni JM, Bayer EA, Wilchek M (1987) Biotin binding to avidin. Oligosaccharide site chain not required for ligand association. Biochem J 248:167−171

Höltke H-J, Kessler C (1990) Non-radioactive labeling of RNA transcripts in vitro with the hapten digoxigenin (DIG); hybridization and ELISA-based detection. Nucleic Acids Res 18, 5843−5851

Höltke H-J, Seibl R, Burg J, Mühlegger K, Kessler C (1990) Non-radioactive labeling and detection of nucleic acids: II. Optimization of the digoxigenin system. Mol Gen Hoppe-Seyler 371:929−938

Hopman AHN, Wiegant J, Tesser GI, Van Duijn P (1986a) A nonradioactive in situ hybridization method based on mercurated nucleic acid probes and sulfhydryl-hapten ligands. Nucleic Acids Res 14:6471−6488

Hopman AHN, Wiegant J, van Duijn P (1986b) A new hybridocytochemical method based on mercurated nucleic acide probes and sulfhydryl-hapten ligands. I. Stability of the mercurysulfhydryl bond and influence of the ligand structure on immunochemical detection of the hapten. Histochem 84:169−178

Jablonski E, Moomaw EW, Tullis RH, Ruth JL (1986) Preparation of oligodeoxynucleotide-alkaline phosphatase conjugates and their use as hybridization probes. Nucleic Acids Res 14:6115−6128

Jones CJ, Mosley SM, Jeffrey IJ, Stoddart RW (1987) Elimination of the nonspecific binding of avidin to tissue section. Histochem J 19:264−268

Keller GH, Huang D-P, Manak MM (1989) Labeling or DNA probes with a photoactivatable hapten. Anal Biochem 177:392−395

Kessler C (1990) Detection of nucleic acids by enzyme-linked immuno-sorbent assay (ELISA) technique: an example for the development of a novel non-radioactive labeling and detection system with high sensitivity. In: Obe G (ed) Advances in Mutagenesis Research, Springer-Verlag, Berlin Heidelberg, Vol 1, pp 105−152

Kessler C (1991) The digoxigenin:anti-digoxigenin (DIG) technology − a survey on the concept and realization of a novel bioanalytical indicator system. Mol Cell Probes 5:161−205

Kessler C, Höltke H-J, Seibl R, Burg J, Mühlegger K (1990) Non-radioactive labeling and detection of nucleic acids: I. A novel DNA labeling and detection system based on digoxigenin:anti-digoxigenin ELISA principle (digoxigenin system). Mol Gen Hoppe-Seyler 371, 917−927

Kumar A, Tchen P, Roullet F, Cohen J (1988) Nonradioactive labeling of synthetic oligonucleotide probes with terminal deoxynucleotidyl transferase. Anal Biochem 169:376−382

Lacey E, Grant WN (1987) Photobiotin as a sensitive probe for protein labeling. Anal Biochem 163:151−158

Langer PR, Waldrop AA, Ward DC (1981) Enzymatic synthesis of biotin-labeled polynucleotides: novel nucleic acid affinity probes. Proc Natl Acad Sci USA 78:6633−6637

Li P, Medon P, Skingle DC, Lanser JA, Symons RH (1987) Enzyme-linked synthetic oligonucleotide probes: nonradioactive detection of Escherichia coli in faecal specimens. Nucleic Acids Res 15:5275−5287

Mühlegger K, Huber E, von der Eltz H, Rüger R, Kessler C (1990) Non-radioactive labeling and detection of nucleic acids: IV. Synthesis and properties of the nucleotide compounds of the digoxigenin system and of photodigoxigenin. Mol Gen Hoppe-Seyler 371:939−951

Nur I, Reinhartz A, Hyman HC, Razin S, Herzberg M (1989) Chemiprobe, a nonradioactive system for labeling nucleic acid. Ann Biol Clin 47:601−606

Oser A, Roth WK, Valet G (1988) Sensitive nonradioactive dot-blot hybridization using DNA probes labeled with chelate group substituted psoralen and quantitative detection by europium ion fluorescence. Nucleic Acids Res 16:1181–1196

Pollard-Knight D, Read CA, Downes MJ, Howard LA, Leadbetter MR, Pheby SA, McNaughton E, Syms A, Brady MAW (1990) Nonradioactive nucleic acid detection by enhanced chemiluminescence using probes directly labeled with horseradish peroxidase. Anal Biochem 185:84–89

Proverenny AM, Podgorodnichenko VK, Bryksina LE, Monastyrskaya G-S, Sverdlov ED (1979) Immunochemical approaches to DNA structure investigation-I. Mol Immunol 16:313–316

Reisfeld A, Rothenberg JM, Bayer EA, Wilchek M (1987) Nonradioactive hybridization probes prepared by the reaction of biotin hydrazide with DNA. Biochem Biophys Res Commun 142:519–526

Renz M, Kurz C (1984) A colorimetric method for DNA hybridization. Nucleic Acids Res 12:3435–3444

Rothenberg JM, Wilchek M (1988) p-Diazobenzoyl-biocytin: a new biotinylating reagent for DNA. Nucleic Acids Res 16:7197

Sakamoto H, Traincard F, Vo-Quang T, Ternynck T, Guesdon JL, Avrameas S (1987) 5-Bromodeoxyuridine in vivo labeling of M13 DNA, its use as a nonradioactive probe for hybridization experiments. Mol Cell Probes 1:109–120

Saman E (1986) A simple and sensitive method for detection of nucleic acids fixed on nylon-based filters. Gene Anal Technol 3:1–5

Schmitz GG, Walter T, Kessler C (1991) Non-radioactive labeling of oligonucleotides in vitro with the hapten digoxigenin (DIG) by tailing with terminal transferase. Anal Biochem 192:222–231

Sheldon EL, Kellogg DE, Watson RE, Levinson CH, Erlich HA (1986) Use of nonisotopic M13 probes for genetic analysis: application to class II loci. Proc Natl Acad Sci USA 83:9085–9089

Sproat BS, Beijer B, Rider P (1987) The synthesis of protected 5'-amino-2',5'-dideoxyribonucleoside-3'-O-phosphoramidites: applications of 5'-amino-oligodeoxyribonucleotides. Nucleic Acids Res 15:6181–6196

Strauss WL (1984) Sulfhydryl groups and disulfide bonds: modification of amino acid residues in studies of receptor structure and function. In: Venter JL, Harrison LC (eds) Membranes, detergents, and receptor solubilization, Alan R Liss, New York, pp 85–97

Syvänen AC, Alanen M, Söderlund H (1985) A complex of single-strand binding protein and M13 DNA as hybridization probe. Nucleic Acids Res 13:2789–2802

Syvänen AC, Tchen P, Ranki M, Söderlund H (1986) Time-resolved fluorometry: a sensitive method to quantify DNA-hybrids. Nucleic Acids Res 14:1017–1028

Takahashi Y, Arakawa H, Maeda M, Tsuiji A (1989) A new biotinylating system for DNA using biotin aminocaproyl hydrazide and glutaraldehyde. Nucleic Acids Res 17:4899–4900

Taub F (1986) An assay for nucleic acid sequences, particularly genetic lesions. PCT Int Appl WO 86/03227

Tchen P, Fuchs RPP, Sage E, Leng M (1984) Chemically modified nucleic acids as immunodetectable probes in hybridization experiments. Proc Natl Acad Sci USA 81:3466–3470

Tomlinson S, Lyga A, Huguenel E, Dattagupta N (1988) Detection of biotinylated nucleic acid hybrids by antibody-coated gold colloid. Anal Biochem 171:217–222

Urdea MS, Warner BD, Running JA, Stempien M, Clyne J, Horn T (1988) A comparison of nonradioisotopic hybridization assay methods using fluorescent, chemiluminescent and enzym labeled synthetic oligodeoxyribonucleotide probes. Nucleic Acids Res 16:4937–4956

Verdlov ED, Monastyrskaya GS, Guskova LI, Levitan TL, Sheichenko VI, Budowsky EI (1974) Modification of cytidine residues with a bisulfite-O-methylhydroxyl-amine mixture. Biochem Biophys Acta 340:153–165

Viscidi RP, Connelly CJ, Yolken RH (1986) Novel chemical method for the preparation of nucleic acids for nonisotopic hybridization. J Clin Microbiol 23:311–317

Wilchek M, Bayer EA (1987) Labeling glycoconjugates with hydrazide reagents. Meth Enzymol 138:429–442

Wilchek M, Bayer EA (1988) The avidin-biotin complex in bioanalytical applications. Anal Biochem 171:1–32

Wilchek M, Bayer EA (1989) Avidin-biotin technology ten years on: has it lived up to is expectations? Trends Biol Sc 14:408–412

Wilchek M, Ben-Hur H, Bayer EA (1986) p-Diazobenzoyl biocytin – a new biotinylating reagent for the labeling of thyrosines and histidines in proteins. Biochem Biophys Res Comm 138:872–879

3 The Digoxigenin:
Anti-Digoxigenin (DIG) System

3.1 Overview

CHRISTOPH KESSLER

The digoxigenin:anti-dioxigenin (DIG) indicator system is based on the specific interaction between the cardenolide steroid DIG, a chemically derived aglycon of digoxin and lanatoside C (see figure), and a high-affinity, DIG-specific antibody (Kessler, 1991).

Structure of digoxigenin
(DIG)

With the DIG system, specific detection of subpicogram levels of DNA or RNA, picogram amounts of proteins, and subnanogram levels of glycans is possible. The high specificity and low amount of background by e.g. unspecific matrix binding of the DIG-labeled probe or the DIG-specific antibody or side reactions of the DIG system by e.g., unspecific cross-reactions of the DIG-specific antibody with compounds structurally related to DIG especially with biological material, reflect the fact that the occurrence of the DIG modification group is limited to *Digitalis* plants and that the number of background reactions is low using the DIG-specific antibody.

Due to the high specificity of the mature polyclonal antibody, isolated from sheep, even structurally related steroids such as the bufadienolide k-strophanthin (cross-reactivity 0.1%) or the sex and suprarenal steroids (cross-reactivity <0.0003%) show only minor side reactions.

In order to avoid nonspecific binding reactions of the Fc portion of the antibody, the Fab fragment of the DIG-specific antibody is used as the binding component when only primary antibodies are applied. The entire DIG-specific antibody, isolated from sheep including the Fc portion is used for signal generation by secondary antibodies. Labeled mouse antibodies whose specificity is directed against the Fc portion of sheep antibodies serve as secondary antibodies in this case.

3.2 Labeling and Detection of Nucleic Acids

HANS-JOACHIM HÖLTKE, RUDOLF SEIBL, GUDRUN G. SCHMITZ,
THOMAS WALTER, RÜDIGER RÜGER, GREGOR SAGNER, JOSEF BURG,
KLAUS MÜHLEGGER, and CHRISTOPH KESSLER

3.2.1 Principle and Applications

The novel, highly sensitive, nonradioactive DIG system allows specific detection of 0.1 pg DNA or RNA within 1−16 h using the optical detection reagents 5-bromo-4-chloro-3-indolyl phosphate (BCIP)/nitroblue tetrazolium salt (NBT) or within 15−30 min using chemiluminescent 3-(2-spiroadamantane-4-methoxy-(3″-phosphoryloxy)-phenyl-1,2dioxetane (Lumigen™PPD/Lumiphos™530)/3-(4-methoxyspiro[1,2-dioxetane-3,2′-tricyclo[3.3.1.1^{3,7}]decan]-4-yl) phenyl phosphate (AMPPD) detection in dot, slot or Southern blots without any significant background on nitrocellulose and nylon membranes (Kessler et al., 1990; Höltke et al., 1990; Seibl et al., 1990; Mühlegger et al., 1990; Kessler, 1991).

DNA is labeled either by random-primed (Feinberg and Vogelstein, 1983; 1984) or polymerase chain reaction-(PCR-)guided incorporation of DIG-labeled dUTP. The PCR labeling approach permits synthesis of vector-free, DIG-labeled probes (Reischl et al., 1992; see also Sect. 16.1). The dUTP is linked via an 11-atom linear spacer arm to DIG (DIG-[11]-dUTP). RNA is labeled by run-off transcription using phage-coded RNA polymerases (Krieg and Melton, 1987) by incorporating DIG-[11]-UTP (Höltke and Kessler, 1990). Oligonucleotides can be labeled by a 3′tailing reaction catalyzed by terminal transferase (Roychoudhury and Wu, 1980) with DIG-[11]-dUTP/dATP or DIG-[11]-ddUTP (Schmitz et al., 1991). In a light-induced reaction DNA or RNA can also be DIG-labeled with photoDIG (Mühlegger et al., 1990). After hybridizazion to the target DNA or RNA, the labeled hybrids are detected by an ELISA reaction using conjugates composed of high-affinity, DIG-specific antibodies coupled to alkaline phosphatase (<DIG>:AP). A detection reaction is obtained by subsequent addition of AP catalyzed either with the color substrates BCIP and NBT or with the chemiluminescent reagents Lumigen™ PPD or Lumiphos™530 substrate (Höltke et al., 1992).

The DNA and RNA labeling methods efficiently modify small (10 ng) and large (up to 3 μg) amounts of linearized DNA in a fast reaction (1 h). Every 20th−25th nucleotide of the newly synthesized DNA or RNA bears a hapten.

DIG oligonucleotide tailing results either in a single addition of a DIG-labeled nucleotide (DIG-[11]-ddUTP) or in the addition of DIG tails

(DIG-[11]-dUTP). The tail length and hapten spacing can be modified by using either different unlabeled nucleotide(s) or none at all.

Hapten-labeled DNA is used under standard hybridization conditions. All tested hybridization buffers including those containing Denhardt's or formamide can be used with the novel system, although a buffer containing only SSC, SDS, sarcosine, and a blocking agent is preferred. Hybridization solutions containing labeled DNA can be reused.

Hybridized filters are either detected immediately or stored dry for later detection. After blocking of the membrane with blocking agent the antibody conjugate is bound to hapten-labeled DNA hybrids. After three washing steps, the color reaction is started at alkaline pH by the addition of either the optical substrates BCIP and NBT or the chemiluminescent substrates Lumigen™PPD or Lumiphos™530. Multiple optical or fluorescent staining protocols have also been developed (Kunz et al., 1990; Lichter et al., 1990; Arnold et al., 1991; Höltke et al., 1992). With the optical substrates the oxidized, blue product starts to precipitate within a few minutes, continuing for up to 16h. Background color development is usually not observed even after 72h because of efficient blocking of the membrane. Rehybridization of the membrane is possible especially after the chemiluminescent detection reaction (Gebeyehu et al., 1987; Hölke et al., 1992).

The advantages of nonradioactive labeling and detection of DNA, RNA, and oligonucleotides with the DIG system − which have high sensitivity and low background − have been utilized in a variety of applications. The DIG system has been used to detect single-copy genes in human DNA, for colony and plaque hybridizations, to detect hepatitis B virus DNA in human sera and Epstein-Barr virus (EBV) DNA in DNA of latently infected cells, and also for in situ hybridizations of EBV DNA and amplified tissue plasminogen activator (t-PA) genes on metaphase chromosome spreads in CHO cells. Moreover, the DIG system has also been applied in the labeling of PCR products (see also chapter 16.1; Kessler, 1990; Rüger et al., 1990; Seibl et al., 1990; Reischl et al., 1992). Further applications of the DIG system are described for Northern blot hybridization (Rüger et al., 1990) and various other techniques in molecular biology (Casacuberta et al., 1988; Manstein et al., 1989; Jessberger and Berg, 1991), to viral DNAs (Dooley et al., 1988; Kimpton et al., 1989), to characterize DNA binding proteins (Suske et al., 1989; Dorward and Garon, 1989), and for histological in situ hybridizations (Tautz and Pfeifle, 1989; Young, 1989; Grega et al., 1989a, 1989b; Baldino et al., 1989; Herrington et al., 1989a, 1989b, Nakano et al., 1989; Cohen, 1990; Lichter et al., 1990; Arnold et al., 1991; Baldino et al., 1991). The DIG system has also been used for DNA fingerprinting with oligonucleotide probes specific for repetitive sequences (Schäfer et al., 1988; Zischler et al., 1989a, 1989b), for nonradioactive restriction mapping of large fragments (Zuber and Schuman, 1991), and DNA sequencing by analyzing either standard sequencing gels or using the direct blotting electrophoresis (DBE) approach (Höltke et al., 1992; for review see also references in Kessler, 1991).

3.2.2 Reaction Scheme

The standard procedure of the nonradioactive DIG system is divided into three parts:

1. Labeling reaction
   - DNA labeling by random-primed or PCR-guided incorporation of DIG-[11]-dUTP
   - RNA labeling by run-off transcription with DIG-[11]-UTP
   - Oligonucleotide labeling by tailing with either DIG-[11]-dUTP/dATP or by incorporation of DIG-[11]-ddUTP
   - DNA or RNA photolabeling with photoDIG
2. Hybridization of DIG-labeled probe to target DNA or DNA analyte
3. Immunological detection
   - <DIG>:AP conjugate binding
   - Optical detection with color substrates BCIP/NBT
   - Chemiluminescent detection with Lumigen™PPD or Lumiphos™530 substrate

The standard reaction scheme for DIG labeling of DNA, RNA, or oligo-nucleotides and BCIP/NBT or Lumigen™PPD/Lumiphos™530 detection is given in the accompanying figure.

3.2.3 Random-Primed DNA Labeling with DIG-[11]-dUTP and Klenow Enzyme*

Standard reagents
- DIG-[11]-dUTP (Boehringer Mannheim)
- dATP, dGTP, dCTP, dTTP (Boehringer Mannheim)
- Klenow enzyme (Boehringer Mannheim)
- Random primer [d(pN)$_6$] (Boehringer Mannheim)
- Tris-HCl (Boehringer Mannheim)
- EDTA, ethanol (99% [v/v]) (Merck)
- Herring sperm DNA (Boehringer Mannheim)
- pBR328 DNA (control DNA) (Boehringer Mannheim)

Standard solutions
- TE buffer: 10mM Tris-HCl; 1mM EDTA; pH 8.0/25°C
- Hexanucleotide mixture: 3mg/ml buffered random primer solution
- dNTP labeling mixture: 1mM dATP; 1mM dGTP; 1mM dCTP; 0.65mM dTTP, 0.35mM DIG-[11]-dUTP; pH 7.0/25°C
- Klenow enzyme solution: 2U/μl Klenow enzyme
- EDTA solution: 0.2M EDTA, pH 8.0/25°C
- LiCl solution: 4M LiCl
- Glycogen solution: 20mg/ml
- Control DNA solution: 20μg/ml pBR328 DNA in TE buffer
- DNA dilution buffer: 50μg/ml herring sperm DNA in TE buffer

Labeling reaction
1. Pipette the following into a microfuge tube on ice:
   10ng – 3μg   linearized purified DNA
                   (Control: 5μl control DNA ≙ 1μg pBR328 DNA)

---

* Available also as DIG DNA Labeling Kit (Boehringer Mannheim)

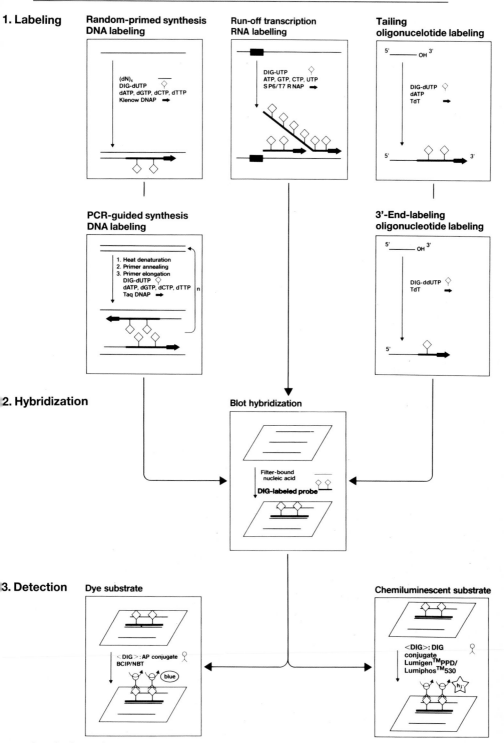

Standard reaction scheme für DIG labeling of DNA, RNA, or oligonucleotide and BCIP/NBT or Lumigen^{TM}PPD/Lumiphos^{TM}530 detection

Denature for 10 min at 95 °C, chill quickly on ice/NaCl for 2 min and add:

2 µl	hexanucleotide mixture
2 µl	dNTP labeling mixture

Make up to:

19 µl	with sterile redistilled H_2O
	(Control: 10 µl)

Add:

1 µl	Klenow enzyme solution (2 units).

2. Centrifuge briefly and incubate for at least 60 min at 37 °C. Longer incubation (up to 20 h) can increase the yield of labeled DNA.

3. Add:

2 µl	0.2 M EDTA solution (pH 8.0) to stop the reaction

Precipitate with:

2.5 µl	4 M LiCl solution
(1 µl	20 mg/ml glycogen solution, if necessary) and
75 µl	prechilled ethanol (−20 °C)

Mix well.

4. Leave the tube for at least 30 min at −70 °C or at −20 °C overnight.

5. Centrifuge (12 000 g) for 10 min, wash pellet with 50 µl cold ethanol (70 % [v/v]), centrifuge again, dry under vacuum and dissolve for approx. 30 min at room temperature in:

50 µl	10 mM Tris-HCl, 1 mM EDTA (pH 8.0)
	(Control: concentration of DNA is 5.2 µg/ml of newly synthesized DNA).

Note: If the DNA has been labeled in the presence of low melting point agarose, the ethanol precipitation (steps 3−5) must be replaced by gel filtration.

3.2.4 PCR-Guided Synthesis of Vector-Free, DIG-Labeled DNA Probe with DIG-[11]-dUTP

Standard reagents
− For details of standard reagents, standard solutions and detailed protocol see also Sects. 3.2.3 and 16.1
− *Taq* DNA polymerase
− PCR thermal cycler
− Reaction tubes adapted to PCR thermal cycler
− PCR oligonucleotides
− For other reagents see Sect. 3.2.3

- 10× PCR buffer: 100 mM Tris-HCl; 500 mM KCl; 15 mM MgCl$_2$; **Standard**
  0.1 mg/ml [w/v] gelatin; pH 8.5/25 °C   **solutions**
- dNTP labeling mixture: 1 mM dATP; 1 mM dGTP; 1 mM dCTP;
  0.65 mM dTTP; 0.35 mM DIG-[11]-dUTP; pH 7.0/25 °C
- *Taq* DNA polymerase solution: 2 U/µl *Taq* DNA polymerase

1. Each 100 µl reaction volume contains 10 µl PCR buffer, 10 µl dNTP **Labeling**
   labeling mixture, 200 ng of each primer, and 2 units *Taq* DNA poly- **reaction**
   merase.
   Mix the compounds in the reaction tube and add the template DNA.
   With samples of known DNA concentration, add a volume corresponding to greater than 2 ng but not exceeding 40 µl. After the addition of the DNA, close each reaction tube before proceeding to the
   next one. Do not vortex or mix. In a negative control reaction add
   sterile autoclaved H$_2$O instead of template DNA. Overlay reactions
   with 100 µl mineral oil to prevent evaporation during thermal cycling.

2a. In three-step PCR, cycling parameters are: 30 s denaturation at 94 °C,
    30 s annealing at temperatures ranging from 40 °C to 60 °C (the anneal-
    ing temperature varies depending on the primers' length and their GC
    content) and a primer extension time of 30 s at 72 °C. Thirty cycles are
    carried out in a DNA thermal cycler, followed by an additional 5 min
    incubation at 72 °C to insure completion of all polymerization pro-
    ducts.

2b. Two-step PCR: For short sequences (<500 bp) to be amplified, a two-
    step PCR reaction works as well as the three-step method described in
    step 2a. Cycling parameters are: 1 min denaturation at 92 °C and 2 min
    annealing and primar extension at 65 °C−75 °C (the temperature var-
    ying depending on the primers' length and their GC content; the opti-
    mal annealing and primer extension temperature has to be evaluated
    for every individual primer pair; see also Part III). Thirty cycles are
    carried out in a DNA thermal cycler, following an additional 5 min
    incubation at 72 °C to insure completion of all polymerization pro-
    ducts.

3. Perform an agarose or polyacrylamide gel electrophoresis on aliquots
   (10 µl) of the PCR-amplified samples to verify that the amplified product is the correct size.

Notes: The denuration and annealing temperatures may need to be
adjusted for various flanking oligomers, depending on their GC content.
Oligomers with high GC content require higher annealing temperatures to
reduce nonspecific binding.

Use pH 7.5 preadjusted dNTP solutions to prevent contamination during
pH adjusting of these small volumes.

3.2.5 RNA Labeling by Run-off Transcription with DIG-[11]-UTP and SP6, T7 or T3 RNA Polymerase*

Standard reagents
- DIG-[11]-UTP (Boehringer Mannheim)
- SP6 RNA polymerase (Boehringer Mannheim)
- T7 RNA polymerase (Boehringer Mannheim)
- T3 RNA polymerase (Boehringer Mannheim)
- ATP, GTP, CTP, UTP (Boehringer Mannheim)
- pSPT18/19 vector DNA (Boehringer Mannheim)
- DNase I, RNase-free (Boehringer Mannheim)
- RNase inhibitor (Boehringer Mannheim)
- Tris-HCl (Boehringer Mannheim)
- EDTA, ethanol (99% [v/v], LiCl) (Merck)
- Diethylpyrocarbonate (Sigma)
- pSPT18-*neo* DNA, cleaved with *Pvu* II (control DNA) (Boehringer Mannheim)
- pSPT19-*neo* DNA, cleaved with *Pvu* II (control DNA) (Boehringer Mannheim)

Standard solutions
- TE buffer: 10 mM Tris-HCl; 1 mM EDTA; pH 8.0/25 °C
- 10× transcription buffer: 400 mM Tris-HCl; 60 mM $MgCl_2$; 100 mM NaCl; 100 mM dithiothreitol; 20 mM spermidine; 1 U/μl RNase inhibitor; pH 8.0/25 °C
- NTP labeling mixture: 10 mM ATP, 10 mM GTP; 10 mM CTP; 6.5 mM UTP; 3.5 mM DIG-[11]-UTP in H_2O
- SP6 RNA polymerase solution: 20 U/μl SP6 RNA polymerase
- T7 RNA polymerase solution: 20 U/μl T7 RNA polymerase
- T3 RNA polymerase solution: 20 U/μl T3 RNA polymerase
- pSPT18-*neo*/*Pvu* II DNA solution (control DNA): 250 μg/ml in TE buffer
- pSPT19-*neo*/*Pvu* II DNA solution (control DNA): 250 μg/ml in TE buffer
- EDTA solution: 0.2 M EDTA; pH 8.0/25 °C
- LiCl solution: 4 M LiCl
- Diethylpyrocarbonate-treated H_2O: Dissolve diethylpyrocarbonate to 1% [v/v] in a 50% ethanol/H_2O mixture; mix redistilled H_2O 1 : 10 with this solution ($\hat{=}$ 0.1% diethylpyrocarbonate); incubate for 30 min at room temperature, then autoclave.

Labeling reaction
Usually 1 μg of linear template DNA is used per standard assay, but smaller or larger amounts can also be used. Circular DNA can also serve as template if one wishes to synthesize „run-around" instead of „run-off" transcripts. The yield of DIG-labeled RNA is approximately 10 μg in the standard reaction starting with 1 μg template DNA. Larger amounts of RNA can be DIG-labeled by scaling up all components and volumes.

The linearized DNA to be transcribed should be purified by phenol/chloroform extraction and ethanol precipitation.

---

* Available also as DIG RNA Labeling Kit (Boehringer Mannheim)

1. Pipette the following into a microfuge tube on ice:
   2 µl   10× transcription buffer
   13 µl   sterile redistilled H_2O (control: 10 µl)
   2 µl   NTP labeling mixture
   1 µg   template DNA (1 µg/µl) (control: 4 µl control DNA $\hat{=}$ 1 µg
           pSPT18/19-*neo/Pvu* II DNA)
   2 µl   SP6, T7 or T3 RNA polymerase solution (40 units).

2. 1 µl of RNase inhibitor can be added to the transcription assay.

3. Centrifuge briefly and incubate for 2 h at 37 °C. Longer incubation does not increase the yield of labeled RNA.

4. The amount of DIG-labeled RNA transcript is far in excess of the template DNA (ratio approximately 10) and it is not usually necessary to remove the template DNA by DNase treatment. If desired, the template DNA can be removed by directly adding 2 µl RNase-free DNase I and incubating for 15 min at 37 °C.

5. With or without prior DNase treatment, add 2 µl EDTA solution to stop the reaction.

6. Precipitate the labeled RNA with 2.5 µl LiCl solution and 75 µl prechilled (−20 °C) ethanol. Mix well.

7. Leave for at least 30 min at −70 °C or 2 h at −20 °C.

8. Centrifuge at 12000 g, wash the pellet with 50 µl cold ethanol (70% [v/v]), dry under vacuum, and dissolve for 30 min at 37 °C in 100 µl diethylpyrocarbonate-treated water. RNase inhibitor (1 µl) can be added to inhibit possible contaminating RNases.

Notes: The amount of newly synthesized labeled RNA depends on the amount, size (site of linearization), and purity of the template DNA. In the control reaction with 1 µg template DNA per assay, linearized to give run-off transcripts of 760 bases with SP6 or T7 RNA polymerase, approximately 37% of the nucleotides are incorporated into about 10 µg of transcribed DIG-labeled RNA.

Scaling up the reaction volume and components while keeping the amount of template DNA constant can improve the yield of DIG-labeled RNA. With 1 µg linear pSPT18-neo DNA as template in a 5× upscaled reaction over 40 µg of DIG-labeled RNA is synthesized after 2 h of incubation at 37 °C.

3.2.6 Oligonucleotide Tailing with DIG-[11]-dUTP/dATP and Terminal Transferase*

- DIG-[11]-dUTP (Boehringer Mannheim)   **Standard**
- dATP (Boehringer Mannheim)   **reagents**
- Terminal transferase (Boehringer Mannheim)

---

* Available also as DIG Oligonucleotide Tailing Kit (Boehringer Mannheim)

3.2.6

- Tris-HCl (Boehringer Mannheim)
- Potassium cacodylate, $CoCl_2$, KCl, EDTA (Merck)
- Bovine serum albumin (BSA) (Boehringer Mannheim)
- Glycogen (Boehringer Mannheim)
- 30-mer oligonucleotide 5'-p TTG GGT AAC GCC AGG GTT TTC CCA GTC ACG OH-3' (control) (Boehringer Mannheim)

- TE buffer: 10 mM Tris-HCl; 1 mM EDTA; pH 8.0/25 °C
- 5× reaction buffer: 1 M potassium cacodylate; 125 mM Tris-HCl; 1.25 mg/ml BSA; pH 6.6/25 °C
- $CoCl_2$ solution: 25 mM $CoCl_2$
- DIG-[11]-dUTP solution: 1 mM DIG-[11]-dUTP; pH 7.0/25°
- dATP solution: 10 mM dATP in 10 mM Tris-HCl; pH 7.5/25 °C
- Terminal transferase solution: 50 U/µl in 0.2 M potassium cacodylate; 1 mM EDTA; 200 mM KCl; 0.2 mg/ml BSA; 50% [v/v] glycerol; pH 6.5/25 °C
- Glycogen solution: 20 mg/ml [w/v] glycogen
- EDTA solution: 0.2 M EDTA; pH 8.0/25 °C
- LiCl solution: 4 M LiCl
- 30-mer oligonucleotide solution: Control, 20 ng/µl ≙ 20 pmol

Tailing reaction With 1 µg of a 30-mer oligonucleotide virtually all of the applied oligonucleotide is tailed. HPLC- or gel-purified oligonucleotides should be dissolved in sterile H_2O.

1. Pipette the following into a microfuge tube on ice:

4 µl	5× reaction buffer
4 µl	$CoCl_2$ solution
100 pmol	oligonucleotide (control: 5 µl 30-mer oligonucleotide ≙ 100 pmol)

   Make up to:

17 µl	with sterile redistilled H_2O

   Add:

1 µl	DIG-[11]-dUTP solution
1 µl	dATP solution
1 µl	terminal transferase solution (50 units).

2. Incubate at 37 °C for 15 min, then place on ice.

3. Mix 1 µl glycogen solution with 200 µl 200 mM EDTA solution; pH 8.0/25 °C and add 2 µl of the dilution to the reaction mixture to stop the reaction. Alternatively, 1 µl of the concentrated glycogen solution can be used. Do not use phenol/chloroform to stop the reaction since this treatment leads to partitioning of DIG-tailed oligonucleotides in the organic phase.

4. Precipitate the tailed oligonucleotide with 2.5 µl LiCl solution and 75 µl prechilled (−20 °C) absolute ethanol. Mix well.

5. Leave for at least 30 min at $-70°C$ or 2 h at $-20°C$.

6. Centrifuge at 12000g, wash the pellet with 50 μl cold ethanol (70% [v/v]), dry under vacuum, and dissolve in an appropriate volume of sterile redistilled H_2O. Store tailed oligonucleotides at $-20°C$ if not used immediately.

Notes: The efficiency of the tailing reaction can be checked by comparison with the tailed, control oligonucleotide in a direct detection reaction without prior hybridization. The tailed oligonucleotide can be analyzed by polyacrylamide gel electrophoresis and subsequent silver staining and compared to the untailed oligonucleotide. DIG tailing of oligonucleotide results in a heterogeneous higher molecular weight shift and is detectable as a smear in polyacrylamide gels. The control oligonucleotide tailed in the standard reaction is completely shifted to the labeled form.

Increasing the amount of oligonucleotide in the tailing reaction is not recommended. Larger amounts of oligonucleotide may be labeled by increasing the reaction volume and all components proportionally.

3.2.7 Oligonucleotide 3′ End Labeling with DIG-[11]-ddUTP and Terminal Transferase*

- DIG-[11]-ddUTP (Boehringer Mannheim)   **Standard reagents**
- For other reagents see Sect. 3.2.6

- DIG-[11]-ddUPP solution: 1 mM DIG-[11]-ddUTP; pH 7.0/25°C   **Standard solutions**
- For other solutions see Sect. 3.2.6

1. With 1 μg of a 30-mer oligonucleotide virtually all of the applied oligo- **End labeling reaction** nucleotide is labeled. HPLC- or gel-purified oligonucleotides should be dissolved in sterile H_2O.

2. Pipette the following into a microfuge tube on ice:
4 μl	5× reaction buffer
4 μl	$CoCl_2$ solution
100 pmol	oligonucleotide (control: 5 μl 30-mer oligonucleotide $\hat{=}$ 100 pmol)

   Make up to:
18 μl	with sterile redistilled H_2O

   Add:
1 μl	DIG-[11]-ddUTP solution
1 μl	terminal transferase solution (50 units).

3–6. Follow the respective instructions of procedure 3.2.6.

---

* Available also as DIG Oligonucleotide 3′-End Labeling Kit (Boehringer Mannheim)

3.2.8 Photolabeling of DNA or RNA with PhotoDIG

Standard reagents
- PhotoDIG (Boehringer Mannheim)
- Tris-HCl (Boehringer Mannheim)
- EDTA, ethanol, LiCl, NaCl, 2-butanol (Merck)
- Dimethylpyrocarbonate (Velcorin) (Bayer Leverkusen)

Standard solutions
- PhotoDIG solution: $10\,\mu g/\mu l$ PhotoDIG in dimethylformamide; stable at 4°C; protect from light
- TE buffer: 10 mM Tris-HCl; 1 mM EDTA; pH 8.0/25°C
- Tris buffer: 100 mM Tris-HCl; 1 mM EDTA; pH 9.0/25°
- LiCl solution: 4 M LiCl
- NaCl solution: 5 M NaCl

Photolabeling reaction
Reaction steps 1−3 should be carried out in the dark. The target DNA or RNA should be free for any organic buffer compounds (e.g., Tris). Denaturation of the DNA is not necessary.

1. Mix $10\,\mu g$ DNA or RNA with $1\,\mu l$ photodigoxigenin solution in a final volume of $40\,\mu l$ H_2O.

2. Place the open tube on ice at a distance of 10 cm from a Philipps HPLR, 400 W lamp (or equivalent).

3. Irradiate the nucleic acid/photoDIG mixture 15 min for labeling of DNA or 10 min for labeling of RNA.

4. Add $60\,\mu l$ of Tris buffer.

5. Add $15\,\mu l$ of NaCl solution.

6. Extract twice with $100\,\mu l$ 2-butanol.

7. Precipitate the photoDIG-labeled nucleic acid with $10\,\mu l$ LiCl solution and $200\,\mu l$ prechilled absolute ethanol. Mix well.

8. Leave for at least 40 min at −70°C or 2 h at −20°C.

9. Centrifuge at 12000 g, wash the pellet with cold ethanol (70% [v/v]), dry under vacuum, and dissolve in $40\,\mu l$ TE buffer.

Notes: The labeled nucleic acids can be electrophoresed and directly blotted on membranes or, alternatively, used as DIG-labeled probes for hybridization assays.

For RNA labeling use dimethylpyrocarbonate treated buffers.

3.2.9 Hybridization with DIG-Labeled DNA or DIG-Labeled Oligonucleotides*

Standard reagents
- Blocking reagent (Boehringer Mannheim)
- SDS (Boehringer Mannheim)

---

* Included also in DIG DNA Labeling and Detection Kit as well as DIG Nucleic Acid Detection Kit (Boehringer Mannheim)

- NaCl, sodium citrate (Merck)
- *N*-lauroylsarcosine (Sigma)
- Nylon membrane, positively charged (Boehringer Mannheim)

- Maleic acid buffer: 100 mM maleic acid; 150 mM NaCl; pH 7.5/25 °C   **Standard**
- Blocking stock solution: blocking reagent is dissolved in maleic acid **solutions** buffer to a final concentration of 10% [w/v] with shaking and heating either on a heating block or in a microwave oven. This stock solution is autoclaved and stored at 4 °C or −20 °C.
- 20× SSC: 3 M NaCl; 0.3 M sodiumcitrate; pH 7.0/25 °C
- Hybridization solution: 5× SSC; 1% [w/v] blocking reagent; 0.1% [w/v] *N*-lauroylsarcosine; 0.02% [w/v] SDS. One can also add formamide up to 50% [v/v] to the hybridization buffer. In this case the concentration of the blocking reagent must be increased to 2% [w/v]. Hybridization with formamide is at 42 °C.
- Washing solution 1: 2× SSC; 0.1% [w/v] SDS
- Washing solution 2: 0.1× SSC; 0.1% [w/v] SDS

1. Prepare nitrocellulose membranes by presoaking in water and then **Hybridiza-** 20× SSC. Nitrocellulose membranes must be dried before loading **tion** with DNA. Nylon membranes can be used without any pretreatment. **reaction**

2. Dot blot: Denature DNA by heating in a boling water bath and chilling quickly on ice. Spot the DNA onto the dry membrane.

3. Southern blot: Load the diluted control DNA (e.g., 100 pg−1 pg per lane) onto an agarose gel, separate the fragments, and subsequently perform a Southern transfer to membrane.

4. Bind the DNA to the nitrocellulose by baking for 2 h in a vacuum oven at 80 °C and to nylon membranes either by baking (vacuum not required) or by UV cross linking with a transillumination device.
   (a) Fixation of nucleic acids to the membrane: 1. Bake for 15−30 min at 120 °C, or 2. Cross-link for 3 min with UV (transilluminator or Stratalinker). UV cross-linking is not recommended after alkaline transfer (see below).
   (b) Alkaline transfer: Transfer DNA using 0.4 M NaOH as transfer buffer from agarose gels directly after electrophoresis without prior denaturation or neutralization steps. Fix DNA by baking at 120 °C after alkaline transfer. UV cross-linking is less effective after alkaline transfer.

5. Prehybridize filters in a sealed plastic bag or box with at least 20 ml hybridization buffer per 100 cm² of filter for at least 1 h under the following prehybridization conditions:
   (a) DNA/without formamide: 68 °C
   (b) DNA/with 50% [v/v] formamide: 42 °C
   (c) Oligonucleotide/without formamide: 54 °C (control 30-mer oligonucleotide) or adequate temperature reflecting length and base composition of probe oligonucleotide (T_m).

Redistribute the solution occasionally. Do not allow the filters to dry out between prehybridization and hybridization.

6. Replace the solution with about 2.5 ml per 100 cm² filter of hybridization buffer containing 25 ng of freshly heat-denatured, labeled DNA per milliliter. Very small filters may require slightly more hybridization solution.

7. Hybridization conditions depend on the nature of the probe and whether formamide is present.
   (a) DNA/without formamide:
       Incubate the filters for at least 6 h at 68°C.
   (b) DNA/with 50% [v/v] formamide:
       Incubate the filters for at least 6 h at 42°C.
   (c) Oligonucleotide/without formamide:
       Incubate the filters for 1−6 h at 54°C (control 30-mer oligonucleotide) or adequate temperature reflecting length and base composition of probe oligonucleotide (T_m).
   Redistribute the solution occasionally.

8. Wash the filters 2 × 5 min at room temperature with at least 50 ml washing solution 1 per 100 cm² filter and 2 × 15 min at 68°C with washing solution 2.

9. Filters can be used directly for detection of hybridized DNA or stored air-dried for later detection.

10. The hybridization solution containing labeled DNA can be stored at −20°C and reused several times. Immediately before use, redenature the probe by heating the hybridization solution at 95°C for 10 min. This step also redissolves any precipitates which may have formed during storage.

Notes: If various oligonucleotides with strongly different nucleotide composition shall be used under comparable hybridization conditions, hybridization solutions containing 3 M tetramethylammonium chloride (TMACl) can be used.

The hybridization behavior of oligonucleotides in 3 M TMACl does not depend on the nucleotide composition of the oligonucleotides. The effect is caused by a stronger interaction of A-T base pairs than of G-C base pairs. Therefore, the T_m value is only a function of the length of the oligonucleotide and the temperature of the stringent wash is generally higher than in conventional procedures.

Hybridization is carried out as described above (steps 1−7) however, the hybridization temperature is lowered to between 37° and 42°C. Probes are hybridized with less stringency in this procedure, and specificity is determined in the TMACl wash. Nitrocellulose membranes are not stable for extended periods of time in TMACl solutions and are not recommended for that reason.

- Rinse the membrane three times with 5× SSC at 4°C and incubate membrane 2 × 30 min in 5× SSC at 4°C. Carefully rinse membrane twice at room temperature with TMACl wash solution (3 M TMACl; 50 mM Tris-HCl; 2 mM EDTA; 0.1% [w/v] SDS; pH 8.0/25°C).
- Incubate the membrane in TMACl wash solution for 20 min at 68°C for stringent washing of the labeled 30-mer oligonucleotide. Wash temperature for oligonucleotides of different lengths are: 50°C/16mer; 55°C/20mer; 68°C/30mer; 75°C/50-mer.
- Repeat the stringent wash. Blot the membrane on chromatography paper to remove excess liquid. Do not allow the filters to dry out if rehybridization is to be performed. The filters are now ready for immunological detection.

3.2.10 Hybridization with DIG-Labeled RNA*

- For reagents see Sect. 3.2.9   **Standard reagents**

- Hybridization solution: 50% [v/v] formamide; 5× SSC; 2% [w/v] blocking reagent; 0.1% [w/v] N-lauroylsarcosine; 0.02% [w/v] SDS   **Standard solutions**
- For other solutions see Sect. 3.2.9

1. Prepare nitrocellulose membranes by presoaking in water and then 20× SSC. Nitrocellulose membranes must be dried before loading with RNA. Nylon membranes can be used without pretreatment.   **Hybridization reaction**

2. Dot blot: Denature RNA by heating in a boiling water bath and chilling quickly on ice. Spot the RNA onto the dry membrane.

3. Northern blot: Load the diluted control RNA (e.g., 100 pg–1 pg per lane) onto an agarose gel, separate the fragments, and then perform a northern transfer to membrane.

4. Bind the RNA to nitrocellulose by baking for 2 h in a vacuum oven at 80°C and to nylon membranes either by baking (vacuum not required) or by UV cross-linking.

5. As RNA:RNA hybrids are more stable than RNA:DNA hybrids, more stringent prehybridization conditions must be chosen.
   (a) RNA:RNA hybrids (RNA probe):
        Prehybridize filters in a sealed plastic bag or box with at least 20 ml hybridization buffer with 50% [w/v] formamide per 100 cm² of filter at 68°C for at least 1 h. Distribute the solution from time to time. Do not allow the filters to dry out between prehybridization and hybridization.
   (b) RNA:DNA hybrids (DNA probe):
        Prehybridize filters in a sealed plastic bag or box with at least 20 ml hybridization buffer per 100 cm² of filter at 50°C for at least 1 h.

---

* Included also in DIG Nucleic Acid Detection Kit (Boehringer Mannheim)

Distribute the solution from time to time. Do not allow the filters to dry out between prehybridization and hybridization.

6. Replace the solution with about 2.5 ml of hybridization buffer per 100 cm² filter containing 100 ng of freshly heat-denatured (boiling water bath) labeled RNA or DNA probe per milliliter. Very small filters may require slightly more hybridization solution.

7. During hybridization, more stringent hybridization conditions must also be chosen for RNA:RNA hybrids than for RNA:DNA hybrids.
   (a) RNA:RNA hybrids: Incubate the filters for at least 6h at 68°C. Redistribute the solution occasionally. Higher RNA probe concentrations (up to 500 ng/ml) in the hybridization solution can be used to shorten the hybridization times down to approximately 2h.
   (b) RNA:DNA hybrids: Incubate the filters for at least 6h at 50°C. Redistribute the solution occasionally. Higher DNA probe concentrations (up to 200 ng/ml) in the hybridization solution can be used to shorten hybridization times down to approximately 2h.

8. Wash the filters 2 × 5 min at room temperature with at least 50 ml of washing solution 1 per 100 cm² filter and 2 × 15 min at 68°C with washing solution 2.

9. Filters can then be used directly for detection of hybridized RNA or stored air-dried for later detection.

3.2.11 Optical Detection of DIG Modification with BCIP/NBT*

Standard reagents
- <DIG>:AP (polyclonal sheep anti-DIG antibody [Fab]:alkaline phosphatase conjugate, 750 U/ml; store at 4°C) (Boehringer Mannheim)
- BCIP (5-bromo-4-chloro-3-indolyl phosphate) (Boehringer Mannheim)
- NBT (nitroblue tetrazolium salt) (Boehringer Mannheim)
- For other reagents see Sect. 3.2.3

Standard solutions
- Buffer 1: 100 mM maleic acid; 150 mM NaCl; pH 7.5/25°C
- Buffer 2: 1% [w/v] blocking reagent in buffer 1; prepare the solution from a 10% [w/v] stock solution containing 10% [w/v] blocking reagent in buffer 1, autoclave, and store at 4°C
- Buffer 3: 100 mM Tris-HCl; 100 mM NaCl; 50 mM MgCl₂; pH 9.5/25°C
- Buffer 4: 10 mM Tris-HCl; 1 mM EDTA; pH 8.0/25°C
- Antibody-conjugate solution for optical detection, freshly prepared: Dilute <DIG>:AP solution to 150 mU/ml (1:5000) in buffer 2 directly prior to optical detection
- BCIP solution: 50 mg/ml BCIP in 100% [v/v] dimethylformamide
- NBT solution: 75 mg/ml NBT in 70% [v/v] dimethylformamide

---
* Available also as DIG DNA Labeling and Detection Kit as well as DIG Nucleic Acid Detection Kit (Boehringer Mannheim)

– Color substrate solution, freshly prepared (35 µl BCIP solution and 45 µl NBT solution are added to 10 ml buffer 3 directly prior to detection)

All the following incubations are performed at room temperature. Except for the color reaction, all incubations require shaking or mixing. The volumes of the solutions are calculated for a membrane size of 100 cm^2 and should be adjusted for other membrane sizes. **Detection reaction**

1. Wash membranes briefly (1 min) in buffer 1.

2. Incubate for 30 min in 100 ml buffer 2.

3. Incubate membranes for 30 min with about 20 ml freshly diluted antibody-conjugate solution for color detection (1 : 5000).

4. Remove unbound antibody conjugate by washing 2 × 15 min with 100 ml buffer 1.

5. Equilibrate the membrane for 2–5 min with 20 ml buffer 3.

6. Incubate membrane in the dark with 10 ml freshly prepared color substrate solution sealed in a plastic bag. The color precipitate starts to form within a few minutes and the reaction is usually complete after 12–16 h. Do not shake or mix while color is developing.

7. When the desired spots or bands are detected, stop the reaction by washing the membrane for 5 min with 50 ml buffer 4.

8. Document the results by photocopying the wet membrane or by photography. Photocopying onto overhead transparencies allows for densitometric scanning. For this purpose, the color reaction can be interrupted for a short time and continued afterwards.

Notes: The membrane may then be dried at room temperature or by baking at 80 °C and stored. Colors fade upon drying. Do not dry the membrane if you intend to reprobe it.

The color can be restored by wetting the membrane with buffer 4. The membranes can also be stored in sealed plastic bags containing buffer 4. The color remains unchanged in this case.

3.2.12 Chemiluminescent Detection of DIG Modification with Lumigen™PPD or Lumiphos™530*

– <DIG>:AP (polyclonal sheep anti-DIG antibody [Fab]:alkaline phosphatase conjugate, 750 U/ml; store at 4 °C) (Boehringer Mannheim) **Standard reagents**
– Lumigen™PPD**, Lumiphos™530** (Boehringer Mannheim)
– Nylon membrane, positively charged (Boehringer Mannheim)
– For other reagents see Sect. 3.2.3

---

* Available also as DIG Luminescent Detection Kit for Nucleic Acids (Boehringer Mannheim)

** Trademark of Lumigen Inc., Detroit, MI, USA. Lumigen™PPD and Lumiphos™530 are the subjects of U.S. patents 4,962,192 and 4,969,182 granted to Lumigen Inc., Detroit, MI, USA

3.2.12

Standard solutions

- Buffer 1: 100 mM maleic acid; 150 mM NaCl; pH 7.5/25°C; the pH is adjusted with solid or concentrated NaOH and autoclaved
- Washing buffer: buffer 1 supplement with 0.3% [v/v] Tween 20
- Buffer 2: 1% [w/v] blocking reagent in buffer 1; prepare the solution from 10% [w/v] stock solution in buffer 1, autoclave and store at 4°C
- Buffer 3: 100 mM Tris-HCl; 100 mM NaCl; 50 mM $MgCl_2$; pH 9.5/25°C
- Antibody-conjugate solution for chemiluminescence detection, freshly prepared: Dilute <DIG>:AP solution to 75 mU/ml (1:10000) in buffer 2 directly prior to chemiluminescence detection
- Lumiphos™530 stock solution: 0.1 mg Lumiphos™530/ml buffer [750 mM 2-amino-2-methyl-1-propanol buffer, pH 9.6/25°C; 0.88 mM $MgCl_2$; 1.13 mM cetyltrimethyl ammonium bromide; 0.035 mM fluorescein surfactant]
- Lumigen™PPD stock solution: 10 mg Lumigen™PPD/ml H_2O ≙ 23.5 mM Lumigen™PPD
- Chemiluminescent substrate solution, freshly prepared: Lumiphos™530 stock solution is ready for use. Dilute Lumigen™PPD stock solution to 0.235 mM (1:100) in buffer 3 directly prior to chemiluminescent detection

Detection reaction

Chemiluminescence detection can be performed on nylon membranes, not on nitrocellulose. All the following incubations are performed at room temperature. Except for the chemiluminescence reaction, all incubations require shaking or mixing. The volumes of the solutions are calculated for membrane sizes of 100 cm² and should be adjusted for other membrane sizes.

1. Wash membrane briefly (1−5 min) in washing buffer.
2. Incubate for 30 min in 100 ml buffer 2.
3. Incubate membrane for 30 min in 20 ml freshly diluted antibody-conjugate solution for chemiluminescent detection (1:10000).
4. Remove unbound antibody conjugate by washing 2 × 15 min with 100 ml washing buffer.
5. Equilibrate the membrane for 2−5 min with 20 ml buffer 3.
6. Incubate the membrane in the dark with 10 ml freshly prepared chemiluminescent substrate solution for 5 min.
7. Let excess liquid drip off the membrane; blot membrane for a few seconds backside down on a sheet of Whatman 3MM paper, but not so that it is completely dry.
8. Seal the damp membrane in a hybridization bag.
9. Preincubate the sealed membrane for 5−15 min at 37°C.
10. Expose for 15−25 min at room temperature to X-ray or Polaroid b/w film. The time of exposure depends on strength of both signal and

background. Multiple exposures may be taken, as luminescence continues for at least 24 h; signal intensity seems even to accumulate with time.

11. For reprobing, the membrane must be wet.

Note: Repeated reprobing is easily achieved by washing the membrane first in sterile redistilled H_2O, then 2×15 min in reprobing solution (200 mM NaOH; 0.1% [w/v] SDS) at 37°C followed by a short wash in $2 \times$ SSC. Then prehybridization and hybridization can be done with a further probe according to the above protocol.

3.2.13 Special Hints for Application and Troubleshooting

It is useful to keep as closely as possible to the listed reagents, solutions, and protocols. Major pitfalls are low sensitivity or increased background. Either problem can be caused by any of the three reactions: labeling, hybridization, or detection.

Low sensitivity

— Check the efficiency of DIG DNA or RNA labeling by comparison to the controls.
— The quality of the membrane used as support for dot, southern, or northern blotting influences sensitivity and speed of detection. Membranes which are not suitable can cause strong background formation. Nitrocellulose membranes cannot be used for chemiluminescence detection unless a modified protocol is applied — with additional special blocking material (Nitro-Block) and longer exposure times.
— Increase the concentration of DIG-labeled DNA or RNA probe in hybridization solution.
— Increase the concentration of antibody conjugate and check the time of substrate reaction (BCIP/NBT: ≤ 72 h; Lumigen™PPD/Lumiphos™530: ≤ 24 h)
— Increase the duration of preincubation prior to chemiluminescence exposure to X-ray film to over 30 min and up to 12 h, and/or increase time of exposure to X-ray or Polaroid film. The type of film may also influence the sensitivity.

High background

— Purify DNA or RNA by phenol/chloroform extraction and/or ethanol precipitation before labeling. Make sure that probe does not contain cross-hybridizing vector sequences.
— Even though the protocol is optimized for the use of charged nylon membranes, some types which are very highly charged can cause background. Also lot-to-lot variations of some membrane types may cause problems. By use of the recommended membrane, these problems can be avoided.
— Important: it may be necessary to decrease the concentration of DIG-labeled DNA or RNA probes. The critical probe concentration limit (concerning background formation) can be determined by hybridization increasing probe concentrations with the unloaded membrane.

Take care that membranes do not dry out between prehybridization and hybridization steps.
- Decrease the concentration of <DIG>:AP conjugate and/or increase volumes of the washing and blocking solution and duration of the washing and blocking steps. Spotty background may be caused by precipitates in the <DIG>:AP conjugate; remove these by a short centrifugation step. Note, that several centrifugation steps can cause a certain loss of material, which must be compensated for by using larger amounts.)
- Shorten the time of preincubation and for shorten the exposure time. Keep in mind that signal intensity increases with time.

References

Arnold N, Seibl R, Kessler C, Wienberg J (1992) Nonradioactive in situ hybridization with digoxigenin-labeled DNA probes. Biotechnic & Histochemistry 67:59−67

Baldino F, Jr, Robbins E, Grega D, Meyers SL, Springer JE, Lewis ME (1989) Nonradioactive detection of NGF-receptor mRNA with digoxigenin-UTP labeled RNA probes. Neurosci Abstr 15:864

Baldino F, Jr, Lewis ME (1992) Nonradioactive in situ hybridization histochemistry with digoxigenin-dUTP labeled oligonucleotides. Methods in Neuroscience, in press

Casacuberta JM, Jardi R, Buti M, Puigdoménech P, SanSegundo B (1988) Comparison of different nonisotopic methods for hepatitis-B virus detection in human serum. Nucleic Acids Res 16:11834

Cohen SM (1990) Specidication of limb development in the Drosophila embryo by positional cues from segmentation genes. Nature 343:173−177

Dooley S, Radtke J, Blin N, Unteregger G (1988) Rapid detection of DNA-binding factors using protein-blotting and digoxigenine-dUTP marked probes. Nucleic Acids Res 16:11829

Dorward DW, Garon CF (1989) DNA-binding proteins in cells and membrane plots of Neisseria gonorrhoeae. J Bacteriol 171:4196−4201

Feinberg AP, Vogelstein B (1983) A technique for radiolabeling DNA restriction endonuclease fragments to high specific activity. Anal Biochem 132:6−13

Feinberg AP, Vogelstein B (1984) A technique for radiolabeling DNA restriction endonuclease fragments to high specific activity. Anal Biochem 137:266−267

Gebeyehu G, Rao PY, SooChan P, Simms DA, Klevan L (1987) Novel biotinylated nucleotide analogs for labeling and colorimetric detection of DNA. Nucleic Acids Res 15:4513−4534

Grega DS, Cavanagh TJ, Grimme S, Martin R, Lewis M, Robbins E, Baldino F, Jr (1989) Localization of neuronal mRNA by in situ hybridization using a nonradioactive detection method. Neurosci Abstr 15:739

Grega DS, Cavanagh TJ, Martin R, Lewis M, Baldino F, Jr (1989) In situ hybridization histochemistry using a new nonradioactive detection method. Advances in Gene Technology: Molecular Neurobiology and Neuropharmacology, ICSU Short Reports 9:69

Herrington CS, Burns J, Graham AK, Evans M, McGee JD (1989) Interphase cytogenetics using biotin and digoxigenin labeled probes. I: Relative sensitivity of both reporter molecules for detection of HPV16 in CaSki cells. J Clin Pathol (Lond) 42:592−600

Herrington CS, Burns J, Graham AK, Bhatt B, McGee JD (1989) Interphase cytogenetics using biotin and digoxigenin labeled probes. II: Simultaneous detection of human and papilloma virus nucleic acids in individual nuclei. J Clin Pathol (Lond) 42:601−606

Höltke HJ, Kessler C (1990) Nonradioactive labeling of RNA transcripts in vitro with the hapten digoxigenin (DIG); hybridization and ELISA-based detection. Nucleic Acids Res 18:5843−5851

Höltke HJ, Seibl R, Burg J, Mühlegger K, Kessler C (1990) Non-radioactive labeling and detection of nucleic acids: II. Optimization of the digoxigenin system. Biol Chem Hoppe-Seyler 371:929−938

Höltke HJ, Sagner G, Kessler C, Schmitz GG (1992) Sensitive chemiluminescent detection of digoxigenin-labeled nucleic acids: a fast and simple protocol and its applications. BioTechniques 12:104−113

Höltke HJ, Ettl I, Finken M, West S, Kunz W (1992) Multiple nuleic acid labeling and rainbow detection. Anal Biochem, in press

Jessberger R, Berg P (1991) Repair of deletions and double-strand gaps by homologous recombination in a mammalian in vitro system. Mol Cell Biol 11:445−457

Kessler C (1990) Detection of nucleic acids by enzyme-linked immuno-sorbent assay (ELISA) technique: an example for the development of a novel non-radioactive labeling and detection system with high sensitivity. In: Obe G (ed) Advances in Mutagenesis Research, Springer-Verlag, Berlin/Heidelberg, pp 105−152

Kessler C (1991) The digoxigenin (DIG) technology − a survey on the concept and realization of a novel bioanalytical indicator system. Mol Cell Probes 5:161−205

Kessler C, Höltke H-J, Seibl R, Burg J, Mühlegger K (1990) Non-radioactive labeling and detection of nucleic acids: I. A novel DNA labeling and detection system based on digoxigen:anti-digoxigen ELISA principle (digoxigenin system). Biol Chem Hoppe-Seyler 371:917−927

Kessler C, Sagner G, Höltke H-J (1991) Nonradioactive dideoxy sequencing using the digoxigenin system. The 1991 San Diego Conference of Nucleic Acids, Poster 27

Kimpton CP, Corbitt G, Morris DJ (1989) Detection of cytomegalovirus DNA using probes labeled with digoxigenin. J Vir Meth 24:335−346

Krieg PA, Melton DA (1987) In vitro RNA synthesis with SP6 RNA polymerase. Meth Enzymol 155:397−415

Lichter P, Tang C-JC, Call K, Hermanson G, Evans GA, Housman D, Ward DC (1990) High-resolution mapping of human chromosome 11 by in situ hybridization with cosmid clones. Science 247:64−69

Manstein DJ, Titus MA, De Lozanne A, Spudich JA (1989) Gene replacement in Dictyostelium: generation of myosin null mutants. EMBO J 8:923−932

Mühlegger K, Huber E, von der Eltz H, Rüger R, Kessler C (1990) Non-radioactive labeling and detection of nucleic acids: IV. Synthesis and properties of the nucleotide compounds of the digoxigenin system and of photodigoxigenin. Biol Chem Hoppe-Seyler 371:953−965

Nakano Y, Guerrero I, Hidalgo A, Taylor A, Whittle JRS, Ingham PW (1989) A protein with several possible membrane-spanning domains encoded by the Drosophila segment polarity gene patched. Nature 341:508−513

Reischl U, Rüger R, Kessler C (1992) Nonradioactive labeling of PCR products. In: White BA (ed) PCR: Selected protocols and applications. The Humana Press, Clifton, NJ, in press

Roychoudhury R, Wu R (1980) Terminal transferase-catalyzed addition of nucleotides to the 3' termini of DNA. Meth Enzymol 65:43−62

Rüger R, Höltke HJ, Sagner G, Seibl R, Kessler C (1990) Rapid labeling methods using the DIG-system: incorporation of digoxigenin in PC reactions and labeling of nucleic acids with photodigoxigenin. Fresenius' Zeitschrift für Analytische Chemie 337:114

Schäfer R, Zischler H, Epplen JT (1988) DNA fingerprinting using nonradioactive oligonucleotide probes specific for simple repeats. Nucleic Acids Res 16:9344

Schmitz GG, Walter T, Seibl R, Kessler C (1991) Non-radioactive labeling of oligonucleotides in vitro with the the hapten digoxigenin by tailing with terminal transferase. Anal Biochem 192:222−231

Seibl R, Höltke H-J, Rüger R, Meindl A, Zachau HG, Raßhofer R, Roggendorf M, Wolf H, Arnold N, Wienberg J, Kessler C (1990) Non-radioactive labeling and detection of nucleic acids: III. Applications of the digoxigenin system. Biol Chem Hoppe-Seyler 371:939−951

Suske G, Gross B, Beato M (1989) Nonradioactive method to visualize specific DNA-protein interactions in the band shift assay. Nucleic Acids Res 17:4405

Tautz D, Pfeifle C (1989) A nonradioactive in situ hybridization method for the localization of specific RNAs in Drosophila embryos reveals translational control of the segmentation gene hunchback. Chromosoma 98:81−85

West S, Schröder J, Kunz W (1990) A multiple staining procedure for the detection of different DNA fragments on a single blot. Anal Biochem 190:254−258

Young WS, III (1989) Simultaneous use of digoxigenin- and radiolabeled oligodeoxyribonucleotide probes for hybridization histochemistry. Neuropeptides (Edinburgh) 12:271−275

Zischler H, Nanda I, Schäfer R, Schmid M, Epplen JT (1989) Digoxigenated oligonucleotide probes specific for simple repeats in DNA fingerprinting and hybridization in situ. Hum Genet 82:227−233

Zischler H, Schäfer R, Epplen JT (1989) Non-radioactive oligonucleotide fingerprinting in the gel. Nucleic Acids Res 17:4411

Zuber U, Schumann W (1991) Tn5cos: a transposon for restriction mapping of large plasmids using phage lambda terminase. Gene 103:69−72

3.3 Labeling and Detection of Proteins and Glycoproteins

ANTON HASELBECK and WOLFGANG HÖSEL

3.3.1 Principle and Applications

The specific labeling and detection of macromolecules by low molecular weight compounds and their respective binding partners (e.g., the biotin/streptavidin system) have proven to be very successful. The labeling and detection system employing DIG and a labeled anti-DIG (<DIG> antibody is, however, better suited for these purposes due to several advantages pointed out in previous chapters. We therefore set out to adapt the DIG/<DIG> system for the detection of proteins and glycoconjugates. For this purpose DIG is introduced onto the particular molecule by specific chemical or enzymatic steps or by digoxigenylated binding proteins (e.g., lectins). The introduced DIG modification can than be detected by the methods previously outlined.

3.3.2 Labeling and Detection of Proteins/Peptides

Proteins or peptides bound to a surface (e.g., blotting membranes) are derivatized with DIG-carboxy-methyl-N-hydroxy-succinimide ester (DIG-ester) (reacting with amino groups) and/or DIG-3-O-succinyl-[2-maleimido)]-ethylamide (DIG-maleimide) (reacting with sulfhydryl groups). The DIG-labeled proteins/peptides are subsequently detected using <DIG> antibody conjugated with alkaline phosphatase (<DIG>:AP). With the two DIG reagents it is possible to label specifically either amino or sulfhydryl groups or both. Regarding the sulfur-containing groups, a distinction between free SH groups and disulfide (S-S) bridges is possible by performing the labeling with and without prior reduction of the S-S bridges, e.g., with 2-mercaptoethanol or dithiothreitol. In order to achieve the highest sensitivity for protein/peptide detection with this system, reduction (e.g., with 2-mercaptoethanol) prior to the incubation with the two DIG reagents is recommended. With the standard procedure described later approximately 0.1 ng of the standard protein rec creatinase from E. coli is detected in a dot blot type of experiment and approximately 1 ng by western blotting.

3.3.3 Labeling and Detection of Glycoconjugates

DIG can be attached specifically to glycans in several ways. For the general labeling and detection of glycoconjugates, oxidation of vicinal diols of sugars by periodate with subsequent covalent attachment of DIG-succinyl-ε-amidocaproic acid-hydrazide (DIG-hydrazide) is the method of choice. The periodate oxidation step can be made specific for the detection of sialic acids by choosing the appropriate conditions (1 mM at 0 °C). For the detection of terminal galactose units, oxidation with galactose oxidase in combination with the incorporation of DIG-hydrazide into the resulting aldehyde groups is suitable. These methods were outlined in detail (Haselbeck and Hösel, 1990). The sensitive detection of glycolipids by these methods was also described (Kniep and Mühlradt, 1990).

By using DIG conjugates of lectins with well known carbohydrate specificity a whole array of reagents for the structural analysis of glycoconjugates on blots became available and has been applied to the analysis of glycoproteins on blots (Haselbeck et al., 1990). Moreover, the DIG-lectin conjugates proved very suitable for histochemical studies of glycoconjugates, above all by using gold-labeled <DIG> antibodies (Sata et al., 1990).

3.3.4 General Labeling and Detection of Proteins/Peptides with DIG-Ester and DIG-Maleimide

The standard procedures consist of two steps: (1) introduction of the DIG label specifically into amino acids or sugars and (2) detection of DIG by an ELISA-type of reaction using <DIG>:AP.

3.3.4

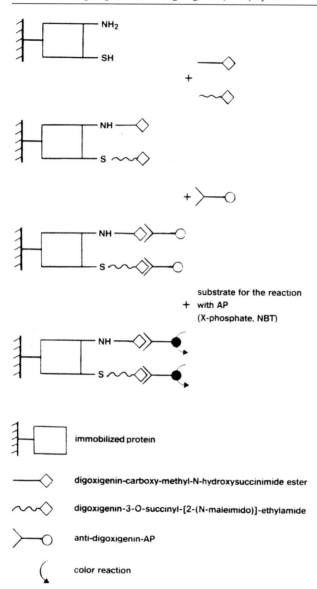

Labeling and detection of proteins/peptides

Labeling reagents
- DIG-carboxy-methyl-*N*-hydroxysuccinimide ester (Boehringer Mannheim)*
- DIG-3-O-succinyl-(2-(N-maleimido))-ethylamide (Boehringer Mannheim)*
- K_2HPO_4 (Merck)
- KH_2PO_4 (Merck)
- Nonidet P-40 (Boehringer Mannheim)

---

* Included in the DIG Protein Detection Kit (Boehringer Mannheim)

- Tween 20 (Boehringer Mannheim)
- 2-Mercaptoethanol (Merck)

- Tris (Boehringer Mannheim)
- NaCl (Merck)
- MgCl$_2$ (Merck)
- Blocking reagent (Boehringer Mannheim)*
- <DIG>:AP (polyclonal sheep anti-DIG antibody[Fab]: alkaline phosphatase conjugate, 750 U/ml) (Boehringer Mannheim)*
- BCIP (5-bromo-4-chloro-3-indolyl phosphate) (Boehringer Mannheim)*
- NBT (4-nitroblue tetrazolium chloride) (Boehringer Mannheim)*
- Dimethylformamide (Merck)

- Creatinase (Boehringer Mannheim)*
- Bovine serum albumin (Boehringer Mannheim)

- Nitrocellulose BA 85 (Schleicher & Schüll)
- Polyvinyl difluoride (Immobilon, Millipore): Both types of membranes are suited for the techniques described below.

Volumes indicated are sufficient for processing one 100 cm^2 blot.
- Phosphate buffer: 50 mM potassium phosphate, pH 8.5/25 °C
- TBS (Tris-buffered saline): 50 mM Tris-HCl, 150 mM NaCl, pH 7.5/ 25 °C
- DIG solution: 5 µl DIG-ester (0.5 mg/ml, dissolved in dimethylformamide) and 5 µl DIG-maleimide (0.5 mg/ml, dissolved in dimethylformamide) in 20 ml phosphate buffer, containing 0.05% Nonidet P-40.
- Wash solution 1: TBS, 0.1% [v/v] Tween 20
- 2-mercaptoethanol solution: 20 ml phosphate buffer, containing 2% [v/v] 2-mercaptoethanol

Volumes indicated are sufficient for processing one 100 cm^2 blot.
- Blocking solution: 0.1 g blocking reagent dissolved in 20 ml phosphate buffer; the dissolution needs heating to 50°−70 °C for approximately 1 h; it can be accelerated by ultrasonication and incubation in a microwave oven. The solution remains turbid. If stored for a longer time, adding sodium azide (final concentration: 0.1%) is recommended.
- Antibody-conjugate solution: 10 µl <DIG>:AP diluted in 10 ml TBS
- Staining buffer: 100 mM Tris-HCl, 100 mM NaCl, pH 9.5.
- BCIP solution: 50 mg/ml BCIP in 100% [v/v] dimethylformamide
- NBT solution: 75 mg/ml NBT in 70% [v/v] dimethylformamide
- Staining solution, freshly prepared: 35 µl BCIP solution and 45 µl NBT solution diluted in 10 ml 100 mM Tris-HCl, 50 mM MgCl$_2$, 100 mM NaCl, pH 9.5

All incubations are carried out by gentle shaking at room temperature.

1. Wash the membrane after the protein transfer 3 × 5 min with 50 ml phosphate buffer (not necessary for dot blot samples).

---

* Included in the DIG Protein Detection Kit (Boehringer Mannheim)

2. Reduction step (not necessary for already reduced protein samples): incubate the membrane for 30 min in the 2-mercaptoethanol solution.
3. Wash 3 × 10 min with approximately 50 ml phosphate buffer.
4. Incubate the membrane for 1 h in the DIG solution.
5. Wash once with approximately 50 ml TBS, 0.1% [v/v] Tween 20 and 2 × 50 min with approximately 50 ml TBS.

All incubations are carried out at room temperature with gentle shaking, except the colour reaction which should be done without shaking.

Detection reaction
1. Incubate the membrane for at least 30 min in the blocking solution.
2. Wash 3 × 5 min with aprroximately 50 ml TBS.
3. Incubate for 1 h in the antibody solution.
4. Wash 3 × 5 min with approximately 50 ml TBS.
5. Immerse the membrane in the staining solution and observe the colour development. This is normally complete within 10–30 min but can take longer if very small amounts of proteins are present.
6. Stop the colour reaction by rinsing the membrane several times with redistilled water and dry the membrane on paper towels.
7. Document the results by photographing or photocopying the membrane; the dry membrane can also be stored for documentation.

3.3.5 Selective Labeling of Sulfhydryl Groups

To perform selective SH group labeling, pH 7.0 phosphate buffer has to be used instead of pH 8.5 phosphate buffer and the labeling is done only with DIG-maleimide. The reduction step is omitted. All other steps are identical to those in the previous section (Sect. 3.3.4).

Standard reagents
Reagents for labeling and detection are identical to those used in the previous procedure (Sect. 3.3.4).

Labeling solutions
Volumes indicated are sufficient for processing one 100 cm^2 blot.

– Phosphate buffer pH 7.0: 50 mM potassium phosphate pH 7.0
– DIG solution: 5 µl DIG-maleimide (5 mM in dimethylformamide) diluted in 10 ml phosphate buffer containing 0.01% [v/v] Nonidet P 40; pH 7.0/25 °C

Detection solutions
Identical with solutions in previous procedure (Sect. 3.3.4).

Labeling reaction
1. Wash the membrane 3 × 5 min with approximately 50 ml phosphate buffer, pH 7.0.
2. Incubate the membrane for 1 h in the DIG solution.
3. Wash 1 × 5 min with approximately 50 ml TBS, 0.1% [v/v] Tween 20 and 2 × 5 min with approximately 50 ml TBS.

Detection reaction
Identical to procedure in Sect. 3.3.4.

3.3.6 Selective Labeling of Disulfides

- *N*-ethylmaleimide (Merck)
- All other required reagents are identical to those in (Sect. 3.3.4).

Labeling reagents

Identical with reagents in Sect. 3.3.4.

Detection reagents

- NEM solution: 10 mM *N*-ethylmaleimide dissolved in phosphate buffer, pH 7.0 (20 ml)
- 2-mercaptoethanol solution: 20 ml phosphate buffer, pH 7.0, containing 2% [v/v] 2-mercaptoethanol
- DIG solution: 5 μl DIG-maleimide diluted in 10 ml phosphate buffer, pH 7.0

Labeling solutions

Identical with solutions in Sect. 3.3.4.
For exclusive labeling of disulfides, the free SH groups present in the protein have to be blocked first and then the disulfides are reduced to the reactive sulfhydryl groups.

Detection solutions

1. Wash the membrane 3 × 5 min with approximately 50 ml phosphate buffer, pH 7.0.
2. Incubate for 30 min in the 2-NEM solution. Note: alternatively, 2-vinyl-pyridine, 1% [v/v] in phosphate buffer, pH 7.0 can be used for blocking the SH groups.
3. Wash the membrane 3 × 10 min with approximately 50 ml phosphate buffer, pH 7.0.
4. Incubate the membrane for 30 min in the 2-mercaptoethanol solution.
5. Wash the membrane 3 × 5 min with approximately 50 ml phosphate buffer, pH 7.0.
6. Incubate for 1 h in the DIG solution.
7. Wash 1 × 5 min with approximately 50 ml TBS; 0.1% [v/v] Tween 20 and 2 × 5 min with approximately 50 ml TBS.

Labeling reaction

Identical to procedure in Sect. 3.3.4.

Detection reaction

3.3.7 Special Hints for Application and Troubleshooting

- It is important to follow the protocols as closely as possible.

- The sensitivity depends on the amount of reactive amino and sufhydryl groups present in a protein.
- The detection limit for most proteins is about 1 ng in one band (transfer blot) or 0.1 ng in one spot (dot blot) in a 30 min color reaction after performing the general labeling procedure described in Sect. 3.3.4.

Sensitivity

- Background staining is usually a light gray but may increase, especially if the incubation periods in the colour solution exceed 30 min. Unwanted high background can be reduced by prolonging the blocking step for up to 15 h.

Background

3.3.8

Labeling — All DIG labeling solutions must be freshly prepared; the DIG-ester and DIG-maleimide solutions in dimethylformamide must always be stored at −20°C.

Artifacts — When applying proteins reduced with 2-mercaptoethanol or dithiothreitol to SDS-PAGE and blotting, bands which do not originate from the sample may be visible in the 60−70kDa range (Bjerrum et al., 1988). These artifacts have been described in the literature (Ochs, 1983) and seem to be associated with skin contaminations (keratins) from the solutions and equipment used (Riches et al., 1988).

3.3.8 General Labeling and Detection of Glycoconjugates with DIG-Hydrazide

This procedure is based on the chemical oxidation of vicinal hydroxyl groups in sugars of glycoconjugates to aldehydes. These aldehyde groups react with DIG-hydrazide, and DIG is then detected in an ELISA-type reaction with the <DIG>:AP conjugate. The detection procedure is identical to that described in Sect. 3.3.4. Oxidation and labeling can be performed in solution (procedure 3.3.8.1) or on a membrane (procedure 3.3.8.2).

The advantages of oxidizing and labeling membrane-bound glycoproteins are: (1) removal of substances in the glycoprotein solution which might interfere with the oxidation and labeling steps resulting in unspecific labeling of nonglycosylated proteins or prevention of labeling of glycoproteins and (2) the electrophoretic mobility of the proteins on SDS-PAGE is not altered; due to the digoxigenylation, glycoprotein bands tend to streak. Furthermore, with this method, sialic acid-containing glycoproteins produce stronger signals than do glycoproteins without sialic acids.

Labeling reagents
- $NaIO_4$ (Merck)*
- $Na_2S_2O_5$ (Merck)*
- DIG-succinyl-ε-amidocaproic acid hydrazide (Boehringer Mannheim)*
- Na-acetate (Merck)
- SDS (Boehringer Mannheim)
- Glycerol (Merck)
- Bromophenol blue (Merck)

Detection reagents Identical with reagents in Sect. 3.3.4.

Control proteins
- rec creatinase (negative control; Boehringer Mannheim)*
- Transferrin (positive control; Boehringer Mannheim)*

---

* Included in the DIG Glycan Detection Kit (Boehringer Mannheim)

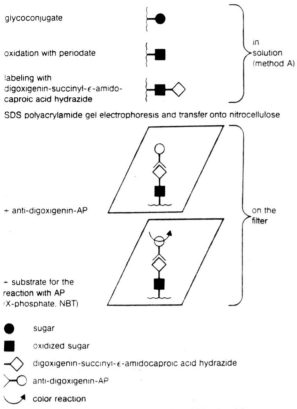

glycoconjugate

oxidation with periodate

labeling with
digoxigenin-succinyl-ε-amidocaproic acid hydrazide

in
solution
(method A)

SDS polyacrylamide gel electrophoresis and transfer onto nitrocellulose

+ anti-digoxigenin-AP

on the
filter

+ substrate for the
reaction with AP
(X-phosphate, NBT)

●   sugar

■   oxidized sugar

◇   digoxigenin-succinyl-ε-amidocaproic acid hydrazide

>–O   anti-digoxigenin-AP

    color reaction

Use of DIG-succinyl-ε-amidocaproic acid hydrazide

— Na-acetate buffer for procedure 3.3.8.1: 100 mM Na-acetate, pH 5.5/ **Labeling**
  25 °C   **solutions**
— PBS (phosphate buffered saline) for procedure 3.3.8.2: 50 mM potassium phosphate, 150 mM NaCl, pH 7.5
— $NaIO_4$ solution for procedure 3.3.8.1: 15 mM $NaIO_4$ dissolved in H_2O
— $NaIO_4$ solution for procedure 3.3.8.2: 10 mM $NaIO_4$ dissolved in Na-
  acetate buffer (10 ml)
— $Na_2S_2O_5$ solution: 20 mM $Na_2S_2O_5$ dissolved in H_2O
— DIG-hydrazide for procedure 3.3.8.1: 5 mM dissolved in dimethylformamide
— DIG-hydrazide solution for procedure 3.3.8.2: 1 μl DIG-hydrazide
  (5 mM) diluted in 5 ml Na-acetate buffer
— 4× concentrated SDS sample buffer: 25 mM Tris-HCl, pH 6.8;
  8% [w/v] SDS; 40% [v/v] glycerol; 20% [v/v] 2-mercaptoethanol; and
  bromophenol blue as tracking dye
— Ponceau S solution: 0.2% Ponceau S dissolved in 3% acetic acid

— Identical with solutions in Sect. 3.3.4.   **Detection**
                                                             solutions

3.3.8.1 Oxidation and Labeling of Glycoconjugates in Solution

Labeling reaction
1. Dissolve $0.1-10 \mu g$ glycoprotein in $20 \mu l$ Na-acetate buffer or dilute the protein solution at least $1:1$ with the Na-acetate buffer.
2. Add $10 \mu l$ NaIO$_4$ solution (15 mM), mix, and incubate for 20 min in the dark at room temperature.
3. Add $10 \mu l$ Na$_2$S$_2$O$_5$ solution to destroy excess periodate. Mix and leave for 5 min at room temperature.
4. Add $5 \mu l$ DIG-hydrazide (5 mM) and incubate for 1 h at room temperature.
5. Add $15 \mu l$ 4× concentrated SDS sample buffer, mix, heat at 100°C for 2 min and load an aliquot onto an SDS gel.

Note: Dot blot samples can be spotted directly onto a nitrocellulose membrane without adding the SDS sample buffer.

Detection reaction — Identical to procedure in Sect. 3.3.4.

3.3.8.2 Oxidation and Labeling of Glycoconjugates Bound to a Membrane

Labeling reaction
1. Wash the membrane 3×5 min with approximately 50 ml PBS; do not use TBS at this stage as Tris interferes with the subsequent DIG labeling.
2. Incubate the membrane for 20 min at room temperature in the 10 mM NaIO$_4$ solution.
3. Wash 3×5 min with approximately 50 ml PBS.
4. Incubate for 1 h at room temperature in the DIG-hydrazide solution ($1 \mu l$ 5 mM in 5 ml Na-acetate buffer).
5. Wash 3×5 min with approximately 50 ml TBS.

Detection reaction Identical to procedure in Sect. 3.3.4.

3.3.9 Special Hints for Application and Troubleshooting

Sensitivity — For oxidation and labeling in solution (procedure 3.3.8.1), after 1 h of developing in the colour substrate solution, different amounts of glycoproteins can be detected on dot blots. The various amounts are: α1-acid glycoprotein, 1 ng; fetuin, 1 ng; transferrin, 5 ng; carboxypeptidase Y, 10 ng.
— Approximate detection limits for membrane-bound glycoproteins (procedure 3.3.8.2) are: α1-acid glycoprotein, 5 ng; fetuin, 5 ng; transferrin, 25 ng; carboxypeptidase Y, 50 ng.

Labeling in Solution The presence of detergents, e.g., Triton X-100, Nonidet P-40, SDS, up to a concentration of 0.25% does not interfere with oxidation and labeling in solution (procedure 3.3.8.1), with the exception of sugar-containing detergents such as octylglucoside. In addition, 2-mercaptoethanol, dithiothreitol ($>0.1\%$), glycerol ($>0.01\%$), and amino group-containing buffers such as Tris and glycine must be strictly avoided.

Positive and negative controls should always be included. When labeling **Controls** in solution (procedure 3.3.8.1), purified *E. coli* proteins, e.g., creatinase, are suitable as negative controls as are glycoproteins such as carboxypeptidase Y or transferrin after complete removal of their carbohydrate chains with N-glycosidase F (Boehringer Mannheim). Some serum-derived proteins, e.g., albumin, contain chemically linked glucose or fructose which will be positive in this method. It should also be taken into account that stabilizers such as sucrose or glycerol, which are often present in enzyme preparations, interfere with this procedure.

In the labeling and detection of membrane-bound glycoproteins (procedure 3.3.8.2), when using nitrocellulose, nonglycosylated proteins appear **Back-ground** as white bands or spots on a slight pink to gray background. This is caused by a certain degree of oxidation and labeling of the membrane itself which does not occur on protein-covered areas.

3.3.10 Sialic Acid-Specific Oxidation

By using 1 mM $NaIO_4$ and performing the reaction at 0°C only sialic acids will be oxidized and then labeled with digoxigenin. Following the above procedures, glycoproteins can be oxidized and labeled either in solution (procedure 3.3.8.1) or bound to a membrane (procedure 3.3.8.2). The necessary modifications are described below.

Identical with reagents in Sect. 3.3.8 (labeling) and in Sect. 3.3.4 (detection). **Standard reagents**

- $NaIO_4$ solution: 3 mM $NaIO_4$ in H_2O (procedure 3.3.8.1)
- $NaIO_4$ solution: 1 mM $NaIO_4$ dissolved in Na-acetate buffer (procedure 3.3.8.2)
- $Na_2S_2O_5$ solution: 4 mM $Na_2S_2O_5$ dissolved in H_2O (procedure 3.3.8.1)

Identical with solutions in Sect. 3.3.4. **Labeling solutions**

Follow the methods outlined above for procedures 3.3.8.1 and 3.3.8.2 **Labeling** using the modified solutions for this procedure. Incubate the tubes **and** (3.3.8.1) or membranes (3.3.8.2) at 0°C for 20 min. **detection**

3.3.11 Special Hints for Application and Troubleshooting

– The sensitivity is about a factor 10 lower than the standard oxidation procedures.

Controls – As positive and negative controls, fetuin and asialofetuin (Boehringer Mannheim) can be used. Only fetuin will be positive under these conditions.

3.3.12 Selective Detection of Terminal Galactose Residues

Terminal galactose residues in glycoconjugates can be oxidized with the enzyme galactose oxidase. DIG-hydrazide is then linked exclusively to galactose residues.

Labeling – Galactose oxidase (Boehringer Mannheim)
reagents – Bovine serum albumin (BSA); Boehringer Mannheim)

Detection Identical with reagents in Sect. 3.3.4.
reagents

Labeling – TBS-BSA: 1% [w/v] BSA dissolved in 20 ml TBS
solutions – Phosphate buffer, pH 6.0: 100 mM potassium-phosphate; pH 6.0
– Galactose oxidase solution: 7.5 U galactose oxidase and 2 µl DIG-hydrazide (5 mM) diluted in 10 ml phosphate buffer; pH 6.0.

Detection Identical with those in Sect. 3.3.4.
solutions

Labeling 1. Incubate the membrane for 30 min with 20 ml TBS-BSA.
and 2. Wash 1 × 5 min with approximately 50 ml phosphate buffer, pH 6.0/
detection   25 °C.
reaction 3. Incubate the membrane for 15 h at 37 °C in galactose oxidase solution.
4. Wash 2 × 5 min with approximately 50 ml TBS and proceed with the blocking step as outlined in Sect. 3.3.4.

3.3.13 Special Hints for Application and Troubleshooting

Sensitivity – The detection limit varies from glycoprotein to glycoprotein, but is usually at about 50–100 ng.

Oxidation – Sialic acid containing glycoprotein can be rendered sensitive to galactose oxidase by incubating with a neuraminidase, e.g., from *Arthrobacter ureafaciens* (Boehringer Mannheim). For this purpose 0.1 U/ml neuraminidase should be included in step 3 of the labeling reaction in Sect. 3.3.12.

glycoprotein:

M-GlcNAc-Gal-[SA]

Asn-GlcNAc₂-M

M-GlcNAc-Gal

SDS polyacrylamide gel electrophoresis and transfer onto nitrocellulose

(specific for SA)

+   NBT/X-phosphate

→ color reaction

■   = SA (sialic acid)   GlcNAc = N-acetylglucosamine

= lectin, digoxigenin-labeled   M   = mannose

= anti-digoxigenin-AP   Gal = galactose

Asn = asparagine

Use of DIG-labeled lectins

3.3.14 Labeling of Glycoconjugates with DIG-Lectins

DIG can be introduced into the carbohydrate part of glycoconjugates by binding of DIG-lectins. The use of different lectins with a well defined substrate specificity allows characterization of the carbohydrate chains.

— DIG-lectins (Boehringer Mannheim); these are available as single reagents. Five lectins from *Datura stramonium* (DSA), *Galanthus nivalis* (GNA), *Maackia amurensis* (MAA), *Arachis hypogäa* (PNA) and

Labeling reagents

Sambucus nigra (SNA) are also available in the Glycan Differentiation Kit.
- $MgCl_2$, $MnCl_2$, $CaCl_2$ (Merck)

Detection reagents Identical with reagents in procedure in Sect. 3.3.4.

Labeling solutions – Incubation buffer: TBS, containing 1 mM $MgCl_2$, $MnCl_2$, $CaCl_2$

Detection solutions Identical with solutions in procedure in Sect. 3.3.4.

Labeling reaction
1. Incubate the membrane for at least 30 min in the blocking solution.
2. Wash 2 × 5 min with approximately 50 ml TBS and 1 × 5 min with approximately 50 ml incubation buffer.
3. Incubate the membrane for 1 h at room temperature in 10 ml incubation buffer containing the desired lectin. The concentration of the DIG-lectin depends on the lectin used and is in the range of 1–10 μg/ml (indicated in the package insert of the respective DIG-lectin or the Glycan Differentiation Kit).
4. Wash 3× 5 min with approximately 50 ml TBS and continue with the <DIG>:AP incubation as outlined in Sect. 3.3.4.

3.3.15 Special Hints for Application and Troubleshooting

Sensitivity – Sensitivity depends greatly on the respective glycoprotein which is analyzed and also varies somewhat between the individual lectins. Generally the detection limit ranges from 1 to 10 ng glycoprotein on dot blot samples.

Developing time – Developing time in the color substrate solution is usually 10–30 min. In order to detect very small amounts of glycoproteins (1 ng or less) longer developing times may be required. In this case it is advisable to increase the blocking time from 30 min to up to 15 h to reduce background staining.

Lectins – The lectins used for this type of analysis are primarily tested with glycoproteins from yeast and animal sources; when analyzing glycoproteins from plants or bacteria different carbohydrate structures may be recognized also.
– Positive and negative control proteins have to be included to ensure the carbohydrate specificities of the individual lectins.

References

Bjerrum OJ, Larsen KP, Heegard NHH (1988) Nonspecific binding and artifacts: specificity problems and trouble shooting with an atlas of immunological artifacts. In: Bjerrum OJ, Heegard NHH (eds) Handbook of immunoblotting of proteins, CRC press Inc, Boca Raton, Florida

Haselbeck A, Hösel W (1990) Description and application of an immunological detection system for analyzing glycoproteins on blots. Glycoconjugate J 7:63−74

Haselbeck A, Schickaneder E, von der Eltz H, Hösel W (1990) Structural characterization of glycoprotein carbohydrate chains by using digoxigenin-labeled lectins on blots. Anal Biochem 191:25−30

Kniep B, Mühlradt PF (1990) Immunochemical detection of glycosphingolipids on thin layer chromatograms. Anal Biochem 188:5−8

Ochs D (1983) Protein contaminants of sodium dodecyl sulfate-polyacrylamide gels. Anal Biochem 135:470−474

Riches PG, Polce B, Hong R (1988) Contaminant bands on SDS-polyacrylamide gel electrophoresis are recognised by antibodies in normal human serum and saliva. J Immunological Methods 110:117−121

Sata T, Zuber C, Roth J (1990) Lectin-digoxigenin conjugates: a new hapten system for glycoconjugate cytochemistry. Histochemistry 94:1−11

4 The Biotin System

4.1 Labeling and Detection of Nucleic Acids

AYOUB RASHTCHIAN AND JESSE MACKEY

4.1.1 Principle and Applications

Nucleic acid hybridization has been one of the most powerful techniques in molecular biology during the past two decades. The applications of nucleic acid hybridization range from determination of overall similarity between organisms (Brenner, 1989) to determination of even a single base mutation in a given gene. Nucleic acid hybridizations are basically performed in three general ways: (1) solution hybridization, (2) hybridization on membrane filters, and (3) in situ hybridization to cytological preparations.

Although radiolabeled probes continue to be used in many applications, the short halflife, problems with disposal and safety and regulatory concerns have resulted in a widespread search for an ideal nonradioactive substitute. Consequently, a large number of nonradioactive labeling methods and appropriate signal generation systems have been developed (See Matthews and Kricka, 1988 for view). The properties of these systems vary greatly with regard to sensitivity of detection as well as the simplicity/complexity and requirements for specialized equipment. These properties, the hybridization format used and ultimately the nature of the biological/biochemical question(s) being addressed by individual researchers affect the choice of the nonradioactive system to be used.

The choice of a strategy for nonradioactive labeling of nucleic acid probes is heavily dependent on the type of probe being used. However, the labeling strategies can generally be divided into two categories:

1. Indirect labeling with reporter groups that require secondary recognition after hybridization, such as biotin and other haptens.
2. Direct primary labels which are covalently attached to the nucleic acid probe and are directly detectable such as enzymes and fluorophores.

In this section we discuss labeling of nucleic acids with biotin and provide protocols for labeling of DNA and RNA probes with biotin (bio) using enzymatic and chemical methods.

4.1.1.1 Labeling of Nucleic Acids with Biotin

The most common nonradioactive labeling method in recent years has been the use of biotin as the reporter group. The high binding affinity of avidin or streptavidin to biotin has made biotin one of the most widely used nonradioactive systems. In this system the probe is labeled with biotin, and the presence of labeled probe after hybridization is detected by streptavidin or avidin conjugated to an enzyme. Of the many enzymes available, alkaline phosphatase conjugated to streptavidin has most often been used for a variety of applications.

The method used for labeling of nucleic acid probes is dependent on the type of probe used as well as the intended use of probe. Labeling methods can be generally divided into (a) enzymatic and (b) chemical categories. Both approaches have been successfully and widely applied to labeling of nucleic acid probes. The most common enzymatic methods of labeling DNA probes are by nick translation (Rigby et al., 1977) or by random primer labeling (Feinberg and Vogelstein, 1983, 1984) using a suitable biotinylated nucleoside triphosphate (Gebeyehu et al., 1987; Langer et al., 1981). Both methods are efficient in incorporating biotin-labeled nucleoside triphosphates into DNA. There are kits available for labeling of DNA with biotin using these procedures (Life Technologies/BRL). An alternative chemical method for biotin labeling is photobiotin. This reagent contains a photoreactive aryl azide group attached to biotin through a charged linker arm. When irradiated with strong visible light in the presence of DNA photobiotin covalently links biotin groups to the DNA (Forster et al., 1985).

The choice of labeling method for preparation of biotinylated probes depends on the specific use of the probe. While all three labeling methods mentioned above are suitable for detection of nucleic acids on membrane filters, only methods which yield probes of small size (50−500 bases) are suitable for use in in situ hybridization (Singer et al., 1987). The most widely used method for preparation of labeled probe for use in situ is nick translation (Singer et al., 1987). Preparation of probes of optimal size for use in in situ requires optimization of the DNase concentration in the nick translation reaction. A kit specifically developed for preparation of probes suitable for in situ hybridization is available (Life Technologies/BRL).

4.1.1.2 Biotinylated Nucleotides

Biotinylated DNA probes can be prepared by enzymatic methods such as nick translation and random primer extension; biotinylated RNA probes can be synthesized by in vitro transcription using an RNA polymerase such as T3, T7, or SP6. A number of biotinylated deoxyribonucleoside triphosphates and ribonucleoside triphosphates have been used for the enzymatic preparation of biotinylated DNA and biotinylated RNA probes. These include derivatives at the 5-position of dUTP as well as the N^4 and N^6 positions of dCTP or dATP, respectively.

Langer and colleagues (1981) first synthesized biotinylated derivatives of dUTP in which the biotin moiety was attached via a linker arm at the 5-position of the deoxyuridine residue. Additional biotinylated dUTP analogs with linker arms of varying lengths attached at the 5-position were prepared by others. The analogs were designated as biotin-n-dUTP where n represents the number of atoms between the carbonyl group of the biotin and the position of attachment to the nucleoside base. Brigati et al. (1983) observed that probes labeled with biotin-11-dUTP and biotin-16-dUTP were detected with a high degree of sensitivity while those labeled with biotin-4-dUTP were not detected as well. These data indicate that sensitivity of detection depends on length of the linker arm and that longer is generally better.

A number of biotin derivatives of dATP and dCTP were prepared by Gebeyehu et al., 1987. In these compounds the biotin was attached through a linker arm to the N^6 position of adenine or to the N^4 position of cytosine. They included biotin-3-dATP and -dCTP, biotin-7-dATP and -dCTP, biotin-10-dATP and -dCTP, biotin-14-dATP and -dCTP, as well as biotin-17-dATP. These deoxyadenosine and deoxycytosine derivatives have the linker attached at a hydrogen bonding position of the purine or pyrimidine base, as opposed to the deoxyuridine derivatives in which the position of attachment of the linker is a non-hydrogen bonding position. The biotin-7-derivatives and the biotin-14-derivatives were incorporated at a significant rate into DNA probes by nick translation and were detected with streptavidin alkaline phosphatase conjugates using a chromogenic substrate. The other derivatives were not incorporated as well.

In general, biotinylated nucleotides with shorter linker arms appear to be incorporated more readily by DNA polymerases than those with longer linker arms. However, the effect of linker length on detection of the biotin is in direct contrast to this. Detection with avidin and streptavidin conjugates appears to increase with increasing length of the linker. In practice, most nucleic acid labeling and detection has been carried out using commercially available biotin-11-dUTP, and biotin-7-dATP, biotin-14-dATP, and biotin-14-dCTP. No significant differences in detection limits have been reported. Biotin-4-dUTP, biotin-11-dUTP, biotin-21-dUTP, biotin-7-dATP, biotin-14-dATP, and biotin-14-dCTP are commercially available. The procedures which follow specify either biotin-11-dUTP, biotin-14-dATP or biotin-14-dCTP, but other nucleotides may be used by adjusting the protocols.

Several studies have looked at the effect of biotin on the behavior of the biotinylated DNA probe. DNA probes with approximately 20 biotin-11-dUMPs per kilobase have approximately the same melting temperature (T_m) as unsubstituted DNA. At about 250 biotin-11-dUMPs per kilobase, the T_m was only decreased by 5 °C. Probes which contained fewer biotinylated dUMPs (14 biotins per kilobase) show similar rates of hybridization in solution to nonbiotinylated probes (Langer et al., 1981). Gebeyehu et al. (1987) compared the hybridization rates of probes containing approximately 12% biotin-11-dUMP or 10% biotin-7-dAMP to the rates for nonbiotinylated probes. Initial reassociation rates were similar, but the time for one-half of the DNA to reassociate was five times longer for the biotin-11-dUMP labeled probe and 10 times longer for the biotin-7-dAMP labeled probe than for the corresponding nonlabeled probe. Maximum detection sensitivities were observed with probes containing 35 or 98 biotin-7-dAMPs per kilobase, while reduced sensitivities were seen with 7 biotins per kilobase probe. Maximum sensitivity was reached with between 7 and 35 biotins per kilobase; additional biotin substitution did not result in increased sensitivity. Perhaps, this reflected a steric hindrance problem in attaching more streptavidin-enzyme conjugate molecules to the biotinylated probe. A steric limit to the number of biotins per kilobase that can be accessed by the detection complex may exist. Cook et al. (1988) have shown that for synthetic oligonucleotide probes, attachment of biotins at or near the termini resulted in maximal signal and increased sensitivity of detection. In most cases biotinylated probes may be used under the same conditions developed for radiolabeled probes with few, if any, minor modifications.

4.1.2 Methods of Labeling DNA

4.1.2.1 Biotin Labeling of DNA by Nick Translation

Labeling of DNA with radioisotopes by nick translation was developed by **Description** Rigby et al. (1977). Single stranded nicks are introduced in the DNA by pancreatic DNase I. *E. coli* DNA polymerase I then attaches at the 3'-OH terminus of a nick and adds nucleotide residues while simultaneously removing nucleotides from the 5'-side of the nick with its 5' →3' exonuclease activity. Initially, radiolabeled nucleoside-5'-triphosphates were supplied to provide radiolabeled probes. Biotinylated nucleoside triphosphates are incorporated in an analogous manner. Biotinylated probes prepared by nick translation generally contain 20−100 biotins per kilobase of DNA probe. These probes are very stable, can be prepared in large quantities, and used for at least 12 months.

− Nucleotide mix: 0.2 mM dCTP; 0.2 mM dGTP; 0.2 mM dTTP; 0.1 mM **Reagents** dATP; 0.1 mM bio-14-dATP; 500 mM Tris-HCl, pH 7.8; 50 mM MgCl$_2$; **and solu-** 100 mM 2-mercaptoethanol; 100 µg/ml nuclease free bovine serum **tions** albumin (BSA)
− Enzyme mix: 0.5 units/µl DNA polymerase I; 0.0075 units/µl DNase I; 50 mM Tris-HCl, pH 7.5; 5 mM magnesium acetate; 1 mM 2-mercaptoethanol; 0.1 mM phenymethylsulfonyl fluoride; 50% [v/v] glycerol, 100 µg/ml nuclease-free BSA
− 300 mM EDTA, pH 7.5
− 3 M sodium acetate, pH 4.5
− 95% ethanol
− TE buffer: 10 mM Tris-HCl, pH 7.5; 1 mM EDTA
− Sephadex G-50 or similar material

4.1.2

- 1× SSC: 0.15M NaCl; 15 mM sodium citrate, pH 7.0; 0.1% [w/v] SDS
- 1% [w/v] blue dextran in water

Procedure 1. Pipette the following into a 1.5 ml microcentrifuge tube on ice:
- 5 µl nucleotide mix
- 1 µg DNA to be labeled (up to 40 µl)
- up to 40 µl distilled water to a total volume of 45 µl
- 5 µl enzyme mix
2. Close the tube, mix well, and centrifuge briefly (15000 g for 5 s).
3. Incubate at 16 °C for 60 min.
4. Add 5 µl 300 mM EDTA to stop the reaction.
5. Unincorporated nucleotides can be separated from biotinylated DNA probe by either ethanol precipitation or gel filtration column chromatographys. Both methods produce suitable probes for filter hybridizations.

Repeated ethanol precipitation. Add 1/10 volume 3 M sodium acetate and 2 volumes cold 95% (or absolute) ethanol to the reaction tube. Mix by inverting the tube. Freeze at −70°C (dry ice) for 15 min or at −20°C for 2 h. Centrifuge at 15000 *g* for 10 min. Carefully remove the supernatant with a pipetter and dry the pellet. Resuspend the probe in 50 µl distilled water and precipitate the probe with sodium acetate and ethanol as described above. Resuspend the probe in TE buffer (10 mM Tris-HCl, pH 7.5, 1 mM EDTA) and store at −20°C.

Column chromatography. Use a 0.5 cm × 5.0 cm (~1 ml) column of Sephadex G-50 equilibrated with 1 × SSC, 0.1% [w/v] SDS. The column can be made in a 6-inch Pasteur pipette, plugged with siliconized glass wool. Add blue dextran to the reaction mixture. The biotinylated DNA elutes in the void volume with the blue dextran.

Notes
1. The reaction can be scaled up linearly for large-scale production of probes. Scaling down is more difficult and may require adjustment of the DNA polymerase I and DNase I concentrations. Smaller quantities of DNA are more readily biotinylated in a random primers labeling reaction (see below).
2. The presence of SDS in the column elution buffer reduces the nonspecific adsorption of biotinylated DNA to the column and the tubes.
3. The reaction conditions above result in probes with a size range of 50−200 nucleotides.
4. Biotinylated probes may be stored in solution at −20°C for at least a year.
5. Complete systems for biotin labeling of nucleic acid probes by nick translation are available commercially from a number of manufacturers.

4.1.2.2 *Random Primer Labeling of DNA with Biotin*

Radiolabeling of DNA probes using random primer extension was initially reported by Feinberg and Vogelstein (1983; 1984). Random oligodeoxy- **Description** nucleotides were annealled to a denatured DNA and extended with Klenow fragment in the presence of labeled nucleoside triphosphates. Nonradioactive biotinylated probes can be prepared by incorporating a biotinylated nucleoside triphosphate in a modification of the original reaction. Random oligonucleotide primers are annealed to a denatured single stranded DNA template and extended by DNA polymerase in the presence of biotinylated nucleoside triphosphates to produce a labeled probe. Since the DNA polymerase should contain no $5' \rightarrow 3'$ exonuclease activity to prevent degradation of the synthesized probe and the primers, Klenow fragment of DNA polymerase I is often used. Random primer biotinylation easily labels relatively small amounts of DNA: $1-500$ ng. Recently we have investigated the various parameters affecting the labeling of DNA probes using random primers (Mackey and Rashtchian, 1992; Hartley, 1991). We have found conditions which allow simultaneous amplification and labeling of DNA probes with random primers. Using this procedure, 10- to 100-fold amplification of DNA has been obtained, and has allowed preparation of large amounts of probes from small amounts of starting template. The details of this protocol is given below.

- TE buffer: 10 mM Tris-HCl, pH 7.5; 1 mM EDTA   **Reagents**
- Random primer buffer: 125 mM Tris-HCl, pH 6.8; 12.5 mM $MgCl_2$;   **and solu-**
  25 mM 2-mercaptoethanol; 1 mg/ml BSA; 750 µg/ml oligodeoxyribo-   **tions**
  nucleotide primers (random octamers)
- Nucleotide mix: 1 mM bio-14-dCTP; 1 mM dCTP; 2 mM dATP;
  2 mM dGTP; 2 mM dTTP (in 10 mM Tris-HCl, pH 7.5; 1 mM EDTA)
- Klenow fragment (large fragment of DNA polymerase I): 40 units/µl in
  100 mM potassium phosphate, pH 7.0; 10 mM 2-mercaptoethanol;
  50% [v/v] glycerol
- 0.2 M EDTA (pH 7.5)
- 3 M sodium acetate, pH 4.5
- 95% [v/v] ethanol
- Sephadex G-50 or similar material
- 1× SSC: 0.15 M NaCl; 15 mM sodium citrate, pH 7.0; 0.1% [w/v] SDS
- 1% [w/v] blue dextran in water

1. Denature 100 ng DNA dissolved in $5-20$ µl of a dilute buffer such as TE **Procedure**
   in a microcentrifuge tube by heating for 5 min in a boiling water bath;
   then immediately cool on ice. (The amount of template per reaction has
   been varied from $1-500$ ng with satisfactory results.)
2. Add the following to the denatured template on ice:
   - 5 µl nucleotide mix

- 20 μl random primer buffer
- distilled water to a total volume of 49 μl
3. Mix briefly
4. Add 1 μl Klenow fragment, mix gently but thoroughly, and centrifuge briefly.
5. Incubate at 37 °C for 60 min.
6. Add 5 μl EDTA.
7. Unincorporated nucleotides can be separated from biotinylated DNA probe by either ethanol precipitation or by gel filtration column chromatography. Both procedures produce satisfactory probes for filter hybridizations.

Repeated ethanol precipitation. Add ¹⁄₁₀ volume 3 M sodium acetate, pH 4.5, and 2 volumes cold 95% (or absolute) ethanol to the reaction tube. Mix by inverting the tube. Freeze at −70 °C (dry ice) for 15 min or at −20° for 2 h. Centrifuge at 15 000 g for 10 min. Carefully remove the supernatant with a pipetter and dry the pellet. Resuspend the probe in 50 μl distilled water and precipitate the probe with sodium acetate and ethanol as described above. Resuspend the probe in TE buffer (10 mM Tris-HCl, pH 7.5, 1 mM EDTA) and store at −20 °C.

Column chromatography. Use a 0.5 cm × 5.0 cm (~1 ml) column of Sephadex G-50 equilibrated with 1× SSC; 0.1% [w/v] SDS. The column can be made in a 6-inch Pasteur pipette, plugged with siliconized glass wool. Add blue dextran to the reaction mixture. The biotinylated DNA elutes in the void volume with the blue dextran.

Notes
1. Net DNA synthesis occurs during random primer labeling reactions. Using varying amounts of linear φX174, the amount of biotinylated probe DNA synthesized was determined by incorporation of radiolabeled nucleotide and is shown below. Amplification of probe with 100 ng template routinely varies from 10- to 40-fold in a 1-h reaction.

Template DNA (ng)	Biotinylated probe (ng)
1	396
2	422
5	620
10	1029
25	1780
50	2244
100	2693
200	3036
300	3274
500	3577

2. Difficulties in random primer labeling of DNA often result from contaminants in the template DNA preparation. To a certain extent these
   problems may be overcome by extending the reaction time to 2−4 h
   and/or by increasing the amount of enzyme to 2 µl per reaction.
3. The reaction may be scaled up proportionately for larger amounts of
   DNA. Smaller amounts of template may be used while still producing a
   significant amount of biotinylated probe.
4. The presence of SDS in the column elution buffer reduces the
   nonspecific adsorption of biotinylated DNA to the column and the
   tubes.
5. The reaction conditions above result in production of probe in the size
   range of 100−700 nucleotides.
6. Biotinylated probes may be stored in solution at −20°C for at least a
   year.
7. A complete system for biotin labeling of nucleic acid probes by random
   primer reactions are available commercially (BioPrime DNA labeling
   system: Life Technologies/BRL).
8. The amount of probe synthesized cannot be easily determined directly
   and will vary considerably depending on the quality of the template
   DNA. Appropriate probe concentrations for hybridizations should be
   determined experimentally.

4.1.2.3 Polymerase Chain Reaction

The polymerase chain reaction (PCR) is a rapid method for amplification **Description**
of specific nucleic acids (Saiki et al., 1985). The method takes advantage
of the specificity of annealing of two oppositely oriented oligonucleotide
primers and their extention using a thermostable DNA polymerase such as
Taq DNA polymerase. The repeated cycling of this process by denaturation of the template DNA for multiple rounds results in exponential
amplification of nucleic acid sequences of interest. PCR has extensively
been used for amplification of target nucleic acids of interest from small
amounts of DNA from variety of sources, making their detection easier
(Erlich et al., 1991).

PCR has recently been used for preparation of DNA probes labeled
with radioisotopes as well as nonradioactive reporter groups. Schowalter
and Sommer (1989) have used PCR to label amplified DNA with ^{32}P. In
this procedure amplification is performed in the presence of ^{32}P-dCTP for
30 cycles. This method is applicable to preparation of probes with high
specific activity ($1-5 \times 10^9$ cpm/µg) and appears to be most useful in cases
where the quantity of the template DNA is limited or with small templates
(<500 bases). In addition, this procedure has been shown to be useful for
generation of labeled probe directly from genomic DNA without any need
for cloning.

Lo et al. (1988) have described incorporation of biotin-11-dUTP by *Taq*
DNA polymerase for generation of short PCR products labeled with
biotin. We have tested incorporation of other biotinylated nucleotide

triphosphates by *Taq* DNA polymerase in PCR. Biotinylated derivatives of dATP and dCTP have been shown to be effectively incorporated by *Taq* DNA polymerase (Rashtchian and Mackey, 1992). Our studies have shown that various biotinylated nucleotides behave differently in PCR reactions. Biotinylated derivatives of dATP and dCTP appear to be incorporated more readily than the dUTP derivative. We have also found that when biotinylated dUTP is used amplification and labeling is limited to fragments <600 bases, and larger fragments are not amplified (Rashtchian, unpublished data). This limitation, however, is related to the incorporation of biotin-11-dUTP and can be overcome by using biotinylated dATP or dCTP. When using biotin-14-dATP or biotin-14-dCTP, PCR can be used for efficient preparation of biotinylated DNA probes up to 3 kb in length. Such probes have been shown to be functional in filter hybridization as well as in situ applications to metaphase chromosome spreads. The following procedure has been used successfully to produce biotinylated DNA probe by PCR.

Reagents
- 10× PCR buffer: 200 mM Tris-HCl, pH 8.3; 15 mM MgCl$_2$; 50 mM KCl
- dNTP mix: 2 mM dTTP; dGTP; and dCTP; and 0.5 mM dATP in 10 mM Tris-HCl, pH 7.8
- 0.4 mM bio-14-dATP
- PCR primers
- *Taq* DNA polymerase (5 units/µl)
- Mineral oil
- Thermal cycler

Procedure The following reaction conditions have been chosen as a standard PCR condition that is applicable to most cases. However, depending on the sequence of template and the primers, optimum reaction conditions may need to be established for efficient amplification (Erlich et al., 1991).

1. Add the following components to a 0.5-ml microfuge tube on ice:
   - 5 µl of 10 × PCR buffer
   - 5 µl dNTP mix
   - 19.2 µl 0.4 mM bio-14-dATP
   - PCR primers at a final concentration of 200 nM
   - 10–1000 pg template DNA
2. Bring the volume to 49 µl with sterile water
3. Add 2.5 units *Taq* DNA polymerase
4. Mix the reaction by vortexing at slow speed and centrifuge briefly.
5. Overlay with 2 drops of mineral oil and place in thermocycler.
6. Denature the template for 5 min at 94 °C and thermocycle for 30 cycles as follows:
   - 1 min at 55 °C
   - 3 min at 72 °C
   - 1 min at 94 °C

The PCR products can be analyzed by agarose gel electrophoresis. Generally, labeling with biotin will result in an increase in the molecular weight of the labeled DNA and will result in somewhat retarded migration in agarose gels compared to the unlabeled DNA.

This procedure has been applied to preparation of probes from plasmid DNA containing various genes as well as amplification of genomic DNA sequences such as the inter-alu sequences. When using genomic DNA as template a larger amount of template DNA ($0.5-1.0\,\mu g$) should be used.

4.1.2.4 Labeling with Photobiotin

In addition to the enzymatic methods of introducing biotin into nucleic **Description** acids described above, biotinylation of both DNA and RNA may be performed chemically using photoreactive biotin compounds known as Photobiotin (Forster et al., 1985). Photobiotin contains a photoreactive aryl azide group attached to biotin through a charged linker arm. When exposed to strong visible light in the presence of nucleic acids, Photobiotin covalently links biotin groups to DNA and RNA. Photobiotin is highly reactive when photoactivated and is nonspecific. Double or single stranded DNA and RNA may be labeled.

- Photobiotin acetate: lyophilized powder   **Reagents**
- TE buffer: 100 mM Tris-HCl, pH 9.0; 1 mM EDTA   **and solutions**
- Distilled water
- 2-Butanol
- 7.5 M Ammonium acetate
- Lamp: A 250- to 500-W mercury vapor discharge lamp with relector is most suitable.

Steps $1-3$ should be carried out in subdued light or in the dark. The **Procedure** remainder of the procedure can be performed in normal room light.

1. Dissolve $100\,\mu g$ photobiotin acetate in $100\,\mu l$ distilled water. Mix vigorously for several minutes under subdued light, preferably in a darkroom.
2. Pipette $2-10\,\mu l$ linearized double stranded or single stranded DNA ($0.5-1.0\,\mu g/\mu l$) into a microcentrifuge tube. Add an equal volume of photobiotin solution. Mix well, close tube, and lay on its side on ice.
3. Irradiate the solution for 20 min with a 250- to 500-W mercury vapor lamp at a distance of $8-10\,cm$. Ensure that the solution remains cold during the irradiation procedure.
4. Dilute the reaction mixture to $100\,\mu l$ by adding an appropriate volume of TE buffer (pH 9.0).
5. Add $100\,\mu l$ 2-butanol and vortex. Centrifuge for 1 min. Remove and discard the upper butanol phase.
6. Repeat the butanol extraction. The aqueous volume will be reduced to $30-40\,\mu l$. Carrier DNA may be added to facilitate the ethanol precipitation.

7. Measure the volume of the aqueous phase. Add 0.5 volume of 7.5 M ammonium acetate and 3 volumes of cold ethanol. Cool at −70°C for 30 min or at −20°C for 2 h. Centrifuge 15 min at 4°C. Dry the brown pellet and redissolve in dilute buffer such as 10 mM Tris-HCl, pH 7.5; 1 mM EDTA. Store frozen at −20°C.

Notes

1. Photobiotin is light sensitive and should be protected from direct light. Steps 1−3 must be performed under low-light conditions: a darkroom is preferred.
2. DNA to be labeled should be highly purified. The reaction is nonspecific and will label contaminants in the DNA preparation as well as the DNA. Phenol and ether extractions are recommended. Since inorganic salts inhibit the reaction, the DNA to be labeled should be dissolved in water. Labeling can also be performed on DNAs in dilute buffers (<10 mM).
3. Mercury vapor lamps are preferred over tungsten lamps due to the high heat output of tungsten lamps.
4. Reactions may be carried out in either glass or transluscent plastic tubes. Tubes without caps should be sealed with parafilm.
5. Using the above procedure, 1−10 μg DNA at concentrations of 0.5−1.0 μg/μl may be biotinylated. Smaller amounts of DNA may be used if the mass ratio of Photobiotin to DNA is maintained at 2:1.
6. Using photobiotin, ~1−10 biotins per kilobase of DNA are introduced. These levels of biotinylation have minimal effect on hybridization.

4.1.3 Labeling of Synthetic Oligonucleotides

In recent years synthetic oligonucleotide probes have been used extensively for detection of nucleic acids by hybridization. Synthetic oligonucleotide probes prepared from known DNA or amino acid sequences can be used as gene probes in a variety of applications. Short oligonucleotide probes are capable of discriminating between two sequences which differ by only a single nucleotide (DiLella and Woo, 1987). In addition, synthetic probes can be prepared with degenerate sequences which allow detection of closely related genes. A number of chemical or enzymatic methods are available for labeling of short oligonucleotide probes with biotin. In the following discussion we provide examples and protocols for each method.

4.1.3.1 Enzymatic Labeling

Description The enzyme terminal deoxynucleotidyl transferase (TdT) can be used for labeling of synthetic oligonucleotide probes with biotin (Riley et al., 1986). TdT can add biotinylated nucleotide triphosphates to the 3′ end of oligonucleotides in a template-independent manner. Flickinger et al.

(1992) have studied the efficiency of labeling of oligonucleotide probes with biotin using TdT. Although all biotinylated nucleotides are incorporated by TdT, there appears to be a marked difference in the efficiency of incorporation depending on the nucleotide triphosphate used. Among the available biotinylated nucleotides the dCTP derivatives are incorporated most efficiently. Flickinger et al. (1992) have described a protocol for rapid and efficient labeling of oligonucleotide probes with biotin which is described below.

Reagents and solutions

- Bio-14-dCTP: 5 mM bio-14-dCTP (in 10 mM Tris-HCl, pH 7.2)
- Terminal deoxynucleotidyl transferase (TdT): 15 units/µl
- 5× tailing buffer: 0.5 M potassium cacodylate, pH 7.2; 10 mM CoCl$_2$; 1 mM DTT
- Stop buffer: 0.2 M Na$_2$EDTA, pH 7.5
- Sterile distilled water

Procedure

1. Thaw components (except enzyme) at room temperature.
2. Pipette the following components *in order* into a 1.5-ml microcentrifuge tube at room temperature:
   - 10 µl 5 × TdT buffer
   - 5 µl bio-14-dCTP
   - 250 pmol synthetic oligonucleotide (up to 30 µl)
3. Bring the volume to 46 µl with dH$_2$O. Add 4 µl of TdT.
4. Mix briefly, but thoroughly. Centrifuge 5–20 s to pellet liquid.
5. Incubate in 37 °C water bath for 15 min.
6. Add 2.5 µl stop buffer.
7. For most filter hybridizations, it is not necessary to separate the labeled probe from the unincorporated nucleotides. For procedures that may require this separation, an ethanol precipitation using carrier nucleic acid is possible.

Notes
1. Add all components in order given.
2. If it is necessary to *decrease* the molar concentration of oligonucleotide in the reaction from the recommended 5 µM, do not scale down the remainder of the reaction components. Do *not increase* the molar concentration of the oligonucleotide in the reaction above 5 µM.
3. The purity of the oligonucleotide will affect the success of the reaction. Generally, it is not necessary to use gel-purified oligonucleotide substrate; however, the presence of contaminating organic compounds or detergents could adversely affect the amount of labeled product.
4. Increasing the time of incubation will not necessarily increase the efficiency of tailing, but longer incubations do not adversely affect the reaction.
5. Unmodified dCTP may be added to increase the tail length.
6. Biotinylated DNA is very stable, and under nuclease-free conditions can be stored for extended periods of time at −20 °C.

4.1.3.2 Chemical Synthesis of Biotinylated Oligonucleotides

A number of techniques have been used to prepare biotinylated oligonuc-leotides by chemical synthesis or by postsynthesis chemical modifications (Goodchild, 1990). Oligonucleotides can be synthesized with biotinylated residues at the 5′ and 3′ ends as well as at internal positions. Considerable effort has been directed toward attachment of biotin at the 5′ end of oligonucleotides (Goodchild, 1990). Labeling of oligonucleotides at the 5′ end has the advantage that the oligonucleotide can be used as hybridiza-tion probes as well as primers in enzymatic reactions such as PCR. Label-ing at other positions can interfere with usage of oligonucleotides as primers. Chemical labeling of oligonucleotides can be accomplished via use of biotinylated phosphoramidites or a phosphodiester linkage. Chollet and Kawashima (1985) labeled synthetic oligonucleotides with biotin in a two-step reaction. These authors provided a method for synthesis of oligonucleotides with an aminoalkylphosphoramide linker arm at the 5′ terminus. The functionalized oligonucleotide was then reacted with biotin-N-hydroxysuccinamidyl ester to generate biotin-labeled oligo-nucleotides at the 5′ terminus. Cook et al. (1988) synthesized a deoxyuridine phosphoramidite containing a protected allylamino sidearm which was used directly in an automated oligonucleotide synthesizer. Biotinylation was achieved by reaction with N-biotinyl-6-aminocaproic acid N-hydroxysuccinimide ester. This procedure allowed placement of the biotin at various positions within the synthetic oligonucleotide.

More recently, biotinylated phosphoramidite derivatives have been synthesized which can be incorporated into the synthetic oligonucleotides during automated chemical synthesis. Roget et al. (1989) demonstrated synthesis of protected nucleoside phosphoramidites bearing a biotinyl group. These labeled phosphoramidites were used for solid-phase oligonucleotide synthesis without any change in the synthetic cycle and deprotection step. Misiura et al. (1990) have also synthesized biotinylated nonnucleosidic phosphoramidites suitable for use in automated DNA synthesizers. Use of such phosphoramidite derivatives has not only simplified the synthesis of biotinylated oligonucleotide probes, but has allowed incorporation of biotin at various positions in the probe. These reagents have also made it possible to add multiple biotinylated residues in a single oligonucleotide probe.

4.1.4 Methods of Labeling RNA

Description In vitro transcription of RNA by phage RNA polymerases is a powerful method for preparation or large amounts of RNA probes. The procedure is rapid and results in the synthesis of strand-specific transcripts with a defined length. RNA probes have a number of characteristics which make them suitable as hybridization probes:

- Since the RNA probes are single stranded, there are no complementary labeled strands that can compete with each other during hybridization to the target sequences.
- Unhybridized RNA probes can be removed by treatment with RNaseA (Melton et al., 1984), which substantially removes nonspecific background.
- RNA/DNA hybrids are more stable than DNA/DNA hybrids (Casey and Davidson, 1977).

RNA probes have been used for hybridization with DNA and RNA blots and in situ studies. They are also useful in ribonuclease protection assays where the transcript is hybridized with target mRNA sequences.

Biotinylated probes can be prepared readily by incorporation of biotinylated ribonucleotides in the transcription reaction. Generally, all three phage RNA polymerases (T7, SP6, and T3) can be used for preparation of RNA probes. T7 and SP6 are used more frequently. There are various plasmid vectors which carry two promoters, allowing use of two different RNA polymerases for preparation of probes of different polarity (pSPORT 1: Life Technologies/BRL). There are kits available for preparation of biotinylated RNA probes (Life Technologies/BRL). The biotinylated ribonucleotide triphosphate used in this kit is biotin-14-CTP, which appears to be incorporated more readily by the RNA polymerase than other nucleotides (unpublished data).

Biotinylated RNA probes are effective in Southern, northern, and in situ hybridization experiments. The sensitivity of RNA probes in filter hybridization formats is similar to those of DNA probes, and as little as 0.25 pg of target nucleic acid can be detected by colorimetric or chemiluminescence methods. For application of biotinylated RNA in situ to cytological specimens RNA probes, RNA was hydrolyzed to an average size of 100 bases and the target DNA was denatured using the conditions of Angerer et al. (1987).

Reagents

- T7 RNA polymerase
- SP6 RNA polymerase
- 10× transcription buffer: 400 mM Tris-HCl, pH 8.0; 80 mM $MgCl_2$; 250 mM NaCl; 20 mM spermidine
- 10× rNTP mixture: 10 mM ATP; 10 mM UTP; 10 mM GTP; 5 mM CTP; 20 mM bio-14-CTP (in 10 mM Tris-HCl, pH 7.0; 0.1 mM EDTA)
- RNase inhibitor: 10000−20000 U/ml
- DEPC-treated dH_2O with 1 U/μl RNase inhibitor
- DEPC-treated distilled water
- DEPC-treated 0.5 M EDTA
- 0.1 M Dithiothreitol

Procedure

The following procedure is for a standard 50-μl reaction using 1 μg linearized template DNA.

1. Place the following components in a microcentrifuge tube in the order listed:
   - DEPC-treated water for final volume of 50 μl
   - 5 μl 10× transcription buffer
   - 2.5 μl 0.1 M dithiothreitol
   - 5 μl 10× rNTP mixture
   - 5 μl linearized template DNA (to give 1 μg)
   - 50 units RNase inhibitor (to give 1 unit/μl)
   - 40 units RNA polymerase (to give 0.8 units/μl)
2. Mix gently but thoroughly and centrifuge briefly.
3. Incubate for 1–2 h at 37°C.
4. Stop reaction by adding 2 μl DEPC-treated 0.5 M EDTA, pH 8.0.

The biotinylated RNA can be separated from unincorporated nucleotides by passage of the reaction mixture through a Sephadex G-50 column equilibrated with DEPC-treated 1× SSC (SSC: 0.15 M sodium chloride; 0.015 M sodium citrate; pH 7.0) containing 0.1% [w/v] SDS.

The typical yields of biotinylated RNA after a 90 min reaction are at least 20 μg for T7 RNA polymerase and at least 15 μg for SP6 RNA polymerase. Since the amount of biotinylated RNA produced is substantially greater than the amount of template DNA in the reaction mixture, it is generally not necessary to remove the template DNA.

4.1.5 Hybridization and Detection of Biotinylated Probes

There are basically three categories for detection of biotinylated probes: (1) chromogenic detection; (2) chemiluminescent detection; and (3) fluorescent detection. All of these detection methods have served as reliable and sensitive methods for nonradioactive detection of nucleic acid probes as well as solid-phase immunochemical assays. The chromogenic and chemiluminescent detection methods are performed using enzyme conjugates of streptavidin (or avidin) or an antibody against biotin. Both alkaline phosphatase and horseradish peroxidase conjugates have been used, with alkaline phosphatase detection being more prevalent. Fluorescent detection of biotinylated probes has been accomplished by streptavidin or anti-biotin antibodies conjugated to fluorescent moieties such as fluorescein, rhodamine or Texas Red.

The choice of the detection methodology is mainly dependent on the hybridization format used. The chromogenic methods are the most versatile and are applicable to membrane hybridization as well as the in situ formats. The chemiluminescent methods have generally been used for membrane applications and fluorescent methods have been most useful in in situ applications.

Both chromogenic detection and chemiluminescent detection of membrane-bound nucleic acid probes have advantages and disadvantages. They differ primarily in potential sensitivity, method of data storage, and methods of stripping and reprobing. Both of these nonradioactive detection procedures require additional steps compared to radioactive procedures. However, the additional steps do not mean additional elapsed time. With chromogenic detection, storage of experimental data is as photographs or preserved membranes. Data for chemiluminescent detection is collected on film as autoradiographs, similar to radioactive detection. The procedure for reprobing chromogenic blots requires boiling the membrane in dimethylformamide, while chemiluminescent blots can be stripped by the same methods used routinely for radiolabeled blots (heating in low salt, detergent solutions). Chromogenic blots may be reprobed a maximum of 3−5 times, while chemiluminescent blots have been reprobed more than 40 times. Chemiluminescent detection is potentially more sensitive than current chromogenic methods but is often limited by high backgrounds. The choice betwen chemiluminescent and chromogenic detection will depend on the sensitivity required, the preferred method of data storage, and the need for multiple probings.

Numerous procedures have been developed for immobilization of nucleic acids on membranes for hybridization (Sambrook et al. 1989). All of the methods are generally acceptable and can be used with biotinylated probes. However, for chemiluminescent detection of biotinylated probes (see below) the choice of membrane is critical (Carlson et al. 1990). For this purpose, the Magnagraph membrane is recommended (Life Technologies/BRL). The detailed procedure for use of biotinylated probes for in situ hybridization has been reviewed recently (Rashtchian, 1992). Below we will limit our discussion to use of biotinylated probes in membrane hybridization assays.

4.1.5.1 Hybridization

Description Numerous procedures for hybridization of biotinylated and radiolabeled probes on nitrocellulose and nylon membranes have been developed. Many of these procedures are interchangeable for most purposes; no attempt will be made to discuss all the variations of hybridization procedures which have been employed. Instead, a single method which has worked well with both chromogenic and chemiluminescent detection will be outlined here.

Reagents
- Prehybridization solution: 6× SSPE, 5× Denhardt's, 1% [w/v] SDS, 50% [v/v] formamide, 200 μg/ml sheared denatured salmon sperm DNA
- Hybridization solution: 6× SSPE, 5× Denhardt's, 1% [w/v] SDS, 50% [v/v] formamide, 200 μg/ml sheared denatured salmon sperm DNA, 10% dextran sulfate
- 5× SSC, 0.5% [w/v] SDS
- 0.1× SSC, 1% [w/v] SDS

Procedure
1. Place membrane in plastic hybridization bag and prehybridize in 250 μl prehybridization solution per cm^2 of membrane surface area for 2h at 42°C.

2. Denature the biotinylated probe by heating in a boiling water bath for 10 min.
3. Remove prehybridization solution from the hybridization bag and replace it with 0.1 ml hybridization solution per cm^2 membrane containing 50 ng/ml denatured probe. Incubate overnight at 42 °C while agitating gently.
4. Wash the membrane twice with 2 ml/cm^2 5× SSC; 0.5% [w/v] SDS at 65 °C, 5 min each wash.
5. Wash the membrane once with 2 ml/cm^2 0.1× SSC; 1% [w/v] SDS at 50 °C for 30 min. The temperature of this wash may be adjusted to achieve the desired level of hybridization stringency.
6. Wash the membranes once with 2 ml/cm^2 2× SSC for 5 min at room temperature.
7. Proceed to Sect. 4.1.5.3.

4.1.5.2 Hybridization of Synthetic Oligonucleotide Probes

The hybridization of oligonucleotides is different from that with longer DNAs as each oligonucleotide requires conditions based on its length and base composition. A detailed discusion of the hybridization conditions for synthetic oligonucleotides can be found in Meinkoth and Wahl (1984). The hybridization kinetics of biotinylated probes are the same as for radioactive probes except that they may exhibit lower melting temperatures. Therefore slightly modified hybridization conditions are necessary.

Reagents and solutions
- TBS-Tween 20: 100 mM Tris-HCl, pH 7.5; 150 mM NaCl, 0.05% [v/v] Tween 20
- Blocking solution: 3% [w/v] BSA in TBS-Tween 20
- Prehybridization solution: 5× SSC; 1% [w/v] SDS; 0.5% [w/v] BSA
- Hybridization solution: 5× SSC; 1% [w/v] SDS; 0.5% [w/v] BSA; 5 pmol/ml oligonucleotide probe
- Washing solution 1: 2× SSC; 0.1% [w/v] SDS
- Wash solution 2: 2× SSC; 0.1% [v/v] Triton X-100
- Washing solution 3: 2× SSC

Procedure **Note:** The hybridization temperature for oligonucleotide probes should be optimized for each probe individully (see Dillela and Woo, 1987; Meinkoth and Wahl, 1984 for review).

1. Preblock membranes at 65 °C for 1 h in (3% [w/v] BSA in TBS-Tween 20).
2. Prehybridize at 37–55 °C for 30 min with shaking in prehybridization solution.
3. Hybridize the membrane with probe (5 pmol/ml hybridization solution) in approximately 1 ml hybridization solution per 9 cm^2 of blot at 37–55 °C for 1 h (overnight is possible, but not necessary).
4. Wash blots with gentle shaking: 1 × 10 min at 37–55 °C in 2× SSC, 1% [w/v] SDS; 1× 10 min at 37–55 °C in 2× SSC, 1% [v/v] Triton X-100; 1× 10 min at room temperature in 2× SSC.
5. Proceed to step 3 in Sect. 4.1.5.3.

4.1.5.3 Binding the Streptavidin-Alkaline Phosphatase Conjugate

The hybridized probe is detected by binding a streptavidin-alkaline phosphatase conjugate to the biotin reporter groups on the probe and then supplying either a chromogenic or chemiluminescent enzyme substrate. Throughout all of these steps it is essential for the solutions to flow freely around each membrane. Make certain the filters do not stick to one another or to container. **Description**

– TBS-Tween 20: 100 mM Tris-HCl, pH 7.5; 150 mM NaCl; 0.05% [v/v] Tween 20 **Reagents and solutions**
– Blocking solution: 3% [w/v] BSA in TBS-Tween 20
– Streptavidin-alkaline phosphatase, 1 mg/ml (SA:AP)

1. Wash the hybridized membranes for 1 min with 1 ml/cm^2 TBS-Tween **Procedure** 20.
2. **Note:** Step 2 must be omitted for short oligonucleotide probes.

   Incubate the membranes in 0.75 ml/cm^2 blocking solution for 1 h at 65 °C in a covered plastic container. Agitate the membranes gently during blocking, making certain the membranes do not stick to one another or to the container.
3. Microcentrifuge the tube of streptavidin-alkaline phosphatase (SA:AP) conjugate for 4 min at room temperature. With a sterile pipet tip, carefully remove 7 µl SA:AP from the supernatant solution for each 100 cm^2 of membrane area. Avoid pipeting any pelleted material. Dilute the supernatant SA:AP 1:1000 in TBS-Tween 20.
4. Incubate the membrane with the SA-AP dilution for 10 min at room temperature. Gently agitate the membranes during binding. Make certain that the membranes are completely covered by the SA-AP solution, and that they do not stick to one another.
5. Wash the membranes twice with 1 ml/cm^2 TBS-Tween 20, for 15 min each at room temperature.
6. Proceed to chromogenic or chemiluminescent detection of biotin (see Sects. 4.1.5.4 and 4.1.5.4).

4.1.5.4 Chromogenic Detection of Biotinylated Probes with Alkaline Phosphatase

After binding of streptavidin-alkaline phosphatase conjugate, probe-target hybrids are then visualized by formation of a insoluble colored precipitate resulting from dephosphorylation of 5-bromo-4-chloro-3-indolyl phosphate (BCIP) by alkaline phosphatase and subsequent reaction with the dye nitroblue tetrazolium (NBT). **Description**

– Washing solution: 0.1 M Tris-HCl, pH 9.5; 0.1 M NaCl; 50 mM MgCl$_2$ **Reagents and solutions**
– NBT: Nitroblue tetrazolium, 75 mg/ml in dimethylformamide

4.1.6

- BCIP: 5-bromo-4-chloro-3-indolyl phosphate, 50 mg/ml in dimethyl-formamide
- Stop solution: 20 mM Tris-HCl, pH 7.5, 5 mM EDTA

Procedure
1. Wash filters for 15 min in washing solution
2. Develop color by placing filters in 330 µg/ml NBT and 167 µg/ml BCIP in washing solution. Allow color to develop at room temperature in the absence of light for 1–3 h.
3. Terminate color reaction by rinsing membrane in 0.25 ml stop solution per cm^2 membrane.
4. Dry membranes in vacuum oven at 80°C for 1–3 min.
5. Store membranes dry and protect from bright light.

4.1.5.5 Chemiluminescent Detection of Biotinylated Probes with Alkaline Phosphatase

Description After binding of the streptavidin-alkaline phosphatase conjugates, probe target hybrids can then be detected using a chemiluminescent 1,2-dioxetane substrate for alkaline phosphatase. After dephosphorylation the 1,2-dioxetane, PPD, forms an unstable intermediate which decomposes at alkaline pH and emits visible light. PPD is commercially available in a premixed formulation, Lumi-Phos 530, containing stabilizers and enhancers. Data is obtained as a stable image on X-ray film.

Reagents and solutions
- Detection buffer: 0.65 M 2-amino-2-methyl-1-propanol, pH 9.6; 0.88 mM $MgCl_2$
- Lumi-Phos 530

Procedure
1. Wash filters for 60 min at room temperature in detection buffer.
2. Dip membrane in Lumi-Phos 530 warmed to room temperature for 15–60 s. Place membrane between two thin plastic sheets. Squeeze out excess reagent by rolling a pipet over plastic sheets. Heat seal.
3. Expose membrane to X-ray film immediately, or incubate at 37°C for 30–45 min or for 3–4 h at room temperature to allow light emission to reach a steady state.

4.1.6 Summary

During the past 10 years a variety of nonradioactive systems have been developed for labeling and detection of nucleic acid probes. Labeling with biotin still remains the major nonradioactive method. In this section we have discussed the major methologies used for labeling of nucleic acid probes with biotin. We have also cescribed the methodologies used for detection of biotinylated probes. Availability of three different detection technologies, chromogenic, chemiluminescent, and fluorescent, for detection of biotinylated probes allows application of biotinylated probes for a variety of hybridization assays. Each of the detection systems has found widespread use in certain applications. The colorimetric methods are the most versatile and are applicable to membrane hybridization as well as to in situ formats. The chemiluminescent methods have generally been used for membrane applications and fluorescent methods have been most useful in in situ applications.

The progress in labeling and detection technologies has been considerable in the last few years. These methods have become simpler and more reliable, and this trend is continuing. These improvements have made nonradioactive detection a reasonable substitute for the radioactive methods in many research and clinical diagnostic applications.

Acknowledgements. We would like to thank all of our colleagues, especially Jeannette Flickinger and Ray Hadley. We also thank James Hartley for sharing his unpublished data regarding random primer amplification with us and Jill Bradshaw for help in typing the manuscript.

References

Angerer LM, Cox KH, Angerer RC (1987) Demonstration of tissue-specific gene expression by in situ hybridization. Meth Enzymol 152:649

Brenner DJ (1989) DNA hybridization for characterization, classification and identification of bacteria. In: Swaminathon B, Prakash G (eds) Nucleic Acid and Monoclonal Antibody Probes, Applications in Diagnostic Microbiology. Marcel Dekker, New York, N.Y.

Brigati D, Myerson D, Leary J, Spalholz B, Travis S, Fong C, Hsiung G, and Ward D (1983) Detection of viral genomes in cultured cells and paraffin-embedded tissue sections using biotin labelled hybridization probes. Virology 126:32

Carlson DP, Superko C, Mackey J, Gaskill E, Hanson P (1990) Chemiluminescent detection of nucleic acid hybridization. Focus 12:9–12

Casey J, Davidson N (1977) Rates of formation and thermal stabilities of RNA:DNA and DNA:DNA duplexes at high concentrations of formamide, Nucleic Acids Res 4:1539

Chollet A, Kawashima EH (1988) Biotin labeled synthetic oligodeoxyribonucleotides: Chemical synthesis and uses as hybridization probes. Nucleic Acids Res 13:1529–1541

Cook A, Vuocolo E, and Brakel C (1988) Synthesis and hybridization of a series of biotinylated oligonucleotides. Nucleic Acids Res 16:4077

DiLella AG, Woo SLC (1987) Hybridization of genomic DNA to oligonucleotide probes in the presence of tetramethylammonium chloride. Meth Enz 152:447–451

Erlich HA, Gelfand D, Sninsky JJ (1991) Recent advances in polymerase chain reaction. Science, 252, 1643

Feinberg AP, Vogelstein B (1983) A technique for radiolabeling DNA restriction endonuclease fragments to high specific activity. Anal Biochem 132:6–13

Feinberg AP, Vogelstein B (1984) A technique for radiolabeling DNA restriction endonuclease fragments to high specific activity. Addendum Anal Biochem 137:266–267

Flickinger JL, Gebeyehu G, Buchman G, Haces A, Rashtchian A (1992) Differential incorporation of biotinylated nucleotides by terminal deoxynucleotidyl transferase. Nucleic Acids Res 20:2382

Forster AC, McInnes JL, Skingle DC, and Symona RH (1985) Nonradioactive hybridization probes prepared by the chemical labeling of DNA and RNA with a novel reagent, photobiotin. Nucleic Acids Res 13:745−761

Gebeyehu G, Rao PY, SooChan P, Simms DA, and Klevan L (1987) Novel biotinylated nucleotide − analogs for labeling and calorimetric detection of DNA. Nucleic Acids Res 15:4513−4534

Goodchild J (1990) Conjugates of oligonucleotides and modified oligonucleotides: A review of their synthesis and properties. Bioconjugate Chem 1:165−187

Hartley JL (1991) United States Patent 5043272

Klevan L and Gebeyehu G (1990) Biotinylated nucleotides for labeling and detecting DNA. Meth Enzymol 184:561

Langer P, Waldrop A, Ward S (1981) Enzymatic synthesis of biotin-labeled polynucleotides: novel nucleic acid affinity probes. Proc Natl Acad Sci USA 18:6633−6637

Lo YMD, Mehal WZ, Fleming KA (1988) Rapid production of vector-free biotinylated probes using the polymerase chain reaction. Nucleic Acids Res 16:8719

Mackey J and Rashtchian A (1992) A method for simultaneous labeling and amplification of DNA using random primers. Focus 14:21−23

Matthews JA, Kricka LJ (1988) Analytical strategies for the use of DNA probes. Anal Biochem 169:1−25

Meinkoth J, and Wahl G (1984) Hybridization of nucleic acids immobilized on solid supports. Biochem 138:267−284

Melton DA, Kreig P, Rebagliti MR, Maniatis T, Zinn K, Green MR (1984) Efficient in vitro synthesis of biologically active RNA and RNA hybridization probes from plasmids containing a bacteriophage SP6 promoter. Nucleic Acids Res 12:7035

Misiura K, Durrant I, Evans MR, Gait MJ (1990) Biotinyl and phosphotyrosinyl phosphoramadite derivatives useful in incorporation of multiple reporter groups. Nucleic Acids Res 18:4345−4354

Rashtchian A, and Mackey J (1992) Efficient synthesis of biotinylated DNA probes using polymerase chain reaction. Focus, 14:64−65

Rashtchian A (1992) Colorimetric detection of alkaline phosphate, In: Kricka L (ed) Nonisotopic DNA Probe Techniques, Academic Press

Rigby P, Dieckmann M, Rhodes C, and Berg P (1977) Labeling deoxyribonucleic acid to high specific activity in vitro by nick translation with DNA polymerase I. J Mol Biol 113:237

Riley LK, Marshall ME, and Coleman MS (1986) A method for biotinylating oligonucleotide probes for use in molecular hybridization. DNA 5, 333

Roget A, Bazin H, Teoule R (1989) Synthesis and use of labeled nucleoside phosphoramidite building blocks bearing a reporter group: biotinyl, dinitrophenyl, pyrenyl, and dansyl. Nucleic Acids Res 17:7643−7651

Saiki RK, Scharf SJ, Faloona FA, Mullis KB, Horn GT, Erlich HA, and Arnheim N (1985) Enzymatic amplification of b-globin genomic sequences and restriction site analysis for diagnosis of sickle cell anemia. Science, 230, 1350−1354

Sambrook J, Fritsch EF, and Maniatis T (1989) Molecular Cloning: A Laboratory Manual, 2nd Edition, Cold Spring Harbor Laboratory, Cold Spring Harbor, New York

Schowalter DB, Sommer SS (1989) The generation of radiolabeled DNA and RNA probes with polymerase chain reaction. Anal Biochem 177, 90

Singer RH, Lawrence JB, and Rashtchian RN (1987) Toward a rapid and sensitive in situ hybridization methodology using isotopic and nonisotopic probes. In: Valentino K, Eberwine J, Barchas J (eds) In situ Hybridization: Application to the Central Nervous System, Oxford University Press, New York

4.2 Labeling and Detection of Proteins and Glycoproteins

EDWARD A. BAYER AND MEIR WILCHEK

4.2.1 Principle and Applications

Historically, the development and extensive application of the avidin-biotin system was a definitive breakthrough in the area of nonradioactive labeling and detection of biologically active molecules. Since the initial group-specific labeling of membranes (Heitzmann and Richards 1974; Bayer et al. 1976a) and labeling of RNA (Manning et al. 1975; Angerer et al. 1976), membrane antigens, and lectin receptors (Bayer et al. 1976b), the applications of this system have expanded drastically to include dozens of different uses for literally thousands of different target biomolecules. Many of these have been described in an ongoing series of reviews, which culminated in a recent book on the subject (Wilchek and Bayer, 1990).

The basis of the avidin-biotin system is the chemical attachment of the low molecular weight vitamin biotin via its carboxy terminus to a biologically active molecule which can then be introduced into an experimental system. The biotinylated biomolecule usually recognizes and binds in selective fashion to a target molecule in the system. The glycoprotein avidin (or its bacterial relative, streptavidin) can then be added to the system to pursue the extraneously added biotinylated molecules. The avidin is usually derivatized or conjugated to a probe of some sort. Thus, the complex formed between the target biomolecule and its biologically active counterpart are combined with the probe of choice by virtue of the mediation afforded by the avidin-biotin interaction. This interaction is recognized as the strongest protein-ligand interaction yet known in nature.

It is interesting to note that the avidin-biotin system was originally developed as an improved, naturally occurring alternative to hapten-antibody systems (for a review, see Wofsy, 1983). For both commercial and scientific design, there has been a trend in recent years to return to such hapten-antibody systems as an alternative to the avidin-biotin system. There are numerous advantages and disadvantages in the use of one system over the other; however, the required brevity of this short chapter prevents further discussion.

The subject matter which follows will be confined to discussing the major crux of the avidin-biotin system, i.e., the chemistry by which the biotin moiety is attached to the protein or glycoprotein of choice. By using an appropriate chemically reactive derivative of the vitamin, the biotin moiety can be inserted quite selectively onto a variety of functional groups of the amino acid side chains of proteins. Thus, N-hydroxysuccinimide ester derivatives of biotin have found widespread use in labeling the amine function (lysines and the amino terminus of the protein). Likewise, maleimido-biotin derivatives are selective for cysteines, diazobenzoyl biotins react with tyrosines or histidines, biotin hydrazides can be rendered selective for glycoproteins or carboxylic acid groups, and so on. In other systems, this basic type of chemistry has more or less been copied, with the derivatized hapten of choice, e.g., digoxigenin (DIG), dinitrophenol (DNP), azobenzene arsonate, fluorescein, being substituted for the biotin group. Throughout the past decade, the principles used for the development of the avidin-biotin system have thus served as a strict theoretical model upon which other approaches have been based.

Once the biotin group has been attached to the desired protein and the resultant derivative has been introduced into the experimental system, an appropriate avidin-probe conjugate or derivative can be employed for the required application. For more details of this particular aspect of the system, refer to the book by Wilchek and Bayer (1990) or any of the reviews on avidin-biotin technology (Bayer and Wilchek, 1978; 1980; Bayer et al., 1979; Wilchek and Bayer, 1984; 1988; 1989).

Biotinylating reagents presented in this chapter are all available commercially from many companies, as are a large variety of both biotinylated proteins and avidin-containing probes.

4.2.2 Biotinylation of Proteins via Lysines

This procedure was the first to have been described for biotinylation of proteins (Bayer et al., 1976b) and remains the preferred method today. Novice users of the avidin-biotin system are typically surprised at how easy it is to biotinylate a protein using BNHS. This is indeed one of the advan-

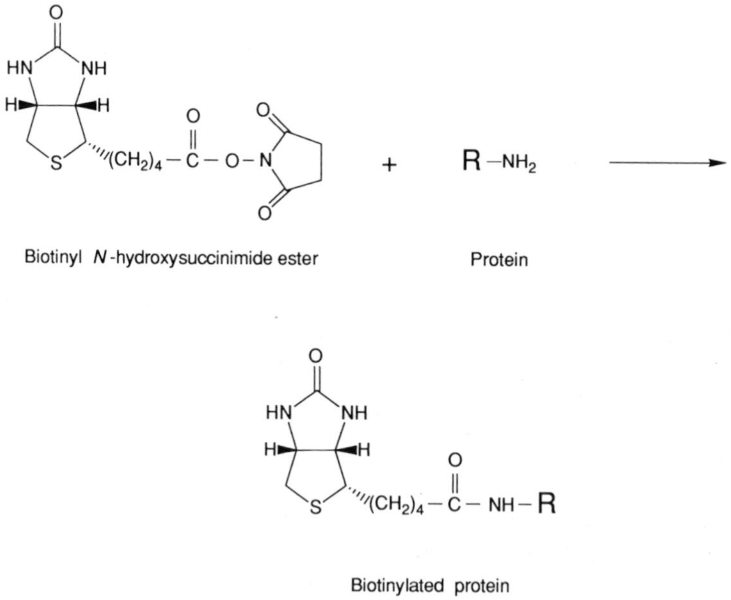

Biotinyl *N*-hydroxysuccinimide ester   +   R—NH₂

Protein

Biotinylated protein

tages of the method, and both students and veteran scientists adjust quickly to the situation. BNHS and a variety of its longer chained analogues are all commercially available from dozens of companies.

- Biotinyl-*N*-hydroxysuccinimide (BNHS)   **Standard reagents**
- Dimethylformamide (DMF)
- Dialysis tubing

- BNHS solution: 1 mg/ml in DMF   **Standard solutions**
- PBS: 0.1 M phosphate-buffered saline, pH 7.4

1. Dissolve protein in PBS at a concentration of 1 mg/ml.   **Biotinyla-**
2. Add rapidly 25 µl BNHS solution for every ml of protein solution; tap   **tion reac-** test tube several times to mix solutions.   **tion**
3. Let the solution stand at room temperature for 1 h without stirring.
4. Dialyze the reaction solution overnight against PBS at 4°C.
5. Store at −20°C.

4.2.3 Special Hints for Application and Troubleshooting

- We usually prepare the BNHS solution fresh although it can be stored   **BNHS** in dry DMF at −20°C.
- Longer chained homologues of BNHS, e.g., biotinyl aminocaproyl *N*-hydroxysuccinimide ester, and water soluble analogues, e.g., sulfo-*N*-hydroxysuccinimidobiotin, can be used instead of BNHS. In many cases, use of a long chained derivative results in improved interaction with the avidin probe. Use of water soluble derivatives precludes exposure of the protein to organic solvents such as DMF, which can be important when biotinylating sensitive proteins such as certain enzymes.

- If the biotinylated protein (after interaction with the experimental sys-   **Biotinyla-** tem) reacts poorly with the avidin probe, the reason is usually either   **tion** overbiotinylation or underbiotinylation. If the protein of choice is abundant, incremented biotinylation (using two to three fold higher and lower levels of BNHS) can be used and the performance of the corresponding products tested empirically. Other methods can be employed to characterize the biotinylated protein more directly, as described previously in detail (Bayer and Wilchek, 1990).
- Biotinylation of lysines alters the pI of the protein. If excessive, the product may interact nonspecifically (on the basis of electrostatic interactions) with other components of the experimental system.

- For stable biotinylated proteins (e.g., most antibodies and antigens), it   **Storage** is usually convenient to store at −20°C in aliquots. In the case of sensitive protein species (e.g., many enzymes), the biotinylated protein can also be stored under sterile conditions at 4°C after passage through a membrane filter.

4.2.4 Biotinylation of Proteins via Cysteines

This procedure is another technically very simple method to biotinylate proteins – in this case, proteins which contain free exposed sulfhydryl groups (cysteines). Proteins which contain disulfides (cystines) can first be reduced using, for example, mercaptoethanol to convert the disulfides to sulfhydryl. Following dialysis to remove the free mercaptoethanol, the newly exposed sulfhydryls of the protein can be biotinylated using 3-(N-maleimidopropionyl) biocytin (MPB). Many different maleimido derivatives of biotin are available commercially.

3-(N-Maleimidopropionyl) biocytin   Protein

Biotinylated protein

Standard reagents
– 3-(N-Maleimidopropionyl) biocytin (MPB)
– Dialysis tubing

Standard solutions
– MPB solution: 1 mg/ml in PBS
– PBS: 0.1 M phosphate-buffered saline, pH 7.4

Biotinylation reaction
1. Dissolve protein in PBS at a concentration of 1 mg/ml.
2. Add rapidly 25 µl MPB solution for every ml protein solution; tap test tube several times to mix solutions.
3. Let the solution stand at room temperature for 2 h without stirring.
4. Dialyze the reaction solution overnight against PBS at 4 °C.
5. Pass solution of biotinylated protein through a membrane filter and store at 4° under sterile conditions, or, alternatively, store at −20 °C.

4.2.5 Special Hints for Application and Troubleshooting

- Unlike BNHS and its analogues, MPB is soluble in aqueous solutions. The MPB solution can be stored either at 4°C or −20°C for limited periods of time, after which the reagent loses its potency.
- Since the extent of biotinylation is dependent on the number of free sulfhydryl groups per protein molecule, and this quality varies greatly among individual proteins, the above-presented procedure may have to be modified regarding the amount of MPB reagent added.
- In one study (Bayer et al. 1985), β-galactosidase was extensively studied as a model sulfhydryl protein for biotinylation using MPB. In this case, the product was unstable upon storage at −20°C but could be stored indefinitely at 4°C under sterile conditions.

4.2.6 Biotinylation of Proteins via Tyrosines or Histidines

Diazobenzoyl biocytin (DBB) selectively modifies both tyrosines and histidines of proteins. The number of biotin groups per protein molecule can be assessed spectroscopically (Wilchek et al. 1986), and of those, the number of labeled tyrosines versus labeled histidines can be determined (Bayer and Wilchek, 1990).

p-Diazobenzoyl biocytin   Protein

Biotinylated protein

Standard reagents
- *p*-Aminobenzoyl biocytin, stable precursor of diazobenzoyl biocytin (DBB)
- HCl
- NaNO$_2$
- NaOH
- Dialysis tubing

Standard solutions
- 2 N HCl (ice-cold)
- NaNO$_2$ solution (7.7 mg/ml of ice-cold double distilled water)
- 1 N NaOH
- Borate buffer (0.1 M, pH 8.4)
- PBS

Preparation of DBB from precursor
1. Dissolve 2 mg *p*-aminobenzoyl biocytin in 40.7 µl of ice-cold 2 N HCl.
2. Add 40.7 µl NaNO$_2$ solution.
3. Let the solution stand at 4 °C for 5 min.
4. Add 35 µl 1 N NaOH to terminate the reaction.
5. Add 12 µl of the reaction solution to 1 ml of borate buffer to prepare the DBB solution (0.2 mg/ml) for the biotinylation step.

Biotinylation of proteins using DBB
1. Add 1 ml of the DBB solution to 1 ml of protein (1 mg/ml).
2. Let the solution stand at room temperature for 2 h without stirring.
3. Dialyze the reaction solution overnight against PBS at 4 °C.
4. Pass solution of biotinylated protein through a membrane filter and store at 4 °C under sterile conditions, or, alternatively, store at −20 °C.

4.2.7 Special Hints for Application and Troubleshooting

- Unlike its precursor, DBB is unstable during storage. Once prepared, the DBB solution must be used immediately.
- For new applications, the exact amount of DBB reagent per protein should be examined experimentally.

4.2.8 Biotinylation of Proteins via Aspartatic and Glutamic Acids

Standard reagents
- Biocytin hydrazide
- Water-soluble carbodiimide (WSC),
  e.g., 1-ethyl-3-(3-dimethyl aminopropyl) carbodiimide hydrochloride
- HCl
- Dialysis tubing

Standard solutions
- Biocytin hydrazide solution: 20 mg/ml in distilled water, brought to pH 5 with 1 N HCl
- HCl solution: 0.5 N
- NaCl: 0.15 M
- PBS

Biocytin hydrazide   +   R—COOH $\xrightarrow{\text{WSC}}$

Protein

Biotinylated protein

1. Dissolve protein in biocytin hydrazide solution to a final concentration of 1 mg/ml.
2. Add solid WSC (2 mg per mg protein).
3. Let the solution stand at room temperature.
4. Check pH of solution every half hour, and, if altered, bring to pH 5 using HCl solution.
5. After 6 h of reaction, dialyze the reaction solution overnight against PBS at 4°C.
6. Pass solution of biotinylated protein through a membrane filter and store at 4°C under sterile conditions, or store at −20°C.

Biotinylation reaction

4.2.9 Special Hints for Application and Troubleshooting

- In order to facilitate the reaction, the biotin hydrazide reagent must be kept in excess over the carboxyl groups of the protein. The reagent may be heated in order to ensure its complete solubilization.
- The optimal amount of WSC should be determined empirically for each protein. Insufficient amounts will lead to partial biotinylation, whereas excessive amounts may inactivate the protein.
- Biotinylation via the carboxyl group alters the pI of the protein. This may result in a highly alkaline product which may lead to high levels of nonspecific binding.

4.2.10 Biotinylation of Glycoproteins via Sugar Residues

Sugar residues are oxidized either chemically using sodium periodate or enzymatically using a mixture of neuraminidase and galactose oxidase. The enzyme mixture first removes terminal sialyl residues and then oxidizes penultimate galactosyl groups from an appropriate glycoprotein. The biocytin hydrazide, present in the reaction mixture, then reacts chemically with the newly oxidized galactose. For glycoproteins which have terminal galactose or N-acetyl galactosamine groups, neuraminidase can be excluded.

Biocytin hydrazide   Oxidized
                                      glycoprotein

Biotinylated glycoprotein

Periodate-mediated oxidation reagents	– Biocytin hydrazide
	– Sodium *meta*-periodate
	– Dialysis tubing

Periodate-mediated oxidation reagents
– Biocytin hydrazide
– Sodium *meta*-periodate
– Dialysis tubing

Enzyme-mediated oxidation reagents
– Biocytin hydrazide
– Galactose oxidase from *Dactylium dendroides* (Sigma)
– Neuraminidase from *Vibrio cholerae* (Behringwerke AF)

Periodate-mediated oxidation solution
– Sodium *meta*-periodate: 0.1 M in double distilled water
– Biocytin hydrazide solution: 20 mg/ml in PBS
– PBS: 0.1 M phosphate-buffered saline, pH 7.4

Enzyme-mediated oxidation solution
– Biocytin hydrazide solution: 20 mg/ml in PBS
– Galactose oxidase solution: 100 units/ml in PBS+

– Neuraminidase solution: 1 unit/ml as supplied by the manufacturer
– PBS+: 0.1M phosphate-buffered saline, pH 7.4, supplemented with 1mM $CaCl_2$ and 1mM $MgCl_2$
– NaCl solution: 0.15M, containing 10mM of EDTA and 0.1% sodium azide

1. Dissolve glycoprotein in PBS to a final concentration of 1mg/ml.   **Periodate-**
2. Add 0.1ml of the periodate solution for every 0.9ml of glycoprotein   **mediated**
   solution.   **oxidation**
3. Let the solution stand for 30min at room temperature.
4. Dialyze the reaction solution against PBS for 4h at 4°C.
5. Add 0.1ml of biocytin hydrazide solution for every ml of dialyzed glycoprotein, and incubate for 2h at room temperature.
6. Dialyze exhaustively against PBS with several buffer changes.
7. Store at −20°C.

1. Dissolve the glycoprotein in PBS+ to a final concentration of 1mg/ml.   **Enzyme-**
2. Add consecutively, 30µl each of neuraminidase, galactose oxidase, and   **mediated**
   biocytin hydrazide solutions.   **oxidation**
3. Let reaction proceed for 2h at 37°C.
4. Dialyze the reaction mixture against the NaCl solution.
5. Store at −20°C.

4.2.11 Special Hints for Application and Troubleshooting

– The above-described reactions are pertinent to glycoproteins which have either vicinal hydroxy groups or bear the sialyl-galactose motif on the termini of the oligosaccharide chains. Other classes of glycoproteins will not be biotinylated using these protocols.

References

Angerer L, Davidson N, Murphy W, Lynch D, Attardi G (1976) An electron microscope study of the relative positions of the 4S and ribosomal RNA genes in HeLa cell mitochondrial DNA. Cell 9:81−90
Bayer EA, Wilchek M (1978) The avidin-biotin complex as a tool in molecular biology. Trends Biol Sci 3:N257−N259
Bayer EA, Wilchek M (1980) The use of the avidin-biotin complex as a tool in molecular biology. Meth Biochem Anal 26:1−45
Bayer EA, Wilchek M (1990) Protein biotinylation. Methods Enzymol 184:138−160
Bayer EA, Skutelsky E, Wynne D, and Wilchek M (1976a) Preparation of ferritin-avidin conjugates by reductive alkylation for use in electron microscopic cytochemistry. J Histochem Cytochem 24:933−939
Bayer EA, Wilchek M, Skutelsky E (1976b) Affinity cytochemistry: the localization of lectin and antibody receptors on erythrocytes via the avidin-biotin complex. FEBS Letters 68:240−244
Bayer EA, Skutelsky E, Wilchek M (1979) The avidin-biotin complex in affinity cytochemistry. Meth Enzymol 62:308−315

Bayer EA, Zalis M, Wilchek M (1985) 3-(*N*-Maleimido-propionyl) biocytin: a versatile thiol-specific biotinylating reagent. Anal Biochem 149:529—536

Heitzmann H, Richards FM (1974) Use of the avidin-biotin complex for specific staining of biological membranes in the electron microscope. Proc Natl Acad Sci USA 71:3537—3541

Manning JE, Hershey ND, Broker TR, Pellegrini M, Mitchell HK, Davidson N (1975) A new method of in situ hybridization. Chromosoma, 53:107—117

Wilchek M, Bayer EA (1984) The avidin-biotin complex in immunology. Immunol Today 5:39—43

Wilchek M, Bayer EA (1988) The avidin-biotin complex in bioanalytical applications. Anal Biochem 171:1—32

Wilchek M, Bayer EA (1989) Avidin-biotin technology ten years on: has it lived up to its expectations? Trends Biol Sci 14:408—412

Wilchek M, Bayer EA (eds) (1990) Avidin-biotin technology. Meth Enzymol, Vol 184, Academic Press, Orlando, Fla

Wilchek M, Ben-Hur H, Bayer EA (1986) *p*-Diazobenzoyl biocytin — a new biotinylating reagent for the labeling of tyrosines and histidines in proteins. Biochem Biophys Res Commun 136:80—85

Wofsy L (1983) Methods and applications of hapten-sandwich labeling. Meth Enzymol 92:472—488

5 In Vivo Labeling of DNA Probes with 5-BrdU

Jean-Luc Guesdon

5.1 Principle and Applications

The thymidine analogue 5-bromodeoxyuridine (BrdU) can specifically and quantitatively replace thymidine in DNA molecules using either an in vivo (Dunn and Smith, 1954; Zamenhof and Griboff, 1954; Hackett and Hanawalt, 1966) or in vitro (Bessman et al., 1958) enzymatic reactions.

BrdU is often used as a substitute for [^{3}H]thymidine in cytochemistry studies. Indeed, BrdU labeling of DNA is widely used to measure DNA replication in cells and chromosomes (Gratzner et al., 1975; 1976), to distinguish DNA synthesizing cells in flow cytometric procedures (Dolbeare et al., 1983), to study human malignant or normal cell kinetics in vivo (Morstyn et al., 1983), to analysze the cell cycle (Dean et al., 1984), to measure DNA replication in stimulated B cells (Porstman et al., 1985; Martinon et al., 1987), to study nuclear incorporation of an S phase-specific chemotherapeutic agent and its effects of DNA synthesis of individual bone marrow cells from patients with acute nonlymphocytic leukemia (Raza et al., 1985), and to identify dividing plasma cell populations in patients with multiple myeloma (Greipp et al., 1985). These techniques, which are considerably less cumbersome and time-consuming than the use of radioactive isotopes of thymidine, are applicable in vitro or in vivo.

BrdU is also used to prepare nonradioactive DNA probes in hybdridization experiments. These BrdU-labeled probes can be made using either in vitro or in vivo enzymatic reactions. In vitro BrdU labeling of DNA is performed by nick translation (Traincard et al., 1983; Niedobitek et al., 1988, 1989) or by complementary strand synthesis from recombinant M13 DNA (Frommer et al., 1988). In vivo incorporation of BrdU was proposed to produce BrdU-labeled plasmid (Kitazawa et al., 1989) or single-stranded M13 DNA (Sakamoto et al., 1987). In order to achieve the best in vivo substitution of thymine by 5-bromouracil in to the nucleic acid of the phage M13, a thymine-requiring strain of *Escherichia coli* was developed (Sakamoto et al., 1987). With this labeling system, BrdU-labeled M13 single-stranded DNA accumulates inside the bacterial cells. Alkaline lysis (Birnboim and Doly, 1979) followed by phenol/chloroform extraction (Sambrook et al., 1990) is used to isolate the BrdU-labeled DNA probe.

In addition, BrdU end-labeled oligonucleotides can be synthesized (Scippo et al., 1989; Jirikowski et al., 1989).

Polyclonal antibodies specific for BrdU were produced and successfully used to detect BrdU incorporated into DNA (Sawicki et al., 1971; Gratzner et al., 1975, 1976). However, the antisera were variable in specificity, depending on the animal immunized; the polyclonal antibodies were heterogenous and difficult to separate from immunoglobulin species that bound to unmodified DNA. These difficulties were resolved by production of mouse monoclonal antibodies highly specific for BrdU (Gratzner, 1982; Traincard et al., 1983; Raza et al., 1984; Vanderlaan and Thomas, 1985; Gonchoroff et al., 1985, 1986). To date, all the anti-BrdU antibodies described and well characterized have low reactivity to double-stranded DNA (generally from $\frac{1}{100}$ to $\frac{1}{1000}$ of single-stranded DNA); thus, BrdU detection is sensitive only in single-stranded DNA molecules. No antibody is available which binds strongly to halogenated nucleotides such as BrdU in double-stranded DNA. Thus the hybridizing region of the probe is not available to the detection system. To counteract this drawback, double-stranded DNA can be denatured with HCl, NaOH, or nuclease before the addition of antibody. However, the requirement to partially denature DNA before addition of antibody may be difficult to control reproducibly. An easier way to overcome this problem is to add poly(BrdU) to DNA by using terminal transferase (Tereba et al., 1979) and to use BrdU-labeled M13 single-stranded DNA (Sakamoto et al., 1987) or BrdU end-labeled oligonucleotide (Scippo et al., 1989; Jirikowski et al., 1989) as probes. With recombinant M13 DNA probe, because the cloning vector is usually bigger than the cloned insert, about 80% of the hybridized probe is in the single-stranded state and thus still available for antibody detection.

Although not necessarily more sensitive or accurate than other non-radioactive techniques, hybridization using BrdU-labeled single-stranded probes has a number of clear advantages. Substitution of a bromine atom for a methyl group in the 5 position of thymine still maintains the keto grouping in the 6 position and an available H in the 1 position for hydrogen-bonding with the 6-amino and 1-N groups, respectively, of adenine. The methyl group in thymine plays no role in hydrogen-bonding of thymine to adenine. As a consequence, the presence of 5-bromouracil in the hybridizing region of the DNA probe causes only a slight decrease in the melting temperature of hybrid molecules and, in contrast to other non-radioactive DNA labels, does not require any modification of hybridization conditions. Moreover, the BrdU in vivo labeling system allows preparation of DNA probes in large quantities. The procedure is simple and cheap and the BrdU-labeled probes are indefinitely stable when stored at $-20°C$.

The BrdU system was used for the detection of viral DNA sequences by in situ hybridization (Niedobitek et al., 1988, 1989), for the demonstration of c-*myc* mRNA expression in leukemic cell lines (Kitazawa et al., 1989), for studying the hormonal dependence of insulin-like growth factor I mRNA expression in the testis of immature hypophysectomized rat

(Scippo et al., 1989), and for the localization of oxytocin mRNA in mouse neurons (Jirikowski et al., 1989), and satellite DNA sequences on human chromosomes (Frommer et al., 1988).

5.2 Reaction Scheme

The standard procedure for nonradioactive in vivo BrdU labeling of DNA probes is comprised of:

1. In vivo incorporation of BrdU into the M13 DNA probe
2. Hybridization of single-stranded BrdU-labeled probe to immobilized target DNA
3. Immunological detection of the BrdU-labeled hybrid molecules by an enzyme immunoassay using a monoclonal anti-BrdU antibody, an alkaline phosphatase-antibody conjugate, and 5-bromo-4-chloro-3-indolyl phosphate/nitroblue tetrazolium salt (BCIP/NBT) as substrate solution.

The standard reaction scheme for in vivo BrdU labeling and BCIP/NBT detection is shown in the figure.

5.3 In Vivo Recombinant M13 DNA Labeling with 5-BrdU, Hybridization, and Immunoenzymatic Detection

- Strain TUC0701, n° I 541 (Collection Nationale de Cultures de Microorganismes, Institut Pasteur, Paris), a thymidine-dependent derivative of *E. coli* JM 103 (Sakamoto et al., 1987)
- 5-Bromouracil (Sigma)
- Recombinant M13 phage
- SDS, lysozyme (Sigma)
- Tris, EDTA, NaOH, ethanol, potassium acetate, glacial acetic acid, chloroform (Merck)
- Salt-saturated phenol
- Prepacked column with superose 6 (Pharmacia)

Labeling reagents

- SDS (Sigma)
- BSA (Boehringer Mannheim)
- Ficoll, type 400 (Pharmacia)
- Polyvinylpyrrolidone (Sigma)
- NaCl, Na citrate (Merck)
- Nitrocellulose membrane BA83 (Schleicher & Schüll)

Hybridization reagents

- Monoclonal anti-5-bromodeoxyuridine antibody, 76-7 (Immunotech, Marseille, France)

Detection reagents

I. Production and labeling of M13 DNA probe

Strain TUC0701 is infected by recombinant M13 phages
and grown in a medium that contains BrdU

Intracellular BrdU-labeled M13 single-stranded DNA probe
accumulates

The BrdU-labeled M13 DNA probe is extracted from
bacterial cells and purified by FPLC chromatography

BrdU-labeled single-stranded M13 DNA probe

II. Hybridization

Solid support-immobilized target DNA or RNA

+ BrdU-labeled single-stranded M13 DNA probe

hybrid target DNA/BrdU-labeled M13 DNA probe

III. Detection

hybrid target DNA/BrdU-labeled M13 DNA probe
+ monoclonal anti-BrdU antibody

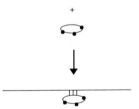

+alkaline phosphatase-labeld anti-mouse IgG antibody

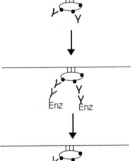

+ BCIP/NBT solution substrate

purple-blue precipitate

Standard reaction scheme for in vivo BrdU labeling and BCIP/NBT detection

- Alkaline phosphatase labeled anti-mouse IgG antibodies (Biosys, Compiègne, France)
- Tween 20 (Merck)
- BCIP (5-bromo-4-chloro-3-indolyl phosphate) (Boehringer Mannheim)
- NBT (nitroblue tetrazolium salt) (Sigma)

Labeling solutions

- Medium 1: M63 medium containing 0.1 mg thymidine/ml
- Medium 2: 2 YT medium containing 0.3 mg 5-bromouracil/ml
- TE buffer: 10 mM Tris-HCl; 1 mM EDTA; pH 8.0/25 °C
- Lysis solution 1: 25 mM Tris-HCl, pH 8.0/25 °C; 50 mM glucose, 10 mM EDTA, 4 mg/ml lysozyme
- Lysis solution 2: 1% [w/v] SDS; 0.2 N NaOH
- K-acetate buffer: 60 ml 5M potassium acetate; 11.5 ml glacial acetic acid; 28.5 ml H_2O; pH 4.8/25 °C

Hybridization solutions

- 20× SSC: 3 M NaCl; 0.3 M Na-citrate, pH 7.0/25 °C
- 50× Denhardt's solution: 1% [w/v] Ficoll 400; 1% [w/v] polyvinylpyrrolidone; 1% [w/v] BSA
- Hybridization solution: 6× SSC; 5× Denhardt's solution; 0.1% [w/v] SDS; 100 µg/ml denatured, fragmented salmon sperm DNA
- Washing solution 1: 2× SSC
- Washing solution 2: 2× SSC; 0.1% [w/v] SDS
- Washing solution 3: 0.1× SSC; 0.1% [w/v] SDS

Detection solutions

- Buffer 1: 10 mM Tris-HCl; 150 mM NaCl; pH 7.5/25 °C, 1‰ [v/v] Tween 20
- Buffer 2: 1% [w/v] BSA in buffer 1
- Buffer 3: 100 mM Tris-HCl; 100 mM NaCl; 20 mM $MgCl_2$; pH 9.5/25 °C
- Monoclonal anti-5-bromodeoxyuridine antibody solution, freshly prepared: 10 µg/ml in buffer 2
- Antibody-conjugate solution: dilute alkaline phosphatase-labeled antibody, $\frac{1}{1000}$ in buffer 2
- BCIP solution: 50 mg/ml in 100% dimethylformamide
- NBT solution: 50 mg/ml in 70% [v/v] dimethylformamide
- Substrate solution, freshly prepared: 30 µl BCIP solution and 60 µl NBT solution are added to 9 ml buffer 3

In vivo labeling reaction

1. Select a bacterial colony (TUC0701 strain) and place in 2 ml of medium 1. Grow overnight at 37 °C with shaking.
2. Add recombinant M13 phages (2 ml) at a multiplicity of infection of 1 and incubate the inoculum for 15 min at 37 °C without agitation.
3. Dilute the mixture 100-fold in medium 2 and incubate at 37 °C with agitation for 1 h.
4. Harvest the bacterial cells by centrifugation at 3000 g for 15 min at 4 °C and resuspend the pellet in 100 µl lysis solution. 1. Incubate for 5 min at room temperature.

5. Add 200 µl of freshly prepared lysis solution 2. Place on ice for 5 min.
6. Add 150 µl of ice cold potassium-acetate solution. Mix and centrifuge at 12000 g at 4°C for 10 min.
7. Pour the supernatant to a new tube. Discard the pellet.
8. Add to the supernatant one volume of salt-saturated phenol plus 1 volume of chloroform. Mix gently and centrifuge at 12000 g for 5 min.
9. Remove the upper aqueous layer and transfer to a new tube. Discard the lower organic layer and the interface.
10. Add 2.5 volumes of prechilled ethanol (−20°C). Mix well and leave the tube 15 min at −70°C or at −20°C overnight.
11. Centrifuge at 12000 g for 10 min, wash the pellet with 0.5 ml cold ethanol (70%, v/v), centrifuge again, dry under vacuum and dissolve in 200 µl TE buffer.
12. Purify the recombinant M13 DNA by gel filtration through a Superose 6 column equilibrated in TE buffer using Fast Protein Liquid Chromatography (FPLC) (Pharmacia). DNA probe is collected in the first peak and is ready for further use.

Hybridization reaction

1. If necessary, prepare dilutions of the DNA to be tested in TE buffer. Add 10 µl of 1 M NaOH to 0.1 ml of each DNA solution, leave the tubes 10 min at 4°C then add 15 µl of 1 M NaH_2PO_4.
2. Prepare nitrocellulose membrane by presoaking in water.
3. Transfer the DNA (or RNA) to be probed to the membrane by dot blot, slot blot, plaque lift, colony lift, Southern or northern blot following a standard protocol (Sambrook et al., 1990).
4. Wash the membrane in washing solution 1. Fix the DNA to nitrocellulose either by baking for 2 h under vacuum or by UV irradiating at 254 nm for 2 min on a transilluminator (Vilber-Lourmat TF 35 M).
5. Place the membrane in a heat-sealable plastic bag and add the hybridization solution (0.2 ml per cm^2 membrane) to the bag. Remove bubbles, fold, and clip end of bag to seal. Incubate the bag at 65°C for 1−3 h.
6. Discard the hybridization solution. Add the 5-bromodeoxyuridine labeled probe (2 µg/ml) in freshly prepared hybridization solution (0.2 ml/cm^2). Carefully remove bubbles and seal the bag.
7. Incubate for at least 6 h submerged in a water bath at 65°C.
8. Wash membrane briefly at room temperature with washing solution 1 (at least 0.5 ml/cm^2 membrane) to remove residual hybridization solution, 3 × 5 min at 65°C with washing solution 2 and 3 × 15 min at 65°C with washing solution 3.
9. Membrane can be used directly for immunoenzymatic detection of the hybridized DNA or stored air-dried for later detection.

Detection reaction

All the following steps are performed at room temperature in a clean plastic box under shaking. The volumes of incubation buffer and washing buffer must be at least 0.2 ml/cm^2 and 1 ml/cm^2 membrane, respectively.

1. Wash the membrane briefly in buffer 1.
2. Incubate for 15 min with buffer 2.

3. Incubate the membrane for 1 h with freshly diluted monoclonal anti-5-bromodeoxyuridine antibody.
4. Remove excess antibody by washing 3 × 10 min with buffer 1.
5. Incubate for 1 h with the alkaline phosphatase-antibody conjugate.
6. Wash the membrane as in step 4 to remove unbound conjugate.
7. Soak the membrane for 2 min in buffer 3.
8. Incubate the membrane in the dark with substrate solution. Color development usually requires approximately 20 min.
9. Stop the enzyme reaction by washing the membrane in distilled water for 15 min. If required, photograph the membrane while still wet to obtain best contrast.
10. Dry the membrane between two sheets of filter paper and store it in the dark.

5.4 Special Hints for Application and Troubleshooting

The major difficulty in the use of nonradioactive probes lies in setting up a procedure which yields the highest signal-to-noise ratio (specific signal vs nonspecific signal). The strength of the specific signal depends upon the specific activity of the probe, the amount of the target DNA spotted on the membrane, the accessibility of the probe to the target, the stability of the probe/target DNA duplex, the affinity of the antibodies used, the accessibility of the immunoenzymatic detection system to the labeled probe, and the sensitivity of the technique used to detect the enzyme activity. The nonspecific signal depends upon the interaction of the major components of the sample with the entire detection system (i.e., nucleic acids and proteins on the one hand, labeled probe and immunoenzymatic detection system on the other). The major problems encountered when using non-radioactive probe technology, and BrdU in vivo labeling in particular, are low sensitivity or high background.

− Do not denature the BrdU-labeled M13 DNA probe prior to use. **Low sensitivity** Autohybridization of the contaminant double-stranded DNA, leading to a hybridization network, may cause low sensitivity because the anti-BrdU antibody recognizes BrdU mainly in the labeled single-stranded DNA rather than in the double-stranded DNA.
− The concentration of labeled DNA in hybridization solution may need to be increased.
− Extend hybridization time.
− Do not wash too stringently after hybridization; this results in loss of specific signal.
− Increase the concentration of anti-BrdU monoclonal antibody (e.g., 30 µg/ml).
− Use an anti-BrdU monoclonal antibody with a high affinity ($K_a > 10^8$ M^{-1}).
− Use the amplifying effect of avidin-biotin interaction to detect anti-BrdU monoclonal antibody (Guesdon et al., 1979; Guesdon, 1988).

- Replace BCIP/NBT detection by a luminescent detection system. Adamantyl-1,2-dioxetane phenyl phosphate substrate can be substituted for alkaline phosphatase. Alternatively, when using peroxidase-antibody conjugate instead of alkaline phosphatase-antibody conjugate, peroxidase-catalyzed enhancement of chemiluminescence can be used in the final detection of nucleic acid hybrids.

High — Generally, nylon membranes give higher background than nitrocel-
background   lulose membranes. If nylon membrane is required, test different membrane batches from various commercial sources. Batch-to-batch variations of some membrane types may be observed.

- Replace the classical baking of nitrocellulose membrane by ultraviolet irradiation. This procedure improves the DNA fixation and lowers the nonspecific signal caused by protein contaminants in the sample.
- Decrease the concentration of labeled DNA in the hybridization solution.
- Purify the anti-BrdU monoclonal antibody. High background may be caused by the presence in ascites fluid of naturally occurring anti-DNA antibodies.
- Use a monoclonal antibody that does not bind to thymidine. Check the specificity of the anti-BrdU monoclonal antibody by ELISA using unlabeled and BrdU-labeled DNA.

References

Bessmann MJ, Lehmann IR, Adler J, Zimmermann SB, Simms ES, Kornberg A (1958) Enzymatic synthesis of deoxyribonucleic acid III. The incorporation of pyrimidine and purine analogues into deoxyribonucleic acid. Proc Natl Acad Sci USA 44:633−640

Birnboim HC, Doly J (1979) A rapid alkaline extraction procedure for screening recombinant plasmid DNA. Nucleic Acids Res 7:1513

Dean PN, Dolbeare F, Gratzner H, Rice GC, Gray JW (1984) Cell-cycle analysis using a monoclonal antibody to BrdUrd. Cell Tissue Kinet 17:427−436

Dolbeare F, Gratzner H, Pallavicini MG, Gray JW (1983) Flow cytometric measurement of total DNA content and incorporated bromodeoxyuridine. Proc Natl Acad Sci USA 80:5573−5577

Dunn DB, Smith JD (1954) Incorporation of halogenated pyrimidines into the deoxyribonucleic acids of Bacterium coli and its bacteriophages. Nature 174:305−306

Frommer M, Paul C, Vincent PC (1988) Localisation of satellite DNA sequences on human metaphase chromosomes using bromodeoxyuridine-labeled probes. Chromosoma (Berl.) 97:11−18

Gonchoroff NJ, Greipp PR, Kyle RA, Katzmann JA (1985) A monoclonal antibody reactive with 5-bromo-2-deoxyuridine that does not require DNA denaturation. Cytometry 6:506−512

Gonchoroff NJ, Katzmann JA, Currie RM, Evans EL, Houck DW, Kline BC, Greipp PR, Loken MR (1986) S-phase detection with an antibody to bromodeoxyuridine. Role of DNase pretreatment. J Immunol Meth 93:97−101

Gratzner HG, Leif RC, Ingram DJ, Castro A (1975) The use of antibody specific for bromodeoxyuridine for the immunofluorescent determination of DNA replication in single cells and chromosomes. Exptl Cell Res 95:88−94

Gratzner HG, Pollack A, Ingram DJ, Leif RC (1976) Deoxyribonucleic acid replication in single cells and chromosomes by immunologic techniques. J Histochem Cytochem 24:34−39

Gratzner HG (1982) Monoclonal antibody to 5-bromo- and 5-iododeoxyuridine: a new reagent for detection of DNA replication. Science 218:474–475

Greipp PR, Witzig TE, Gonchoroff NJ (1985) Immunofluorescent plasma cell labeling indices (LI) using a monoclonal antibody (BU-1). Am J Hematol 20:289–292

Guesdon J-L, Ternynk T, Avrameas S (1979) The use of avidin-biotin interaction in immunoenzymatic techniques. J Histochem Cytochem 27:1131–1139

Guesdon J-L (1988) Amplification systems for enzyme immunoassay. In: Ngo TT (ed) Nonisotopic immunoassay, Plenum Publishing Corporation, New York, pp 85–106

Hackett P, Hanawalt P (1966) Selectivity for thymine over 5-bromo-uracil by thymine.requiring bacterium. Biochim Biophys Acta (Amst) 123:356–363

Jirikowski GF, Ramalho-Ortigao JF, Lindl T, Sekinger H (1989) Immunocytochemistry of 5-bromo-2′-deoxyuridine labeled oligonucleotide probes. A novel technique for in situ hybridization. Histochemistry 91:51–53

Kitazawa S, Takenaka A, Abe N, Maeda S, Horio M, Sugiyama T (1989) In situ DNA-RNA hybridization using in vivo bromodeoxyuridine-labeled DNA probe. Histochemistry 92:195–199

Martinon F, Rabian C, Loiseau P, Ternynck T, Avrameas S, Colombani J (1987) In vitro proliferation of human lymphocytes measured by an enzyme immunoassay using an anti-5-bromo-2-deoxyuridine monoclonal antibody. J Clin Lab Immunol 23:153–159

Morstyn G, Hsu SM, Kinsella T, Gratzner H, Russo A, Mitchell JB (1983) Bromodeoxyuridine in tumors and chromosomes detected with monoclonal antibody. J Clin Invest 72:1844–1850

Niedobitek G, Finn T, Herbst H, Bornhöft G, Gerdes J, Stein H (1988) Detection of viral DNA by in situ hybridization using bromodeoxyuridine-labeled DNA probes. Am J Pathol 131:1–4

Niedobitek G, Finn T, Herbst H, Stein H (1989) Detection of viral genomes in the liver by in situ hybridization using ^{35}S-, bromodeoxyuridine-, and biotin-labeled probes. Am J Pathol 134:633–639

Porstman T, Ternynck T, Avrameas S (1985) Quantitation of 5-bromo-2-deoxyuridine incorporation into DNA: an enzyme immunoassay for the assessment of the lymphoid cell proliferative response. J Immunol Meth 82:169–179

Raza A, Preisler HD, Mayers GL, Bankert R (1984) Rapid enumeration of S-phase cells by means of monoclonal antibodies. N Engl J Med 310:991

Raza A, Ucar K, Preisler HD (1985) Double labeling and in vitro versus in vivo incorporation of bromodeoxyuridine in patients with acute nonlymphocytic leukemia. Cytometry 6:633–640

Sakamoto H, Traincard F, Vo Quang T, Ternynck T, Guesdon J-L, Avrameas S (1987) 5-bromodeoxyuridine in vivo labeling of M13 DNA, and its use as a nonradioactive probe for hybridization experiments. Mol Cell Probes 1:109–120

Sambrook J, Fritsch EF, Maniatis T (1990) Molecular cloning. A laboratory manual. Cold Sping Harbor Laboratory Press

Sawicki DL, Erlanger FB, Beiser SM (1971) Immunochemical detection of minor basis in nucleic acids. Science 174:70–71

Scippo ML, Dombrowicz D, Igout A, Closset J, Hennen G (1989) A nonradioactive method to detect RNA or DNA using an oligonucleotide probe with bromodeoxyuridine free ends, a monoclonal antibody against bromodeoxyuridine and immunogold silver staining. Arch Int Physiol Biochimie 97:279–284

Tereba A, Lai MMC, Murti KG (1979) Chromosome 1 contains the endogenous RAV-0 retrovirus sequences in chicken cells. Proc Natl Acad Sci USA 76:6486–6490

Traincard F, Ternynck T, Danchin A, Avrameas S (1983) An immunoenzymatic procedure for the detection of nucleic acid hybridization. Ann Immunol (Inst Pasteur) 134 D:399–405

Vanderlaan M, Thomas CB (1985) Characterization of monoclonal antibodies to bromodeoxyuridine. Cytometry 6:501–505

Zamenhof S, Griboff G (1954) Incorporation of halogenated pyrimidines into the deoxyribonucleic acids of Bacterium coli and its bacteriophages. Nature 174:306–307

6 The Sulfone System

Israel Nur and Max Herzberg

6.1 Principle and Applications

The novel use of nonradioactive sulfone labeling of DNA has been utilized in the highly sensitive ChemiProbe system, enabling the detection of 0.1 pg of DNA in less than 16 h in dot, slot, and Southern blot hybridization techniques (Herzberg et al., 1987; Nur et al., 1989).

Labeling is performed by treating denatured DNA with sodium bisulfite and methoxyamine, resulting in the conversion of cytosine groups into N^4-methoxy-5,6-dihydro-cytosine-6-sulfonate (Sverdlov et al., 1974). As estimated by HPLC analysis of DNA labeled by the ChemiProbe procedure, sulfonation of about 10%−15% of the cytosine groups of DNA takes place, which is equivalent to 2%−3% of the nucleotide residues in DNA with a G-C content of 47%. The simple procedure enables efficient labeling of both small amounts of single-stranded DNA in the range of 1 pg/ml and large amounts in the range of 1 mg/ml. The system is well suited in particular for detecting minute quantities of DNA in biological fluids (Corti et al., 1990; Richardson and McAvoy 1989).

Labeled DNA is detected using an immunoenzymatic „sandwich" procedure: specific monoclonal antibodies bind to sulfonated DNA and an enzyme-anti-immunoglobulin antibody conjugate binds to the monoclonal antibodies. The complex is visualized using the chemiluminescent substrate for alkaline phosphatase 3-(4-methoxyspiro [1,2-dioxetane-3,2'-tricyclo [3.3.1^{3,7}] decan]-4-yl) phenyl phosphate (AMPPD), whereby dephosphorylation results in the production of a light signal which can be recorded on photographic film, e.g., X-ray films commonly used in autoradiography. Alternatively, the antibody complex can be visualized by the chromogenic reaction with 5-bromo-4-chloro-3-indolyl phosphate (BCIP) and nitroblue tetrazolium salt (NBT).

The ChemiProbe system has been successfully used for the detection of DNA from *Chlamydia trachomatis* (Dutilh et al., 1988), *Mycoplasma pneumoniae* and *M. genitalium* (Hyman et al., 1987), and human papillomavirus (Melki et al., 1988), HIV-1 (Pezzella et al., 1989), and herpes simplex virus (Mullink 1989). The sulfone system has been found to enhance the sensitivity of the polymerase chain reaction (PCR) technique (Paper et al., 1991; Uchimura et al., 1992).

6.2 The Photo-ChemiProbe Procedure: Labeling by Sulfonation, Hybridization, and Chemiluminescence Visualization

- Sodium bisulfite (4 M) (Orgenics) **Labeling reagents**
- Methoxyamine (1 M) (Orgenics)
- Tris-HCl, EDTA, ethanol (99% and 70% [v/v], Merck)

- Salmon sperm DNA (0.5 mg/ml, Orgenics) **Control reagent**

- Blocking powder (Orgenics) **Hybridization reagents**
- Nonhomologous DNA: salmon sperm DNA (Orgenics),
  E. coli DNA or calf thymus DNA (Sigma)
- Formamide, Tris-HCl, EDTA, NaCl (Merck)
- SDS, dextran sulfate (Sigma)
- Poly(acrylic acid), 25%, molecular weight: 90000 (Aldrich)
- Nylon membrane (Orgenics)

- Anti-sulfonated DNA (monoclonal mouse antibodies; Orgenics; **Detection reagents**
  store at 4°C)
- Anti-mouse Ig:enzyme (polyclonal goat antibodies to mouse immunoglobulin: alkaline phosphatase conjugate; Orgenics; store at 4°C)
- PBS (phosphate-buffered saline) powder (Orgenics)
- Tween 20 (Orgenics)
- 10× assay buffer (0.5 M diethanolamine, 10 mM MgCl$_2$; Orgenics)
- AMPPD (Tropix, Inc.; 10 mg/ml solution; Orgenics)
- BCIP (5-bromo-4-chloro-3-indolyl phosphate) (Orgenics)
- NBT (nitroblue tetrazolium salt) (Orgenics)

- TE buffer: 10 mM Tris-HCl; 1 mM EDTA; pH 7.0/25°C **Hybridization solutions**
- Hybridization solution: 50% deionized formamide; 0.7 M NaCl; 50 mM Tris-HCl; 1% [w/v] SDS; either 5% [w/v] dextran sulfate or 1.5% [w/v] poly(acrylic acid); 100 µg/ml freshly denatured nonhomologous carrier DNA; pH 7.5/25°C
- 20× SSC: 3 M NaCl; 0.75 M Na-citrate, pH 7.0/25°C
- 2× SSC; 0.1% [w/v] SDS
- 0.1× SSC; 0.1% [w/v] SDS

- Blocking buffer, freshly prepared: 1 ml/l Tween 20 in conjugate buffer **Detection solutions**
- Conjugate buffer, freshly prepared: 2 g/l blocking powder in 1× PBS
- 10× PBS: 0.58 M Na$_2$HPO$_4$, 0.17 M NaH$_2$PO$_4$·H$_2$O, 0.68 M NaCl; pH 7.3/25°C
- Wash solution: 0.3% [v/v] Tween 20 in 1× PBS
- 10× assay buffer: 0.5 M diethanolamine; 10 mM MgCl$_2$
- Chemiluminescent substrate solution, freshly prepared: dilute 0.1 ml of AMPPD solution in 10 ml 1× assay buffer

6.2

Labeling reaction

1. Pipette the following into a microcentrifuge tube:
   1 volume DNA ($\leq 0.5\,\mu g/\mu l$) to be labeled (Control: $50\,\mu l$ salmon sperm DNA)
   Denature for 5 min in boiling water, chill quickly on ice and add:
     1 volume 4 M sodium bisulfite
   0.5 volume 1 M methoxyamine
2. Mix with a vortex and incubate overnight at room temperature ($20°-28°C$).

Note: Control DNA can be labeled separately as a trial run: there is no need to perform removal of modification chemicals, and dot-blots of serial dilutions ($1000-0.1\,pg/\mu l$) on nylon filter strips can be made directly. Store dotted nylon strips with control DNA at room temperature.

Removal of modification chemicals

1. Precipitate with 2.5 volumes cold absolute ethanol.
2. Maintain in an ice bath for 15 min.

Note: If small quantities of DNA are present, incubate in the ice bath for $1-18\,h$.

3. Centrifuge ($15000\,g$) for 10 min, rinse the pellet with 70% ethanol, repeat rinsing with absolute ethanol, dry under vacuum and dissolve the pellet containing modified DNA in $400-500\,\mu l$ 10 mM Tris-HCl, 1 mM EDTA (pH 7.0)

Hybridization reactions

1. Transfer the DNA to be probed to a nylon filter by dot blot, plaque lift, colony lift or Southern blot transfer. For Southern transfer, proceed as follows:
2. Perform size separation of a DNA digest on an agarose gel. Stain and photograph the gel. Depurinate large DNA fragments, denature and neutralize the gel (Sambrook et al., 1989).
3. Perform Southern transfer onto a nylon filter, rinse the filter in $2\times$ SSC and dry in air on a sheet of filter paper at room temperature for 10 min. Bake the filter in an oven at $80°C$ for 2 h or at $120°C$ for $15-30$ min.
4. Prehybridize in a sealed plastic bag or box of suitable size with at least 20 ml hybridization solution per $100\,cm^2$ at $42°C$ for $2-4\,h$.
5. Discard the hybridization solution. Replace with hybridization solution containing freshly denatured sulfonated probe ($100-200\,ng/ml$) and incubate at $42°C$ for $12-24\,h$.
6. Wash membrane filters using $50-70$ ml per cm^2 of filter. The appropriate washing procedure should be adapted for the specific hybridization system and probe used. As a general procedure, rinse the filter 2×5 min in $2\times$ SSC/0.1% SDS at room temperature and 2×15 min in $0.1 \times$ SSC/0.1% SDS at $42°C$.
7. Filters can then be used directly for detection of hybridized DNA or stored air-dried for later detection.

Membrane filters are placed in sealed plastic bags for incubation with: **Detection** (1) monoclonal antibodies to sulfonated DNA, (2) anti-mouse Ig:enzyme **reaction** conjugate, and (3) chemiluminescent substrate; filters are placed in suitable plastic or glass containers in the washing steps.

1. If using a dried membrane filter, rehydrate in $1\times$ SSC.
2. Transfer the filter to a plastic bag. Add 5 ml of blocking buffer per $100\,cm^2$ of filter. Seal the bag and incubate for 30 min at room temperature.
3. Dilute the anti-sulfonated DNA 1:1000 in conjugate buffer. Add 5 ml of freshly diluted solution per $100\,cm^2$. Seal the bag and incubate for 30 min at 37 °C.
4. Wash the filter 3×10 min in the washing solution (0.3% Tween 20 in $1\times$ PBS) using 300 ml per $100\,cm^2$.
5. Insert the filter into a new plastic bag. Dilute the anti-mouse Ig:enzyme 1:2500 in conjugate buffer. Prepare 5 ml of freshly diluted conjugate per $100\,cm^2$ of filter. Add the diluted conjugate to the bag and incubate for 30 min at 37 °C.
6. Wash the filter 3×10 min in the washing solution (0.3% Tween 20 in $1\times$ PBS) using 300 ml per $100\,cm^2$.
7. Equilibrate the filter in $1\times$ assay buffer for 2 min at room temperature. Remove excess buffer by blotting. Incubate the filter at room temperature in diluted chemiluminescent substrate buffer in a sealed plastic bag using $1-2$ ml per $100\,cm^2$. Spread the substrate solution uniformly using a 10 ml pipet. The optimal exposure time may vary from 10 to 90 min.
8. Record results by exposing X-ray film (Kodak XAR film, Agfa-Gevaert Curix, or equivalent).

6.2.1 BCIP/NBT as an Alternative Chromogenic Systems

As an alternative to the chemiluminescent substrate, a chromogenic substrate system containing 5-bromo-4-chloro-3-indolyl phosphate (BCIP) and nitroblue tetrazolium (NBT) can be used.

6.3 Special Hints for Application and Troubleshooting

The above procotol is applicable to most of the commonly used techniques in molecular biology. However, in some cases modifications should be made to compenaate for low sensitivity and excessive background.

- Repeat precipitation of DNA in 70% ethanol.   **Excessive**
- Use another type of nylon filters.   **background**
- Prolong the duration or increase number of the prehybridization, post-hybridization, and immunological washes.

6.4

Low sensitivity
- Optimize the hybridization conditions by increasing the probe concentration or by lengthening the time of the hybridization reaction.
- Increase the time of incubation with the enzyme-antibody conjugate or with the chemiluminescent substrate.

Low specificity
- Repeat precipitation of DNA in ethanol.
- Increase the duration and stringency of washes following the hybridization and the enzyme-antibody conjugate reaction.

6.4 Labeling of Oligonucleotide Primers for Use in PCR and the Use of Amplified Products as a Labeled Probe

In the following protocol, oligonucleotide primers synthesized with a 5'-polycytidine end (C-tailed primer) are sulfonated and used to amplify a designated sequence on a DNA template by means of PCR. The enzymatically amplified, labeled PCR products may then be used in hybridization experiments as a labeled DNA probe.

Preparation of the sulfonated C-tailed primers
1. Synthesize primers with a polycytidine tail of about 15−25 bases (depending on the base length of the primer) at the 5'-end of the oligodeoxyribonucleotide. For best results, primer sequences with repetitive cytidines, particularly at the 3'-end, should be avoided.
2. Label 100 µl of primers (0.2−0.5 mg/ml) with an equal volume of 4 mM sodium bisulfite and a half volume of 1 M methoxyamine and incubate for 2h at 37°C.
3. Desalt the primers by centrifuging 250 µl of the sulfonated oligonucleotide mixture through a 3 ml bed of Sephadex G-50 spin column. Alternatively, primers can be purified by using a reverse phase column (e.g., DuPont Nensorb, or Waters SEP-PAK C_{18} cartridge).
4. Perform amplification according to standard PCR technique.
5. Sample 10 µl of PCR products and analyze for enzymatic amplification of target DNA by electrophoresis.
6. Add 0.6 volume of PEG solution (20% [w/v] polyethylene glycol 6000 and 2.5 M NaCl) to 1 volume of the PCR products and store the mixture on ice for 1h.
7. Centrifuge for 5 min at 14000 g. Discard the supernatant, rinse the pellet with 80% [v/v] ethanol, and dry in a desiccator at reduced pressure.
8. Dissolve the dried pellet in 100 µl of TE buffer and estimate the DNA concentration by measuring the absorbance at 260 nm. Store the probe at 4°C or, for long-term storage, at −20°C avoid repeated freezing and thawing.

Hybridization experiments with labeled PCR products can be performed as described in Sect. 6.2 with minor modifications.

References

Corti A, Tripputi P, Cassani G (1990) A rapid method for monitoring DNA labeling reactions with haptens. J Immunol Meth 134:81−86

Dutilh B, Bebear C, Taylor-Robinson D, Grimont PAD (1988) Detection of Chlamydia trachomatis by in situ hybridization with sulfonated total DNA. Ann Inst Pasteur/Microbiol 139:115−128

Herzberg M, Reinhartz A, Ritterband M, Twizer S, Smorodinski NI, Fish F (1987) New concept in diagnostic procedures based on nonradioactive DNA probes. Chimica Oggi (Nov):69−71

Hyman HC, Yogev D, Razin S (1987) DNA probes for detection and identification of Mycoplasma pneumoniae and Mycoplasma genitalium. J Clin Microbiol 25:726−728

Lebacq P, Squalli D, Duchenne M, Pouletty P, Joannes M (1988) A new sensitive nonisotopic method using sulfonated probes to detect picogram quantities of specific DNA sequences on blot hybridization. J Biochem Biophys Meth 15:255−266

Melki R, Khoury B, Catalan F (1988) Nucleic acid spot hybridization with nonradioactive labeled probes in screening for human papillomavirus DNA sequences. J Med Virol 26:137−143

Morimoto H, Monden T, Shimano T, Higashiyama M, Tomita N, Murotani M, Matsuura N, Nakamori S, Furukawa J, Okuda H, Mori T (1987) Use of sulfonated probes for in situ detection of amylase mRNA in formalin-fixed paraffin sections of human pancreas and submaxillary gland. Laboratory Investig 57:737−741

Mullink H, Walboomers JMM, Raap AK, van der Ploeg M, Meyer CJLM (1989) Two colour DNA in situ hybridization for the detection of two viral genomes using nonradioactive probes. Histochemistry 91:195−198

Nur I, Reinhartz A, Hyman HC, Razin S, Herzberg M (1989) ChemiProbe, a nonradioactive system for labeling nucleic acid − principles and applications. Ann Biol Clin 47:601−606

Paper T, Friedman M, Nur I (1991) Use of sulfonated primers to detect and type papillomavirus in cell cultures and cervical biopsies. Gene 103:155−161

Pezzella M, Rossi P, Lombardi V, Gemelli V, Mariani Constantini R, Mirolo M, Fundaro C, Moschese V, Wigzell H (1989) HIV viral sequences in seronegative people at risk detected by in situ hybridisation and polymerase chain reaction. Brit Med J 298:712−716

Richardson NA, McAvoy JW (1989) An enzyme-linked immunosorbent assay for the quantitation of DNA. J Immunol Meth 125:287−289

Sambrook J, Fritsch JF, Maniatis T (1989) Molecular cloning − a laboratory manual, Second Edition. Cold Spring Harbor Laboratory Press

Sverdlov ED, Monastyrskaya GS, Guskova LI, Levitan TL, Sheichenko VI, Budowsky EI (1974) Modification of cytidine residues with a bisulfite-o-methyl-hydroxylamine mixture. Biochim Biophys Acta 340:153−165

Uchimura Y, Ishida H, Asada K, Mukai H, Kato I (1992) Nonradioactive labeling with chemically modified cytosine tails by the polymerase chain reaction. Gene 108:103−108

7 Colloidal Gold as a Marker in Molecular Biology: The Use of Ultra-Small Gold Probes

PETER F. E. M. VAN DE PLAS and JAN L. M. LEUNISSEN

7.1 Principle and Applications

Since its introduction by Faulk and Taylor in 1971, the immunogold staining method has undergone an enormous evolution. Methods were developed to prepare gold sols in a wide range of sizes, with a narrow size distribution enabling double or even triple labeling in electron microscopy (EM). A large variety of macromolecules other than antibodies (e.g., streptavidin, protein A, lectins, enzymes, and ligands such as insulin) were coupled to colloidal gold particles. Gold probes have not only proven to be efficient markers for EM, but in combination with silver enhancement, their use has been extended to the light microscopic and macroscopic level (immunoblotting).

The success of gold particle-based immunodetection systems is determined by the following:

1. Colloidal gold particles can be prepared as monodisperse sols in a wide range of sizes, from 2–3 nm to 150 nm (Baschong and Lucocq 1985, Horisberger 1981, Van Bergen en Henegouwen and Leunissen 1986). This allows double or even triple labeling at the EM level (Geuze et al., 1981).
2. Macromolecules coupled to colloidal gold generally retain their bioactivity, and the probes are stable over a long period of time.
3. Colloidal gold particles are highly electron dense and therefore clearly visible in heavy metal ion-contrasted biological structures in transmission EM.
4. Gold particles are useful markers in scanning EM since they can be excited to emit secondary electrons and to back-scatter electrons (De Harven et al., 1984).
5. Large conglomerates of gold particles are intensely colored and as such visible in the light microscope (De Mey et al., 1981) and on blotting membranes (Brada and Roth 1984, Moeremans et al., 1984).
6. Gold particles reflect and depolarize light. These properties are used to attain improved detectability in, e.g., epipolarization microscopy (De Waele et al., 1988).
7. Since colloidal gold particles can act as nuclei for the deposition of silver (Danscher and Rytter-Nörgaard 1983), the sensitivity of detection at

the light microscopic (Van den Brink et al., 1990) and blotting level (Moeremans et al., 1984) can be substantially enhanced.

8. Colloidal gold is preferred over radioactive or enzyme markers due to its particulate nature, resulting in improved resolution. In addition gold probes are nonhazardous reagents.

The different aspects of gold sols and probes, their preparation, and use in immuno(cyto)chemical studies are covered in several reviews (Beesley 1985, De Mey 1983, Hayat 1989, Roth 1983, Verkley and Leunissen 1989).

7.1.1 Limitations of the Gold Marker System

Although the colloidal gold marker system is well-suited for postembedding and surface labeling, there are limitations, e.g., with respect to probe penetration and steric hindrance. When labeling of intracellular components in intact cells is to be achieved, severe permeabilization procedures are required using, for instance, Triton X-100 before aldehyde fixation (Langanger et al., 1984). Stierhof et al. (1986) showed that only the surface layer of semi-thin cryosections was immunomarked with colloidal gold, although such sections should have the best accessibility since they are not embedded and fully hydrated.

7.1.2 Improvement by Reduction of Particle Size

It has been reported by several authors (Ghitescu and Bendayan 1990, Van Bergen en Henegouwen and Leunissen 1986, Yokata 1988) that the labeling density increases when the particle size of the gold probe decreases. A reduction of the gold particle size leads to a reduced probe size and a decreased net negative charge. In this way both steric hindrance and charge determined repulsion will be reduced. Thus, colloidal gold probes were prepared with gold particles having an average diameter of 1 nm or less. These *ultra-small gold probes* (Leunissen et al., 1989) have been commercially available for several years now. Ultra-small gold probes differ in their characteristics and performance compared to probes built around larger particles. Whereas the „classical" probes can be considered as particles coated with proteins, ultra-small gold probes can be considered as proteins coated with one or more gold particle(s).

Ultra-small gold particles can hardly be visualized directly by transmission EM. However, the gold signal can be detected by increasing the particle diameter by silver enhancement as, e.g., described by Danscher (1981). The improved marking characteristics of ultra-small gold probes have been described for immunocytochemistry (De Graaf et al., 1991, Moeremans et al., 1989, Nielsen and Bastholm 1990, Van Bergen en Henegouwen 1989, Van de Plas et al., 1989). In combination with controlled silver enhancement, ultra-small gold particles are excellent markers for

EM, light microscopy (LM), and immunoblotting. As such, ultra-small gold probes are valuable tools for integrated immunolabeling studies.

Although the incubation protocol for the indirect immunodetection of antigens described in the following sections has been especially developed for ultra-small gold probes, it can be universally applied, independent of the probe size.

7.2 Reaction scheme

In the accompanying figure, schematic representation of the indirect immunogold silver staining method, using ultra-small gold-antibody conjugates for immunodetection, is shown.

1) Blocking of non-specific binding
   sites

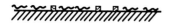

Legend:

	BSA
~	CWF gelatin
	Normal Serum

primary antibody

ultra small gold conjugate

2) Primary antibody incubation

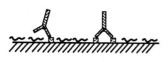

3) Secondary incubation with the
   ultra small gold conjugate

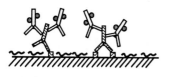

4) Visualization using silver
   enhancement

7.3 Indirect Immunogold Silver Staining

The immunolabeling protocol described has been especially developed for ultra-small gold probes but is generally applicable independent of the particle size. However, for most EM applications silver enhancement is omitted when „classical" probes are used. It is advisable to exactly follow the described procedure.

Immuno-detection reagents
- Specific primary antibody, preferably affinity-purified
- Ultra-small gold-labeled secondary antibodies (Amersham/Aurion/Biocell)

Silver enhancement reagents
- IntenSE (Amersham)
- Aurion R-gent (Aurion)
- Biocell Silver Enhancement Kit (Biocell)
- Gum arabic (for EM) (Merck/Fluka/Aurion)

Silver enhancement solutions
The silver enhancement procedure (for EM) is according to Danscher (1981).

- Protecting colloid: Dissolve 100 g gum arabic in 200 ml distilled water (this lasts several days). Filter the solution through layers of gauze. The solution can be stored frozen.
- Citrate buffer: Dissolve 2.55 g citric acid·H_2O (Merck) and 2.35 g sodium citrate·$2H_2O$ (Merck) in distilled water to make 10 ml.
- Reducing agent: Dissolve 0.57 g hydroquinone (Merck) in 10 ml distilled water. Prepare immediately before use and protect from light.
- Silver ion supply: Dissolve 0.073 g silver lactate (Fluka) in 10 ml distilled water. Prepare immediately before use and protect from light.

Buffer solutions
- PBS: 10 mM phosphate buffer. 150 mM NaCl, pH 7.6
- PBS, pH 7.6, supplemented with 0.05 M glycine (Merck) (LM/EM) or 0.1% $NaBH_4$ (Merck) (LM), freshly prepared
- Incubation buffer: PBS, pH 7.6, supplemented with 0.8% bovine serum albumin fraction V (Sigma), 0.1% cold water fish skin gelatin (Sigma), 20 mM NaN_3 (Merck); check the pH and adjust to 7.6 if necessary.
- Block buffer: for blotting (LM/EM): Incubation buffer supplemented with 5% bovine serum albumin
- 5% normal serum prepared from the same species as the ultra-small gold-labeled secondary antibody (Amersham/Aurion/Biocell)
- Detection buffer: Incubation buffer supplemented with 1% normal serum

Postfixation solution
- 2% glutaraldehyde (BDH) in PBS, pH 7.6

Light microscopy
- Counterstaining, e.g., hematoxylin/eosin (Fluka)
- Ethanol (Merck)
- Xylene (Merck)
- Pertex (Histolab)

Electron microscopy reagents
- Heavy metal contrast, e.g., uranyl acetate/Reynold's lead citrate
- Methylcellulose (Merck) for ultra-thin cryosections

7.3

Blocking reaction

1. Aldehyde-fixed specimens are incubated in PBS with 0.05 M glycine (LM/EM) or 0.1% NaBH$_4$ (LM) for 15 min to inactivate residual aldehyde groups. Rinse in PBS 2 × 1 min before proceeding.
2. For LM/EM, incubate in block buffer at room temperature for 30 min; for blots, incubate in block buffer at 45 °C for 30 min.

Primary antibody reaction

1. Incubate the specimens with primary antibody (1–5 µg/ml) in detection buffer for 60 min.
2. Remove excess primary antibody by washing in incubation buffer for 3 × 10 min. The washing should be extended to 6 × 5 min for EM on-grid marking.

Ultra-small gold labeling and fixation

1. Incubate with the appropriate ultra-small gold conjugate reagent, diluted according to the manufacturers instructions (1/50–1/100) in detection buffer for 120 min. It is recommended to test the optimal dilution for each new localization study.
2. Remove excess gold probe by washing in incubation buffer for 4 × 10 min. The washing steps should be extended to 6 × 5 min for EM on-grid marking.
3. Wash in PBS 3 × 5 min.
4. Postfix in 2% glutaraldehyde in PBS for 15 min.
5. Wash in distilled water 3 × 5 min.

Silver enhancement and staining

1. Visualize the gold signal using the silver enhancement as follows:
   (a) For LM and immunoblotting: Mix the enhancer and the initiator/developer in a 1:1 ratio immediately before use. The specimens should be fully covered by the mixture. Typical enhancement times are between 15 and 30 min. For LM, the ongoing process can be monitored using an (inverted) light microscope with dimmed light conditions.
   (b) To obtain a homogeneous enhancement for EM, gum arabic (50% w/v in distilled water) is added to the enhancement mixture in a 1:2 ratio. Typical enhancement time is between 10 and 20 min.
   (c) An alternative for EM visualization is the silver enhancement method described by Danscher (1981). Preparation of the enhancement solution: Mix 6 ml gum arabic solution, 1 ml citrate buffer, and 1.5 ml hydroquinone solution. Add 1.5 ml silver lactate solution and mix carefully (the mixture has to be protected from light). Use the enhancement solution immediately. The enhancement procedure should be performed shielded from direct light. After 20 min at room temperature the ultra-small gold-silver particles have reached a diameter between 5 and 10 nm.
2. Wash the specimens with excess distilled water for 3 × 5 min. For complete removal of the gum arabic, more but shorter wash steps are necessary (e.g., 10 × 2 min).
3. For LM, specimens can be counterstained in, e.g., hematoxylin/eosin or basic fuchsin. The specimens are dehydrated in a graded series of ethanol, xylene, and embedded in a *water incompatible* mounting medium, e.g., Pertex. This results in a permanent gold-silver signal.

For EM uranyl acetate and lead citrate (Reynold's) can be used for heavy metal contrasting. Plastic sections are air dried. Ultra-thin cryosections are protected from drying artifacts by mounting in a thin layer of methyl cellulose. Immunoblots are air dried.

7.4 Special Hints for Application and Troubleshooting

Immuno-blotting

- Depending on the size of blots or strips, incubations can be carried out in sealed plastic bags, Petri dishes, or in disposable, screw cap, sealed tubes.
- It is important to use a rocking table or tilting apparatus for all incubation steps except for blocking.

Light microscopy

- Sections mounted on microscope slides are washed on a magnetic stirrer in 250 ml color trays with separate slide rack.
- Monolayers (or sections mounted) on coverslips can be easily incubated using 6-well culture plates (Falcon). Washing steps are carried out on a rocking table, using about 2 ml of washing medium/per well.
- Cell suspensions are gently pelleted after each incubation step. The pellets are resuspended in the medium used in the next step. Incubations are carried out on a rocking table.
- Living cells are preferably incubated at $0°-4°C$ or in the presence of $0.05\%-0.2\%$ NaN_3 to avoid internalization of reagents.
- For immunodetection of intracellular antigens, permeabilization steps may be required. The degree of penetration of immunoreagents depends on the specimen characteristics, the type of fixation, and the incubation time. For the detection of cytoskeletal antigens or antigens associated with the cytoskeleton, the use of $0.1\%-0.5\%$ Triton X-100 (Sigma) (*only* after aldehyde fixation) is recommended. Successful labeling of intracellular antigens in nerve tissue in the absence of detergent has been reported (Van Lookeren Campagne 1991).

Electron microscopy

- Pre-embedding immunolabeling: The set-up for marking in light microscopy is used. If penetration problems occur Triton X-100 may be applied (*only* after aldehyde fixation) but the ultrastructure may suffer from this treatment. For some specimens a compromise between ultrastructural preservation and penetration can be obtained with saponin (Sigma) or by the use of a graded series of ethanol followed by acetone (100%) and rehydration in the ethanol series.
- The use of nickel grids covered with a carbon coated formvar or parlodion film is recommended since silver enhancement procedures will be used. For most applications, grids are floated on top of drops of (immuno)reagents on a sheet of Parafilm.
- If necessary osmium tetroxide fixations may be used. Osmification must be done before silver enhancement since osmium tetroxide removes metallic silver by oxidation.

7.4

Back-
ground
staining

- Background staining is due to unintended reactions in immuno(cyto)chemical localization studies. To unravel the cause, it is advisable to include the following controls in every new experimental set-up: (1) Omission of the primary antibody; specimens are incubated with the gold-labeled secondary antibody and silver enhanced. (2) Omission of primary and secondary antibody; specimens are silver enhanced only.

- If background occurs and both controls are negative, then the primary antibody has caused an unintended staining pattern. To evaluate the performance and usability of every new primary antibody test a dilution series.

- In sera many clones of antibodies are present and part of them may be directed to other antigens. A test using pre-immune serum may elucidate this. Affinity purify the serum or use a different well characterized serum or primary antibody.

- Polyclonal antibodies recognize a whole panel of different epitopes, some of which are present in structures other than the one under investigation. In this case cross-adsorption may substantially improve specificity. If the same problem occurs with a monoclonal antibody, it is necessary to use a clone recognizing a different epitope.

- If background staining is caused by the gold-labeled secondary antibody the following suggestions may be helpful:
  (1) If possible avoid the use of positively charged adhesives such as poly-L-lysine. If charge-based interactions occur between the gold probe and the specimen itself negatively charged additives in the incubation and detection buffer can prevent this type of background. In practice 0.1%−0.2% acetylated BSA (Aurion) in PBS, pH 7.4 has been found to be the most effective
  (2) Increase the number of washing steps after the secondary antibody incubation.
  (3) When the positive signal is strong the gold reagent may be diluted further to obtain a better signal to noise ratio

- If background staining is present in both controls, the following is then recommended:
  If it is necessary to apply silver enhancement for longer than 25 min, it is advisable to follow a two-step procedure to avoid autonucleation.
  After a first enhancement period of 25 min, wash the specimen extensively in distilled water and subsequently proceed with the second enhancement step in freshly mixed reagents.

- More information on the subject of background prevention can be found in Aurion Newsletter, October 1990.

— It is advised to include a positive control. If this control gives satisfactory results it can be concluded that the followed procedure is reliable. In the specimens under investigation, the antigen is either not present or only present in low amounts.

— Troubleshooting should be directed to the activity of the immunogold/silver reagents, the reactivity of the primary antibody, and/or the fixation or the embedding medium/procedure.

— The activity of the silver enhancement reagents can be easily tested. Apply 1 μl of a 1 : 10 dilution of immunogold reagent in PBS on a strip of nitrocellulose membrane. Allow the spot to dry and soak the strip in distilled water. Wash once in distilled water and proceed with silver enhancement. The spot should be clearly visible after 15−20 min by its brown/black to black color.

— The bioactivity of the immunogold probe can be assessed in a dot-spot test (Moeremans et al., 1984): 1 μl drops of a serial dilution (250−0.1 μg/ml) of the primary antibody in PBS along with 50 μg/ml BSA are spotted on a strip of nitrocellulose paper. The spots are allowed to dry. The strip is washed in PBS. Proceed as for immunoblotting (see procedure in Sect. 7.3). After silver enhancement 1 ng of primary antibody has to be detected.

— When the problem of low signal intensity cannot be explained by the poor reactivity of the immunogold silver reagents, the reactivity of the primary antibody is suboptimal or the fixation and/or embedding procedure have to be adapted.

— Chemical fixation is commonly used to stabilize the (ultra)structure of tissue. It has a self-contradicting goal of protecting the tissue components from loss or displacement during dehydration and embedding without affecting antigenicity. The effect of chemical fixation on immunoreactivity can be tested using the antigen-spot test (Moeremans 1984). In this way the effect of a fixation protocol on preservation of antigenicity can be evaluated without the use of time-consuming preparation and microscopy techniques.

— For EM applications, the influence of the preparation technique and the immunodetection procedure should be evaluated with LM, e.g., by optimizing the labeling conditions on semi-thin cryosections (LM) and extrapolating these conditions for EM (see following example).

Low (or absent) signal intensity

7.5 Selected Example: Immunodetection of Gonodatropic Hormone in Catfish Pituitary

Rabbit-anti-*Clarias* gonadotropic hormone (GTH) serum diluted 1 : 20 000 and Aurion goat-anti-rabbit conjugated to ultra-small gold particles 1 : 100 were used as primary and secondary antibodies, respectively. At the light microscope level, Aurion R-Gent was used for 15 min to visualize the gold signal by silver enhancement. Sections were counterstained with hematoxilin and eosin. For EM, the ultra-small gold particles were silver enhanced with the Danscher method (Danscher 1981) for 20 min. Ultra-thin sections were heavy metal-contrasted with 2% aqueous uranyl acetate and Reynold's lead citrate (see figure).

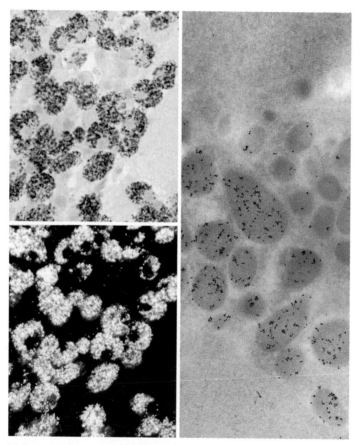

Light and electron microscopic detection of gonadotropic hormone (GTH) on H_2O_2-etched Epon sections of catfish pituitary gland. *Top left:* Bright field mode. The brown to black silver deposit in the cytoplasm identifies the GTH-positive cells. *Bottom left:* Epipolarization image of the same section. *Right:* Electron microscopic demonstration of GTH-containing granules. The diameter of the gold-silver particles is between 5 and 10 nm. With special thanks to Mr. Zandbergen and Dr. J. Peute (Department of Experimental Zoology, State University Utrecht, The Netherlands) for providing embedded specimens and primary antiserum

References

Baschong W, Lucocq JM (1985) Thiocyanate gold: small (2—3 nm) colloidal gold for affinity cytochemical labeling in electron microscopy. Histochemistry 83:409—411

Beesly JE (1985) Colloidal gold: A new revolution in marking cytochemistry. Proc RMS 20:187—197

Brada D, Roth J (1984) Golden blot detection of polyclonal and monoclonal antibodies bound to antigens on nitrocellulose by protein-A gold complexes. Anal Biochem 142:79—83

Danscher G (1981) Localization of gold in biological tissue. A photochemical method for light and electron microscopy. Histochemistry 71:81—88

Danscher G, Rytter-Nörgaard (1983) Light microscopic visualization of colloidal gold on resin-embedded tissue. J Histochem Cytochem 31:1394—1398

De Graaf J, Van Bergen en Henegouwen PMP, Meijne AML, Van Driel R, Verkley AJ (1991) Ultrastructural localization of nuclear matrix proteins in HeLa cells using silver enhanced ultra small gold probes. J Histochem Cytochem 39:1035—1046

De Harven E, Leung R, Christensen H (1984) A novel approach for scanning electron microscopy of colloidal gold labeled cell surfaces. J Cell Biol 99:53—57

De Mey J (1983) Colloidal gold probes in immunocytochemistry. In: Polak S, Van Noorden S (eds) Immunocytochemistry. Practical Application in Pathology and Biology, Wright, London, pp 82—112

De Mey J, Moeremans M, Geuens G, Nuydens R, De Brabander M (1981) High resolution light and electron microscopic localization of tubulin with the IGS (immuno gold staining) method. Cell Biol Int Rep 5:889—899

De Waele M, Renmans W, Segers E, Jochmans K, Van Camp B (1988) Sensitive detection of immunogold-silver staining with darkfield and epi-polarization microscopy. J Histochem Cytochem 36:679—683

Faulk WP, Taylor GM (1971) An immunocolloid method for the electron microscope. Immunochemistry 8:1081—1083

Geuze HJ, Slot JW, Van der Ley P, Schuffer R, Griffith J (1981) Use of colloidal gold particles in double labeling immuno electron microscopy on ultrathin frozen sections. J Cell Biol 89:653—665

Ghitescu L, Bendayan M (1990) Immunolabeling efficiency of protein A-gold complexes. J Histochem Cytochem 38:1523—1530

Hayat MA (ed) (1989) Colloidal Gold. Principles, Methods and Applications. Academic Press, New York, Vol 1—2

Holgate CS, Jackson P, Cowen PN, Bird C (1983) Immunogold-silver staining: new method of immunostaining with enhanced sensitivity. J Histochem Cytochem 31:938—944

Horisberger M (1981) Colloidal gold: a cytochemical marker for light and fluorescent microscopy and for transmission and scanning electron microscopy. Scan Electron Microsc 2:9—28

Langanger G, De Mey J, Moeremans M, Daneels G, De Brabander M, Small JV (1984) Ultrastructural localization of α-actinin and filamin in cultured cells with the immunogold (IGS) staining method. J Cell Biol 99:1324—1334

Leunissen JLM, Van de Plas PFEM, Borghgraef PEJ (1989) Auroprobe One: a new and universal ultra small gold particle based (immuno)detection system for high sensitivity and improved penetration. In: Janssen Life Sciences (ed) Aurofile 2:1—2

Moeremans M, Daneels G, Van Dijck A, Langanger G, De Mey J (1984) Sensitive visualization of antigen-antibody reactions in dot and blot immuno overlay assays with the immunogold and immunogold/silver staining. J Immunol Methods 74:353—360

Moeremans M, Daneels G, De Raeymaeker M, Leunissen JLM (1989) Auroprobe One in immunoblotting. In: Janssen Life Sciences (ed) Aurofile 2:4—5

Nielsen MH, Bastholm L (1990) Improved immunolabeling of ultrathin cryosections using antibody conjugated with 1-nm gold particles. In: Proc 12th Int Congr Electron Microsc, p 928

Roth J (1983) The colloidal marker system for light and electron microscopic cytochemistry. In: Bullock GR, Petrusz P (eds) Techniques in immunocytochemistry. Vol 2, Academic Press, New York, pp 217–284

Stierhof Y-D, Schwarz H, Hermann F (1986) Transverse sectioning of plastic embedded immuno-labeled cryosections: morphology and permeability to protein A-gold complexes. J. Ultrastruct Mol Struct Res 97:187–196

Van Bergen en Henegouwen, Van Lookeren Campagne (1989) Subcellular localization of phosphoprotein B-50 in isolated presynaptic nerve terminals and in young adult rat brain using a silver-enhanced ultra-small gold probe. In: Janssen Life Sciences (ed) Aurofile 2 pp 6–7

Van Bergen en Henegouwen, Leunissen JLM (1986) Controlled growth of colloidal gold particles and implications for labeling efficiency. Histochemistry 85:81–87

Van den Brink W, Van der Loos C, Volkers H, Lauwen R, Van den Berg F, Houthoff H-J, Das PK (1990) Combined B-galactosidase and immunogold/silver staining in immunohistochemistry and DNA in situ hybridisation. J Histochem Cytochem 38:325–329

Van de Plas PFEM, Leunissen JLM (1989) Immunocytochemical detection of tubulin in whole mount preparations of PtK2-cells: improved penetration characteristics of Auroprobe One. In: Janssen Life Sciences (ed) Aurofile 2 pp 3–4

Van Lookeren Campagne M (1991) Ph.D. thesis, University of Utrecht

Verkley AJ, Leunissen JLM (eds) (1989) Immuno-gold labeling in cell biology. CRC Press, Boca Raton, Florida

Yokata S (1988) Effect of particle size on labeling density for catalase in protein A-gold immunocytochemistry. J Histochem Cytochem 36:107–109

8 Direct Peroxidase Labeling of Hybridization Probes and Chemiluminescence Detection

IAN DURRANT

8.1 Principle and Applications

The labeling of long nucleic acid probes is based on the work first described by Renz at the European Molecular Biology Laboratories (EMBL) (Renz and Kurz, 1984). Single-stranded nucleic acid (RNA or DNA) is labeled with a positively charged, modified, horseradish peroxidase (HRP) complex in a rapid, reliable, and simple reaction process to produce a stable probe. In conjunction with enhanced chemiluminescence (ECL), a light-based detection system, the HRP-labeled probes can be used to detect single copy genes in as little as 1 µg of a restriction enzyme digest of human DNA blotted onto either nylon or nitrocellulose membranes (Stone and Durrant, 1991). The light output of the ECL reaction is captured on X-ray film; for most high sensitivity applications the exposure time required is less than 60 min, although up to 4 h exposures are possible. For high target applications the associated rapid hybridization and rapid detection procedures enable the whole process to be completed in 1 working day.

The labeling reagent is made by cross-linking HRP to polyethyleneimine to produce a stable labeling reagent. The detection system is based on chemiluminescence, using the HRP label as a catalyst for the enhanced luminol light-producing reaction (Thorpe and Kricka, 1987). The labeling reaction takes 20 min only and the labeled probe does not require further purification. The probe can be stored in 50% glycerol at $-20\,°C$ for at least 6 months (Durrant, 1990). The reaction is a reliable chemical process and can be performed with as little as 100 ng or as much as 2 µg with equal efficiency. The optimal template is greater than 300 bases in length and it is recommended that cloned sequences are purified away from the vector. The reaction leads to the introduction of an average of one HRP every 25 bases. The amount of labeled probe produced is easy to calculate; it is exactly that put into the labeling reaction.

Hybridization takes place in a special buffer that is optimized for enzyme stability and hybridization efficiency. Recent improvements of the buffer have given ten fold better sensitivity due to better enzyme stability and an improved rate enhancement system (Cunningham, 1991). The hybridization is performed at $42\,°C$ in the presence of $6\,M$ urea; this denaturant does not inhibit the enzyme label and has properties similar to those of traditional buffers containing 50% formamide (Hutton, 1977). The buffer is stable at room temperature and is clear and nonviscous. A proteinaceous blocking agent is added before use and the buffer can be stored in aliquots for up to 6 months at −20 °C. After hybridization, no antibody incubation steps are required. Blots are detected immediately, although they can be stored, wetted in 2× SSC, at 4 °C overnight if required.

Oligonucleotides (usually <50 bases in length) are labeled by a separate process. The oligonucleotide is synthesized with the addition of a reactive thiol group attached at the 5′ end (Connolly and Rider, 1985). The thiol group reacts with freeze-dried HRP which has been previously derivatized with a bifunctional cross-linking agent to leave a suitable reactive group exposed. The labeling reaction is optimized to label 5 μg of oligonucleotide (Fowler et al., 1990). This is a reliable and very efficient chemical process that labels over 85% of the oligonucleotide molecules; no further probe purification is required.

Hybridization takes place in a simple buffer and the presence of the HRP does not appear to affect the stringency of the hybridization; probes can still distinguish mismatched from perfectly matched target sequences. The high molar concentrations of oligonucleotide probes used means that the hybridization is rapid, typically only 1 h is required, enabling the whole process to be completed in as little as 4 h (Simmonds et al., 1991).

The ECL detection process is common to both HRP-labeled long probes and HRP-labeled oligonucleotide probes. The hybridized and stringently washed blots are simply covered in the ECL substrate and left for only 1 min. The blots are then drained and exposed to X-ray film. The length of exposure varies with the target system but is usually in the 10−60 min range, even for sensitive applications such as the detection of single copy genes (Pollard-Knight et al., 1990). The ECL light output can also be collected on a charge-coupled device (CCD), i.e., a photon counting camera (Boniszewski et al., 1990). The signal dies away with time (over 5−6 h) as the HRP label becomes gradually, irreversibly inhibited by the free radicals generated by the ECL reaction process. In addition, subsequent hybridizations do not appear to be affected by the presence of the previous probe so there is no need to physically strip the previous probe before going on with the next hybridization. This saves time and avoids the use of harsh stripping protocols which may cause damage to the target. It is possible to reprobe a nylon membrane over ten times using the same or different probes at each stage (Evans et al., 1990). These direct labeling systems have found application in all areas of nucleic acid detection in molecular biology. The long probes can be used in particular for the detection of single copy genes (Pollard-Knight, 1990; Lambalk, 1991) and for other sensitive Southern and northern blotting applications (Tonjes, 1991; Moore, 1991). Both systems (long and oligonucleotide probes) have been demonstrated as effective in applications such as fingerprinting (Amersham, 1990), colony and plaque lifts (Durrant et al., 1990), and PCR product detection (Sorg et al., 1990; Grimsley et al., 1991). It is also possible to use the ECL detection system in in situ hybridization (Hawkins and Cumming, 1990) and it is particularly useful in conjunction with either densitometry or CCD cameras for techniques requiring quantification of the results (Misiura et al., 1990). The ECL system can also be used to great effect when HRP is used as the label (such as HRP-labeled antibody molecules); examples include the detection of western blots (Leong and Fox, 1988; Wisdom and Winter, 1990), in immunoassay (Thorpe and Kricka, 1989), and for hapten-based nucleic acid labeling and detection systems, such as those based upon fluorescein-dUTP (Cunningham et al., 1991).

8.2 Reaction Scheme

The standard procedures for use of the direct HRP labeling systems can be split into three parts (see figure):

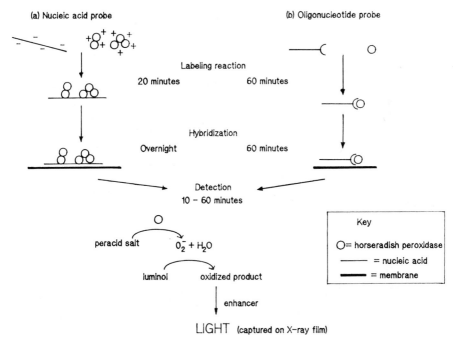

Direct labeling of hybridization probes with horseradish peroxidase

1. Introduction of the HRP enzyme label onto probe molecules
2. Use of HRP-labeled probes, under conditions designed to retain enzyme activity and promote hybridization, to hybridize to immobilized target sequences
3. Light-based detection of the hybrid molecules to produce a hard-copy of the result on X-ray film

8.3

8.3 Directly Labeled Nucleic Acid (Long) Probes

Labeling — Labeling reagent (Amersham)
reagents — Glutaraldehyde solution (Amersham)

Hybridiza- — NaCl (BDH)
tion — Blocking agent (Amersham)
reagents — Hybridization buffer (Amersham)
— Hybond-N$^+$ (nylon) membrane (Amersham)
— Hybond-ECL (nitrocellulose) membrane (Amersham)
— Urea (BDH)
— Sodium citrate (Sigma)
— Sodium dodecyl sulphate (Sigma)

Detection — ECL detection reagent 1 (Amersham)
reagents — ECL detection reagent 2 (Amersham)
— Hyperfilm-ECL (X-ray film) (Amersham)
— Saranwrap (Dow Chemical Company)

— Labeling reagent (Prediluted solution of HRP cross-linked to polyethylene imine, PEI, in deionized water)
— Glutaraldehyde (1.5% [v/v] solution in deionized water)

Hybridiza- — Hybridization buffer — working stock: buffer as supplied with the addi-
tion solu- tion of NaCl to a concentration of 0.5 M and blocking agent to 5% [w/v]
tions — Primary wash solution: 6 M urea, 0.4% [w/v] SDS, 0.5× SSC
— Secondary wash solution: 2× SSC

Detection Detection reagent — working stock (equal volume mix of ECL detection
solutions reagents 1 and 2)

Probe 1. Dilute the probe to be labeled to a concentration of 10 ng/μl in
labeling deionized water; 200 ng should be labeled for 100 cm^2 of membrane.
2. If the probe is double-stranded, boil for 5 min at 100°C in a vigorously
boiling water bath. Cool immediately on ice for 5 min. Add:
20 μl labeling reagent
20 μl glutaraldehyde solution
Mix gently by pipetting and incubate at 37°C for 10 min.
Store labeled probe in 50% glycerol at −20°C or place on ice if to be
used immediately.

Note: to label alternative quantities of probe scale the volume of the reactants accordingly; the volume of the three solutions used to make the labeled probe should always be equal to one another.

Hybridiza- 1. Prepare a working stock of the hybridization buffer as described below.
tion Allow approximately 1 h for this process initially.

- 100 ml hybridization buffer
- 2.92 g NaCl
- 5.00 g blocking agent

2. Prehybridize membrane in $0.125-0.25 \, ml/cm^2$, depending on membrane size and the type of container, for a minimum of 15 min and preferably for 1 h.
3. Add labeled probe to the prehybridization buffer to a concentration of 10 ng/ml. Mix and hybridize for 17 h at 42 °C.
4. Wash membrane in primary wash buffer for 2×20 min at 42 °C, followed by washes in the secondary wash buffer for 2×5 min at room temperature.

1. Prepare sufficient working detection reagent to cover the membrane at **Detection** $0.125 \, ml/cm^2$. Place the membrane on a clean surface. Cover with detection reagent and leave at room temperature for 1 min.
2. Drain the membrane well, wrap in Saranwrap, and expose to X-ray film in a cassette. The initial exposure is often timed for 1 min in order to assess the level of signal so that subsequent optimum exposure times can be deduced.

8.4 Oligonucleotide Probes

- Thiolmodifier (Amersham)   **Labeling**
- Freeze-dried derivatized HRP (Amersham)   **reagents**
- DNA synthesizer (ABI, Pharmacia)
- Spin columns (Sephadex G25, Pharmacia)

- Blocking agent (Amersham)   **Hybridiza-**
- Trisodium citrate (Sigma)   **tion**
- Lauroyl sarcosine, sodium salt (Sigma)   **reagents**
- Sodium dodecyl sulphate (Sigma)
- Hybond-ECL (nitrocellulose) membrane (Amersham)
- Hybond-N+ (nylon) membrane (Amersham)

- ECL detection reagent 1 (Amersham)   **Detection**
- ECL detection reagent 2 (Amersham)   **reagents**
- Saranwrap (Dow Chemical Company)
- Hyperfilm-ECL (X-ray film) (Amersham)

- Hybridization buffer: $5 \times$ SSC, 0.1% [w/v] *N*-lauroylsarcosine,   **Hybridiza-**
  0.02% [w/v] SDS, 5% [w/v] blocking agent   **tion solu-**
- Wash solution 1: $3 \times$ SSC, 0.1% SDS   **tions**
- Wash solution 2: $2 \times$ SSC

- Detection reagent − working stock: equal volume mix of detection re- **Detection**
  agents 1 and 2   **solutions**

Probe labeling

1. Synthesize an oligonucleotide bearing a thiol group at the 5′end using the protocols provided by the supplier of the thiolmodifier. Purify as specified by the supplier of the thiolmodifier; usually by reverse phase HPLC.

2. Prepare a 1 ml spin column equilibrated in deionized water. Use the column to remove the dithiothreitol (thiol group stabilizing agent) from 5 μg of thiol-linked oligonucleotide. Immediately add the freshly desalted oligonucleotide to freeze-dried derivatized HRP. Redissolve the enzyme and incubate for 1 h at room temperature. Store labeled probes in 50% glycerol at −20°C.

Hybridization

1. Prehybridize in the hybridization buffer for 15 min at 42°C. Hybridize at a probe concentration of 20 ng/ml for 1 h at 42°C.

2. Wash the membrane for 2 × 15 min in wash buffer 1 at 42°C followed by washes in wash buffer 2 for 2 × 5 min at room temperature.

Note: stringency control can be achieved by any of the usual methods including alteration in salt concentration, the inclusion of formamide or urea, or by alteration of the wash temperature. The values given are guidelines only, and the correct conditions will have to be calculated empirically.

Detection

1. Prepare sufficient working detection reagent to cover the membrane at 0.125 ml/cm^2. Place the membrane on a clean surface. Cover with detection reagent and leave at room temperature for 1 min.

2. Drain the membrane well, wrap in Saranwrap, and expose to X-ray film in a cassette. The initial exposure is often timed for 1 min in order to assess the level of signal so that subsequent optimum exposures times can be deduced.

8.5 Special Hints for Application and Troubleshooting

– This section highlights the areas of the protocols where particular attention is required. Adhere as closely as possible to the recommended protocols and the specified reagents.

Nucleic acid (long) probes

– Ensure that probe concentration is correct and that the template is dissolved in water with a NaCl concentration of less than 10 mM.

– Double-stranded DNA must be fully denatured. Do not use a heating block for this step. Do not boil labeled probes prior to adding to the hybridization buffer.

– Check that both the blocking agent and the correct amount of NaCl have been added to the hybridization buffer.

– Ensure that the temperature of both the hybridization procedure and the stringency washes does not exceed 42°C. The stringency can be con-

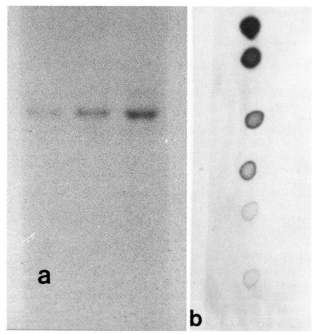

Examples of the use of directly labeled probes. **a** Single copy gene detection using a long probe. Human genomic DNA digested with *Eco*RI, blotted onto Hybond-N$^+$ (loadings of 1, 2, and 5 µg), and hybridized with 10 ng/ml of a 1.5 kbp HRP-labeled probe specific for the N-*ras* proto-oncogene overnight at 42 °C; enhanced chemiluminescence (ECL) detection; 30 min exposure **b** Dot blot analysis using an oligonucleotide probe. M13 mp8 single-stranded DNA dot, blotted onto Hybond-N$^+$ (loadings of 10, 5, 2, 1, 0.5, and 0.1 ng), and hybridized with 20 ng/ml of a specific 16 bp HRP-labeled probe at 42 °C for 1 h; ECL detection; 30 min exposure

trolled by alteration of the salt concentration of the primary wash buffer in the range $0.1 \times - 2 \times$ SSC.
- Drain the detection reagents from the membrane and see that the Saranwrap and other surfaces are kept clean.
- When rehybridizing, ensure that blocking agent is still added to the hybridization buffer.

Ensure that the oligonucleotide synthesizer is operating efficiently.   **Oligo-nucleotide probes**

- The 5'end trityl group should be removed, after HPLC purification, by treatment with DTT and silver nitrate. The oligonucleotide should be stored in DTT. However, the DTT must be removed immediately prior to labeling to achieve the optimum coupling efficiency.
- The hybridization temperature should not exceed 42 °C.

References

Amersham International (1990) Detection of nucleic acids and protein with light: ECL technical manual

Boniszewski ZAM, Comley JS, Hughes B, Read CA (1990) The use of charge-coupled devices in the quantitative evaluation of images, on photographic film or membranes, obtained following electrophoretic separation of DNA fragments. Electrophoresis 11:432−440

Connolly BA, Rider P (1985) Chemical synthesis of oligonucleotides containing a free sulphydryl group and subsequent attachment of thiol specific probes. Nucleic Acids Res 13:4485−4502

Cunningham M (1991) Nucleic acid detection with light. Life Science 6:2−5

Cunningham M, Harvey B, Benge L, Wheeler C (1991) ECL random prime system: protocol variations. Highlights 2:9−10

Durrant I (1990) Light based detection of biomolecules. Nature (London) 346:297−298

Durrant I, Benge LCA, Sturrock C, Devenish AT, Howe R, Roe S, Moore M, Scozzafava G, Proudfoot LMF, Richardson TC, McFarthing KG (1990) The application of enhanced chemiluminescence to membrane-based nucleic acid detection. BioTechniques 8:564−570

Evans MR, Benge LCA, Devenish AT, Durrant I, Fowler SJ, Harding ER, Howe R, Richardson TC, Scozzafava G, Sturrock C, Proudfoot LMF (1990) Chemiluminescence: nucleic acid detection for the future. In Proceedings of the 6th International Congress on Rapid Methods and Automation in Microbiology and Immunology, Helsinki

Fowler SJ, Harding ER, Evans MR (1990) Labeling of oligonucleotides with horseradish peroxidase and detection using enhanced chemiluminescence. Technique 2:261−267

Grimsley G, Witt C, Saveracker G, Dawkins RL (1991) HLA DRB typing of polymerase chain reaction (PCR) products using horseradish peroxidase labeled sequence specific oligonucleotides (SSOs) and enhanced chemiluminescence. Highlights 3:1−3

Hawkins E, Cumming R (1990) Enhanced chemiluminescence for tissue antigen and cellular viral DNA detection. J Histochem Cytochem 38:415−419

Hutton JR (1977) Renaturation kinetics and thermal stability of DNA in aqueous solutions of formamide and urea. Nucleic Acids Res 4:3537−3555

Lambalk JJM (1991) ECL direct system: single copy gene detection in tomato plants. Highlights 2:3−4

Leong MM, Fox GR (1988) Enhancement of luminol-based immunodot and western blotting assays by iodophenol. Analyt Biochem 172:145−150

Misiura K, Durrant I, Evans MR, Gait MJ (1990) Biotinyl and phosphotyrosinyl phosphoramidite derivatives useful in the incorporation of multiple reporter groups on synthetic oligonucleotides. Nucleic Acids Res 18:4345−4354

Moore M (1991) ECL direct system: stringency washes without the use of urea in a northern blotting application. Highlights 2:5−6

Pollard-Knight D (1990) Current methods in nonradioactive nucleic acid labeling and detection. Technique 2:113−132

Pollard-Knight D, Read CA, Downes MJ, Howard LA, Leadbetter MR, Pheby SA, McNaughton E, Syms A, Brady MAW (1990) Nonradioactive nucleic acid detection by enhanced chemiluminescence using probes directly labeled with horseradish peroxidase. Analyt Biochem 185:84−89

Renz M, Kurz C (1984) A colorimetric method for DNA hybridization. Nucleic Acids Res 12:3435−3444

Simmonds AC, Cunningham M, Durrant I, Fowler SJ, Evans MR (1991) Enhanced chemiluminescence in filter-based DNA detection. Clinical Chemistry 37:1527−1528

Sorg R, Enczmann J, Sorg U, Kogler G, Schneider EM, Wernet P (1990) Specific nonradioactive detection of PCR-amplified HIV-sequences with enhanced chemiluminescence labeling (ECL) − an alternative to conventional hybridization with radioactive isotopes. Life Science 2:3−4

Stone T, Durrant I (1992) Enhanced chemiluminescence for the detection of membrane bound nucleic acid sequences. Genetic Analysis: Techniques and Applications 8:230−237

9 A Highly Sensitive Method for Detecting Peroxidase in In Situ Hybridization or Immunohistochemical Assays

JAMES G. LAZAR and FLOYD E. TAUB

9.1 Principle and Applications

The sensitivity of peroxidase detection in tissue sections and cellular samples has been limited by nonspecific background, poor contrast of chromogen with counterstain, fading of chromogen with time or exposure to solvents, and inability to detect signals present at low levels. A highly sensitive procedure for peroxidase detection in tissue sections and cellular samples is described (Taub and Higgs 1991). The peroxidase label may be attached directly to a nucleic acid probe, avidin, streptavidin, antibodies, or antibody fragments. After reaction with the peroxidase-labeled probe or conjugate, the sample is washed, a colorimetric development solution is applied, and a precipitate forms where peroxidase is present. The initial colorimetric precipitate is then subjected to several subsequent chemical reactions that reduce background and greatly enhance the signal intensity. The resulting product is permanent, black, and insoluble in most aqueous and organic solvents. In contrast to current colorimetric detection methods' used for detection of peroxidase in in situ hybridization or immunohistochemical assays, the silver enhancement procedure for the detection of peroxidase is extremely sensitive. Moreover, high sensitivity detection allows for the use of lower probe concentrations in hybridization assays, thus improving specificity and reducing cross-reactivity. In immunohistochemical assays, a sensitive detection system allows the use of more dilute antibodies or antibody conjugates.

Highly sensitive methods for the detection of peroxidase in solution and in blot formats are currently available (Hosoda et al., 1986; Gehle and Lazar B, 1990; IBI, Inc. 1991). However, these methods are not well suited to the detection of peroxidase in situ. Detection of peroxidase in situ has traditionally been accomplished with the use of peroxidase substrates that produce colored precipitates (De Jong et al., 1985). Limitations of these substrates are low sensitivity, product solubility, color instability over time, and lack of contrast with common stains. In addition to high sensitivity, optimal colorimetric substrates for in situ analyses should give precipitates that are insoluble in both aqueous and organic solvents to allow flexibility in mounting procedures. An optimal peroxidase substrate should also given a precipitate that does not fade over time and that provides a signal with high contrast to a variety of common tissue stains.

A high contrast signal is especially important if the sample is to be photographed. Traditional peroxidase substrates give products of several colors, ranging from red to brown. The colors and signal intensity produced by these substrates may not provide sufficient contrast with a number of commonly used counterstains. The silver enhancement procedure provides a highly sensitive, colorimetric method of detecting peroxidase in situ that overcomes the major drawbacks of traditional in situ colorimetric detection methods.

Silver amplification techniques have been used to increase the intensity of peroxidase signals in a number of different systems (Gallyas et al., 1982; Newman et al., 1983; Rodriguez et al., 1984). Since these procedures are nonspecific, however, background and signal are amplified, resulting in only a marginal increase in detection sensitivity. Moreover, these reactions are often difficult to control and often require monitoring of the progress of the reaction to achieve optimals results. The method described below includes a step that has been found to drastically reduce background without reducing specific signal. The overall procedure is easily controlled, does not require careful monitoring, and gives uniform and reliable results.

In an in situ hybridization or immunohistochemical assay, peroxidase labeled probe or conjugate is reacted with an appropriate tissue or cellular sample. After an incubation period, the sample is washed extensively to remove excess probe or conjugate. The first step of the described method for sensitive detection of peroxidase is the peroxidase catalyzed oxidation of 3,3′-diaminobenzidene (DAB). The product of this reaction is a brown polymeric precipitate (Josephy et al., 1983). The DAB precipitate can be further oxidized by a number of reagents including ferricyanide, Ni(II), and Co(II) (Lazar 1991; Scopsi and Larsson 1986). The two oxidation steps can occur either simultaneously or in sequence. The product of DAB peroxidation in the presence of Ni(II) or Co(II) is a dark purple-black precipitate containing metallic nickel or cobalt. If the steps are performed in sequence, the initial light brown DAB precipitate immediately turns dark purple-black upon exposure to a solution of Ni(II) or Co(II) as a result of the precipitation of metallic nickel or cobalt. The sample is rinsed thoroughly to remove unreacted DAB and metals prior to the next step.

The next step of the procedure is critical for minimizing background. The sample is washed in a low pH buffer to remove small particles of DAB reaction product that would lead to high background if amplified. A potassium phthalate buffer, pH 2.2, has been found to be most effective (Taub and Higgs 1991). The phthalate wash also removes a portion of the DAB-nickel precipitate that constitutes real signal. However, the final signal strength and background levels can be optimized by adjusting the incubation time in the phthalate wash buffer.

Following the phthalate wash, the sample is rinsed and exposed to a solution of hydrogen tetrachloroaurate ($HAuCl_4$). The remaining DAB-nickel precipitate reacts with $AuCl_4^-$ and precipitates metallic gold onto the DAB-nickel complex. The sample must be rinsed thoroughly at this stage as any remaining AuCl$_4^-$ will react with the subsequent silver amplification reagent.

The final step of the method is silver amplification, the deposition of metallic silver onto the DAB-nickel-gold complex. The silver amplification solution, containing silver acetate, hydroquinone, and gum arabic, is prepared just before use by mixing two stock solutions (Taub and Higgs 1991). Silver acetate is used as the silver source rather than silver nitrate or silver chloride because it is much less light sensitive. Hydroquinone is used as a hydrogen donor to reduce Ag(l) to metallic silver. Gum arabic is used to stabilize the reaction. When the sample contacts the silver amplification solution, the gold surface on the DAB-nickel precipitate catalyzes the reduction of silver by hydroquinone, resulting in the deposition of metallic silver onto the DAB-nickel-gold complex. The reaction is autocatalytic, so that the initial deposit of silver provides an even better catalyst for the reaction. The amount of precipitate and the signal intensity are controlled by the length of time that the sample is exposed to the silver solution. The reaction is stopped by rinsing the sample in deionized water or photographic fixative. The final reaction product is a dense black precipitate. The sample may now be stained and mounted with virtually any stain and mounting medium. The dark silver precipitate provides an intense, long-lasting, high-contrast signal.

The DAB/silver enhancement method for detecting peroxidase has been used in a variety of systems and applications. Its high sensitivity has made it the system of choice when the target analyte is present in low amount, highly specific detection is required, or the amount of detection reagent available is small.

The DAB/silver enhancement system was used for the type-specific detection of human papillomavirus in formalin-fixed, paraffin-embedded sections using DNA probes directly labeled with horseradish peroxidase (HRP) (Higgs et al., 1990; Park et al., 1991). In comparison to a biotinylated DNA probe/streptavidin-alkaline phosphatase system, the HRP system was found to be superior due to the higher levels of signal achieved in sections containing very low levels of target sequences. The DAB/silver enhancement procedure has also been used for the in situ detection of *Leishmania* parasites in sandflys (Schoone et al., 1991). The use of the DAB/silver enhancement procedure with DNA probes directly labeled with HRP provided a rapid and specific assay for various *Leishmania* taxa and gave good sensitivity and low background. The DAB/silver enhancement procedure has also been used for the detection of specific RNA transcripts using peroxidase-labeled cDNA probes (McClintock et al., 1992). The usefulness of the enhancement procedure is not limited to hybridization assays. Preliminary experiments have shown excellent reults with a biotin/streptavidin-peroxidase-system and in immunohistochemical assays utilizing peroxidase − labeled antibodies. The enhancement procedure has been used successfully in in situ assays on tissue sections, smears, ctyospins, and cell culture monolayers. The DAB/silver enhancement procedure is a general method for the sensitive detection of peroxidase in in situ assays.

9.2 Reaction Scheme

The procedure for the sensitive in situ detection of peroxidase is shown graphically in the accompanying figure. The procedure is separated into four parts:

1. Peroxidase-catalyzed oxidation of DAB and Ni(II) producing a purple-black precipitate
2. Background and signal reduction with the phthalate wash buffer
3. Reaction of the DAB-nickel complex with $AuCl_4^-$ producing a gold precipitate
4. Autocatalytic reduction of silver acetate to metallic silver by hydroquinone; catalyzation of the initial reaction by the gold precipitate.

9.3 In Situ Detection of Peroxidase

Reagents and buffers
- $40\times$ DAB: 10 mg/ml in deionized H_2O
- DAB buffer: 25 mM Tris, 125 mM NaCl, 2 mM $NiCl_2$, pH 7.6
- 3% [v/v] hydrogen peroxide
- Phthalate buffer: 25 mM potassium hydrogen phthalate, pH 2.2
- $20\times$ Gold solution: 3 mM hydrogen tetrachloroaurate hydrate
- Silver solution I: 12 mM silver acetate
- Silver solution II: 4.5 mM hydroquinone, 50 mM citrate, 1.8% [w/v] gum arabic, 10% [v/v] ethanol

Reagent and buffer preparation
In general, all solutions and reagents should be made up without metal utensils, as metal contaminants can cause high background in the silver amplification step.

- $40\times$ DAB stock solutions (25 ml):
  Dissolve 250 mg of 3,3'-diaminobenzidene tetrachloride dihydrate in 25 ml of deionized water. Separate into 5.0 ml aliquots and store frozen at $-20\,°C$. The frozen DAB stocks should be stable for 1 year at $-20\,°C$.
- DAB reaction buffer (1000 ml):
  Dissolve 3.03 g Tris base; 7.3 g sodium chloride; 0.475 g nickel(II) chloride hexahydrate in 950 ml deionized water. Adjust pH to 7.6 with 1 M HCl. Adjust volume to 1 l with deionized water. Filter sterilize. Stable for 1 year at room temperature.
- 3% [v/v] Hydrogen peroxide:
  Dilute 30% [v/v] hydrogen peroxide to 3% [v/v] in deionized water just prior to use.
- Phthalate wash buffer (1000 ml):
  Add 5.1 g potassium hydrogen phthalate to 950 ml deionized water. Adjust pH to 2.2 with 12 N HCl. Adjust volume to 1 l with deionized water. Filter sterilize. Stable for 1 year at room temperature.

I. Detection of HRP label with diaminobenzidene and Ni(II)

$$DAB \ + \ Ni(II) \ + \ H_2O_2 \ \xrightarrow{\text{Peroxidase}} \ \text{DAB-Ni complex (dark purple)}$$

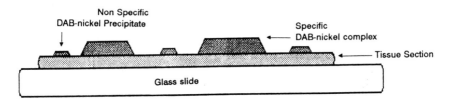

II. Phthalate wash to reduce background

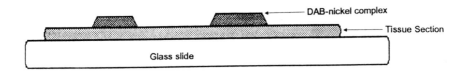

III. Catalytic deposition of metallic gold

$$AuCl_4^- \ + \ DAB\text{-}Ni \ \xrightarrow{\text{DAB-Ni}} \ Au\text{-}DAB\text{-}Ni$$

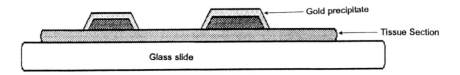

IV. Silver Enhancement

$$2 \ Ag(I) \ + \ \text{hydroquinone} \ \xrightarrow{\text{Au/Ag}} \ 2 \ Ag \ + \ \text{benzoquinone}$$

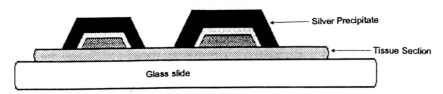

Reaction scheme for the highly sensitive detection of peroxidase

9.3

- 20× Gold solution (100 ml):
  Dissolve 100 mg of hydrogen tetrachloroaurate hydrate in 100 ml of deionized water. Store at 4°C. Stable for 1 year at 4°C.
- Silver solution I (500 ml):
  Dissolve 1 g of silver acetate in 500 ml deionized water. Store in opaque amber bottle at 4°C. Keep from light. Stable for 6 months at 4°C.
- Silver solution II (500 ml):
  Add 9 g of gum arabic to 50 ml of absolute ethanol. Stir vigorously until the gum arabic is a homogeneous slurry in the ethanol (no clumps). Add 425 ml deionized water and stir until the gum arabic has dissolved. Add 3.28 g citric acid (monohydrate) and 2.38 g sodium citrate (dihydrate), and 2.5 g hydroquinone. Stir until dissolved, bring volume to 500 ml. Store in an opaque amber bottle at 4°C. Stable for 6 months at 4°C.

DAB stain
1. Wash sections thoroughly to remove excess probe or peroxidase conjugate. Thaw one 5 ml aliquot of DAB stock solution. Shake DAB reaction buffer, and pour 200 ml into an opaque staining dish. Empty the aliquot of DAB stock solution into the DAB reaction buffer and stir until mixed. Add 100 μl of 3% [v/v] hydrogen peroxide and stir until mixed. This solution must be used within 15 min.
2. Place slide in the DAB stain, cover container, and incubate at room temperature for 15 min.
3. Stop color development with four, consecutive, 1 min, deionized water rinses. Agitate the slides in each rinse step.

Note: DAB stain may now have a beige or light brown color. This color is normal and is caused by exposure to air. Dispose of DAB stain in accordance with laboratory and government guidelines.

Phthalate wash buffer
1. Transfer the slides to a fresh container containing 200 ml of the phthalate wash buffer. Dip the slides into the solution several times, and then let them incubate in the solution for 10 min.
2. Rinse the slides in two changes of deionized water.

Gold deposition
1. Dilute the gold solution to the working concentration by adding 10 ml of the concentrate to 190 ml of deionized water in a fresh staining dish.
2. Transfer the slides from deionized water to the 1X gold solution. Incubate the slides for 6 min at room temperature.
3. Remove the slides from the gold solution and rinse in deionized water. Then wash the slides in three consecutive 3 min deionized water washes.

Silver amplification
Staining dishes for silver enhancement must be scrupulously clean.
1. *Immediately before use,* remove silver solutions I and II from 4°C. Pour 100 ml of each solution into a precleaned opaque staining container and mix well. *Do not cross-contaminate the remaining silver solutions.* Return the stock silver solutions to 4°C.

2. Immerse the slides in the silver enhancement solution, cover, and incubate for 8 min. The silver reaction is mildly light sensitive. Although the procedure can be done on the benchtop, the solutions should be protected from direct sunlight or bright lights.
3. Rinse the slides in two changes of deionized water.

The silver-enhanced DAB precipitate is opaque and black in color. It is **Counter-** insoluble in water, and most organic solvents. Therefore, almost any coun- **stain** terstain can be used. In our laboratory, we routinely counterstain with nuclear fast red.

Dehydrate and mount by standard laboratory practice or as follows: **Dehyd-**
— 70% ethanol (20 dips) **ration and**
— 95% ethanol (20 dips) **mounting**
— 100% ethanol (20 dips in each of two baths)
— 100% xylene (20 dips in each of two baths)
Coverslip with Pro-Texx or Permount

Note: Do not use toluene to clear, as signal fading may occur.

9.4 Special Hints for Application and Troubleshooting

Low sensitivity and/or high background are potential pitfalls of in situ assays. These difficulties may be caused by the assay conditions and/or the silver amplification procedure. Thus, optimization of signal and background should include adjustment to both the assay conditions and the silver amplification procedure.

— Increase probe/conjugate concentration. **Low sen-**
— Increase hybridization/antibody reaction time. **sitivity**
— Decrease incubation time in phthalate wash buffer.
— Increase incubation time in silver enhancement solution.

— Decrease probe/conjugate concentration. **High**
— Be sure that post-probe or post-conjugate washes are effective in **background** removing all unbound probe/conjugate.
— Increase incubation time in phthalate buffer.
— Increase frequency and length of deionized water rinses between all color development steps.
— Keep all utensils and staining dishes scrupulously clean.
— Decrease incubation time in silver enhancement solution.
— DAB stain was allowed to stand for more than 15 min before use.
— Silver stain was allowed to warm above 8 °C before use. Mix cold silver solutions just before use. If necessary, keep the silver enhancement solution on ice during use.

References

De Jong ASH, van Kessel-van Vark M, and Raap AK (1985) Sensitivity of various visualization methods for peroxidase and alkaline phosphatase activity in immunoenzyme histochemistry. Histochemical Journal 17:1119−1130

Gallyas F, Gorcs T, and Merchenthaler I (1982) High-grade intensification of the end-product of the diaminobenzidene reaction for peroxidase. Histochemistry 30(2): 183−184

Gehle WD and Lazar BS (1990) An enhanced-response system for performing chemiluminescent and bioluminescent tests. American Biotechnology Laboratory, March 1990

Higgs TE, Moore NJ, Badawi DY, and Taub FE (1990) Type-specific human papillomavirus detection in formalin-fixed, paraffin-embedded tissue sections using non-radioactive deoxyribonucleic acid probes. Methods in Laboratory Investigation 63(4):557−567

Hosoda H, Takasaki W, Oe T, Tsukamoto R and Nambara T (1986) A comparison of chromogenic substrates for horseradish peroxidase as a label in steroid enzyme immunoassay. Chem Pharm Bull 34(10):4177−4182

IBI Incorporated, Enzygraphic Web for Colorimetric Detection of Peroxidase Linked Probes. Product Insert

Josephy PD, Eling TE, and Mason RP (1983) Co-oxidation of benzidene by prostaglandin synthase and comparison with the action of horseradish peroxidase. Journal of Biological Chemistry 258(9):5561−5569

Lazar JG (1991) Unpublished results: The brown peroxidation product of diaminobenzidene gives a positive result with Schmorl's ferric/ferricyanide reduction test (See: Sheehan, D.C., and Hrapchak, B. BH., *Theory and practice of Histotechnology,* C. V. Mosby, St. Louis, MO (1980), page 223)

McClintock JT, Chan I-J, Thaker SR, Katial A, Taub FE, Aotaki-Keen AE, and Hjelmeland LM (1992) Detection of c-sis proto-oncogene transcripts by direct enzyme-labeled cDNA probes and in situ hybridization. In Vitro Cellular and Developmental Biology 28A:102−108

Newman GR, Jasani B, and Williams ED (1983) The visualization of trace amounts of diaminobenzidene (DAB) polymer by a novel gold-sulphide-silver method. J Microsc 132:RP1−RP2

Park JS, Kurman RJ, Kessis TD, and Shah KV (1991) Comparison of peroxidase-labeled DNA probes with radioactive RNA probes for detection of human papillomavirus by in situ hybridization in paraffin sections. Modern Pathology 4(1):81−85

Ridriguez EM, Yulis R, Peruzzo B, Alvial G, and Andrade R (1984) Standardization of various applications of methacrylate embedding and silver methenamine for light and electron microscopy immunocytochemistry. Histochemistry 81:253−263

Schoone GJ, van Eys GJ, Ligthart GS, Taub FE, Zaal J, Mebrahtu Y, and Lawyer P (1991) Detection and identification of *Leishmania* parasites by in situ hybridization with total and recombinant DNA probes. Experimental Parasitology 73(3):345−353

Scopsi L and Larsson L-I (1986) Increased sensitivity in peroxidase immunocytochemistry: A comparative study of a number of peroxidase visualization methods employing a model system. Journal of Histochemistry 84:221−230

Taub FE and Higgs TH (1991) A Highly Sensitive Method for Detecting Peroxidase, US Patent number 5,116,734, issued May 26, 1992

10 The SNAP System

JAMES E. MARICH and JERRY L. RUTH

10.1 Principle and Applications

Since their advent, the use of synthetic nucleic acid probes for detection of nucleotide sequences has offered specificity and handling advantages over cloned DNA probes. Highly specific oligonucleotides can distinguish between sequences with as few as one nucleotide difference (Wallace et al., 1979). Automated chemistry has enabled the routine production of high purity oligonucleotides using phosphoamidite chemistry. The development of SNAP (synthetic nucleid acid probe) technology has provided a safe and convenient method for labeling these oligomer probes and detecting their targets (Ruth et al., 1985). The SNAP system uses synthetic oligonucleotides which have been covalently coupled to a nonisotopic reporter molecule, most typically alkaline phosphatase (AP). Addition of AP substrates to these conjugated oligomers, AP probes, results in the enzymatic production of a detectable signal (Jablonski et al., 1986; Ruth and Jablonski, 1987). Similar to unconjugated oligomers, AP-labeled probes show the same high degree of specificity achievable with radioactively labeled oligomer probes (Podell et al., 1991). Unlike isotopically labeled probes, these enzyme-labeled probes are more sensitive and show remarkable stability.

SNAP probes are used in a variety of hybridization applications. As with other enzyme labeled probes, detection may be accomplished by using substrates which generate precipitating, fluorescent or chemilumi-

Functionalized thymidine base for incorporation of amines into synthetic oligonucleotides. (DMT = dimethoxytrityl; TFA = trifluoroacetyl; $EtCN$ = cyanoethyl protecting group)

C-5 Amino-dThd

---

* SNAP is a registered trademark of Syngene, Inc.

10.2

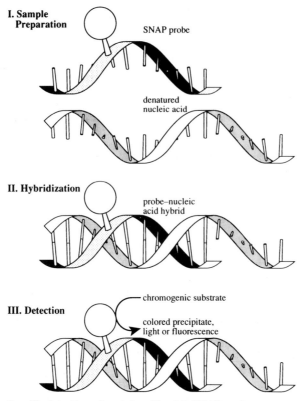

I. Sample Preparation

SNAP probe

denatured nucleic acid

II. Hybridization

probe–nucleic acid hybrid

chromogenic substrate

III. Detection

colored precipitate, light or fluorescence

Specific detection of nucleic acids with SNAP probes

nescent products. These detection methodologies are used in blotted filter hybridizations (dot-, slot-, Centri-Dot[1]- and Southern blots), in situ (viral DNA/RNA and somatic mRNA) and in solution hybridization methods (sandwich and direct capture). Because each detection system is best suited for specific applications, selection of format depends on the type of information required. The following is a discussion of applications which incorporate these various detection methodologies.

10.2 Colorimetric Detection of *Mycobacteria* on Filters

Detection and identification of *Mycobacteria* species is accomplished through a colorimetric assay of nucleic acids bound to nylon membranes, utilizing the SNAP *Mycobacteria* culture identification kit (Syngene, Inc.) (Lim et al., 1991; Huang and Jungkind, 1991). Partially purified nucleic acids extracted from lysed cultures are applied directly to membranes by

---

[1] Centri-Dot is a trademark of Syngene, Inc.

means of a Centri-Dot (a six-chamber device for blotting by centrifuga-tion). Nucleic acids from cell lysates are bound to nylon by centrifugation through the Centri-Dot containing the membrane. This membrane is sub-sequently removed and hybridized with SNAP probes specific for *M. tuberculosis, M. avium,* or *M. intracellulare.*

1. *Lysis:* Add 1 ml SNAP *Mycobacterium* lysis buffer, 300 μl chloroform, and bacterial colonies ($\sim$5 $\times$ 10^7 – 5 $\times$ 10^9 cells, 1 to several colonies depending on size) to a lysis tube containing lysis beads. Agitate at high speed for 3 min in Minibead Beater (Biospec Products, Bartlesville, OK, USA). Centrifuge lysis tube for 5 min at 1000–3000 g.
2. *Denaturation:* Add 300 μl 0.5 N NaOH and 300 μl of supernatant from lysis tube to Centri-Dot. Centrifuge Centri-Dot 5 min at 1000 g. Discard filtrate and remove membrane from Centri-Dot. Rinse membrane in deionized H$_2$O for 15 s. Blot excess liquid.
3. *Hybridization:* Add 1.0 ml of SNAP *Mycobacterium* hybridization buf-fer, 30 μl SNAP *Mycobacterium* probe *(M. tuberculosis, M. avium,* or *M. intracellulare)* and Centri-Dot membrane to sealable reaction bag. Mix, express air bubbles, and seal. Submerge reaction bag in 49±1°C water bath and incubate for 15 min.
4. *Washing:* Transfer membrane to a vessel containing 200 ml HI TEMP membrane wash buffer 1 preheated to 49±1°C. Incubate 15 min 49±1°C. Transfer membrane to 200 ml of preheated HI TEMP mem-brane wash buffer 2. Incubate 15 min, 49±1°C.
5. *Detection:* Add 1 ml of alkaline phosphatase substrate buffer, 5 μl nitro-blue tetrazolium salt (NBT) substrate and 5 μl 5-bromo-4-chloro-3-indolyl phosphate (BCIP) substrate to a reaction bag and mix. Blot excess liquid from membrane and add to reaction bag. Seal and incubate at 37°C in water bath for 1 h in the dark. Rinse membrane in deionized H$_2$O, 15 s. Interpret results. Purple color on membrane in sample wells indicates positive results. Maximum sensitivity is 2–5 $\times$ 10^6 molecules (3–8 $\times$ 10^{-18} mol) of target nucleic acid.

10.3 Colorimetric Detection of Human Papillomavirus In Situ

In situ detection of human papillomavirus (HPV) nucleic acids is accomplished by fixing cellular material directly on glass slides, treating cellular membranes to render them permeable to reagents, and denaturing nucleic acids within the cell matrix. Hybridization and detection take place directly on the slide. SNAP probes are added in concentrations ranging from 0.1 to 10 nM depending on hybridization time. Unbound excess probe is removed by washing. Addition of BCIP and NBT results in the deposit of a purple stain on cellular features containing homologous nuc-leic acids. SNAP probes have been successfully used to detect low-abundance mRNA (Kiyama et al., 1990; Kiyama and Emson, 1990). Compared to radiolabeled probes, this method results in less background, greater cellular resolution, and shorter detection times (Baldino et al., 1989). The following is a method for the in situ detection of HPV nucleic acid in cervical smears.

1. *Sample preparation:* Apply cell smear to slide. Dehydrate in 100% MeOH, 10 min, room temperature. Air dry.
2. *Denaturation:* Cover slide with 70% formamide, 1× SSC, 0.5% [w/v] bovine serum albumin (BSA). Heat to 95 °C for 10 min. Rinse in 1× SSC, 1 min.
3. *Hybridization:* Incubate in 5× SSC; 0.5% [w/v] SDS; 0.5% [w/v] BSA, 120 min, 50 °C with 5 nM specific SNAP probe. Wash 2 × 6 min, 45 °C in 1× SSC, 1% [w/v] SDS. Wash 2 × 6 min, 45 °C in 1× SSC; 0.5% [v/v] Triton X-100. Wash 2 × 6 min, 45 °C in 1× SSC.
4. *Detection:* Incubate in 0.1 M Tris-HCl (9.5), 0.1 M NaCl, 50 mM $MgCl_2$, 15% polyvinyl alcohol, 0.33 mg/ml NBT, 0.16 mg/ml BCIP, 2 h, 37 °C. Rinse with H_2O, 2 min, room temperature. Counterstain in 1% eosin B, 4 min, room temperature. Mount and examine slides under microscope for purple staining. Maximum sensitivity is 20−50 copies of target nucleic acid per cell.

10.4 Chemiluminescent Detection of Human DNA Fingerprints

Human DNA fingerprinting is used chiefly in two applications, paternity and identity testing. Identification of inherited alleles of specific genetic loci is used to determine or exclude paternity. Comparison of allelic patterns between tissues and individuals is used in forensics to confirm or preclude identity. SNAP DNA fingerprinting probes are used to detect variable number tandem repeat alleles (VNTRs) from single polymorphic genetic loci (Nakamura et al., 1987; Odelberg et al., 1989). Allelic frequencies for VNTR loci can be established, enabling statistical analysis of electrophoretic profiles (Baird et al., 1986; Balazs et al., 1989). The following analysis uses probes from two human loci, D2S44 (YNH24) and D10S28 (TBQ7), to create a DNA fingerprint profile.

1. *Restriction enzyme digestion of genomic DNA:* Digest 10 µg of purified genomic DNA with desired restriction enzyme, for example *Hae* III, at 3 units enzyme/µg DNA for a minimum of 2 h to overnight. Add 1/10 volume 10× gel loading buffer and heat at 65 °C for 3 min. Load digested DNA into a 1% agarose gel, 16 cm long.
2. *Electrophoresis:* Electrophorese the DNA for 20 min at 80 V in 40 mM Tris base; 20 mM NaAc; 17 mM NaCl; 2 mM EDTA; pH 8.0 until the DNA has entered the gel. Continue at 30 V for 12−16 h. Recirculate buffer.

3. *Southern transfer:* Soak gel in 0.1 N HCl for 10 min to depurinate DNA. Soak gel 2 × 15 min in 0.5 N NaOH; 1.5 M NaCl to denature. Neutralize gel 2 × 15 min in 1.5 M NaCl; 1 M Tris-HCl, pH 7.0. Cut uncharged nylon membrane to size, i.e., MagnaGraph (Micro Separations, Inc., Westborough, MA, USA). Prewet membrane in deionized H_2O. Equilibrate in 10× SSC for 10 min. Rinse gel in deionized H_2O and set up Southern blot. Transfer overnight. Fix DNA to membrane by baking for 30 min at 80 °C or UV crosslink ($120000\,\mu J/cm^2$, 30 s).

4. *Blocking:* Slowly wick the fixed membrane in freshly prepared 0.2% casein, 0.5× SSC ($1\,ml/cm^2$). Incubate for 30 min at room temperature. Blot excess liquid and transfer to hybridization bag.

5. *Hybridization:* Add 1 nM SNAP human locus probe, i.e., SNAP D2S44 (YNH24) or D10S28 (TBQ7), in hybridization buffer ($50\,\mu l/cm^2$). Incubate for 45 min at 60 ± 1 °C.

6. *Washing:* Wash 10 min in 1% [w/v] SDS; 1× SSC at 60 ± 1 °C, $1.5\,ml/cm^2$. Wash 10 min in 1% [v/v] Triton X-100, 1× SSC at 60 ± 1 °C, $1.5\,ml/cm^2$. Wash 10 min in 1× SSC at room temperature, $1.5\,ml/cm^2$. Wash 10 min in Tropix assay buffer (Tropix, Inc., Bedford, MA, USA), room temperature. Blot excess liquid. Do not allow to dry.

7. *Detection:* Insert membrane in sealable plastic bag or folder. Add $10\,\mu l/cm^2$, 0.26 mM 3-(4-methoxyspiro [1,2-dioxetane-3,2'-tricyclo $[3.3.1.1^{3,7}]$ decan]-4-yl) phenyl phosphate (AMPPD) in Tropix assay buffer. Express bubbles and seal detection bag. Expose bag to X-ray film in light-tight cassette for 1 h to overnight. Develop X-ray film and interpret results. Maximum sensitivity is $5-10 \times 10^4$ molecules ($8-15 \times 10^{-20}$ mol) of target nucleic acid or 30−60 ng human genomic DNA.

8. *Reprobing membrane:* Heat 0.1% [w/v] SDS to 65 °C. Pour solution over membrane and agitate for 45 min at 65 °C. Wash 5 min in 1× SSC at room temperature. Repeat hybridization, beginning with blocking as described above.

10.5 Fluorescence Detection of Human Immunodeficiency Virus Using Sandwich Hybridization

Detection of nucleic acid sequences by sandwich hybridization technology can offer several advantages over filter and in situ based assays. These previously mentioned methods require manual manipulation for either pretreating the sample or performing the basic assay steps. Sample pretreatment is required when interfering substances disrupt the assay or cause high backgrounds. Sandwich hybridization techniques can generally use samples with minimal pretreatment (Nicholls and Malcolm, 1989). Hybridization is performed in solution followed by capture of hybrids on a solid support and washing. Sample impurities and excess probe are eliminated. Sandwich hybridization with SNAP probes is readily automated, eliminating manual processing. For analytes which are present in low numbers in patient samples, i.e., many viruses, the polymerase chain reaction (PCR) may first be performed, followed by sandwich hybridization. Capture of PCR amplicons is performed with probes directly linked to any of several types of solid supports: beads (Albretsen et al., 1990), microtiter plates (Keller et al., 1990), or membranes (Ranki et al., 1983). After hybridization, signal generated by SNAP probes is quantitated in a fluorimeter. Presented here is a method for detecting human immunodeficiency virus (HIV) DNA in blood using PCR, sandwich hybridization, and SNAP probes.

1. *Sample preparation:* Isolate and PCR amplify nucleic acid (Ou et al., 1988; Sninsky and Kwok, 1990).
2. *Denaturation:* Heat nucleic acid to 95°C for 10 min in 20 µl 10 mM Tris-HCl, pH 7.5; 1 mM EDTA (7.0); 15 nM biotin-labeled capture oligonucleotide, followed by chilling in an ice bath for 2 min.
3. *Hybridization:* Add 100 µl hybridization buffer (10% deionized formamide, 6× SSC; 0.05% [w/v] SDS; 0.05% [w/v] sodium lauryl sarkosyl; 0.1 mg/ml BSA; 3.0 nM SNAP HIV probe: pH 8.0. Heat to 37°C for 30 min with agitation.
4. *Capture:* Add 15 µg avidinylated magnetic beads (Advanced Magnetics, Inc., Cambridge, MA, USA), 1 mg/ml in hybridization buffer. Incubate at 37°C for 15 min with agitation. Clear solution with magnet and aspirate supernatant.
5. *Wash:* Wash three times in 200 µl 4× SSC; 0.1% [w/v] SDS, 37°C, 5 min with agitation. Clear solution with magnet and aspirate supernatant after each wash.
6. *Detection:* Resuspend beads in 150 µl 30 µM 4-methylumbelliferyl phosphate; 5 mM MgCl$_2$; 100 mM diethanolamine; pH 9.1. Incubate 60 min at 37°C with agitation. Transfer 110 µl of solution to microtiter plate. Stop reaction with 40 µl 100 mM EDTA. Detect reaction product in fluorescence plate reader (excitation 363 nm, emission 447 nm). Maximum sensitivity is 2×10^5 molecules (3×10^{-19} mol) or target nucleic acid generated by PCR.

References

Albretsen C, Kalland K-H, Haukanes B-I, Håvarstein L-S, Kleppe K (1990) Applications of magnetic beads with covalently attached oligonucleotides in hybridization: isolation and detection of specific measles virus mRNA from crude cell lysate. Anal Biochem 189:40–50

Baird M, Balazs I, Giusti A, Miyazaki L, Nicholas L, Wexler K, Kanter E, Glassberg J, Allen F, Rubinstein P, Sussman L (1986) Allele frequency distribution of two highly polymorphic DNA sequences in three ethnic groups and its application to the determination of paternity. Am J Hum Genet 39:489–501

Balazs I, Baird M, Clyne M, Meade E (1989) Human population genetic studies of five hypervariable DNA loci. Am J Hum Genet 44:182–190

Baldino FB, Ruth JL, Davis LG (1989) Nonradioactive detection of vasopressin mRNA with in situ hybridization histochemistry. Exp Neuro 104:200–207

Huang CH, Jungkind DL (1991) Nonradioactive DNA probe for the rapid identification of Mycobacterium avium complex from clinical specimens. Mol Cell Probes 5:277−280

Jablonski E, Moomaw EW, Tullis RH, Ruth JL (1986) Preparation of oligodeoxy-nucleotide-alkaline phosphatase conjugates and their use as hybridization probes. Nucleic Acids Res 14:6115−6128

Keller GH, Huang D-P, Shih JW-K, Manak MM (1990) Detection of hepatitis B virus DNA in serum by polymerase chain reaction amplification and microtiter sandwich hybridization. J Clin Micro 28:1411−1416

Kiyama H, Emson PC (1990) Distribution of somatostatin messenger RNA in the rat nervous system as visualized by a novel nonradioactive in situ hybridization histo-chemistry procedure. Neuroscience 38:223−244

Kiyama H, Emson PC, Ruth JL (1990) Distribution of tyrosine hydroxylase mRNA in the rat central nervous system visualized by alkaline phosphatase in situ hybvridization histochemistry. Euro J Neuro 2:512−524

Lim SD, Todd J, Lopez J, Ford E, Janda JM (1991) Genotypic identification of pathogenic Mycobacterium species by using a nonradioactive oligonucleotide probe. J Clin Micro, 29:1276−1278

Nakamura Y, Leppert M, O'Connell P, Wolff R, Holm T, Culver M, Martin C, Fujimoto E, Hoff M, Kumlin E, White R (1987) Variable number of tandem repeat (VNTR) markers for human gene mapping. Science 235:1616−1622

Nicholls PJ, Malcolm ADB (1989) Nucleic acid analysis by sandwich hybridization. J Clin Lab Anal 3:122−135

Odelberg SJ, Plaetke R, Eldridge JR, Ballard L, O'Connell P, Nakamura Y, Leppert M, Lalouel J-M, White R (1989) Characterization of eight VNTR loci by agarose gel elec-trophoresis. Genomics 5:915−924

Ou C-Y, Kwok S, Michell SW, Mack DH, Sninsky JJ, Krebs JW, Feorino P, Warfield D, Schochetman G (1988) DNA amplification for direct detection of HIV-1 in DNA of peripheral blood mononuclear cells. Science 239:295−297

Podell S, Maske W, Ibañez E, Jablonski E (1991) Comparison of solution hybridization efficiencies using alkaline phosphatase-labeled and ^{32}P-labeled oligodeoxynucleotide probes. Mol Cell Probes, 5:117−124

Ranki M, Palva A, Virtanen M, Laaksonen M, Söderlund H (1983) Sandwich hybridiza-tion as a convenient method for the detection of nucleic acids in crude samples. Gene 21:77−85

Ruth JL, Jablonski E (1987) Synthesis and hybridization characteristics of oligodeoxynucleotide-alkaline phosphatase conjugates, Nucleosides & Nucleotides 6:541−542

Ruth JR, Morgan C, Pasko A (1985) Linker arm nucleotide analogs useful in oligo-nucleotide synthesis, DNA 4:93

Sninsky JJ, Kwok S (1990) Detection of human immunodeficiency viruses by the polymerase chain reaction. Arch Pathol Lab Med 114:259−262

Wallace RB, Shaffer J, Murphy RF, Bonner J, Hirose T, Itakura K (1979) Hybridization of synthetic oligodeoxyribonucleotides to ϕ_χ 174 DNA: the effect of a single base pair mismatch. Nucleic Acids Res 6:3543−3557

II Specialized Nonradioactive Detection Systems

11 Overview of Colorimetric, Chemiluminometric, and Fluorimetric Detection Systems

H.-J. GUDER and H.-P. JOSEL

11.1 Principle and Applications

A large number of detection systems for biomolecules have been described in the literature; therefore, only a brief overview of the most important nonradioactive methods will be given here. More details can be found in the other chapters of this book or in the cited literature.

For detection and quantitation of biomolecules, e.g., nucleic acids in blotting systems or haptens and proteins in immunoassays, a detection step, in which the antibody must be attached to one or more labeling moieties, follows the specific recognition reaction.

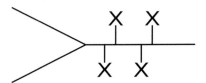

(X = marker enzyme or direct label)

One of the most frequently used methods for labeling and detection of biomolecules is conjugation with marker enzymes (alkaline phosphatase, β-galactosidase, horseradish peroxidase) and subsequent detection with a chromogenic substrate. This procedure has the advantage that even weak signals can be detected with the help of enzymatic amplification.

In the last few years new techniques have been developed for labeling biomolecules directly with certain highly sensitive detection systems that do not need the amplification effect of an enzyme to reach the same detection limits.

Detection in both cases can be accomplished by a variety of measurement techniques, depending on the nature of the enzyme substrate and products or the type of direct label. Methods for the analytical detection of biomolecules are based on the determination of, e.g., colorimetric, fluorimetric, luminometric, potentiometric, or amperometric properties. The most frequently used methods for the detection of biomolecules are still the above mentioned spectrophysical ones, but electrophysical methods are finding increasing importance, especially in sensor applications (Cammann, 1991).

The following overview deals predominantly with spectrophysical detection by colorimetry, fluorescence, and luminescence. In these techniques a detectable compound is formed from a substrate with different spectral properties, e.g., a red dye from a yellow enzyme substrate. The detectable species can be a direct product of the enzyme reaction or it can be formed by a second additional reaction.

Application of the following detection principles depends on the assay format and its measurment possibilities and on the marker enzyme, i.e., alkaline phosphatase, β-galactosidase, or horseradish peroxidase. A photometric or fluorimetric test in a cuvette with instrumental detection has different requirements than a visual colorimetric test on a blot.

For visual detection brilliant dyes with intens red or blue color are required. Ideally they should be created from colorless substrates so that no mixed colors or colored backgrounds occur to disturb the visual interpretation of test results (see Sects. 12.1 and 12.2).

In the case of instrumental detection there are fewer such problems because the photometer can of course differentiate between narrowly separated absorptions.

Another problem in identifying biomolecules can be interference from the biological material, e.g., the color of hemoglobin or the fluorescence of bilirubin. In these cases special dyes with distinct spectral properties are recommended (see Sect. 14.1).

Further aspects to consider are the water solubility of the substrate or the chromogen. For a test measured in a tube or cuvette, a sifficiently high solubility of both compounds is necessary, whereas for matrix bound tests (blots, test strips) the chromogen should precipitate quantitatively to avoid bleeding of the dye (see Sects. 12.1 and 12.2). Nonetheless, for most applications, appropriate substrates exist.

11.2 Substrates for Hydrolases

11.2.1 Direct Substrates

These compounds form a detectable species directly by an enzymatic reaction without a second reaction step or reagent. To this group belong substrates such as:

- Resorufin-β-D-galactopyranoside
- Chlorophenolred-β-D-galactopyranoside
- 2-Nitrophenyl-β-D-galactopyranoside
- 4-Nitrophenyl-β-D-galactopyranoside
- 2-Chloro-4-nitrophenyl-β-D-galactopyranoside
- 4-Methylumbelliferyl-β-D-galactopyranoside

- 4-Nitrophenyl-phosphate
- 4-Methylumbelliferyl-phosphate

All of these substrates release water soluble dyes and are therefore suitable for chromogenic or fluorogenic tests in solution. They are not appropriate for blots.

11.2.2 Indirect Substrates

A detectable compound is formed after enzymatic cleavage of the substrate by an additional reaction step. This can be, e.g., oxidation (Sect. 12.1), azocoupling (Sect. 12.2), decomposition (Sect. 13.1.2) or pH change. Members of this group are:

- 5-bromo-4-chloro-3-indolyl-β-D-galactopyranoside (X-Gal)
- 3-(4-methoxyspiro [1,2-dioxetane-3,2'-tricyclo [3.3.1^{3,7}] decan]-4-yl) phenyl-β-D-galactopyranoside (Lumigen PPG, AMPGD)
- 5-bromo-4-chloro-3-indolyl phosphate (BCIP, X-phosphate)
- Naphthol-AS-phosphate
- 1-Naphthyl-phosphate
- 3-(4-methoxyspiro [1,2-dioxetane-3,2'-tricyclo [3.3.1^{3,7}] decan]-4-yl) phenyl phosphate (Lumigen PPD, AMPPD)

X-Gal and X-phosphate are well known systems that form water insoluble indigoid dyes after oxidative dimerization. High sensitivity can be achieved by adding a tetrazolium salt as oxidant (Sect. 12.1).

Naphthol-AS-phosphate and 1-naphthyl-phosphate are also used for blot detection. After azocoupling of the released naphthol derivative with an appropriate diazonium salt different colors and high sensitivities can be achieved (Sect. 12.2).

AMPGD and AMPPD are substrates for luminescence detection on blots or in solution. The enzymatic cleavage is followed by a decay process of the adamantly moiety with emission of light (Sect. 13.1.2).

11.3 Substrates for Peroxidase

Oxidation of the substrates occurs via a reduction of hydrogen peroxide under the catalytic effect of the enzyme horseradish peroxidase. Favored substrates for biomolecule detection are:

- 2,2'-Azino-di-(3-ethyl-benzthiazoline-sulfonic acid) (ABTS)
- Tetramethylbenzidine (TMB)
- o-Phenylenediamine
- o-Dianisidine
- Diaminobenzidine
- 4-Methoxy-1-naphthol

- 4-Chloro-1-naphthol
- Aminoethylcarbazol
- Luminol

For blot formats the well known substrate TMB is most frequently used. Other precipitating substrates are 4-methoxy-1-naphthol and 4-chloro-1-naphthol, which form blue indigoid dyes through oxidative dimerization (Guilbault, 1964; Elias, 1980; Kyowa Medex Co., 1986; Poor, 1988).

ABTS and *o*-phenylenediamine are highly soluble in water and are often used in cuvette tests.

Optimal compositions of substrate and buffer solution, additives such as detergents or inorganic salts as accelerators, and stopping solutions for activity measurement of marker enzymes are described by Porstmann (1988).

For chemiluminescence detection luminol is the most frequently used horseradish peroxidase substrate (see Sect. 13.1; Thorpe, 1985; Merenyi, 1990).

11.4 Applications of Direct Labels

Direct labels possess the advantage that no substrate reaction step is necessary. The measurement can be done directly after the specific recognition step. To reach an appropriate sensitivity, fluorescence or chemiluminescence measurement is done in most cases. In blotting systems, gold-labeled antibodies are used as direct labels for visual detection. Additional silver staining can improve the sensitivity of the system (Poor, 1988). Very popular direct labels, especially for immunofluorescence, are low molecular weight activated fluorophores. These labels (e.g., FLUOS, FITC, RESOS, Texas Red, RHODOS) can easily be conjugated to antibodies according to standard protocols. If higher sensitivities are needed phycoerythrins or fluorescence latices can be used but require a more laborious coupling procedure (see Sect. 14.1 and Herrmann, 1989; Haugland, 1982).

In some cases background fluorescence, e.g., from serum, interferes with the label emission. One way to circumvent this problem is to use time-resolved fluorescence detection with europium labels. After the decline of background fluorescence, the long-lived metal luminescence is measured (see Sect. 14.1 and Hemmilä, 1984; Diamandis, 1988).

The acridinium ester derivatives are widely used as chemiluminescent direct labels. The light-producing reaction is easily triggered by the addition of H_2O_2. The resulting „flash" signal requires more sophisticated instruments, but these are currently available in several commercial immunoassay systems (Weeks, 1986).

(Iso-)luminol derivatives can also be coupled to antibodies and used as direct labels. Signal generation requires the addition of horseradish peroxidase/H_2O_2 (De Boever, 1990).

11.4

A new system recently described in the literature is the so-called electrogenerated chemiluminescence. In this detection system ruthenium labels are oxidized on a gold electrode ($Ru^{2+} \rightarrow Ru^{3+}$). During subsequent reduction of the strongly oxidized ruthenium with tripropylamine a chemiluminescence signal is generated (see Sect. 13.3 and Blackburn, 1991).

References

Abbott Laboratories (1986) Substrate formulation in 2-amino-2-methyl-1-propanol buffer for alkaline phosphatase assays. European Patent Application EP228663

Altman F (1976) Tetrazolium salts and formazans. Progr Histochem Cytochem 9:1−57

Beck S, Köster H (1990) Applications of dioxetane chemiluminescent probes to molecular biology. Anal Chem 62:2258−2270

Blackburn GF et al. (1991) Electrochemiluminescence detection for developement of immunoassays and DNA probe assays for clinical diagnosis. Clin Chem 37/9:1534−1539

Boehringer Mannheim GmbH (1969) Verfahren und diagnostische Mittel zur Bestimmung von Hydroperoxiden und peroxidatisch wirksamen Substanzen. Deutsche Patentschrift DE1917996C3

Boehringer Mannheim GmbH (1983) Phenolsulfonphthaleinyl-β-D-galactoside, Verfahren zu deren Herstellung sowie deren Verwendung zur Bestimmung der β-Galactosidase. Europäische Patentschrift EP146866

Boehringer Mannheim GmbH (1984) Glycoside von Resorufin-Derivaten, Verfahren zu deren Herstellung sowie deren Verwendung zur Bestimmung der Aktivität von Glycosidasen. Europäische Patentanmeldung EP156347

Bronstein I, Edwards B, Voyta JC (1989) 1,2-Dioxetanes: novel chemiluminescent enzyme substrates. Applications to immunoassays. J Biolumin Chemilumin 4:99−111

Bronstein I, Kricka LJ (1989) Clinical applications of luminescent assay for enzyme and enzyme labels. L Clin Lab Anal 3:316−322

Buonocore V, Sgambati O, De Rosa M, Esposito E, Gambacorta A (1980) A constitutive β-galactosidase from extreme thermoacidiphile archaebacterium Caldariella acidophila. J Appl Biochem 2:390−397

Cammann K (1991) Chemo- und Biosensoren, Grundlagen und Anwendungen. Angew Chem 103:519−541

Ci Y-X, Chen L, Wei C (1989) Fluorescence reaction of the system mimetic peroxidase (Mn-T(4-TAP)P)-homovanillic acid-hydrogen peroxide. Fresenius Z Anal Chem 334:34−36

Coutlee F, Viscidi R, Yolken R (1989) Comparison of colorimetric, fluorescent and enzymatic amplification substrate systems in an enzyme immunoassay for detection of DNA-RNA hybrids. J Clin Mikrobiol 27:1002−1007

Czerkinsky C et al. (1988) A novel two colour ELISPOT assay. J Immunol Meth 115:31−37

De Boever J, Kohen F, Leyseele D, Vandekerckhofe D (1990) Isoluminol as a marker in direct chemiluminescence immunoassays for steroid hormones. J Biolum Chemilum 5:5−10

Diamandis EP (1988) Immunoassays with time-resolved fluorescence spectroscopy: principles and applications. Clin Biochem 21:139−150

Elias J (1980) A rapid, sensitive myeloperoxidase stain using 4-chloro-1-naphthol. Am J Clin Pathol 73:797−799

Garen A, Levinthal C (1960) A fine structure genetic and chemical study of the enzyme alkaline phosphatase of E. coli. Biochim Biophys Acta 38:470−483

Guilbault G (1964) 4-Methoxy-alpha-naphthol as a spectrophotometric reagent substrate for measuring peroxidatic activity. Anal Chem 36:2492−2496

Guilbault GG, Brignac P, Zimmer M (1966) Homovanillic acid as fluorimetric substrate for oxidative enzymes. Anal Chem 40:190−199

Haugland EP (1982) Covalent fluorescent probes in excited states of biopolymers. In: Steiner RF (ed) Plenum Press, New York, London

Hauber R, Geiger R (1987) A new, very sensitive, biochemiluminescence-enhanced detection system for protein blotting. J Clin Chem Clin Biochem 25:511−514

Hemmilä I, Dakubu S, Mukkala V-M, Siitari H, Lövgren T (1984) Europium as a label in time-resolved immunofluorometric assays. Anal Biochem 137:335−343

Herrmann R, Josel H-P, Wörner W, Fetterhoff TJ (1989) Conjugation of proteins to various fluorescence labels. 19th FEBS meeting, Rome, Abstract Nr FR 511

Kronick MN, Grossman PD (1983) Immunoassay technique with fluorescent phycobili-protein conjugates. Clin Chem 29/9:1582−1586

Kyowa Medex Co. Ltd. (1986) Method of assaying a biocomponent. European Patent Application EP167340

Lojda Z, Slaby J, Kraml J, Kolinska J (1973) Synthetic substrates in the histochemical demonstration of intestinal disaccharidases. Histochemie 34:361−369

Merenyi G, Lind J, Eriksen TE (1990) Luminol chemiluminescence: chemistry, excitation, emitter. J Biolum Chemilum 5:53−56

Nugel E, Porstmann B, Porstmann T, Evers U, Schmechta H (1986) Vergleichende Untersuchungen zur Peroxidase, alkalische Phosphatase und β-Galactosidase als Markerenzyme. Z Med Lab Diagn 27:145−153

Poor M, Santa P, Sittampalam S (1988) Visualization of multiple protein bands on the same nitrocellulose membrane by double immunoblotting. Anal Biochem 175:191−195

Porstmann B, Porstmann T (1988) Chromogenic substrates for enzyme immunoassay. In: Ngo T (ed) Nonisotopic immunoassays, Plenum Press, New York, London

Porstmann B, Evers U, Nugel E, Schmechta H (1991) Tetramethylbenzidin − ein chromogenes Substrat für Peroxidase im Enzymimmunoassay. Z Med Lab Diagn 32:3−8

Sauer M, Foulkes J, O'Neil P (1989) A comparison of alkaline phosphatase, β-galactosidase and peroxidase used as labels for progesterone determination in milk by heterologous microtitre plate enzymeimmunoassay. J Steroid Biochem 33:423−431

Schaap AP, Sandison MD, Handley RS (1987) Chemical and enzymatic triggering of 1,2-dioxetanes. Alkaline phosphatase-catalyzed chemiluminescence from an aryl phosphate substituted dioxetane. Anal Biochem 192:222−231

Technicon Instruments Corporation (1987) Substrates for β-galactosidase. Eur Pat Appl EP292169

Thorpe GHG, Kricka LJ, Moseley SB, Whitehead TP (1985) Phenols as enhancers of the chemiluminescent horseradish peroxidase-luminol-hydrogen peroxide reaction. Clin Chem 31:1335−1341

Vaidya HC et al. (1988) Quantification of lactate dehydrogenase-1 in serum with use of an M-subunit-specific monoclonal antibody. Clin Chem 34:2410−2414

Weeks I, Sturgess M, Brown RB, Woodhead JS (1986) Immunoassays using acridinium esters. Meth Enzymol 133:366−406

West S, Schröder J, Kunz W (1990) A multiple-staining procedure for the detection of different DNA fragments on a single blot. Anal Biochem 190:254−258

White JC, Stryer L (1987) Photostability studies of phycobiliprotein fluorescent labels. Anal Biochem 161:442−452

12 Colorimetric Systems

12.1 Indigo/Tetrazolium Dyes

H.-J. GUDER

12.1.1 Principle and Applications

One of the most sensitive and widespread colorimetric detection systems in matrix-based application formats is the redox couple 5-bromo-4-chloro-3-indolyl phosphate (BCIP)/nitroblue tetrazolium chloride (NBT). It is used for the detection and localization of alkaline phosphatase activity.

Indolyl derivatives of type (I) are usually stable over a wide pH range, but after enzymatically catalyzed hydrolysis to the indolyl derivative (II) they rapidly form an indigoid dye (III) through atmospheric or chemical oxidation (see figure).

One advantage of the system appears to be that in the case of substrate (I) the solubility in water, and correspondingly the availability for enzymatic cleavage, is relatively high because of the polar phosphate group and the small lipophilic indolyl moiety. The final indigoid dye, however, is formed after cleavage of the phosphate group by intermolecular dimerization of two indolyl residues. This results in a dye having a much lower solubility relative to the substrate from which it is derived. Thus, once formed, it precipitates very rapidly.

In comparison to unsubstituted indigo, improved staining results are obtained with indolyl derivatives that carry halogen atoms, especially bromine and chlorine in the 4,5 position. This compound leads to the formation of the extremely insoluble 5,5'-dibromo-4,4'-dichloro-indigo with excellent staining properties on a matrix.

However, for the requirements of many tests the sensitivity of even these modified indigoid dyes is often not sufficient because of their generally low extinction coefficients. Amplification of the entire color formation process can be achieved by the addition of a tetrazolium salt. This compound serves as an oxidant for the indolyl and itself forms a highly sensitive formazan dye through reduction. In the case of NBT (IV) the intensity of the formazan (V) exceeds that of the indigo dye by far. The tetrazolium salt is easily soluble in water because of its ionic properties. Reduction, however, destroys the salt structure and the resulting formazan becomes lipophilic and insoluble in water.

The table shows the absorption characteristics of the two dyes and the synergistic effect of both compounds resulting from overlapping absorption bands.

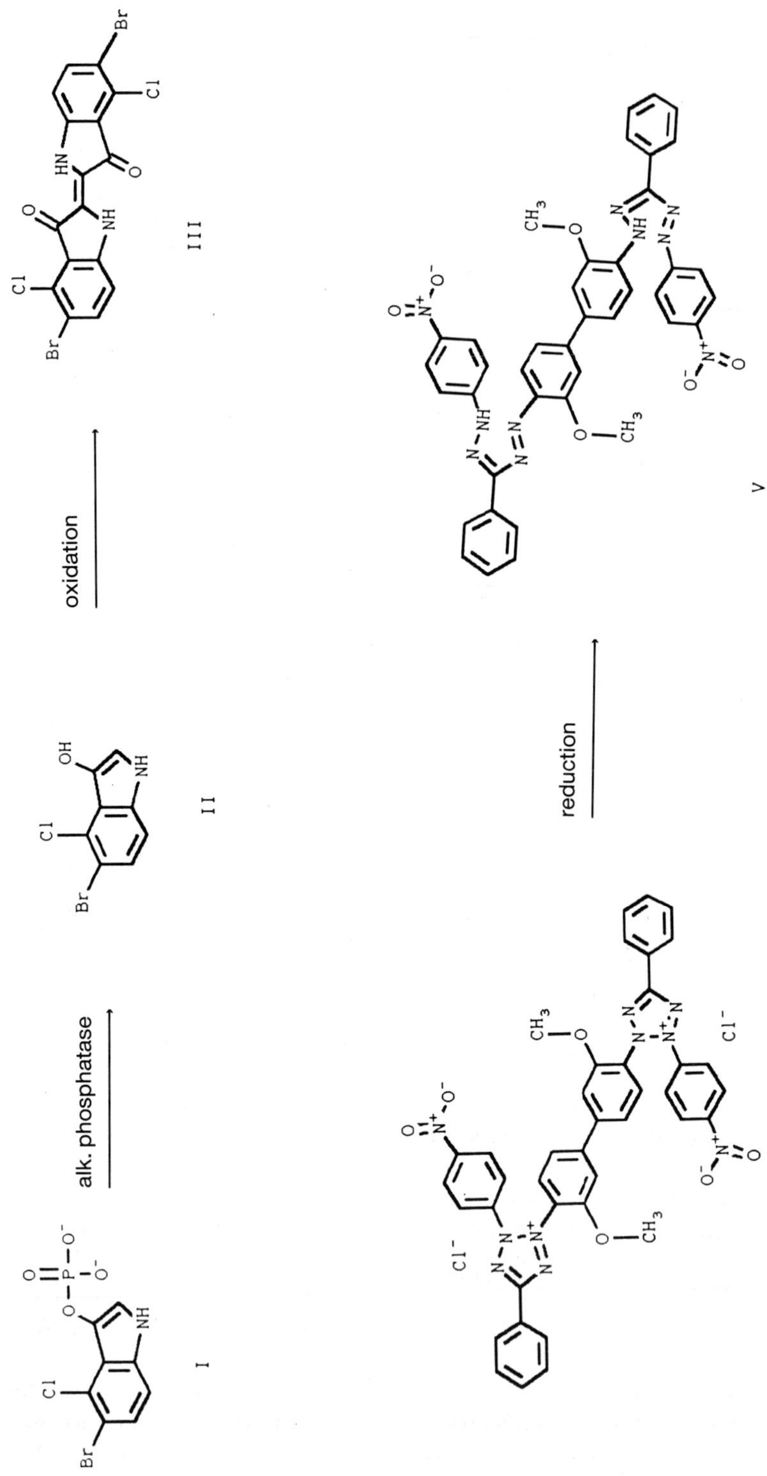

I

alk. phosphatase

II

oxidation

III

reduction

IV

V

Absorption characteristics of BCIP and NBT

	Absorption maximum (nm)	Molar extinction coefficient	Reference
BCIP	620 (DMSO)	4000	Guder, unpublished
NBT-formazan	605 (ethanol)	40200	Altman, 1976

BCIP, 5-bromo-4-chloro-3-indolyl phosphate; NBT, nitroblue tetrazolium salt.

Two dyes are formed in the redox step. Both of them are blue and insoluble in water with good precipitation properties. The rapid redox reaction results in discrete deposition of the dyes with no bleeding effects which would decrease the sensitivity of the test. For this reason BCIP/NBT is ideal for use in matrix-based systems.

12.1.2 Detection with BCIP/NBT

Techniques for the application of BCIP/NBT in nonradioactive DNA detection are described in Sect. 3.2.11.

Reference

Altman F (1976) Tetrazolium salts and formazans. Progr Histochem Cytochem 9:1−57

12.2 Azo Dyes

Werner Kunz and Sabine West

12.2.1 Principle and Applications

In order to detect biomolecules such as nucleic acids or proteins on blots or in situ, enzyme-linked immunoassays are used. In most of these applications, alkaline phosphatase (AP) is the enzyme of choice. AP is either conjugated to an antibody or to streptavidin which then binds to specific antigens or hapten/biotin-labeled target molecules, respectively. In all these examples, AP can be visualized by its ability to catalyze specific color reactions.

Conventionally, BCIP and NBT are employed as substrates yielding a blue precipitate when the phosphoryl group is released (McGadey, 1970).

Here, we describe a new substrate for AP that presents several advantages compared with the BCIP/NBT system: naphthol AS phosphates together with diazonium salts (Fast salts).

Naphthol AS is the 3-carboxylic acid anilide of 2-naphthol (or, in other terms, 2-hydroxy-3-naphthoic acid anilide). A variety of naphthol AS derivatives are available which differ from each other in the side groups on the benzene ring of the anilide residue. Removal of the phosphoryl group from the naphthol AS phosphate by AP in the presence of a diazonium salt results in the formation of an azo dye. Depending on the side groups of the different diazonium salts and on the type of the naphthol AS phosphate partner, varying azo dyes are created that differ in color. Combinations of eight naphthol AS phosphates and eight diazonium salts have been tested. The following table summarizes the five best combinations and indicates differences in color quality and intensity between nylon and nitrocellulose membranes.

Dye	Nylon (intensity)	Nitrocellulose (intensity)
AS-GR/Fast Blue B	Turquoise-blue (+++)	Turquoise-blue (++)
AS/Fast Red TR	Red (++++)	Red (+++)
AS/Fast Blue B	Violet-blue (+++++)	Violet-blue (++)
AS/Fast Brown RR	Dark violet (+++++)	Pink (+)
AS-GR/Fast Brown RR	Turquoise-blue (++++)	Violet-black (+++)

Azo dyes have traditionally been used for AP detection in histological preparations (Cordell et al., 1984), but we have shown that they can more widely be applied in a variety of different nucleic acid or protein screening procedures (Schröder et al., 1989; West et al., 1990):

1. Detection of hapten- or biotin-labeled nucleic acids via antibodies or streptavidin conjugated to AP in: Southern blots of cloned or genomic DNA, northern blots, Dot blots of DNA or RNA, colony hybridization, plaque lifts of phage clones, and in situ hybridization.
2. Detection of fusion proteins or cellular proteins via AP-conjugated specific antibodies in: plaque lifts of expressed phage clones, dot blots of expressed phage clone lysates, western blots, and immunohistology.

Azo dye precipitates adhere directly to the membrane or tissue. Thus, it is possible to remove nucleic acid probes or antibodies from the filter or tissue while the dye remains fixed. The filter or tissue is then ready to be rehybridized with a second or third nucleic acid probe or screened for proteins with further antibodies, respectively. Consecutive probes can be developed in a variety of very distinct colors, making the system suitable for multiple color detection of different probes on one membrane or specimen. It is even possible to visualize different target molecules within one band, colony, or plaque due to the reproducible intermediate colors formed when two different probes are superimposed.

12.2.2

Azo dyes additional advantages. The substrates are cheaper than BCIP and NBT. Furthermore, they are extremely tolerant against background, whereas BCIP/NBT yields background when the membrane is too strongly charged, the probe concentration is too high, or when incubation in staining solution is too long. Complete destaining of the membrane is much easier than with the BCIP/NBT system since azo dyes are ethanol soluble. A simple ethanol incubation is sufficient, whereas BCIP/NBT filters have to be washed in heated dimethylformamide which is toxic and aggressive to plastic dishes. Azo dyes are very stable when exposed to light, and stained filters can be stored for years in aqueous solution or air-dried.

A limitation of azo dyes is that they are a factor of ten less sensitive than BCIP/NBT and are, therefore, not the first choice for single copy gene detection in genomic blots of mammalian DNA.

12.2.2 Detection with Azo Dyes

Detection reagents

— DMF (dimethylformamide) (Aldrich)
— Naphthol AS-GR phosphate; free acid, crystalline (Sigma)
— Naphthol AS phosphate; free acid, crystalline (Sigma)
— Fast Blue B (zinc chloride complex); 20% dye content (Sigma)
— Fast Red TR (hemi-zinc chloride salt); >90% dye content (Sigma)
— Fast Brown RR (hemi-zinc chloride salt); >90% dye content (Sigma)

Detection solutions

— Incubation buffer: 100 mM Tris-HCl; 100 mM NaCl; 5 mM MgCl$_2$; pH 9.5/25°C
— Naphthol AS phosphate stock solutions: 20% [w/v] in 100% DMF
— Fast Blue B stock solution: 10% [w/v] in 70% DMF
— Fast Red TR stock solution: 10% [w/v] in 100% DMF
— Fast Brown RR stock solution: 10% [w/v] in 70% DMF
All stock solutions have to be stored at −20°C.

Preparation of stock solutions

Fast Blue B dissolves incompletely in DMF. Vortex frequently; after 30 min spin down undissolved contaminants and keep supernatant.

Preparation of color substrate solution

1. Dilute stock solution of naphthol AS phosphates in incubation buffer to a final concentration of 0.02% [v/v].
2. Add stock solution of Fast Red TR or Fast Brown RR to a final concentration of 0.006% [v/v] of Fast Blue B to a final concentration of 0.03% [v/v].

Staining reaction

1. Equilibrate membrane or incubate microscope slide (chromosome or tissue preparation) for 2 min in incubation buffer.
2. Incubate filter or slide in freshly prepared color substrate solution. Azo dye formation starts immediately and is complete after 1 h. Do not shake or mix during color precipitate formation.
3. Wash for 2 × 2 min in distilled water.

12.2.3

4. Dry filter at room temperature. Microscope slides either have to be dried or mounted in an aqueous medium (see below).

Removal of probes or dyes For multiple color detection experiment, prior to each further detection reaction on the same blot or microscope slide.

1. Remove hybrids in $0.2\,N$ NaOH; 0.1% [w/v] SDS for $2 \times 15\,min$ at $37\,°C$, or remove antibodies in $4\,M$ MgCl$_2$, $2 \times 15\,min$ at room temperature, or inactivate AP in $0.05\,M$ EDTA, pH 8.0 for 15 min at $65\,°C$.
2. To remove azo dyes, rinse nylon membrane repeatedly in 95% ethanol until it is completely decolorized (usually $3 \times 10-30\,min$); soak membrane in distilled water before reuse or storage.

Note: Do not treat nitrocellulose with ethanol.

12.2.3 Special Hints for Application and Troubleshooting

- If color signals are weak and develop too slowly, replace color substrate solution with fresh solution after 1 h.
- If many target molecules are to be detected in one color substrate solution, increase Fast salt concentration by a factor of two.
- A slight background may arise after long staining times. This will fade after complete drying of the membrane.
- Ethanol solubility of the azo dyes limits their application to histological or cytological preparations, since stained tissues or chromosomes cannot be processed through an ascending ethanol series for the purpose of mounting in nonaqueous media.

References

Cordell JL, Falini B, Erber WN, Gosh AK, Abdulaziz Z, MacDonald S, Pulford KAF, Stein H, Mason DY (1984) Immunoenzymatic labeling of monoclonal antibodies using immune complexes of alkaline phosphatase and monoclonal anti-alkaline phosphatase. J Histochem Cytochem 32:219−239
McGadey J (1970) A tetrazolium method for nonspecific alkaline phosphatase. Histochemie 23:180−184
Schröder J, Symmons P, Kunz W (1989) Immunenzymatische Farbkodierung zur Erkennung von DNA-Klonen in Expressionsvektor-Bibliotheken. BioEngineering 5:24−30
West S, Schröder H, Kunz W (1990) A multiple-staining procedure for the detection of different DNA fragments on a single blot. Anal Biochem 190:254−258

13 Luminescence Systems

13.1 Chemiluminescence: Luminol

HANS-PETER JOSEL

13.1.1 Principle and Applications

The chemiluminescence (CL) of luminol and related compounds is a well known reaction and has been studied intensively (Gundermann, 1974). The mechanism of the CL reaction of the so-called diacyl hydrazides is very complex and depends on several conditions (Gundermann and McCapra, 1987), e.g., whether the reaction is carried out in water or in aprotic solvents. In DMSO luminol exists as an activated aminophthalic acid, which gives off a bright blue-green light in the presence of a strong base together with hydrogen peroxide.

Normally, a catalyst, especially horseradish peroxidase (HRP), is used to trigger the CL reaction but hemin and potassium hexacyanoferrate (III) give the same reaction.

For the nonradioactive detection of biomolecules, in most cases an HRP: antibody conjugate and luminol+ H_2O_2 as substrate are employed (Kricka, 1991). Nevertheless, luminol labels coupled to proteins in which the CL reaction is started with the addition of HRP/H_2O_2 have also been described in the literature (Kricka, 1991).

Under normal conditions, 7-dimethylamino-naphthalene-1,2-dicarbonic acid hydrazide (see figure) gives a better quantum yield than does luminol (Allen, 1982; Gundermann, 1965).

Luminol *(left)* and 7-dimethylamino-naphthalene-1,2-dicarbonic acid hydrazide *(right)*

A great advantage in luminol/HRP-related reactions has been the introduction by Whitehead and coworkers of so-called enhancers, which force the flash signal of the unenhanced reaction into a glow, allowing easier handling and improving the quantum yield of the reaction (Thorpe, 1985).

Typical enhancer compounds are substituted phenols, the most popular being p-iodine-phenol, but many others exist; for some, a better sensitivity has been claimed, e.g., p-phenyl-phenol (Thorpe, 1985; Motsenbocker, 1988).

The results of the enhanced luminol/HRP reaction strongly depend on the reaction conditions: pH, relative luminol/enhancer concentration, HRP concentration, etc.

A standard protocol, giving an example of how to use this detection system is given below, but it should be kept in mind that the specific conditions must be carefully adapted to the particular problem.

The luminol reaction is most often applied in immunoassay systems (Kricka, 1988) but some application in DNA probe hybridization assays (Pollard-Knight, 1991) and blotting have also been reported (Pollard-Knight, 1991; Schneppenheim, 1987).

13.1.2 Chemiluminescent Detection with Horseradish Peroxidase and Luminol

Standard reagents
- Luminol (Boehringer Mannheim)
- 7-Dimethylamino-naphthalene-1,2-dicarbonic acid hydrazide (Boehringer Mannheim)
- p-Iodine-phenol (Aldrich)
- Horseradish peroxidase (Boehringer Mannheim)
- Hydrogen peroxide (Merck)

Standard solutions
- 10 nM Luminol solution in Tris-HCl buffer, pH 8.5
- 5 nM 4-Iodine-phenol solution in Tris-HCl buffer, pH 8.5
- 20 nM Aqueous H_2O_2 solution
- 0.25 mU HRP solution in 10 mM KPO_4 buffer, pH 7.8

Standard equipment
- Standard luminometers, e.g., Berthold: Biolumat, Hamilton: Lumicon

Standard protocol
Pipette 20 µl of the luminol, 4-iodine-phenol, and H_2O_2 solutions and 100 µl of the HRP solution into a tube and vortex for a few seconds. The tube is immediately put into the luminometer and the measurement is started at once.

With most luminometers, the luminol-HRP reaction can be started directly in the instrument by automatic pipetting of the appropriate quantity of H_2O_2 solution.

- Typical results are presented in the accompanying figure.

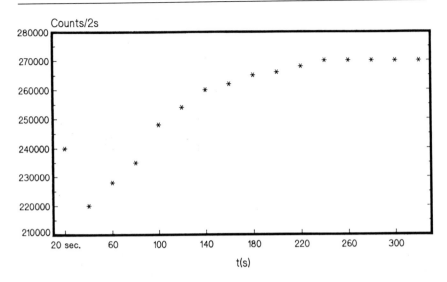

13.1.3 Special Hints for Application and Troubleshooting

— As stated above, the reaction is influenced by the relative concentration of reaction partners and pH. Therefore for every application a careful evaluation of all the reaction components has to be done.

References

Allen RC (1982) Chemical and biological generation of excited states. In: Adam W, Cilento G (eds) Academic Press, New York, pp 309−344

Gundermann KD (1965) Liebigs Ann Chem 682:127−141

Gundermann KD (1974) Topics in Current Chemistry, Springer Verlag 46:61

Gundermann KD, McCapra F (1987) Chemiluminescence in Organic Chemistry, Springer Verlag:77−108

Kricka LJ (1991) Chemiluminescent and bioluminescent techniques. Clin Chem 37: 1472−1481

Kricka LJ, Stott RAW, Thorpe GHG (1988) Enhanced chemiluminescence enzyme immunoassays. In: Colins WP (ed) Complementary Immunoassays, Wiley, Chichester, p 169−179

Motsenbocker MA (1988) Sensitivity limitations encountered in enhanced horseradish peroxidase catalysed chemiluminescence. J Biolumin Chemilumin 2:9−16

Pollard-Knight DV (1991) Rapid and sensitive luminescent detection methods for nucleic acid detection. In: Stanley PE, Kricka L (eds) Bioluminescence and Chemiluminescence: Current Status, Wiley, Chichester, pp 83−90

Schneppenheim R, Rautenberg P (1987) A luminescence western blot with enhanced sensitivity for antibodies to human immunodeficiency virus. Eur J Microbiol 6:49−51

Thorpe, Kricka, Moseley, Whitehead (1985) Phenols as enhancers of the chemiluminescent horseradish peroxidase-luminol-hydrogen peroxide reaction: application in luminescence-monitored enzyme immunoassays. Clin Chem 31:1335−1341

13.2 Chemiluminescence: Properties of 1,2-Dioxetane Chemiluminescence

IRENA BRONSTEIN and LARRY J. KRICKA

The first synthesis of a dioxetane was described in 1969 by Kopecky and Mumford, and the first thermally stable dioxetane, adamantylidene adamantyl 1,2-dioxetane, was described in 1972 by Wierynga et al. The resulting product decomposed chemiluminescently only when heated above 165 °C. McCapra and coworkers (1977) reported on the synthesis of an unsymmetrically substituted adamantyl 1,2-dioxetane (9-(2-adamantylidene)-N-methylacridan 1,2-dioxetane) which, when heated, generated chemiluminescence exclusively from the excited singlet of N-methylacridone. It has been shown that the decomposition of this unsymmetrically substituted 1,2-dioxetane leads to the formation of the excited state carbonyl-based product with the lowest singlet energy state. McCapra also proposed that dioxetanes which are substituted with electron donating groups decompose via charged intermediates in an electron transfer process (McCapra et al., 1977). In 1983, Adam and coworkers (Adam et al., 1983) synthesized and studied the stability of several unsymmetrically substituted adamantyl 1,2-dioxetanes and concluded that the stabilization mechanism is complex and depends on several factors, such as conformational isomerism (McCapra, 1977). More recently a theoretical investigation on the decomposition modes of 1,2-dioxetanes has been reported (Reguero et al., 1991).

The first important application of 1,2-dioxetanes to clinical analysis was described by Hummelen et al. (Hummelen et al., 1987). Suitably derivatized adamantylidene-adamantyl 1,2-dioxetanes were utilized as labels in thermoluminescent immunoassays. In 1986 the first enzyme-activated adamantyl-substituted 1,2-dioxetane substrates were described. These substrates, coupled with the existing ELISA test formats, led to the development of highly sensitive assays for a variety of clinically important analytes (Voyta et al., 1988; Bronstein and Kricka, 1989; Schaap et al., 1989; Beck and Köster, 1990; Bronstein, 1990; Bronstein and Dimond, 1990; Bronstein and Kricka, 1990; Beck and Koster, 1991; Bronstein and Sparks, 1991; Bronstein et al., 1991a). Currently a new generation of dioxetane substrates and enhancers has been developed that have superior properties in a number of bioassay formats. These substrates and enhancers are further described in this and the following sections.

Detailed study of the properties of a large number of 1,2-dioxetane substrates has permitted the intelligent design of such compounds with specific properties, such as increased stability, ease of use, and intense chemical signal. The basic design of 1,2-dioxetane substrates such as AMPPD, shown in the accompanying figure, includes several components.

AMPPD

Disodium 3-(4-methoxyspiro[1,2-dioxetane-3,2′-tricyclo[3.3.1.1^{3,7}]decan]-4-yl) phenyl phosphate

The stability of the compounds is influenced by the presence of the bulky adamantyl group (Hummelen et al., 1987; Schaap et al., 1987). A single adamantyl group provides sufficient thermal stability in the unsymmetric 1,2-dioxetane. The energy required to populate the emitting state of the breakdown product is generated from scission of the oxygen-oxygen single bond in the 1,2-dioxetane ring. This weak bond (23 kcal/mole) thus allows the ring to behave as an "energy store-house" which releases energy when it decomposes to strongly bonded carbonyl products. The alkoxy substituent is an important intermediate in the synthesis and ensures rapid and efficient photooxygenation of the electron-rich enol ether to form a 1,2-dioxetane. The emitting moiety is methyl 3-oxybenzoate anion, generated after the protecting group — the phosphate attached to the aromatic ring via an oxygen atom — is removed by the enzyme alkaline phosphatase. The orientation of the protecting group relative to the attachment point of the dioxetane ring is referred to as "disjoint" or "odd" and is key to the performance of the compound as an enzyme substrate in all applications. (An odd pattern is one in which the donor group's point of attachment to the ring in relation to the ring's point of attachment to the acceptor is such that the total number of ring carbon atoms separating these points, including the atoms at the point of attachment, is an odd whole number.) Our detailed investigations of the substitution pattern leads to the preferred dioxetane properties in the phenyl-based systems. AM1 molecular orbital calculations for the 1,3-carboxyphenolate and 1,4-carboxyphenolate anions have shown that the HOMO-LUMO gap for the disjoint system is 7.14 ev* and 7.97 ev for the conjugated system, and the net charge transfer is 0.77 for the odd and 0.63 for the conjugated systems (Edwards et al., 1990a, b). We found that the *para* analogo of AMPPD exhibited poor thermal stability, very high nonspecific background, and a detection limit for alkaline phosphatase approximately six orders of magnitude below that for AMPPD (unpublished results).

In 1,2-dioxetanes which contain the naphthalene-based aromatic substituent, the position of the donor O$^-$ relative to the position of the accep-

---

* ev = electron volt

tor, or the point of the attachment to the dioxetane ring, also affects the characteristics of the excited ester product (Edwards et al., 1990a, b; Schaap et al., 1991a). The odd-substituted (2,7, 1,3, and 1,6) naphthalene dioxetane phosphates have longer half-lives for their respective oxyanions. In addition, the odd-patterned dioxetanes generate greater levels of chemiluminescence, lower backgrounds, and bathochromically shifted emissions compared to the the even-substituted 2,6-naphthalene compound. Theoretical AM1 molecular orbital calculations indicate that the greatest amount of charge transfer from donor to acceptor occurs in the odd substituted oxynaphthoic acid model systems. The phenomenon of disjoint substitution suggests that dioxetane decomposition occurs along several reaction coordinates which strongly depend on the aromatic substitution pattern.

AMPPD, in aqueous buffer solutions in the presence and absence of polymeric enhancers, exhibits a relatively long delay before reaching a constant level of light emission. The inherent amphiphilic property leads to the aggregation of AMPPD and its dephosphorylated anion in aqueous solutions. In an aggregated state, the signal produced from the nonenzymatically generated excited state emitter is amplified as it resides in a hydrophobic environment, which leads to an undesirably high nonspecific signal. Based on these findings, we have designed a new group of derivatized adamantyl 1,2-dioxetane phosphates with improved analytical performance. The premise of the new design involves conversion of the passive adamantyl-stabilizing moiety, which does not interfere with the chemiluminescent process, to an active substituent which can influence the kinetics of the light process and the hydrophobic/hydrophilic balance of this system and control aggregation. The new generation of chemiluminescent dioxetane substrate contains an adamantyl ring appended with various substituents at the 5 position.

A number of derivatized, adamantyl-substituted, 1,2-dioxetane substrates have been synthesized, in which R = hydroxy-, methoxy-, chloro-, bromo-, iodo- and others, via Wittig-Horner coupling reaction of 5-substituted adamant-2-ones with diethyl 1-methoxy-1-(3-pivaloyloxy) phenyl

CSPD

Disodium 3-(4-methoxyspiro[1,2-dioxetane-3,2'-(5'-chloro)tricyclo[3.3.1.1^{3,7}]decan]-4-yl) phenyl phosphate

Application of adamantyl 1,2-dioxetanes

DNA fingerprinting	Reference
Single and multilocus	Beck and Köster, 1990
DNA sequencing	
Nonhybridization	Beck and Köster, 1990;
	Beck et al., 1989; Creasey et al., 1991;
	Martin et al., 1991; Richterich and Church,
	1992
Hybridization (multiplex)	Beck and Köster, 1990; Creasey et al., 1991;
	Gillevet, 1990; Tizard et al., 1990
Immunoassay	
Alpha-fetoprotein	Bronstein et al., 1990c; Thorpe et al., 1989
Carcinoembryonic antigen	Hummelen et al., 1986
Heptatitis B surface antigen antibody	Ashihara et al., 1991
Hair bundle proteins	Gillespie and Hudspeth, 1991b
HIV-1 antibody	Yamamoto et al., 1991
HTLV-1 antibody	Yamamoto et al., 1991
Human chorionic gonadotropin	Bronstein et al., 1989a
Human growth hormone	Albrecht et al., 1991
Human luteinizing hormone	Bronstein et al., 1989a; Honda et al., 1990
IL-6	Yamamoto et al., 1991
Myeloperoxidase	Wood, 1990
Photoreceptor outer segments	Gillespie and Hudspeth, 1991a
Thyrotropin	Bronstein et al., 1989b, 1991b
Nucleic acid hybridization assay	
Chlamydia trachomatis	Clyne et al., 1989
Hepatitis B	Urdea et al., 1990; Bronstein et al., 1989c
Herpes simplex I	Bronstein and Voyta, 1989
Neisseria gonorrhoeae	Pescador et al., 1989
pBR 322	Bronstein et al., 1990b
Northern blotting	
Angiotensin-converting enzyme	Lanzillo, 1991
HMG-CoA reductase	Höltke et al., 1991
LDL receptor	Höltke et al., 1991
Reporter gene assay	
lac Z gene (β-galactosidase)	Jain and Magrath, 1991
Southern blotting	
Angiotensin-converting enzyme	Lanzillo, 1991
Human beta-globin genes	Carlson et al., 1990
mos proto-oncogene	Pollard-Knight et al., 1990
pBR 322	Bronstein et al., 1990a
pBR 328	Höltke et al., 1991
raf-1 proto-oncogene	Pollard-Knight et al., 1990
t-Pa	Cate et al., 1991, Holtke et al., 1991
Western blotting	
Mouse IgG	Bronstein et al., 1992
Transferrin	Bronstein et al., 1992

methane phosphonate to yield enol ether phenolate salts which are subsequently phosphorylated and photooxygenated to generate the final products. One such compound, namely, the chloro-derivatized adamantyl-1,2-dioxetane (CSPD; see figure) exhibits several important benefits compared to AMPPD, such as faster light emission kinetics, greater sensitivity and higher resolution of images when used to detect proteins and nucleic acids on membranes (described further in Sect. 19.7).

Applications of 1,2-dioxetane labels and substrates are summarized in the table. Among other 1,2-dioxetanes included in this table are adamantylidene adamantyl-1,2-dioxetane labels in which light emission is achieved thermally (thermoluminescence) and can be improved by energy transfer to an appropriate fluorophore (Hummelen et al., 1986). The high temperature for signal generation has deterred extensive development of these labels. However, recently a series of 1,2-dioxetane labels has been prepared that can be triggered chemically (Schaap et al., 1991b). The majority of applications incorporating dioxetanes have utilized adamantyl 1,2-dioxetane aryl phosphates as substrates for alkaline phosphatase labels in immunoassays, blotting assays, and nucleic acid hybridization assays. In all cases these dioxetane substrates have provided significant improvements in assay performance when compared to conventional substrates for alkaline phosphatase labels.

In summary, the study of various properties of 1,2-dioxetane enzyme substrates has permitted the creation of molecules of commercial importance. On the basis of the above results, coupled with empirical and theoretical investigations in progress, excited states with greater charge transfer character will be created which are expected to lead to the design of chemiluminescent systems with greater excitation yields, lower nonspecific backgrounds, and more stable intermediates preceding the formation of the excited state ester products. Finally, it is anticipated that many new commercial applications of this technology for the ultrasensitive detection of substances will appear in the future.

References

Adam W, Encarnacion L, Zinner K (1983) Thermal stability of spiro[adamantane-[1,2]dioxetanes]. Chem Ber 116:839–846

Albrecht S, Ehle H, Schollberg K, Bublitz R, Horn A (1991) Chemiluminescent enzyme immunoassay of human growth hormone based on adamantyl dioxetane phenyl phosphate substrate. In: Stanley PE, Kricka LJ (eds) Bioluminescence and Chemiluminescence: Current Status, Wiley, Chichester, pp 115–118

Ashihara Y, Sakurabayashi Y, Nishizono I, Sone T, Shirane H, Hikata A, Yamauchi S, Saito T, Okada M (1991) Automation of chemiluminescent enzyme immunoassay using ferrite particles. Clin Chem 37:1031

Beck S, Köster H (1990) Applications of dioxetane chemiluminescent probes to molecular biology. Anal Chem 62:2258–2270

Beck S, Köster H (1991) Applications of dioxetane chemiluminescent probes to molecular biology [correction] Anal Chem 63:848

Beck S, O'Keeffe T, Coull JM, Koster H (1989) Chemiluminescent detection of DNA; applications for DNA sequencing and hybridization. Nucleic Acids Res 17:5115–5123

Bronstein I, Kricka LJ (1989) Clinical applications of luminescent assays for enzymes and enzyme labels. J Clin Lab Analysis 3:312−322

Bronstein I, Voyta JC (1989) Chemiluminescent detection of herpes simplex virus I DNA in blot and in-situ hybridization assays. Clin Chem 35:1856−1860

Bronstein I, Edwards B, Voyta JC (1989a) 1,2-Dioxetanes: novel chemiluminescent enzyme substrates, applications to immunoassays. J Biolumin Chemilumin 4:99−111

Bronstein I, Voyta JC, Thorpe GHG, Kricka LJ, Armstrong G (1989b) Chemiluminescent assay for alkaline phosphatase: application in an ultrasensitive enzyme immunoassay for thyrotropin. Clin Chem 35:1441−1446

Bronstein I, Voyta JC, Edwards B (1989c) A comparison of chemiluminescent and colorimetric substrates in a hepatitis B virus DNA hybridization assay. Anal Biochem 180:95−99

Bronstein I (1990) Chemiluminescent 1,2-dioxetane-based enzyme substrates and their applications. In CRC Press, Boca Raton, FL, Van Dyke K, Van Dyke R. Luminescence immunoassay and molecular applications, pp 256−274

Bronstein I, Dimond P (1990) Chemiluminescent compounds for diagnostic tests. Diagn Clin Testing 28:36−39

Bronstein I, Kricka LJ (1990) Chemiluminescence: a new end-point for clinical assays. Clin Lab Management Rev, July/Aug, 114−116

Bronstein I, Voyta JC, Lazzari K, Murphy OJ, Edwards B, Kricka LJ (1990a) Rapid and sensitive detection of DNA in Southern blot with chemiluminescence. BioTechniques 8:310−314

Bronstein I, Voyta JC, Lazzari K, Murphy OJ, Edwards B, Kricka LJ (1990b) Improved chemiluminescent detection of alkaline phosphatase. BioTechniques 9:160−161

Bronstein I, Thorpe GHG, Kricka LJ, Edwards B, Voyta JC (1990c) Chemiluminescent enzyme immunoassay for alpha-fetoprotein. Clin Chem 36:1087−1088

Bronstein I, Sparks A (1991) Sensitive enzyme immunoassay with chemiluminescent detection. In: American Society for Microbiology (ed) Immunochemical assays and biosensor technology for the 1990s, pp 229−250

Bronstein I, Juo RR, Voyta JC, Edwards B (1991a) Novel chemiluminescent adamantyl 1,2-dioxetane enzyme substrates. In: Stanley PE, Kricka LJ (eds) Bioluminescence and chemiluminescence: current status, Wiley, Chichester, pp 73−82

Bronstein I, Voyta JC, Vant Erve Y, Kricka LJ (1991b) Advances in ultrasensitive detection of proteins and nucleic with chemiluminescence: novel derivatized 1,2-dioxetane enzyme substrates. Clin Chem 37:1526−1527

Bronstein I, Voyta JC, Murphy OJ, Bresnick L, Kricka LJ (1992) Improved chemiluminescent western blotting procedures. BioTechniques 12:748−753

Carlson DP, Superko C, Mackey J, Gaskill ME, Hansen P (1990) Chemiluminescent detection of nucleic acid hybridization. Focus 12:9−12

Cate RL, Ehrenfels CW, Wysk M, Tizard R, Voyta JC, Murphy OJ, Bronstein I (1991) Genomic Southern analysis with alkaline phosphatase-conjugated oligonucleotide probes and the chemiluminescent substrate AMPPD. GATA 8:102−106

Clyne JM, Running JA, Stempien M, Stephens RS, Akhaven-Tafti H, Schaap AP, Urdea MS (1989) A rapid chemiluminescent DNA hybridization assay for the detection of Chlamydia trachomatis. J Biolumin Chemilumin 4:357−366

Creasey A, D'Angio LJ, Dunne TS, Kissinger C, O'Keeffe T, Perry-O'Keefe H, Moran L, Roskey M, Schildkraut I, Sears LE, Slatko B (1991) Application of a novel chemiluminescent-based DNA detection method to single-vector and multiplex DNA sequencing. BioTechniques 11:102−109

Edwards B, Sparks A, Voyta JC, Bronstein I (1990a) Unusual luminescent properties of odd- and even-substituted naphthyl-derivatized dioxetanes. J Biolumin Chemilumin 5:1−4

Edwards B, Sparks A, Voyta JC, Strong R, Murphy OJ, Bronstein I (1990b) Naphthyl dioxetane phosphates: synthesis of novel substrates for enzymatic chemiluminescent assays. J Org Chem 55:6225−6229

Gillespie PG, Hudspeth A (1991) Chemiluminescence detection of proteins from single cells. Proc Natl Acad Sci USA 88:2563−2567

Gillespie PG, Hudspeth AJ (1991b) High purity solution of bullfrog hair bundles and subcellular and topological localization of constituent proteins. J Cell Biol 112: 625–640

Gillevet PM (1990) Chemiluminescent multiplex DNA sequencing. Nature 348:657–658

Höltke H, Ettl J, Obermaier J, Schmitz G (1991) Sensitive chemiluminescent detection of digoxigenin (DIG) labeled nucleic acids. A fast and simple protocol and its applications. In: Stanley PE, Kricka LJ (eds) Bioluminescence and Chemiluminescence: Current Status, Wiley, Chichester, pp 179–182

Honda M, Kitamura K, Mizutani Y, Oishi M, Arai M, Okura T, Igarahi K, Yasukawa K, Hirano T, Kishimoto T, Mitsuyasu R, Chermann J-C, Tokunaga T (1990) Quantitative analysis of serum IL-6 and its correlation with increased levels of serum IL-2R in HIV-induced diseases. J Immunol 145:4059–4064

Hummelen JC, Luider TM, Wynberg H (1986) Stable 1,2-dioxetanes as labels for thermochemiluminescent immunoassay. Meth Enzymol 133:531–537

Hummelen JC, Luider TM, Wynberg H (1987) Functional adamantylidene adamantane 1,2-dioxetanes: Investigations of stable and inherently chemiluminescent compounds as a tool for clinical analysis. Pure Appl Chem 59:639–650

Jain VK, Magrath IT (1991) A chemiluminescent assay for quantitation of β-galactosidase in the femtogram range: application to quantitation of β-galactosidase in lac Z-transfected cells. Anal Biochem 199:119–124

Kopecky KR, Mumford C (1969) Luminescence in the thermal decomposition of 3,3,4-trimethyl-1,2-dioxetane. Can J Chem 47:709–711

Lanzillo J (1991) Chemiluminescent nucleic acid detection with digoxigenin-labeled probes – a model system with probes for angiotensin converting enzyme which detect less than one attomole of target DNA. Anal Biochem 194:45–53

Martin C, Bresnick L, Juo RR, Voyta JC, Bronstein I (1991) Improved chemiluminescent DNA sequencing. BioTechniques 11:110–112

McCapra F (1977) Alternative mechanism for dioxetan decomposition. J Chem Soc Chem Commun, 946–948

McCapra F, Beheshti I, Burford A, Hann RA, Zaklika KA (1977) Singlet excited states from dioxetane decomposition. J Chem Soc Chem Commun, 944–946

Pollard-Knight D, Simmonds AC, Schaap AP, Akhaven H, Brady MA (1990) Non-radioactive DNA detection on Southern blots by enzymatically triggered chemiluminescence. Anal Biochem 185:353–358

Richterich P, Church GM (1992) DNA sequencing with direct transfer electrophoresis and nonradioactive detection. Meth Enzymol, Recombinant DNA, Volume H, in press

Reguero M, Bernardi F, Bottoni A, Olivucci M, Robb M (1991) Chemiluminescent decomposition of 1,2-dioxetanes – an MC-SCF-MP2 study with VB analysis. J Am Chem Soc 113:1566–1571

Sanchez-Pescador R, Running JA, Stempien MM, Urdea MS (1989) Rapid nucleic acid assay for detection of baceria with tetM-mediated tetrycycline resistance. Antimicrob Agents Chemother 33:1813–1815

Schaap AP, Handley RS, Giri BP (1987) Chemical enzymatic triggering of 1,2-dioxetanes. Aryl esterase-catalyzed chemiluminescence from a naphthyl acetate-substituted dioxetane. Tetrahedron Lett, 28, 935–938

Schaap AP, Akhaven H, Romano LJ (1989) Chemiluminescent substrates for alkaline phosphatase: applications to ultrasensitive enzyme-linked immunoassays and DNA probes. Clin Chem 35:1863–1864

Schaap AP, DeSilva R, Akhaven-Tafti H, Handley R (1991a) Chemical and enzymatic triggering of 1,2-dioxetanes; structural effects on chemiluminescence efficiency. In: Stanley PE, Kricka LJ (eds) Bioluminescence and Chemiluminescence: Current Status, Wiley, Chichester, pp 103–106

Schaap AP, Goudar JS, Romano LJ (1991b) Chemically trigerable 1,2-dioxetanes as chemiluminescent labels for biological molecules. J Biolumin Chemilumin 6:281–282

Thorpe GHG, Bronstein I, Kricka LJ, Edwards B, Voyta JC (1989) Chemiluminescent enzyme immunoassay of alpha-fetoprotein based on an AMPPD substrate. Clin Chem 35:2319–2321

Tizard R, Cate RL, Ramanchandran KL, Voyta JC, Murphy OJ, Bronstein I (1990) Imaging DNA sequences with chemiluminescence. Proc Natl Acad Sci USA 87:4514−4518

Urdea MS, Kolberg J, Warner BD, Horn T, Clyne J, Ku L, Running JA (1990) A novel method for the rapid detection of hepatitis B virus in human serum samples without blotting or radioactivity. In: Van Dyke K, Van Dyke R (eds) Luminescence Immunoassay and Molecular Applications, CRC Press, Boca Raton, pp 275−292

Voyta JC, Edwards B, Bronstein I (1988) Ultrasensitive detection of alkaline phosphatase activity. Clin Chem 34:1157

Wierynga JH, Strating J, Wynberg H (1972) Adamantylidene-adamantane peroxide: a stable 1,2-dioxetane, Tetrahedron Lett 2:169−172

Wood WG (1990) Spiroadamantane dioxetanes substrates − stable labels for luminescence-enhanced enzyme immunoassays. J Clin Chem Clin Biochem 28:481−484

Yamamoto K, Higashimoto K, Minagawa H, Okada M, Kasahara Y (1991) Simultaneous detection of antibodies to HTLV-1 and HIV-1 by chemiluminescent enzyme immunoassay. Clin Chem 37:1031

13.3 Electrochemiluminescence: Ruthenium Complexes

JOHN H. KENTEN

13.3.1 Principle and Applications

Electrochemiluminescence (ECL) is the generation of emitting excited states from electron transfer reactions of species that are generated electrochemically at the surface of an electrode on application of applied potentials [1]. These reactions have been the subject of theoretical study for a number of years and have largely centered on the use of very clean organic solvents requiring, in many cases, redistillation, recrystallization, deoxygenation, extensive drying of the reagents, and repeated polishing of the electrode surfaces used in the study [1, 2]. Prior studies in aqueous solutions have been based on the use of organic acids at pHs from 3 to 6; however, the desired pH for detection of most biomolecules is between 6 and 8 [3]. Of the ECL species which have been studied, most efforts to date have been directed at $Ru(bpy)_3^{2+}$ (bpy is 2,2'-bipyridine) due to its electrochemical properties and quantum yield, and this is the ECL label we have made use of to develop our assays.

Development of the necessary chemistry and instrumentation which allows for ECL in aqueous buffers at biological pHs (6 through 9) with great accuracy over wide concentration ranges was pivotal in designing the assays [4]. These buffer chemistries have led to novel rapid, sensitive, and quantitative assays using electrochemiluminescent labels based on the requirement for electrode proximity to achieve this type of light production [5, 6, 7]. These labels also demonstrate modulation due to molecular environment, which enables the development of assays based directly on

13.3.2

binding or other similar molecular interactions, i.e., ligand receptor interactions and hybridization [5]. We have made use of the Ru(bpy)$_3^{2+}$ ECL species to generate two labeling reagents Ru(bpy)$_3^{2+}$ n-hydroxysuccinamide ester (Origen label) and Ru(bpy)$_3^{2+}$ phosphoramidite (Origen phosphoramidite) [5, 6]. With these two labeling reagents were are able to modify many biomolecules, using Origen label to couple via the primary amines of antibodies, proteins, and amino-modified nucleic acids and Origen phosphoramidite to carry out both automated direct labeling of synthetic nucleic acids during automated synthesis and labeling of molecules via hydroxyl, sulfhydryl, and amino groups.

Assays based on the use of the labels are covered in detail in a number of publications [5, 6, 7]. The most interesting of these is the application of ECL assays to the polymerase chain reaction (PCR). The Origen ECL labels are very stable and are able to withstand the conditions of the thermocycling reaction without any problems [5, 7]. Stability of these labels allows direct incorporation of the ECL-modified oligonucleotides within the PCR and direct detection of the resultant products. Significant to the value of the Origen ECL assay system is the ability to carry out assays without the need for any user wash steps or multiple reagent additions. When applied to the immunoassay of thyroid stimulating hormone, the sensitivity and simplicity of the system resulted in an excellent assay performance [6].

These extraordinary features of the Origen ECL system are exemplified in the recently described assays for the cystic fibrosis (CF) and HIV 1 *gag* genes, in which both the extreme stability and assay simplicity were used [8]. In outline a PCR for the HIV 1 *gag* gene was carried out using standard conditions, with one of the primers biotinylated. Included within the PCR was a hybridization probe for detection of the deletion. This probe was labeled at a 3′amino group using the Origen label. At the end of the PCR, samples were loaded on to an Origen 1.5 ECL analyzer for analysis.

In conclusion use of the Origen ECL system allows rapid development of assays for all biomolecules in a format with no user washes and results in 15 min [5–8]. This new assay system is unlike the complex, temperamental, and slow nonradioactive assays previously available [9–11].

The following is an outline of one of the many assay formats possible which combines the speed, simplicity, sensitivity, and stability of the Origen ECL assays for the detection of PCR products.

13.3.2 Detection of PCR Products Using an Electrochemiluminescence Assay

Probe synthesis and labeling reagents	– Origen N-hydroxysuccinamide (IGEN) – Origen phosphoramidite (IGEN) – Dimethyl sulfoxide (VWR) – Phosphoramidites (American Bionetics, Inc.) – DNA synthesizer (Applied Biosystems, Inc.)

- Biotin X-*N*-hydroxysuccinimide ester (Clontech)
- Amino modifiers (Clontech)

- *Taq* DNA polymerase and reagent kit (Perkin Elmer Cetus)
- Thermocycler (Perkin Elmer Cetus)

PCR reagents and equipment

- Modified streptavidin beads (IGEN)
- Formamide (VWR)
- Origen 1.5 ECL analyzer (IGEN)
- Origen assay buffer (IGEN)
- Origen cell cleaner (IGEN)

Assay reagents and equipment

1. The oligonucleotides are synthesized using standard methods. Amino groups are introduced into the 5′ end of one of the PCR primers and into the 3′ end of the probe oligonucleotide sequence, which is a sequence able to hybridize to the PCR product of the 5′ biotinylated primer. These amino additions are also made following standard methods.

Probe synthesis and labeling

2. Oligonucleotides are prepared for labeling by Biogel P6 column chromatography in 0.3 M NaCl, followed by precipitation of the excluded oligonucleotide peak. Typically, 0.1 µmole make of oligonucleotide is reacted with 0.5 µmole make of Origen label in 80% dimethyl sulfoxide/phosphate buffered saline, pH 7.4.
3. Biotinylation of the oligonucleotides is performed essentially as above except biotin X-*N*-hydroxysuccinimide ester (Clontech) in 50% dimethyl sulfoxide is used for labeling.
4. The labeled oligonucleotides are precipitated with ethanol and washed to remove unincorporated label. The pellet from ethanol precipitation of the Origen label reaction should be orange-red.

Note: The Origen phosphoramidite can also be used to directly label oligonucleotides during synthesis on automated DNA synthesizers, but the modification occurs at the 5′ end of the sequence. These 5′ ECL-labeled probes are valuable for hybridizations separate from the PCR or for labeling the PCR product directly during the reaction.

1. The PCRs are carried out as normal but should include: (a) a biotinylated primer in place of one of the primer pairs to allow the synthesis of biotinylated PCR products and (b) a probe (typically 20 nM) blocked at its 3′ end to prevent its nonspecific incorporation into the PCR side products thus maintaining specificity. The probe at the end of the PCR is available for hybridization to the biotinylated PCR product during the last cycle extension step.

PCR and assay

2. At the end of the PCR thermocycles 2 µl of sample are added to 15 µg of modified streptavidin beads in 240 µl of Origen assay buffer; the reaction is inserted into the Origen 1.5, shake-incubated for 15 min, and analyzed for ECL. Analysis is carried out automatically by the Origen

1.5 after the 15 min shaking incubation. Control of the specificity of the hybridization can be carried out using the addition of up to 30% formamide in the Origen assay buffer, in addition to the sequence size and base composition. Signals from the Origen 1.5 analyzer are proportional to the amount of specific PCR product produced and, proper controls and a standard curve, can be used for quantitation.

13.3.3 Special Hints for Application and Troubleshooting

The most significant problems which arise in the PCR assays are those which are related to the PCR or to the oligonucleotides.

- If no signal is seen or if signals are low a quick check of the Origen 1.5 using its reference solutions will determine if any ECL instrumentation failure has occurred. Most likely the problem is the PCR, this should be checked by gel electrophoresis for the production of the appropriate bands from the high copy number standard in the assay.
- The oligonucleotides can be checked, by TBE-PAGE; the labeled oligonucleotides are readily separated from the unlabeled ones and analyzed by shadowing the gel with UV light against a fluorescent TLC plate. Typically over 90% will be labeled and two bands can be identified. If only one band is visible then labeling has most likely not occurred. The presence of the ECL moiety can also be confirmed by photography of the gel on a transilluminator as its fluorescence is similar to that of ethidium-stained DNA.

References

Abruna HD (1985) Electrochemiluminescence of osmium complexes, spectral and mechanistic studies. J Electrochem Soc 132:842–849

Arnold LJ, Hammond PW, Weise WA, Nelson NC (1989) Assay formats involving acridinium-ester-labeled DNA probes. Clin Chem 35:1588–1594

Blackburn GF, Shah HP, Kenten JH, Leland J, Kamin RA, Link J, Peterman J, Shah A, Talley DB, Tyagi SK, Wilkins E, Wu T-G, Massey RJ (1991) Electrochemiluminescence detection for development of immunoassays and DNA-probe assays for clinical diagnostics. Clin Chem 37:1534–1539

Casadei J, Powell MJ, Kenten JH (1990) Expression and secretion of aequorin as a chimeric antibody using a mammalian expression vector. Proc Natl Acad Sci USA 87:2047–2051

Ege D, Becker WG, Bard AJ (1984) Electrogenerated chemiluminescent determination of Ru(bpy)$_3^{2+}$ at low levels. Anal Chem 56:2413–2417

Faulkner LR, Bard AJ (1977) Techniques of electrogenerated chemiluminescence. In: Bard AJ (ed) Electroanalytical chemistry, vol 10. New York: Marcel Dekker, 1–95

Kenten JH, Casadei J, Link J, Lupold S, Willey J, Powell M, Ress A, Massey R (1991a) Rapid electrochemiluminescence assay of polymerase chain reaction products. Clin Chem 37:1626–1632

Kenten JH, Gudibande S, Link J, Friedman K (1991b) One step, fifteen minute, non-separation assay for HIV1 gag and cystic fibrosis genes using electrochemiluminescence and PCR, AACC Meeting, Nov 1991, San Diego

Kenten JH, Gudibande S, Link J, Willey J, Curfman B, Major EO, Massey R (1992) Improved electrochemiluminescent label for DNA probe assays: Rapid quantitative assays for HIV1 polymerase chain reaction products. Clin Chem 38:873−879

Leland JK, Powell MJ (1990) Electrogenerated chemiluminescence. An oxidative-reductive type ECL reaction sequence using tripropyl amine. J Electrochem Soc 137:3127−3133

Urdea MS, Running JA, Horn T, Clyne J, Ku J, Warner BD (1987) A novel method for the rapid detection of specific nucleotide sequences in crude biological samples without blotting or radioactivity; application to the analysis of hepatitis B virus in human serum. Gene 61:253−264

13.4 Bioluminescence: Luciferin

Reinhard Erich Geiger and Eva Schneider

13.4.1 Principle and Applications

Bioluminescence is a natural phenomenon found in many lower forms of life (Deluca, 1978; Deluca and McElroy, 1986; Herring, 1987). Naturally-occurring bioluminescent systems differ with regard fo the structure and function of enzymes and cofactors as well as in the mechanism of the light-emitting reactions (Burr, 1985). Due to its high sensitivity, firefly *(Photinus pyralis)* bioluminescence had been used for many years for the sensitive determination of ATP (Lundin et al., 1976). More recently, further highly sensitive bioluminescent and chemiluminescent methods have become available for many different analytes (Kricka et al., 1984; Wood, 1984; Gould and Subramani, 1988; Kricka, 1988). Recently bioluminescent detection of nucleic acid hybridization has been reported (Hauber and Geiger, 1988; 1989).

Several new enzyme substrate, all based on luciferin derivatives, which are highly sensitive for the corresponding enzymes used in detection systems, are described here. These new substrates (Geiger and Miska, 1987; Miska and Geiger, 1987) can be used for unmodified enzymes and for enzyme conjugates and applied in enzymatic activity test systems, in reporter gene tests, in enzyme immunoassays, in protein blot analysis, and in nucleic acid hybridization tests.

The test principle of these new substrates is the release of D-luciferin from D-luciferin derivatives by the action of hydrolytic enzymes. Released D-luciferin can be quantified by a luminometric detection system. The high sensitivity of these bioluminogenic substrates is obtained by both the amplification which occurs in the releasing step (e.g., one molecule of alkaline phosphatase can convert 1000 molecules of D-luciferin-O-phosphate to D-luciferin per second) and by the very sensitive bioluminescence system (*Photinus pyralis;* concentrations of 5×10^{-13} mol/l of D-luciferin can be detected (Miska and Geiger, 1987).

The bioluminescent detection system described can be applied in nucleic acid hybridization experiments using all types of nucleic acid labeling systems (DIG, biotin, BrdU, sulfone, and all other systems involving enzymatic labels; see corresponding chapters of this book). Using the bioluminescence system antibodies have to be labeled with enzymes for example, in biotin-labeled nucleic acids, either avidin has to be labeled by the respective enzyme or the enzyme used has to be labeled by biotin, forming a complex together with avidins and the labeled probe.

Bioluminescence-enhanced detection systems require convenient and reliable light-measuring instruments. In the last few years established manufacturers of photometric microtiter plate readers have also introduced readers for chemiluminescent or bioluminescent measurements. These can now bes used for measuring bioluminescence- or chemiluminescence-enhanced enzyme immunoassays in microtiter plates. If hybridization experiments can be performed in microtiter plate wells (e.g., microtiter plate wells with bottoms of nitrocellulose membranes), convenient and highly sensitive measuring equipment is available. For electronic measurement of emitted photons produced on nitrocellulose sheets, photon-counting camera systems are recommended. These innovative and precise instruments can detect photons within a very short time at an extreme sensitivy ($1 \text{ photon} \times \text{cm}^{-3} \times \text{s}^{-1}$). Magnetic storage of pictures on diskettes for later data analysis is also accomplished.

Detection of nucleic acid hybridization can also be performed using photographic films for light detection, (e.g., Kodak Tri X pan, 380 ASA or Polaroid films). The films should be developed according the procedures given by the manufacturers.

13.4.2 Reaction Scheme

The standard scheme of the bioluminescence-enhanced detection is performed as follows (also see accompanying figures):

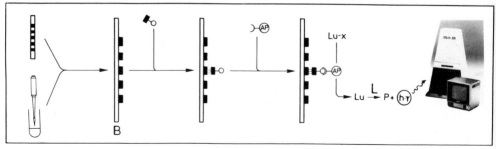

The bioluminescent detection system. *AP,* alkaline phosphatase (or β-galactosidase); *Lu-x,* corresponding D-luciferin derivative; *Lu,* D-luciferin; *L,* luciferase; *P,* oxyluciferin; *B,* nitrocellulose sheet with nucleic acids; ⊱AP , antibody enzyme conjugate; ■–○, labeled probe; h.v, emitted light

1) D-Luciferin-β-D-Galactoside $\xrightarrow{\text{β-Galactosidase}}$ D-Luciferin + β-D-Galaktose

2) D-Luciferin + O_2 + ATP $\xrightarrow[\text{Mg}^{2+}]{\text{Luciferase}}$ Oxyluciferin + h·v + AMP + PP_i

The bioluminescence detection reaction. *1*, release of luminometrically active D-luciferin from D-luciferin-*O*-β-galactoside by the action of β-galactosidase; *2*, light production by oxidation of D-luciferin by firefly luciferase *(Photinus pyralis)*

1. Incorporation of the label into the DNA probe and hybridization of labeled probe to immobilized DNA
2. Binding of the enzymatically labeled probe to the nucleic acid
3. Incubation of the labeled probe-DNA complex with a bioluminegenic substrate and simultaneous measurement of emitted light.

13.4.3 Bioluminescence-Enhanced Detection

13.4.3.1 Enzymatic Labeling with Alkaline Phosphatase

— DNA hybridization kits (purchased or self-composed)   **Standard**
— Nitrocellulose or nylon filters   **reagents**
— ATP (disodium salt), $MgCl_2$, dithiothreitol (DTT)
— *Photinus pyralis* luciferase, native or recombinant
— D-luciferin-*O*-phosphate (Novabiochem AG, CH-4448 Läufelfingen, Switzerland, or Medor GmbH, W-8036 Herrsching, FRG)

— D-luciferin-*O*-phosphate solution: 2 mmol/l D-luciferin-*O*-phosphate   **Standard**
— Buffer solution: 41 mmol/l HEPES; 2.6 mmol/l ATP (added shortly **solutions** before use); 7.8 mmol/l diethanolamine; 5 mmol/l $MgCl_2$; 3.5 mmol/l DTT; pH 8.0
— Luciferase solution: 1 mg luciferase *(Photinus pyralis)*/ml; 0.5 mol/l Tris-succinate buffer, pH 7.7; 3 mmol/l DTT
— Alkaline phosphatase label (antibody:alkaline phosphatase conjugate or biotinylated alkaline phosphatase avidin complex)
— Light detection solution: 1 ml buffer solution, 0.01 ml luciferase solution; 0.01 ml D-luciferin-*O*-phosphate solution

After hybridization of nucleic acids and binding of alkaline phosphatase label, the filter is washed with phosphate-buffered saline, dipped for a few seconds into light detection solution (about 6 ml of solution in a petri dish) and transferred to a transparent plastic tube. The tube is placed, under light exclusion, into a light-tight chamber of a photon counting camera. Photons are counted and integrated for 5 s, 5 min, and 20 min, depending on the photon counting camera system used. Hybridized DNA visualized as bright spots. Semiquantification can be performed using computer programs designed for the photon counting camera systems.

13.4.3.2 Enzymatic Labeling with β-Galactosidase

Standard reagents — Identical to above procedure

Standard solutions
— D-luciferin-O-β-galactoside solution: 0.25 mmol/l D-luciferin-O-β-galactoside
— Buffer solution: 41 mmol/l HEPES; 5 mmol/l MgCl$_2$; 2.6 mmol/l ATP (added shortly before use); pH 7.75
— Luciferase solution: 1 mg luciferase *(Photinus pyralis)*/ml; 0.5 mol/l Tris-succinate buffer, pH 7.7; 3 mmol/l DTT
— β-Galactosidase label (antibody: β-galactosidase conjugate
— Light detection solution: 1 ml buffer solution, 0.005 ml luciferase solution; 0.05 ml D-luciferin-O-β-galactosidase solution

Using β-galactosidase, two detection methods are possible: (1) detection of nucleic acids on filters using a photon counting camera and (2) detection of nucleic acids in microtiter plates using a luminometer.

Filters/photon counting camera detection After hybridization of nucleic acids and binding of β-galactosidase label, the filter is washed with phosphate-buffered saline, dipped for a few seconds into the light detection solution (about 6 ml of solution in a petri dish) and transferred to a transparent plastic tube. The tube is placed under light exclusion into a light-tight chamber of a photon counting camera. Photons are counted and integrated for 5 s, 5 min, and 20 min, depending on the photon counting camera system used. Hybridized DNA can be visualized as bright spots.

Microtiter plates/Luminometer detection After isolation of DNA from cells, blotting of nucleic acids to small nitrocellulose filters which can be placed into microtiter plate wells, and binding of β-galactosidase label (binding of mouse monoclonal anti BrdU antibody to labeled nucleic acid followed by binding of rabbit anti-mouse IgG β-galactosidase conjugate to nucleic acid-bound monoclonal antivodies), the filters are washed with phosphate-buffered saline and placed into microtiter plate wells. Thereafter 0.1 ml of D-luciferin-O-β-galactoside solution is added and incubated for 5 min. After incubation 0.1 ml light detection solution (without D-luciferin-O-β-galactoside) is added. The microtiter plate is placed under light exclusion into a light-tight chamber of a luminogenic microtiter plate reader. Photons are counted and integrated for 5 and 20 min. DNA can be visualized as bright spots in the microtiter plate wells, depending on the concentration applied. Calculation of nucleic acid concentration is performed using computer programs designed for the luminometer.

13.4.4 Special Hints for Application and Troubleshooting

— Using the methods described, experiments have been performed by labeling pBR322 probes either by biotin or by sulfonylation. Alkaline phosphatase or β-galactosidase was bound to biotinylated nucleic acid by a streptavidin: alkaline phosphatase or β-galactosidase complex.

Using sulfonylated nucleic acids for hybridization, antibody (directed against sulfonated nucleotides): alkaline phosphatase or β-galactosidase conjugates were added.

- Using BrdU as a label, growing cells incorporated BrdU from the culture medium. Thereafter cellular DNA (labeled with BrdU) was isolated. Bioluminescent detection of nucleic acids was then performed by detecting the incorporated BrdU using BrdU-specific antibodies labeled with β-galactosidase. For hybridization with either biotinylated or sulfonylated probes and for detection of BrdU-labeled nucleic acids, the bioluminescent detection systems have been the same.

- Many advantages in handling and sensitivity have been obtained using a photon counting camera system instead of photographic films (Hauber and Geiger, 1989). The detection limits can be lowered by powers of ten even at a shorter detection time 5−10 min). The method is relatively simple to perform, but one has to be careful when using nitrocellulose, since different types of nitrocellulose purchased from different distributors may contain substances which interfere with firefly luciferase *(Photinus pyralis)*. A reducing effect on luciferase activity was obtained by adding buffer to the test system in which nitrocellulose was soaked or stored only for a short time. The luciferase may be inhibited or may be denatured by compounds existing in the nitrocellulose sheets.

- The use of high quality water is recommended.

- In earlier syntheses of D-luciferin-*O*-phosphate, very small amounts of free D-luciferin could be detected in the preparations after purification using the highly sensitive bioluminescence reaction. These traces of D-luciferin sometimes influenced the blank values. By improving the purification methods D-luciferin-*O*-phosphate is now available in a highly purified grade and has a very low blank value.

- After dissolving D-luciferin-*O*-phosphate in highly pure water or 0.05 mol/l ammonium acetate, pH 6.5, aliquots should be taken and stored at −80 °C until use. For each experiment a fresh aliquot should be used.

- To purify D-luciferin-*O*-phosphate, methods using high performance liquid chromatography have been published (Miska and Geiger, 1987).

References

Burr GJ (1985) Chemi- and Bioluminescence. Marcel Dekker, New York

Deluca MA (1978) Bioluminescence and Chemiluminescence. Meth Enzymol 57, Academic Press, New York

Deluca M, McElroy WD (1986) Bioluminescence and Chemiluminescence, Part B. Meth Enzymol 133, Academic Press, New York

Geiger R, Miska W (1987) II. The bioluminescence-enhanced immunoassay. New ultrasensitive detections for enzyme immunoassays. J Clin Chem Clin Biochem 25:31−38

Gould SJ, Subramani S (1988) Review. Firefly luciferase as a tool in molecular and cell biology. Anal Biochem 175:5−13

Hauber R, Geiger R (1988) A sensitive, bioluminescence-enhanced detection method for DNA dot-hybridization. Nucleic Acids Res 16:1213

Hauber R, Geiger R (1989) The applications of a photon-counting camera in a sensitive, bioluminescence-enhanced detection system for nucleic acid hybridization. Ultrasensitive detection systems for protein blotting and nucleic acid hybridization, III. J Clin Chem Clin Biochem 27:361–363

Herring PJ (1987) Systematic distribution of bioluminescence in living organisms. J Biolumin Chemilumin 1:146–163

Kricka LJ (1988) Review. Clinical and biochemical applications of luciferase and luciferins. Anal Biochem 175:14–21

Kricka LJ, Stanley PE, Thorpe GHG, Whitehead TP (1984) Analytical Applications of Bioluminescence and Chemiluminescence. Academic Press, New York

Lundin A, Richardsson A, Thorpe A (1976) Continous monitoring of ATP-converting reactions by purified firefly luciferase. Anal Biochem 75:611–620

Miska W, Geiger R (1987) I. Synthesis and characterization of luciferin derivatives for use in bioluminescence enhanced enzyme immunoassays. New ultrasensitive detection systems for enzyme immunoassays. J Clin Chem Clin Biochem 25:23–30

Wood WG (1984) Luminescence immunoassays: problems and possibilities. J Clin Chem Clin Biochem 22:905–918

14 Fluorescence Systems

14.1 Labeling of Biomolecules with Fluorescein, Resorufin, and Rhodamine

HANS-PETER JOSEL

14.1.1 Principle and Application

Fluorescein, resorufin, and rhodamine derivatives are three important fluorescence labels which can be used for labeling of biomolecules. Although there are many more fluorescence labels, which differ in excitation/emission wavelengths, e.g., coumarins, this chapter concentrates mainly on the most popular labels because of their widespread application.

Due to their high sensitivity fluorescence methods are used in many areas of biochemical analysis and clinical chemistry: fluorescence-activated cell sorting (FACS), fluorescence microscopy and fluorescence immunoassays (Hemmilä, 1985; Raffael, 1988). In principle fluorescent probes can be covalently bound to all biomolecules: to antibodies for the already mentioned applications; to nucleic acids and DNA probes for sensitive DNA detection; to lipids for metabolism studies or for use as membrane probes; and to oligosaccharides for a variety of purposes (Haugland, 1982; Nederlof, 1990). Since biomolecules possess different functional groups, e.g., amino or thiol groups, the appropriate label derivative must be chosen.

The table includes a survey of fluorescein, resorufin, and rhodamine labels, their reactivity, and some spectroscopic data.

Perhaps the most important applications are: (a) the labeling of antibodies for direct immunofluorescence detection and (b) the labeling of streptavidin for indirect immunofluorescence. Thus, the labeling of proteins will be discussed here in further detail. Proteins have a multitude of freely accessible amino groups (lysine) to which correspondingly activated fluorescence labels can be bound. While isothiocyanates are still used very frequently, they have certain disadvantages. For example, for conjugation they need relatively basic conditions and the formed thio-urea bonds are sometimes unstable (Rypacek, 1980).

Properties of Fluorescence labels

Label	Corresponding reactive group	Excitation (nm)	Emission (nm)	Absorption coefficient
5(6)-Carboxyfluorescein-N-hydroxysuccinimide ester (FLUOS)	$-NH_2$	494	518	75 000
Fluorescein isothiocyanate (FITC)	$-NH_2$	494	520	72 000
5-Iodoacetamidofluorescein	$-SH$	490	520	>60 000
Fluorescein-5-thiosemicarbazide	$-CO$	492	518	78 000
N-(Resorufin-4-carbonyl)piperidine-4-carbonic acid-N-hydroxysuccinimide ester (RESOS)	$-NH_2$	576	587	60 000
N-(Resorufin-4-carbonyl-N'-iodacetyl-piperazine (RESIAC)	$-SH$	574	591	65 000
5(6)-Carboxyrhodamine-101-N-hydroxysuccinimide ester (RHODOS)	$-NH_2$	577	600	>50 000
Texas red (TM)	$-NH_2$	596	615	85 000

N-Hydroxysuccinimide esters (NHS esters), e.g., 5(6)-carboxyfluorescein NHS, (FLUOS), N-(resorufin-4-carbonyl)piperidine-4-carbonic acid NHS (RESOS), and 5(6)carboxyrhodamine-101-NHS (RHODOS) (see figure) do not have these disadvantages. Reaction takes place under very mild physiological conditions (pH 7–8), with high coupling efficiency. After each reaction a stable amide bond is formed as in proteins (Herrmann et al., 1989).

FLUOS

RESOS

RHODOS

Fluorescent labels can also be covalently bound to biomolecules via thiol groups. Iodacetyl derivatives have proved to be particularly useful in such instances, as they allow conversion under physiological conditions and lead to formation of stable thioether bonds (Haugland, 1982). Furthermore, apart from the labeling of high molecular weight biomolecules, they are also suited for coupling to haptens for use in the appropriate assays (fluorescence polarization immunoassays, Hemmilä, 1985).

14.1.2 Fluorescent Labeling of Proteins

− FLUOS, RESOS, RHODOS (Boehringer Mannheim)   **Standard**
− Sephadex G-50 (Pharmacia)   **reagents**
− Tween 20 (Boehringer Mannheim)
− Dimethylsulfoxide (DMSO) (Merck)

− Buffer for conjugation procedure/chromatography: Phosphate buffer, **Standard**
  0.1 mol/l; pH 7.5−8   **solution**

1. Dissolve 10 mg protein in 1 ml phosphate buffer. A solution consisting of a 10 molar excess of activated label in 500 μl DMSO is then added. Shake for 2 h at room temperature.
2. The conjugate is separated from free dye by gel chromatography, using phosphate buffer containing 0.1% [v/v] Tween 20 as solvent for removing the unspecifically bound portion.

3. Dialyzed and lyophilized.
4. The degree of conjugation can be determined by measuring the absorption at the λ max with the aid of a standard curve using free fluorescence dye.

14.1.3 Special Hints for Application and Troubleshooting

− The amount of label per biomolecule can be influenced by the molar excess of the fluorescence label and the pH.
− Due to the different reactivities of the activated labels, the molar ratio of fluorochrome to protein also varies.
− The procedure can easily be adapted to the conjugation of iodacetyl derivatives, e.g., iodacetamido-fluorescein or N-(resofurin-4-carbonyl-N'-iodacetyl-piperazine (Boehringer Mannheim) to thiol groups using a higher excess of label.

References

Haugland EP (1982) Covalent fluorescent probes. In: Excited States of Biopolymers. Steiner RF (ed), Plenum Press, New York, London
Hemmilä I (1985) Clinical Chemistry 31:359
Herrmann R, Josel H-P, Wörner W, Fetterhof TJ (1989) Conjugation of proteins to various fluorescence labels, 19th FEBS meeting, Rom, Abstract Nr FR 511
Nederlof PM, van der Flier S, Wiegant J, Raap AK, Tanke HJ, Ploem JS, van der Ploeg M (1990) Multiple fluorescence in in situ hybridization. Cytometry 11:126−131
Raffael A (1988) GIT Labor-Medizin, 89−97
Rypacek F, Drobnik J, Kalal J (1980) Fluorescence labeling method for estimation of soluble polymers in living material. Analytical Biochemistry 104:141−149

14.2 Time-Resolved Fluorescence

ELEFTHERIOS P. DIAMANDIS and THEODORE K. CHRISTOPOULOS

14.2.1 Principle and Applications

The fluorescent rare-earth chelates, and the europium chelates in particular, are used frequently as labels in time-resolved fluorometry because they possess certain advantages in comparison to conventional fluors. The fluorescent europium chelates exhibit large Stokes shifts (~290 nm) with no overlap between excitation and emission spectra and very narrow (10 nm bandwidth) emission spectra at 615 nm. Additionally, their long fluorescence lifetimes (600−1000 μs for Eu^{3+} compared with 5−100 ns for conventional fluorophores) allow use of microsecond time-resolved fluorescence measurements, which reduce the observed background signals (Diamandis, 1988; Diamandis and Christopoulos, 1990). The principle of the time-resolved fluorometric measurement is as follows: When a mixture of fluorescent compounds is excited with a short pulse of light from a laser or flash lamp, the excited molecules emit either short- or long-lived fluorescence. Although both types of fluorescence decay follow an exponential curve, short-lived fluorescence dissipates to zero in $<100\,\mu s$. If no measurements are taken during the first $100-200\,\mu s$ after excitation, all short-lived fluorescence background signals and scattered excitation radiation are completely eliminated, and the long-lived fluorescence signals can be measured with very high sensitivity. In practice, the only background signal observed when using europium chelate labels is that produced the nonspecific binding of the labeled reagents used.

Time-resolved fluorometry with europium chelates as labels has already been used extensively in immunological assays (Diamandis, 1991). More recently, the method has been used for Southern (Christopoulos et al., 1991) and western blotting (Diamandis et al., 1991). Labeling of the biospecific probe (e.g., antibody or nucleotide) can be accomplished by linking it with either Eu^{3+} or a Eu^{3+} chelator, as described in recent reviews (Diamandis and Christopoulos, 1990). Alternatively, the probe can be biotinylated; streptavidin carrying Eu^{3+} or the Eu^{3+} chelator can then be used for detection. The latter method is preferable, affords better sensitivity and will be described in detail below.

14.2.2 Reaction Scheme

The standard procedure for nucleic acid hybridization involves the following steps:

1. Incorporation of biotin into the probe (e.g., linearized plasmid pBR 328)
2. Hybridization of the probe to immobilized DNA (e.g., plasmid pBR 328 fragments)
3. Detection of the biotinylated hybrid by using a streptavidin reagent multiply labeled with the europium chelate of 4,7-*bis* (chlorosulfophenyl) 1,10 phenanthroline-2,9-dicarboxylic acid (BCPDA).

14.2.3 DNA Labeling with Biotin, Hybridization, and Detection

An example of this procedure, using linearized pBR 328 as probe and pBR 328 fragments (both from Boehringer Mannheim) as the immobilized DNA, is given below.

– Prime-it random primer kit, incorporating bio-11-dUTP as the biotinylating nucleotide (Stratagene)

Labeling reagents

14.2.3

Control reagents	– Biotinylated DNA molecular weight markers (HindIII lambda DNA digests) (Vector Laboratories)

Membranes – Hybond-C-Extra supported nitrocellulose (Amersham)

Hybridization solutions
– Prehybridization buffer: 5× Denhardt's solution: 0.1% [w/v] SDS; 100 μg/ml denatured salmon sperm DNA
– 10× Denhardt's solution: 2% [w/v] polyvinylpyrrolidone; 2% [w/v] bovine serum albumin; 2% [w/v] Ficoll
– 20× SSPE: 174 g NaCl; 27.6 g $NaH_2PO_4H_2O$; 7.4 g EDTA, pH adjusted to 7.4 with NaOH and volume adjusted to 1 liter
– 20× SSC: 175.3 g NaCl; 88.2 g trisodium citrate·$2H_2O$; pH adjusted to 7.0 with HCl and volume adjusted to 1 liter

Detection reagents and solutions
– Streptavidin-based macromolecular complex (SBMC) labeled with the europium chelate of BCPDA (CyberFluor Inc.); 15 mg/l stock solution.
– SBMC diluent (CyberFluor Inc.)

Labeling reaction Biotinylation of the linearized pBR 328 plasmid (100 ng) is accomplished with the random primer method available as a kit by Stratagene. No probe purification is required.

Hybridization reaction
1. Nucleic acid targets are separated with agarose gel electrophoresis and transferred to nitrocellulose using standard procedures (Sambrook et al., 1989).
2. The membranes are then baked for 2 h at 80°C and prehybridized in prehybridization buffer for 2 h at 65°C.
3. Hybridizations are performed in the hybridization buffer (exactly as the prehybridization buffer but with 1× Denhardt's) at 65°C, overnight. Both prehybridizations and hybridizations are performed in 20 ml solutions using a hybridization incubator.
4. Biotinylated probes are boiled for 5 min before they are added to the hybridization solution at a concentration of 10 ng/ml.
5. After hybridization, the membranes are washed as follows: 3 × 5 min with 2× SSC, 0.1% [w/v] SDS at room temperature; with vigourous shaking 3 × 5 min as above but with 0.2 × SSC, 0.1% [w/v] SDS; and 2 × 15 min 0.2× SSC, 0.1% [w/v] SDS at 65°C.

Detection The membranes are blocked in 6% [w/v] bovine serum albumin solution for 1 h. The biotinylated nucleic acid hybrids are visualized with 50-fold diluted SBMC in SBMC diluent for 3 h at room temperature with continuous rotational shaking (~1 ml per cm^2 of membrane). At the end of the incubation, the strips are washed 3× with a wash solution (50 mM Tris, pH 7.2 containing 0.05% [v/v] Tween 20 and 9 g/l of NaCl) and soaked with shaking for 1 h in the same solution. The strips are then dried with a hair dryer.

Evalua-
tion
of data
- Visual inspection: Membranes can be evaluated by observation in a UV transilluminator (side containing bands facing down).
- Instant photography: As above but photograph with a Polaroid camera with exposure time of ~ 13 s. Filters and films are identical to those used for ethidium bromide photography.
- Quantitative evaluation: Scan membranes on the Cyberfluor 615 Immunoanalyzer using the software CONT, as described elsewhere (Christopoulos et al., 1991).

14.2.4 Special Hints for Application and Troubleshooting

Here we have described the use of a novel reagent, SBMC, in Southern blotting. The SBMC consists of streptavidin covalently and noncovalently linked to BCPDA-labeled bovine thyroglobulin. Eu^{3+} ions are chelated to BCPDA to form the fluorescent complex. The details of the preparation of this reagent are given elsewhere (Morton and Diamandis, 1990). A schematic of the SBMC is shown in the accompanying figure.

SBMC binds biotin thus linking the biotinylated probe with the fluorescent Eu^{3+} chelate. No enhancement of the Eu^{3+}-BCPDA fluorescence is needed for detection, so fluorescence can be measured on dry solid-phases, e.g., polystyrene plastic or nitrocellulose membranes. Quantification of the Eu^{3+}-BCPDA fluorescence can be achieved by using a versatile

The streptavidin-based macromolecular complex. Streptavidin is covalently linked to one molecule of BCPDA-labeled bovine thyroglobulin (TG) and noncovalently linked to another two BCPDA-labeled bovine thyroglobulin molecules. Eu^{3+} acts as a bridge between components. The molar ratio of the components of this complex is 1:3.3:480 for streptavidin:TG:BCPDA. More details are given in Morton and Diamandis (1990)

14.2.4

Current and future applications of the streptavidin-based macromolecular complex

Application	Reference or comment
Immunological assays	Diamandis (1988)
Southern and dot blots	Christopoulos et al. (1991)
Western blots	Diamandis et al. (1991)
Polymerase chain reaction	Done successfully but not reported*
Northern blots	Not done
Plaque, colony lifts	Not done
Sequencing	Not done
Flow cytometry	Not done
Immunohistochemistry	Not done

* Chan A, Diamandis EP, Krajden M (1992) Submitted

time-resolved fluorometer (CyberFluor 615 immunoanalyzer) which scans the solid-phase with high resolution.

We have briefly described the application of this reagent for Southern blotting. However, SBMC has been used in many different applications and could be used for others not tested as yet. In the table, we summarize some possible applications. Biotinylated reagents are used as complementary reagents. The final step, involving SBMC binding to biotin, can be performed as described in the detailed protocol given above. About 1 and 5 pg of proteins or DNA can be detected with this reagent, respectively. Problems with this procedure, as with any Southern blot procedure may occur during the labeling reaction, transfer, hybridization, or detection.

Labeling reaction — Denature the probe before the labeling reaction.

Southern transfer — Neutralization is needed after the alkaline denaturation of DNA if nitrocellulose membranes are used.
— Remove any bubbles trapped between gel and membrane for a successful transfer.
— Cover the area around the gel with parafilm, so that the transfer buffer passes only through the gel.

Hybridization — Complete denaturation of the probe is required.
— Optimize the probe concentration.
— Do not allow membranes to dry out between prehybridization and hybridization.
— Extensive washing after hybridization reduces the background.

Detection — Extensive washing is required to remove the excess of Eu^{3+}-labeled streptavidin from the membrane. Do not leave the membrane to dry out before the washing.

14.2.4

References

Christopoulos TK, Diamandis EP, Wilson G (1991) Quantification of nucleic acids on nitrocellulose membranes with time-resolved fluorometry. Nucleic Acids Res 19: 6015−6019

Diamandis EP (1988) Immunoassays with time-resolved fluorescence spectroscopy. Principles and applications. Clin Biochem 21:139−150

Diamandis EP (1991) Multiple labeling and time-resolvable fluorophores. Clin Chem 37:1486−1491

Diamandis EP, Christopoulos TK (1990) Europium chelate labels in time-resolved fluorescence immunoassays and DNA hybridization assays. Anal Chem 62:1149A−1157A

Diamandis EP, Christopoulos TK, Bean CC (1992) Quantitative western blot analysis and spot immunodetection using time-resolved fluorometry. J Immunol Methods 147:251−259

Morton RC, Diamandis EP (1990) Streptavidin-based macromolecular complex labeled with a europium chelator suitable for time-resolved fluorescence immunoassay applications. Anal Chem 62:1841−1845

Sambrook J, Fritsch EF, Maniatis T (1989) Molecular cloning. A laboratory manual. 2nd Edition

III Enhanced Systems

15 Overview of Amplification Systems

CHRISTOPH KESSLER

15.1 Introduction

In order to detect as low as single molecules in various analytes by the described nonradioactive labeling and detection systems (see Chap. 2) it is necessary to include an amplification step.

The coupling of such amplification reactions with nonradioactive detection systems is essential in the case of low concentration target molecules; a well-known example is the AIDS virus HIV which must be detectable in extremely low concentrations ($10^2 - 10^3$ virus molecules ml^{-1} serum) for early-stage diagnosis.

Three different in vitro amplification principles can be distinguished: target amplification, signal amplification, and target-specific signal amplification; furthermore, there are also in vivo amplifications.

Target-specific amplification reactions are only known for nucleic acids. The detection of proteins may be enhanced by signal amplification or by measurement of the in vivo immunoreactive protein-specific antibodies; the same holds true for glycoproteins and haptens.

The table lists the already developed in vitro and in vivo systems for increasing the sensitivity of nucleic acid detection through target, signal, and target-specific signal amplification and protein detection through signal amplification. Cross-reference to the various amplification systems in Chaps. 16−18 is also given.

15.2 Target Amplification

In this in vitro amplification reaction (see figure) the target molecule itself is amplified. This reaction is only possible with nucleic acids through in vitro replication (exponential amplification: 2^n-fold), combined in vitro reverse transcription/transcription (mixed amplification factor: x^n-fold) or in vitro transcription (linear amplification: x-fold).

15.2.1 Replication

The best known example of a replication amplification is the polymerase chain reaction (PCR) by repeated three-step thermocycles (heat denatura-

Amplification systems

Mode of amplification	Example(s)	Cross reference	Reference(s)
Target amplification			
In vitro amplification: replication			
Elongation temperatur cycles	Polymerase chain reaction; oligonucleotide ligation assay	16.1	Saiki et al. (1985; 1988); Li et al. (1988); Nickerson et al. (1990)
cDNA synthesis/elongation temperature cycles	Polymerase chain reaction on RNA basis	16.1	Murakawa et al. (1987)
Elongation ligation temperature cycles	Repair chain reaction	16.2	Segev (1990); Segev et al. (1990)
Elongation isothermal replacement reactions	Strand displacement amplification	16.3	Alexander et al. (1991)
	RecA enhanced amplificaton	–	Zarling et al. (1991)
In vitro target amplification: ligation			
Ligation temperature cycles	Ligase chain reaction	16.4	Orgel (1989); Barringer et al. (1990)
	Ligation and amplification reaction	–	Wu and Wallace (1989)
In vitro target amplification: transcription			
cDNA synthesis/ ds-promoter-dependent transcription cycles	Nucleic acid sequence based amplification	16.5	Davey and Malek (1988); Compton (1991); Kievits et al. (1991)
	Self-sustained sequence replication	–	Guatelli et al. (1990)
cDNA synthesis/ ds-promoter-dependent transcription	Transcription-based-amplification	–	Gingeras et al. (1988); Joyce (1989); Kwoh et al. (1989)
In vitro target amplification: transcription/ligation			
cDNA synthesis/ligation/ ds-promoter-dependent transcription amplification	Promoter ligation-activated	–	Berninger et al. (1991)
In vitro amplification: increased rRNA copy numbers			
Bacterial rRNA detection	16S/23S rRNA probes	16.6	Fox et al. (1980); Roussau et al. (1986); Yehle (1987); Stull (1988); Kohne (1990)
Target-specific signal amplification			
In vitro indicator amplification: replication			
Replication cycles	Qβ replication	–	Lizardi et al. (1988)
In vitro indicator amplification: hydrolysis			
Restriction cleavage system	Restriction amplification	–	George (1991)
RNA hydrolysis system	Target cycling amplification	–	Duck and Bender (1989)
In vivo amplification of antibodies	Detection of antigen-specific antibodies	–	Malvano (1980); Avrameas et al. (1983)

Signal amplification

Coupling of binding partners	Basic macromolecules as cross-linker	–	Sodja and Davidson (1976); Renz (1983); Al-Hakim and Hull (1986)
Tree structures			
Network of indicator molecules	Probe network (trees, brushes)	–	Urdea et al. (1987); Fahrlander and Klausner (1988); Segev (1991)
	Primary/secondary/tertiary antibody trees	–	Nicholls and Malcolm (1989); Oellerich (1983)
	Peroxidase: antiperoxidase Alkaline phosphate: anti-alkaline phosphatase	–	Mason et al. (1982); Mason (1985)
Enzyme catalysis			
Enzyme-catalyzed signal generation	ELISA	–	Vogt (1978); Maggio (1980); Ishikawa et al. (1981, 1983); Kemeny and Challacombe (1988)
Conjugates with precoupled marker enzymes (hedgehog conjugates)	ELISA with polymeric enzyme conjugates	–	Ward et al. (1987)
Coupled signal cascades			
Cyclic NAD/NADH + H^+ redox reaction	NADH + H^+-coupled reduction of INT violet by ADH/DP	17	Self (1985); Johannsson et al. (1985); Stanley et al. (1985)

ADH, alkohol dehydrogenese; DP, diaphorase; ELISA, enzyme-linked immuno-sorbent assay; INT violet, *p*-iodo-nitro-tetrazolium purple

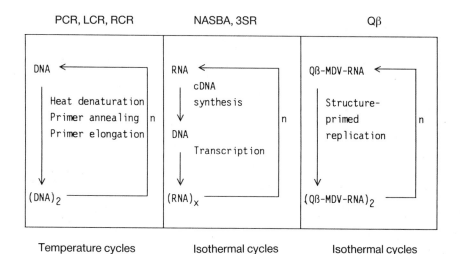

The figure categorizes the various target amplification systems according to thermocyclic reactions (PCR: polymerase chain reaction, LCR: ligase chain reaction, RCR: repair chain reaction), isothermal transcription reactions (NASBA: nucleic acid sequence based amplification, 3SR: self-sustained sequense replication) and isothermal replication reactions (Qβ: Qβ replication). An reactions are target-dependent and are characterized by exponential amplification rates.

tion, primer annealing, primer elongation) with the help of heat-stable *Taq* DNA polymerase from *Thermus aquaticus* (Saiki et al., 1985; 1988). By repeating the temperature cycle up to 60 times, the original target DNA is exponentially amplified. Since the number of target molecules is doubled in each replication cycle, a 2^n-fold amplification rate is reached in n cycles.

By using a preceding reverse transcription step, RNA can also be incorporated as a target molecule into the temperature cycles (Murakawa et al., 1987).

Alternative thermocycling reactions are the ligase chain reaction (LCR) and the repair chain reaction (RCR). In the LCR, adjacent template-bound oligonucleotides are ligated: the ligated product serves as template in the following cycle (Wu and Wallace, 1989). The RCR combines properties of PCR and LCR by limited gap-filling between two oligonucleotides with either dGTP/dCTP or dATP/dTTP and subsequent ligation of the extended primer oligonucleotide and the second stop oligonucleotide. Again, the created product serves as template in the following cycle (Segev, 1990). Strand displacement amplification (SDA) substitutes the heat denaturation step between primer elongation by isothermal strand displacement reactions starting from nicks in the primer regions (Alexander et al., 1991). RecA enhanced amplification has also recently been reported (Zarling et al., 1991).

15.2.2 Transcription

Sequence-specific amplification can also be accomplished by repeated cycles of sequential reverse transcription and transcription steps. In this type of amplification, the RNA components are selectively hydrolyzed from intermediary DNA:RNA hybrids with the enzyme RNaseH; a promoter element is then incorporated in the subsequent cDNA second strand synthesis for the final transcription reaction (NASBA: nucleic acid sequence based amplification; 3SR self-sustained sequence replication) (Davey and Malek, 1988; Guatelli et al., 1990; Compton, 1991). This amplification cycle results in the x^n-fold amplification of the target molecule; x stands for the transcription rate, n for the number of cycles.

Promotor-dependent transcription amplification (TAS: transcription-based amplification system) (Gingeras et al., 1989) takes advantage of the repeated start of the transcription reaction; x-fold transcription thus results in the x-fold linear amplification of the target sequences.

15.3 Signal Amplification

In this in vitro amplification reaction, the final signal generation is amplified by additional coupled reaction steps; this reaction can be used for the detection of nucleic acids, proteins, and glycans. As illustrated in the accompanying figure, there are four different amplification levels:

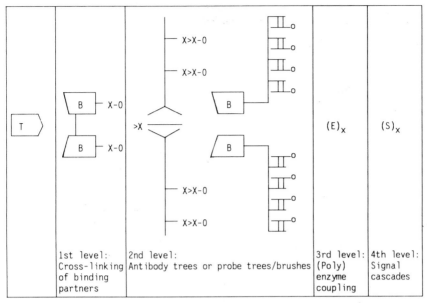

1st level: Cross-linking of binding partners	2nd level: Antibody trees or probe trees/brushes	3rd level: (Poly) enzyme coupling	4th level: Signal cascades

The figure shows various levels of signal amplification: 1. Cross-linking of binding partners (probes); 2. Generation of antibody trees (primary/secondary/tertiary antibodies) or probe trees/brushes (primary/secondary/tertiary probes); 3. use of enzymes or polyenzymes (alkaline phosphatase/peroxidase/β-galactosidase) for catalytic substrate reaction (EIA/ELISA principle); 4. coupled signal cascade (e.g., NAD^+/$NADH+H^+$ cycle). Abbserations: T = target, B = binding partner, X = modification groups, O = reporter group, E = enzyme, $(E)_x$ = polyenzyme, S = substrate, $(S)_x$ = substrate cyclic reaction

15.3.1 Coupling of the Binding Partners

This is accomplished by coupling several binding partners with bifunctional reagents and basic macromoles; the amplification factors are in the tenfold range (Renz, 1983).

15.3.2 Antibody Trees and Probe "Xmas Trees"/Probe "Brushes"

This type of signal amplification is based on the application of secondary or tertiary antibody trees; five- to tenfold amplification factors can be obtained (Oellerich, 1983). The probe Xmas tree/brush concept of signal amplification uses the binding of secondary and tertiary DNA probes resulting in branched probe structures; amplification factors up to 10^2-fold can be achieved (Urdea, 1987; Fahrlander and Klausner, 1988). An analogous concept is described by Segev (1991).

15.3.3 Enzyme/Polyenzyme Amplification

Enzyme catalysis (enzyme-immuno assay, EIA; enzyme-linked immuno-sorbent assay, ELISA) as part of the detection reaction yields up to 10^3-fold amplification factors; when polymeric marker enzymes are additionally used, a further three- to five-fold signal amplification is reached (Vogt, 1978; Maggio, 1980).

15.3.4 Coupled Signal Cascades

Amplification cascades have been described for coupled $NADH+H^+$ redox reactions (Self, 1985). These cyclic amplification reactions are frequently used to obtain a 10- to 100-fold increase in sensitivity or a marked reduction in the overall reaction time.

15.4 Target-Specific Signal Amplification

There are two different modes of target-specific signal amplification. The Qß system uses the structure-primed replication of target-bound Qß RNA; the target cycling reactions are based on repeated selective hydrolysis of target-bound probes.

15.4.1 Replication

This in vitro reaction allows for amplification of the measuring signal, which is analogous to the signal amplification reactions. However, in contrast to these, the signal is not amplified independently from the target molecule but is selectively generated from specific complexes between signal probe and target molecule. These are achieved by a specific interaction between signal-generating probe component and target molecule during the amplification reaction: only the signal-generating components bound in the complex are selectively amplified. An example of this type of amplification is the Qß amplification system (Lizardi et al., 1988). Only those Qß-specific RNA molecules which have been specifically linked to target molecules are selectively amplified. In the first amplification cycle, only the Qß-specific RNA sequences of the probe, but not the target-specific probe sequences, are replicated; the resulting Qß RNA molecule is selectively amplified through further replication steps. Qß replication thus leads to a 2^n-fold amplification of the signal RNA molecule.

15.4.2 Selective Complex Separation

Target cycling amplification is an alternative to Qß replication (Duck and Bender, 1989). In this in vitro reaction, the modified binding component is repeatedly released by selective separation from the complex between target molecule and modified binding component and thus solubilized. Detection via these target-specific amplification cycles results in a linear signal amplification (n-fold); n stands for the number of release cycles.

Thus far, target cyling has so far only been successfully applied to nucleic acid target-specific signal amplification only, using enzymes such as T4 DNA ligase.

As an alternative, cyclic cleavage of oligonucleotides hybridized to target DNA has been described (George, 1991).

15.5 In Vivo Amplification

Aside from the above in vitro amplification reactions, there are others in which specific in vivo situations are used for the detection of target molecules with increased sensitivity:

1. *Detection of rRNA sequences:* in vivo target amplification for the detection of an increased copy number of rRNA target molecules in bacteria (10^3- to 10^4-fold) with the help of rRNA-specific probes (Kohne, 1990).
2. *Detection of antigen-dependent antibodies:* in vivo target-specific signal amplification for the indirect detection of antigens (proteins, glycoproteins, haptens) via target-specifically amplified primary antibodies in sera with the help of secondary labeled antibodies (Linke and Küppers, 1988).

The two described in vivo amplification reactions are not generally applicable; they reflect specific conditions within the cell and the serum. However, in the area of protein detection, generation of amplified signals via secondary antibodies allows for the detection of low concentrations of the antigens of interest, despite the fact that direct target amplification of proteins is not possible.

References

Alexander A, Fraiser M, Little M, Malinowski D, Nadeau J, Schram J, Shank D, Walker T (1991) Isothermal, in vitro amplification of DNA by a novel restriction enzyme/DNA polymerase system – strand displacement amplification (SDA). The San Diego Conference on Nucleic Acids: The Leading Edge, San Diego, CA, Abstract 17

Al-Hakim AH, Hull R (1986) Studies towards development of chemically synthesized nonradioactive biotinylated nucleic acid hybridization probes. Nucleic Acids Res 14:9965–9976

Avrameas S, Druet P, Masseyeff R, Feldmann G (1983) Immunoenzymatic techniques. Elsevier Science Publishers, Amsterdam

Barringer KJ, Orgel L, Wahl G, Gingeras TR (1990) Blunt-end and single-strand ligations by Escherichia coli ligase: influence on an in vitro amplification scheme. Gene 89:117–122

Berninger MS, Schuster DM, Rashtchian A (1991) Promoter ligation activated transcription amplification of nucleic acid sequences. PCT Int Appl WO 91/18115

Compton J (1991) Nucleic acid sequence-based amplification. Nature 350:91−92

Davey C, Malek LT (1988) Nucleic acid amplification process. Eur Pat Appl 0329822

Duck P, Bender R (1989) Methods for detecting nucleic acid sequences. PCT Int Appl WO 89/10415

Fahrlander PD, Klausner A (1988) Amplifying DNA probe signals: a „Christmas tree" approach. BioTechnology 6:1165−1168

Fox GE, Stackebrandt E, Hespell RB, Gibson J, Maniloff J, Dyer TA, Wolfe RS, Balch WE, Tanner RS, Magrum LJ, Zablen LB, Blakemore R, Gupta R, Bonen L, Lewis BJ, Stahl DA, Luehrsen KR, Chen KN, Woese CR (1980) The phylogeny of prokaryotes. Science 25:457−463

George AL Jr (1991) Restriction amplification assay. Eur Pat Appl 0455517

Gingeras TR, Merten U, Kwoh DY (1988) Transcription-based nucleic acid amplification/detection systems. PCT Int Appl WO 88/10315

Guatelli JC, Whitfield KM, Kwoh DY, Barringer KJ, Richman DD, Gingeras TR (1990) Isothermal, in vitro amplification of nucleic acids by a multienzyme reaction modeled after retroviral replication. Proc Natl Acad Sci USA 87:1874−1878

Ishikawa E, Imagawa M, Hashida S, Yoshitake S, Hamaguchi Y, Ueno T (1983) Enzyme labeling of antibodies and their fragments for enzyme immunoassay and immunohistochemical staining. J Immunoassay 4:209−327

Ishikawa E, Kawai T, Miyai K (1981) Enzyme immunoassay. Igaku-Shoin, Tokyo

Johannsson A, Stanley CJ, Self CH (1985) A fast highly sensitive colorimetric enzyme immunoassay system demonstrating benefits of enzym amplification in clinical chemistry. Clin Chim Acta 148:119−124

Joyce GF (1989) Amplification, mutation and selection of catalytic RNA. Gene 82:83−87

Kemeny DM, Challacombe SJ (1988) ELISA and other phase immunoassays. Theoretical and practical aspects. John Wiley and Sons, New York

Kievits T, van Gemen B, van Strijp D, Schukkink R, Dircks M, Adriaanse H, Malek L, Sooknanan R, Lens P (1991) NASBA™ isothermal enzymatic in vitro nucleic acid amplification optimized for the diagnosis of HIV-1 infection. J Virol Meth 35:273−286

Kwoh DY, Davis GR, Whitfield KM, Chappelle HL, DiMichelle LJ, Gingeras TR (1989) Transcription-based amplification system and detection of amplified human immunodeficiency virus type I with a bead-based sandwich hybridization format. Proc Natl Acad Sci USA 86:1173−1177

Li H, Gyllensten UB, Cui X, Saiki RK, Erlich HA (1988) Amplification and analysis of DNA sequences in single human sperm and diploid cells. Nature 335, 414−417

Linke R, Küppers R (1988) Nicht-isotopische Immunoassays − ein Überblick. In: Borsdorf R, Fresenius W, Günzler H, Huber W, Kelker H, Lüderwald I, Tölg G, Wisser H (eds) Analytiker-Taschenbuch, Springer-Verlag, Berlin, pp 127−177

Lizardi PM, Guerra CE, Lomeli H, Tussie-Luna I, Kramer FR (1988) Exponential amplification of recombinant-RNA hybridization probes. Biotechnology 6:1197−1202

Maggio ET (1980) Enzyme-immunoassay. CRC Press, Boca Raton, Florida

Malvano R (1980) Immunoenzymatic assay techniques. Martinus Nujoff Publishers, The Hague, The Netherlands

Mason DY (1985) Immunocytochemical labeling of monoclonal antibodies by the APAAP immunoalkaline phosphatase technique. In: Bullock GR, Petrusz P (eds) Techniques of immunocytochemistry, Vol 3, Academic Press, London, pp 25−42

Mason DY, Cordell JL, Abdulaziz Z, Naiem M, Bordenave G (1982) Preparation of peroxidase-antiperoxidase (PAP) complexes for immunohistological labeling of monoclonal antibodies. J Histochem Cytochem 30:1114−1122

Murakawa GJ, Wallace BR, Zaia JA, Rossi JJ (1987) Method for amplification and detection of RNA sequences. Eur Pat Appl 0272098

Nicholls PJ, Malcolm ADB (1989) Nucleic acid analysis by sandwich hybridization. J Clin Lab Anal 3:122−135

Nickersen DA, Kaiser R, Lappin S, Steward J, Hood L, Landgren U (1990) Automated DNA diagnostics using an ELISA-based oligonucleotide ligation assay. Proc Natl Acad Sci USA 87, 8923−8927

Oellerich M (1983) Principles of enzyme-immunoassays. In: Bergmeyer HU, Bergmeyer J, Grassl M (eds) Methods of enzymatic analysis, Vol 1, Verlag Chemie, Weinheim, Germany, pp 233−260

Orgel LE (1989) Ligase-based amplification method. PCT Int Appl WO 89/09835

Renz M (1983) Polynucleotide-histone H1 complexes as probes for blot hybridization. EMBO J 2:817−822

Rossau R, van Landschoot A, Mannheim W, De Ley J (1986) Intergeneric and intrageneric similarities of ribosomal RNA cistrons of the Neisseriaceae. Int J Syst Bacteriol 36:323−332

Saiki RK, Gelfand DH, Stoffel S, Scharf SJ, Higuchi R, Horn GT, Mullis KB, Erlich HA (1988) Primer-directed enzymatic amplification of DNA with a thermostable DNA polymerase. Science 239:487−491

Saiki RK, Scharf S, Faloona F, Mullis KB, Horn GT, Erlich HA, Arnheim N (1985) Enzymatic amplification of beta-globin genomic sequences and restriction site analysis for diagnosis of sickle cell anemia. Science 230:1350−1354

Segev D (1990) Amplification and detection of target nucleic acid sequences − for in vitro diagnosis of infectious disease, genetic disorders and cellular disorders, e.g. cancer. PCT Int Appl WO 90/01069

Segev D (1991) DNA probe signal amplification. Eur Pat Appl 0450594

Segev D, Zehr S, Lin P, Park-Turkel HS (1990) Amplification of nucleic acid sequences by the repair chain reaction (RCR). 5th San Diego Conference on Nucleic Acids, AACC, Abstract Poster 44

Self CH (1985) Enzyme amplification − a general method applied to provide an immunoassisted assay for placental alkaline phosphatase. J Immunol Methods 76:389−393

Sodja A, Davidson N (1978) Gene mapping and gene enrichment by the avidin-biotin interaction: use of cytochrome-c as a polyamine bridge. Nucleic Acids Res 5:385−401

Stanley CJ, Johannsson A, Self CH (1985) Enzyme amplification can enhance both the speed and the sensitivity of immunoassays. J Immunol Methods 83:89−95

Stull TL, LiPuma JJ, Edling TD (1988) A broad spectrum probe for molecular epidemiology of bacteria: ribosomal RNA. J Inf Diseases 157:280−286

Urdea MS, Running JA, Horn T, Clyne J, Ku L, Warner BD (1987) A novel method for the rapid detection of specific nucleotide sequences in crude biological samples without blotting or radioactivity; application to the analysis of hepatitis B virus in human serum. Gene 61, 253−264

Vogt W (1978) Enzymimmunoassay. Georg Thieme Verlag, Stuttgart

Ward DC, Leary EH, Brigati DJ (1987) Visualization polymers and their application to diagnostic medicine. US Pat 4687732

Wu DY, Wallace RB (1989) The ligation and amplification reaction (LAR) − amplification of specific DNA sequences using sequential rounds of template dependent ligation. Genomics 4:560−569

Yehle CO, Patterson WL, Boguslawski SJ, Albarella JP, Yip KF, Carrico RJ (1987) A solution hybridization assay for ribosomal RNA form bacteria using biotinylated DNA probes and enzyme-labeled antibody to DNA:RNA. Mol Cell Probes 1:177−193

Zarling DA, Sena EP, Green CJ (1991) Process for nucleic acid hybridization and amplification. PCT Int Appl WO 91/17267

16 Enhanced Signal Generation by Target Amplification

16.1 Polymerase Chain Reaction Amplification

RÜDIGER RÜGER

16.1.1 Principle and Applications

The polymerase chain reaction (PCR) represents the most common and widespread method for the direct amplification of specific sequences of nucleic acid target molecules. The basic reaction is comprised of three steps:

1. Denaturation of the target DNA
2. Annealing of sequence specific primers
3. Template-specific elongation of these primers with a DNA polymerase and desoxynucleotides

The nucleic acids synthesized in the first step of polymerization serve again as target molecules in a new cycle starting with a denaturation step. Thus an exponential increase of specific target sequences is achieved (Saiki et al., 1985; Mullis et al., 1986). The use of thermostable DNA polymerase (Saiki et al., 1988) and the introduction of automated thermocyclers generally facilitated the method for application in both molecular biology and diagnostic laboratories.

PCR products can be detected in gel electrophoresis performed directly after the amplification reaction. Alternatively, the PCR product can be spotted on a membrane or, after gel electrophoresis, transferred to a membrane and detected by hybridization with a probe specific for the amplified nucleic acid. Another way to visualize the amplification products is labeling of the newly synthesized DNA strands during the PCR. Labeled PCR products can be detected directly after gel electrophoresis and Southern transfer without the need of hybridization. Besides Southern blotting, dot spotted or slot blotted labeled PCR products can also be detected after purification from the nonintegrated nonradioactively modified mononucleotides or primer sequences. This can be achieved by gel filtration or spin chromatography.

Labeled PCR products can also be applied as hybridization probes. PCR probes are strongly reduced in the content of labeled vector sequences and are defined in length and base pair composition. Thus vector sequences and target nucleic acids cross-hybridize only to a very low extent. To achieve totally vector-free PCR probes the DIG-labeled PCR products can be gel electrophoresed and purified from the gel. Another advantage of PCR labeled products is the generation of probes from sequences which are not yet molecularly cloned.

Labeling of the PCR products is achieved either by incorporation of nonradioactively modified mononucleotides during the PCR or by the use of labeled primer sequences. Nonisotopic groups used for this purpose might be, e.g., digoxigenin, biotin, fluorescein. In the case of labeled primers, the nonradioactive group is preferably positioned in the 5' region of the sequence (Levenson and Chang, 1990).

Here, the incorporation of the nonradioactive modification group digoxigenin (DIG) into the amplification products during PCR is described in more detail. The digoxigenin system for nonradioactive labeling and detection of nucleic acids is highly specific and sensitive (Kessler, 1991). A detailed description of the basic system is given in Chap. 3.

The thermostable DNA polymerase from *Thermus aquaticus (Taq)* accepts DIG-[11]-2'-desoxyuridine-5'-triphosphate (DIG-[11]-dUTP). If DIG-[11]-dUTP is added to the PCR in the first cycle, the yield of newly synthesized DNA is about half that in amplification reactions without DIG incorporation (Rüger et al., 1990). Alternatively, the nonisotopically modified mononucleotide may also be added during a later PCR cycle. The migration of DIG-labeled PCR products in gel electrophoresis is slightly reduced compared to amplification products without integration of DIG-[11]-dUTP. Direct Southern transfer and detection of DIG-PCR products resulted in the detection of single molecules (Rüger et al., 1991). Using this method it was, for example, possible to directly visualize 3.5 altogram of cloned HBV sequences. This corresponds to one to two viral genome equivalents (Rüger et al., 1991). The application of DIG incorporation during PCR was also shown for various cellular target genes (Lion and Maas, 1990; Lanzillo, 1991).

DIG-[11]-dUTP-labeled PCR products can be used as hybridization probes under standard conditions and yield sensitivities of 0.1 pg (Rüger et al., 1991).

Both the Southern-blotted DIG-labeled PCR products and membranes hybridized with DIG-labeled PCR products are detected with DIG-specific antibody conjugates (as described in Sect. 3.2). The indicator reaction is mediated either by the color substrates BCIP and NBT, resulting in blue precipitating products or by the use of the chemiluminescent substrate Lumiphos™ 530* or Lumigen PPD* (see Sect. 3.2).

---

* Trademark of Lumigen Inc., Detroit, MI, USA. Lumiphos™530 and Lumigen™PPD are the subject of U.S. patents 4,962,192 and 4,969,182 granted to Lumigen Inc., Detroit, MI, USA

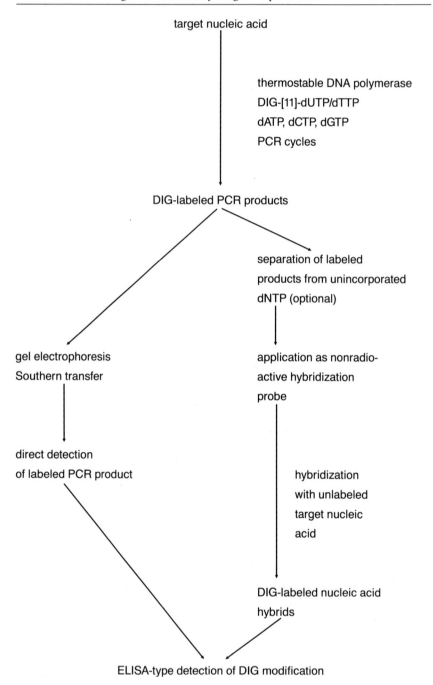

DIG labeling during PCR and BCIP/NBT or Lumiphos^{TM}530/Lumigen^{TM}PPD detection

A flow diagram summarizing the application of DIG-labeled PCR products and the appropriate detection reaction is shown in the figure on page 208.

The PCR can be carried out according to standard protocols. Within the regularly applied temperature and time profiles (including an initial denaturation step of 10 min at 92°–94 °C) and cycle numbers, the stability of DIG-[11]-dUTP will not be reduced.

As an alternative to DIG-labeling of PCR products, biotin-[16]-dUTP can be incorporated during PCR. For detection of biotin-labeled products please see Sect. 4.1. PCR with short time steps from 30 s to 2 min proved to be more advantageous compared to PCR profiles with long time steps (>2 min) with respect to the specifity of the reaction product. For short target sequences (≤ 500 bp) a two step PCR protocol combining primer annealing and elongation at one temperature (64°–75 °C) yielded highly specific products at an extensively shortened reaction time.

In conclusion, labeling of nucleic acids with DIG during PCR combines the advantage of PCR with the highly sensitive nonradioactive DIG technology in nucleic acid detection and probe generation.

16.1.2 DIG Labeling by PCR Amplification

— *Thermus aquaticus (Taq)* DNA polymerase **Standard**
— dATP, dCTP, dGTP, dTTP (Boehringer Mannheim) **reagents**
— DIG-[11]-dUTP (Boehringer Mannheim)
— Sequence-specific primers
— Mineral oil (light) (Sigma)
— Thermal cycler (e.g., Perkin Elmer Cetus)
— Reaction tubes adapted to PCR thermal cycler

— 10× PCR buffer: 100 mM Tris-HCl; 500 mM KCl; 15 mM MgCl$_2$; 0.1 mg/ **Standard**
ml gelatin; pH 8.5/25 °C **solutions**
— 10× dNTP labeling mixture: 1 mM dATP; 1 mM dGTP; 1 mM dCTP; 0.65 mM dTTP, 0.35 mM DIG-[11]-dUTP (alternatively, 0.35 mM bio-[16]-dUTP); pH 7.0/25 °C
— 10× PCR primers: 2 µM primer 1 or 2 in TE (10 mM Tris-HCl; pH 8.5/ 25 °C; 0.1 mM EDTA)

Mix in a thermal cycler-adapted reaction tube: **Reaction**
— 10 µl 10× PCR buffer **mixture**
— 10 µl 10× dNTP labeling mixture
— 10 µl of each primer stock solution respectively
— DNA template solution
— 2 U *Taq* DNA polymerase

Adjust total volume to 100 µl with sterile water. You may overlay reaction volume with 100 µl mineral oil to prevent evaporation during thermocycling.

16.1.3

Note: Use sterile pipette tips, change tips for every pipetting step and close reaction tubes before proceeding to the next reaction tube.

Negative (primer reaction) control: add sterile water instead of template DNA solution.

Reaction profile

Three step PCR:

1. Denaturation: 30s−2min/92°−94°C (according to complexitiy of template nucleic acid). An initial denaturation step of 5−10min can be carried out if, e.g., highly complex nucleic acid serves as template; nucleotides and DNA polymerase are added after this step.
2. Annealing of primers: 1−2min/40°−60°C (conditions depending on primer length/GC content)
3. Primer extension: 1−2min (depending on length of nucleic acid sequence to be amplified)/72°C

Cycles: 30 cycles followed by an additional 5′ incubation at 72°C to insure completion of all polymerization products.

Two step PCR:

This PCR profile is used for shorter target sequences (< 500bp).

1. Denaturation: 30s−1min 92°C
2. Primer annealing/primer extension: 30s−1min/65°−75°C (depending on primer length/GC content)

Cycles: 30 cycles followed by a 5′ incubation at 72°C

Note: The PCR reaction profile has to be tested for each template/primer pair.

16.1.3 Further Processing of DIG-Labeled PCR Products

For direct detection of PCR products, a 10 μl aliquot is gel electrophoresed and Southern blotted on a nylon or nitrocellulose membrane according to standard protocols (e.g., Sect. 19). Conditions for gel electrophoresis depend on size of amplified product. The blotted and immobilized PCR products are immunologically detected according to the procedure described in Sect. 3.2.11 (optical detection) or in Sect. 3.2.12 (chemiluminescent detection).

DIG-labeled PCR products can be used as hybridization probes achieving a sensitivity of at least 0.1 pg with as little as 10 μl of the total PCR reaction. To estimate the efficiency of the DIG PCR, a serial dilution of 1 μl of the total reaction from 1 to 1:1000 might be compared in a dot blot experiment with a dilution series of a standard DIG label [DIG DNA Labeling Kit (Boehringer Mannheim)].

Hybridization and detection is carried out according to the procedures described in Sects. 3.2.9 and 3.2.10.

16.1.4 Special Hints for Application and Troubleshooting

- The PCR profile may need to be adjusted for various primer sequences according to length and GC content.
- For direct detection of DIG- or biotin-labeled PCR products, DIG- and biotin-photolabeled molecular weight markers are available (Mühl-egger et al., 1990). These markers can be directly transferred to a membrane after gel electrophoresis.
- Denaturation of gel electrophoresed DIG PCR products is not necessary, yet binding of denatured single-stranded nucleic acid to the membranes might be more efficient.

References

Kessler C (1991) The digoxigenin: anti-digoxigenin (DIG) technology — a survey on the concept and realization of a novel bioanalytical indicator system. Mol Cell Probes 5:161−205

Lanzillo JJ (1991) chemiluminescent nucleic acid detection with digoxigenin-labeled probes: a model system with probes for angiotensin converting enzyme which detect less than one attomole of target DNA. Anal Biochem 194:45−53

Levenson C, Chu-an, Chang (1990) Nonisotopically labeled probes and primers. In: Innis MA et al. (eds) PCR protocols, Academic Press, Inc, London

Lion T, Haas OA (1990) Nonradioactive labeling of probe with digoxigenin by polymerase chain reaction. Anal Biochem 188, 335−337

Mühlegger K, Huber E, von der Eltz H, Rüger R, Kessler C (1990) Nonradioactive labeling and detection of nucleic acids: IV. Synthesis and properties of the nucleotide compounds of the digoxigenin system and of photodigoxigenin. Biol Chem Hoppe-Seyler 371:953−965

Mullis KB, Faloona FA, Scharf SJ, Saiki RK, Horn GT, Erlich HA (1986) Specific enzymatic amplification of DNA in vitro: the polymerase chain reaction. Cold Spring Harbor Symp Quant Biol 51:263−273

Rüger R, Höltke HJ, Sagner G, Seibl R, Kessler C (1990) Rapid labeling methods using the DIG-system: incorporation of digoxigenin in PC reactions and labeling of nucleic acids with photodigoxigenin. Fresenius' Z Anal Chem 337:114

Rüger R, Höltke H-J, Reischl U, Sagner G, Kessler C (1991) Labeling of specific DNA sequences with digoxigenin during polymerase chain reaction. In: Rolfs A et al. (ed) PCR Topics, Springer-Verlag, Berlin

Saiki R, Scharf S, Faloona F, Mullis KB, Horn GT, Erlich HA, Arnheim N (1985) Enzymatic amplification of β-globin genomic sequences and restriction site analysis for diagnosis of sickle cell anemia. Science 230:1350−1354

Saiki RK, Gelfand DH, Stoffel S, Scharf SJ, Higuchi R, Horn GT, Mullis KB, Erlich HA (1988) Primer-directed enzymatic amplification of DNA with a thermostable DNA polymerase. Science 239:487−491

16.2 Amplification of Nucleic Acid Sequences by the Repair Chain Reaction

DAVID SEGEV

16.2.1 Principle and Applications

Several techniques have been introduced to achieve senstivie DNA target amplification. The most dramatic has been the polymerase chain reaction (PCR) [1]. Other technologies include self-sustained sequence replication [2] and the ligase amplification reaction (LAR) [3, 4]. A combination of PCR and allele-specific oligonucleotide hybridization can detect single-base mismatches. Several DNA sequences polymorphisms, including those that cause certain genetic diseases [5, 6], can be detected with PCR and denaturing gradient gel electrophoresis [7]. Landergren et al. [8] have developed an oligonucleotide ligation assay (OLA) to circumvent the need for either electrophoresis or precise hybridization conditions. Similar assays have been developed by Wu et al. [9] and Alves et al. [10].

A novel technology called the repair chain reaction (RCR) has been developed for DNA target amplification of specific DNA sequences. RCR is performed using two sets of complementary oligonucleotide probe pairs that are directed against specific regions of target DNA for amplification. In each pair, there are 3' overhung base(s). The 5' and 3' ends are protected with an amino linker arm, which is useful for nonradioactive labeling. The four oligonucleotide probes hybridize to the target DNA of interest, leaving two short gaps of bases which are composed of either AT or GC. These gaps are filled with the complementary triphosphate bases by a heat stable DNA polymerase and then joined by a heat stable ligase. The target DNA and the joined probe pairs products are denatured and used as templates of second cycle gap filling. Repetition of the cycle results in a geometric amplification of the specific target region (see figure). RCR is well-suited for the detection of specific point mutations. For the detection of RCR products, we have used gel analysis and a nonisotopic capture assay with a microtiter plate format. The high efficiency of RCR is demonstrated by our studies on the detection of HPV 16 in both Caski cell line DNA and clinical biopsy targets.

16.2.2 Detection of HPV16 in Caski Cell Line DNA and in Clinical Biopsy Targets

Probe preparation Into a 500 μl microfuge tube (USA Scientific) was placed a solution containing 2.8 μg of each of the four probes in 240 μl of 20 mM Tris-HCl; 10 mM MgAc; 25 mM KAc; 0.6 mM NAD; 10 mM DTT; 0.2 mM dATP;

1. Construct two complementary oligonucleotide pairs that are complementary to the HPV 16 genome in the 6631-6287 region.

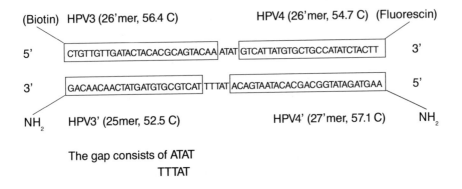

(Biotin) HPV3 (26'mer, 56.4 C)   HPV4 (26'mer, 54.7 C) (Fluorescin)

5'   CTGTTGTTGATACTACACGCAGTACAA ATAT GTCAT TATGTGCTGCCATATCTACTT   3'

3'   GACAACAACTATGATGTGCGTCAT TTTAT ACAGTAATACACGACGGTATAGATGAA   5'

NH₂   HPV3' (25mer, 52.5 C)   HPV4' (27'mer, 57.1 C)   NH₂

The gap consists of ATAT
TTTAT

2. Denature DNA at 94°C. Anneal oligonucelotides at 45°C.

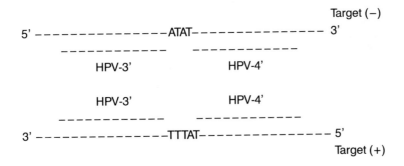

                                                        Target (−)
5' ───────────────ATAT─────────────── 3'
    ─ ─ ─ ─ ─ ─ ─ ─ ─ ─   ─ ─ ─ ─ ─ ─ ─ ─ ─ ─
            HPV-3'   HPV-4'

            HPV-3'   HPV-4'
    ─ ─ ─ ─ ─ ─ ─ ─ ─ ─   ─ ─ ─ ─ ─ ─ ─ ─ ─ ─
3' ───────────────TTTAT─────────────── 5'
                                                        Target (+)

3. Fill the gap with dATP and dTTP using *Taq* DNA polymerase, and seal with thermostable ligase at 62°C.

5'   ─────────────────ATAT──────────────── 3'
            ───────────────TATA────────────
            ───────────── AAATA ───────────
3'   ─────────────── TTTAT ────────────── 5'

4. Repeat cycles 30 or 40 times.

The repair chain reaction (RCR) of HPV16

0.2 mM dTTP; 400 units of heat stable ligase (Epicenter); 20 units of *Taq* DNA polymerase (Perkin Elmer); and 1.75 μg of human placenta DNA (Sigma). This solution is used as a master mix.

Thirty microliters of this solution was placed in a separate tube to be used as a control, and covered with 40 μl of light mineral oil (Sigma).

Target preparation Three biopsy targets were treated with a solution of 11 μl of 10 mg/μl proteinase K (Boehringer Mannheim) for 2 h at 64 °C, and were boiled for 10 min. The target solutions were treated twice with phenol/chloroform (1 : 1) and once with chloroform. The nucleic acids in the water phase were precipitated by ethanol and quantitated by their absorption at 260 nm. The total nucleic acids concentration were: (a) biopsy #41, 92 ng/μl; (b) biopsy #63, 222 ng/μl; (c) biopsy #59, 93 ng/μl. The Caski cell line targets were treated with proteinase K without further purification.

Repair chain reaction Thirty microliters of the master mix solution was placed in each of the six other tubes. The content of each tube is shown in the table:

Tube	Specific experiment
1	Biopsy #41,92 ng
2	Biopsy #63,22 ng
3	Biopsy #59,93 ng
4	Caski DNA
5	HeLa DNA: negative control
6	Human placenta DNA: negative control
7	Probe pairs only

The contents of each tube were covered with 40 μl of light mineral oil. The temperature cycle was programmed as follows: 94 °C, 3 min, 1 cycle. (90 °C, 1 min; 45 °C, 3 min; 62 °C, 1 min) 30 cycles.

The results are summarized in the accompanying figure.

Detection reaction Two approaches to detection have been used: (1) Size separation of the RCR products using 12% polyacrylamide gels, followed by staining the gels with ethidium bromide and (2) a nonradioactive capture assay format using fluorescein and biotin (see figures).

1. Coats wells of microtiter plate (Immulon II) with 10 μg of egg white avidin (Precision Chemicals) in 100 μl of PBS and incubate overnight at 4 °C.
2. Wash plates two times with PBS/0.5% Tween (PBST). Add 100 μl/well of 2% BSA in PBS. Incubate for 30 min at room temperature.
3. Bring RCR reaction product to a volume of 100 μl in PBS. Wash off 2% BSA with PBST. Place 50 μl of the RCR product solution in each well. Incubate for 1 h at room temperature.

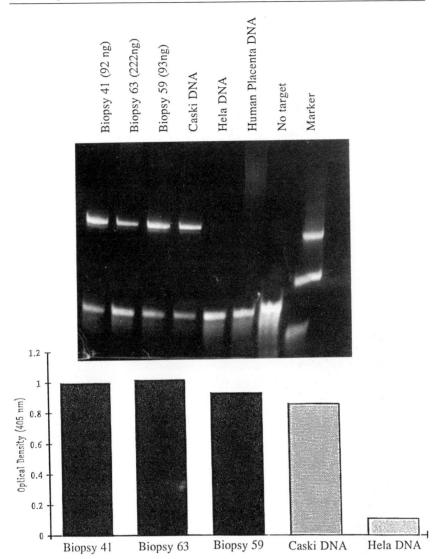

Testing HPV16 biopsy samples with RCR

4. Wash plates five times with PBST. Add 100 μl/well of a rabbit anti-fluorescein alkaline phosphatase-conjugate antibody (diluted 1:100 in PBS, from Biodesign, ME). Incubate for 1 h at room temperature.
5. Wash plates five times with PBST. Add 100 μl/well of pNpp substrate. Incubate for 15 min at room temperature. Add 100 μl/well of 2N NaOH. Read at 405 nm.

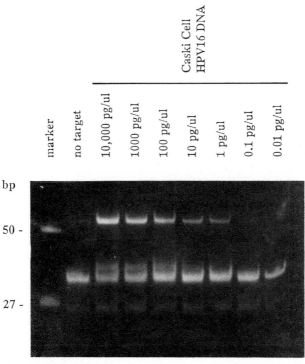

Titration of HPV16 DNA: detection with PAGE and ethidium bromide staining

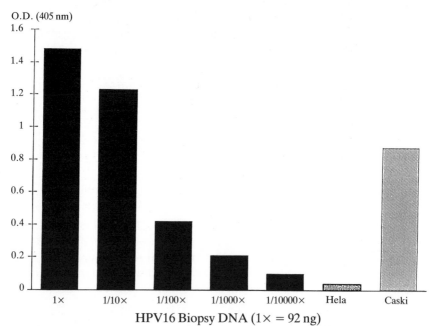

Titration of HPV16 DNA: detection using a colorimetric capture assay

The specificity of the HPV16 RCR was tested as follows: **Specificity**
1. Negative DNAs:
   − HBVp3.2 in PBR322
   − HIV(kX2, cloned in plasmid PBR322)
   − PBR322
   − *Chlamydia trachomatis*
   − HeLa (cell line that contains HPV18)
   − MT2 (cell line that contains HTLV)
   − U037 (a normal human cell line)
   − HPV6 (HPV6 positive by Southern)
2. Positive DNAs:
   − Caski (cell line that contains HPV16)
   − Biopsy samples (HPV16 positive by Southern)

16.2.3 Summary

In this study, we have developed an RCR system for the amplification of
HPV16 sequences using the Caski DNAs isolated from cervical biopsy
samples as targets. In addition, a fast (4h) and efficient nonradioactive
capture and detection assay was also developed for the quantitation of the
RCR products.

The results obtained from this HPV16 RCR study are:

1. The Caski target titration experiment showed that after 40 cycles of
   RCR, 1 pg (or approximately 180 copies) of Caski DNA target can be
   amplified and resulted in a clearly discernible band in an ethidium
   bromide-stained gel.
2. The results on RCR using clinical samples as targets showed that all
   three Southern hybridization-typed HPV16-positive samples tested
   also gave strong positive signals in our nonradioactive capture assay
   and strong bands in an ethidium bromide-stained gel.
3. Titration of the cervical biopsy DNA targets showed that approximately 10 pg of the phenol-extracted biopsy genomic DNA can be specifically amplified using 40 cycles of HPV16 RCR followed by our direct
   capture color assay.
4. A panel negative DNA targets, which includes HPV6 and HPV18, was
   tested using the HPV16 RCR in this study. All the negative DNAs
   tested did not give false positive signals after 40 cycles of RCR.

References

1. Saiki RK, Scharf S, Falcona F, Mullis K, Horn GT, Erlich HA, Arnheim N (1985)
   Enzymatic amplification of beta-globin genomic sequences and restriction site
   analysis for diagnosis of sickle cell anemia. Science 230:1350

2. Guatelli JC, Whitfield KM, Kwoh DY, Barringer KJ, Richman DD, Gingeras TR (1990) Isothermal, in vitro amplification of nucleic acids by multienzyme reaction modeled after retroviral replication. Proc Natl Acad Sci USA 87:1874
3. Wu DY, Wallace RB (1989) The ligation amplification reaction (LAR) − amplification of specific DNA sequences using sequential rounds of template-dependent ligation. Genomics 4:560
4. Barringer K, Orgel L, Wahl G (1990) Blunt-end and single-stranded ligations by E. coli ligase: influence on an in vitro amplification scheme. Gene 89:117
5. Saiki RK, Bugawan TL, Horn GT, Mullis KB, Erlich HA (1986) Analysis of enzymatically amplified beta-globin and HLA-DQ alpha DNA with allele-specific oligonucleotide probes. Nature (London) 324:163
6. Chebab FF, Doherty M, Cai S, Kan YW, Cooper S, Rubin EM (1987) Detection of sickel cell anemia and thalassemia. Nature (London) 329:293
7. Orita M, Iwahara H, Kanazawa H, Hayashi K, Sekiya T (1989) Detection of polymorphisms of human DNA by gel electrophoresis as single-strand conformation polymorphisms. Proc Natl Acad Sci USA 86:2766
8. Landegren U, Kaiser R, Sanders J, Hood L (1988) A ligase-mediated gene detection technique. Science 241:1077
9. Wu DY, Nozari G, Schold, Conner BJ, Wallace RB (1988) Direct analysis of single nucleotide variation in human DNA and RNA using in situ dot hybridizing synthetic oligonucleotide probes. Nucleic Acids Res 16:8723
10. Alves AM, Carr FJ (1988) Dot-blot detection of point mutations with adjacently hybridizing synthetic oligonucleotide probes. Nucleic Acids Res 16:8723

16.3 Isothermal Amplification of DNA Targets by Strand Displacement Amplification

MICHAEL C. LITTLE, JAMES G. NADEAU, G. TERRANCE WALKER,
JAMES L. SCHRAM, MELINDA S. FRAISER, AMY ALEXANDER,
AND DOUGLAS P. MALINOWSKI

16.3.1 Principle and Applications

Strand displacement amplification (SDA) is an isothermal method for amplifying DNA targets (optimally 50−100mer) at 37°C. The approach employs a unique combination of two enzymes − a restriction enzyme capable of nicking a hemimodified restriction site and a DNA polymerase possessing strand displacing activity but lacking 5′ → 3′ exonuclease activity [2, 3]. The method exploits the ability of certain restriction enzymes to create a nick on one DNA strand within the hemimodiefied restriction site. This nick-sensitive site is created by a target extension reaction that places thiolated dNTPs within the restriction site on the amplification primer. Following the creation of the nick and dissociation of the restriction enzyme, the afforded 3′-OH is extended by a DNA polymerase to create a copy of part of the restriction site and the target DNA sequence. The strand that is displaced is captured by its complementary primer and the

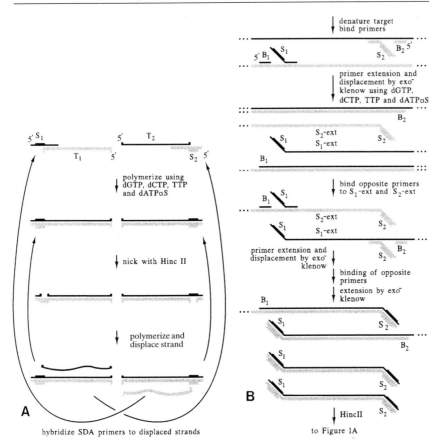

Strand displacement amplification (SDA). **A** Mechanics of SDA reaction. **B** Target generation in SDA

process repeats. By allowing this process to occur with both strands of a target DNA, products of one reaction feed into the other, creating an exponential amplification of target (see figure).

The early practice of SDA using DNA targets with defined 3′ends utilized restriction enzymes to cleave the target sequence from a longer DNA sequence [2]. While this method of generating target sequences for amplification was satisfactory for single and multicopy target sequences present within genomic *Mycobacterium tuberculosis* and *M. bovis* DNA, it presented an obstacle to the simple workflow required for clinical utility. A newer method of generating target sequences (see figure), which occurs ab initio in the SDA reaction, requires only the addition of another set of primers (B1, B2) to serve as bumpers. The exquisite simplicity of this approach is that no additional enzymes are needed and that it exploits the strand displacing property of the DNA polymerase.

The choice of target and primer sequences is of key importance in the practice of SDA, since the system operates at low stringency at 37 °C (without temperature cycling). Primer-dimer hybrids between all four primers, but especially the amplification primers, should be avoided. The choice of specific target sequences within a longer sequence should be made based upon the lowest possible secondary structure within the desired target sequences. It is best to choose target sequences of low G+C content that do not fold into secondary structures which tie-up the primer-binding regions at the 3′ end. To select and screen these primer and target sequences we use computer programs such as the Zuker RNAFOLD program in PC-GENE (Intelligenetics) and OLIGO (National Biosciences). In general, bumper oligos (13−15 mers) are positioned approximately five bases 5′ to the first base of the target sequence which hybridizes to the amplification primer. A generic 5′ → 3′ design of the amplification primers might include 18 bases of noncritical sequence, followed by the restriction site for the enzyme of choice, and 13−15 bases of target-binding sequence. Of course, the absolute lengths of target-binding sequence in the bumper oligos and in the amplification primers may change with the G+C content to allow annealing to occur at 37°C.

16.3.2 Reaction Scheme

The SDA reaction shown in the figure is divided into two parts for clarity: (1) Amplification of DNA sequences from the displaced products of the previous reaction cycle (see part **B** of figure) and (2) target generation using bumper oligonucleotides.

16.3.3 Strand Displacement Amplification

The protocol described below is for the amplification of a *M. tuberculosis* complex insertion element, IS6110 [1]. The region of the IS6110 that is used for target amplification ranges from nucleotide positions 954−966 and 1032−1044 for bumper primers B1 and B2, while the amplification primers S1 and S2 anneal at positions 972−984 and 1011−1023. In the example below, SDA products are detected by extension of a ^{32}P-labeled primer to diagnostic lengths, although nonradioactive detection is also possible.

Reaction reagents
- *Hinc*II restriction endonuclease (New England Biolabs; Life Technologies/BRL), 50 U/μl
- Exonuclease-free (exo⁻) Klenow DNA polymerase (US Biochemical), 5 U/μl
- dCTP, dGTP, TTP, 5′-α(S)dATP (Pharmacia)
- Acetylated bovine serum albumin (BRL)
- Dithiothreitol, molecular biology grade (Sigma)
- KH_2PO_4 (Fisher)
- $MgCl_2$ (Sigma)
- SDA primer S1
  (5′dTTGAATAGTCGGTTACTTGTTGACGGCGTACTCGACC)

- SDA primer S2
  (5'dTTGAAGTAACCGACTATTGTTGACACTGAGATCCCCT)
- Bumper primer B1 (5'dTGGACCCGCCAAC)
- Bumper primer B2 (5'dCGCTGAACCGGAT)
- Human placental DNA (Sigma)
- 1-Methyl-2-pyrollidinone (Sigma)
- *Mycobacterium tuberculosis* genomic DNA (ATCC or Becton Dickinson Bactec-grown cultures)

- Detector primer (5'dCGTTATCCACCATAC)   **Detection**
- γ-^{32}P-ATP 3000 Ci/mmol: 10 mCi/ml (New England Nuclear)   **reagents**
- T4 Polynucleotide kinase: 10 U/μl (New England Biolabs)
- Tris HCl (Sigma)
- $MgCl_2$
- TBE: Tris-borate-EDTA (Sigma)

Prealiquot and store at −20°C. Open and use only once.   **Reaction**
- dNTP stock solution: 10 mM each dCTP, dGTP, TTP, 5'-α(S)-dATP in **solution**
  water
- 60 mM $MgCl_2$: in water
- Bovine serum albumin: 1 mg/ml in water
- Human placental DNA: 0.2−1 mg/ml in water
- Dithiothreitol: 10 mM in water
- SDA buffer: 500 mM $KiPO_4$, pH 7.6
- 1-Methyl-2-pyrollidinone: NMP, 30% [v/v] in water
- Primer stock: 5 μM each S1 and S2; 500 nM each B1 and B2 in water
- Target dilution mix: 5 mM KPO_4; 0.01 μg/μl placental DNA

- Detector probe: 10 μM in water   **Detection**
- 10× Kinase buffer: 500 mM Tris-HCl, pH 8; 100 mM $MgCl_2$ **solutions**
- 0.5× TBE: TBE salts diluted in water

- High voltage power supply (constant power to 55 W)   **Standard**
- Vertical sequencing gel box   **equipment**
- 0.2 mm Sequencing gel plates, combs, spacers

1. Target DNA is diluted serially in target dilution mix and set aside. Prepare fresh daily.
2. A master reaction mix is prepared fresh as below for each reaction;
   5 μl each of:
   - 10 mM dNTP stock solution
   - 10 mM DTT
   - 1 mg/ml BSA
   - 60 mM $MgCl_2$
   - SDA buffer
   - 30% [v/v] NMP
   - primer stock

16.3.4

3. To 35 µl master reaction mix is added diluted target and water to a final volume of 46 µl in a 0.5 ml microfuge tube.
4. Incubate the tubes from step 3 at 95 °C for 3 min; transfer to a 37 °C water bath for at least 3 min.
5. Add 150 U *Hinc*II (3 µl) and 5 U exo⁻ Klenow DNA polymerase (1 µl) and incubate at 37 °C for approximately 2 h. Terminate the reaction by incubating at 95 °C 2 min or 65 °C 10 min.

³²P-labeled primer The following reaction is incubated at 37 °C 20 min and terminated at 65 °C for 10 min:
  − 2 µl 10× kinase buffer
  − 2 µl 10 µM detector probe
  − 2 µl 10 U/µl polynucleotide kinase
  − 14 µl γ-³²P-ATP

Detection reaction
1. To 10 µl of each SDA reaction product is added 1 µl of ³²P-labeled probe in a 0.5 ml microfuge tube. The tube is heated at 95 °C for 2 min and then at 37 °C for at least 2 min.
2. To each tube in step 1, add 1 U exo⁻ Klenow and incubate at 37 °C for 10 min.
3. Terminate the reaction by the addition of 12 µl 50% urea in 0.5× TBE, containing 0.01% each xylene cyanol and bromophenol blue.
4. Place at 95 °C for 2 min. Load onto 10% acrylamide gels containing 50% urea, 0.5× TBE. Electrophorese at 55 W until bromophenol blue dye reaches bottom of gel.
5. Remove gel using exposed film as support and cover with plastic wrap. Transfer to X-ray cassette containing intensifying screens, and place a piece of Kodak AR film over the gel. Expose at −70 °C overnight and develop the film.
6. SDA products appear as 35 and 56 nucleotide bands on the film.

16.3.4 Special Hints for Application and Troubleshooting

SDA reaction To minimize contamination from carryover of high concentrations of target, or from amplicons, several suggestions are given.

Carryover contamination
  − A separate set of pipettors that are never exposed to target or amplicon should be used for all reagent preparation and access to enzymes.
  − The use of aerosol resistant tips (ART, Fisher Scientific) should be used for all reagent deliveries.
  − The frequent changing of gloves and the physical segregation of pre- and post-SDA activities is also helpful.
  − Always include a negative control (no target) to monitor false positive signals and reagent purity.

No amplification products The SDA reaction is sensitive to high levels of extraneous DNA (>10 ug), Mg concentration (2 mM excess over nucleotide-chelated Mg is optimal), and exact reaction temperature (37°C ± 1°C). Close attention to these parameters is important.

Signals present in the negative controls sometimes accumulate over time due to the presence of amplification product (amplicon) contamination. These can be minimized by attention to lab hygiene and following the suggestions above regarding aerosol tips and gloves.

Signals in negative controls

References

1. Eisenach KD, Cave MD, Bates JH, Crawford JT (1990) Polymerase chain reaction amplification of a repetitive DNA sequence specific for *Mycobacterium tuberculosis*. J Infectious Diseases 161:977−981
2. Walker GT, Little MC, Nadeau JG, Shank DD (1992) Isothermal in vitro amplification of DNA by a restriction enzyme/DNA polymerase system. Proc Nat Acad Sci USA, 89:392−396
3. Walker GT, Fraiser MS, Schram JL, Little MC, Nadeau JG, Malinowski DP (1992) Strand displacement amplification − an isothermal, in vitro DNA amplification technique. Nucleic Acids Res. 20:1692−1696

16.4 Ligase Chain Reaction

GEORGE H. SHIMER JR. AND KEITH C. BACKMAN

16.4.1 Principle and Applications

Ligase chain reaction (LCR), employing just oligonucleotide probes and DNA ligase, is capable of detecting approximately 1000 copies of a specific target DNA sequence in the presence of a vast excess of other DNA sequence information. Since the first description in 1989 (Backman and Wang, 1989; Royer et al., 1989; Wallace, 1989; Wu and Wallace, 1989; Orgel, 1989; Richards and Jones, 1989) LCR has been improved by the employment of a thermostable DNA ligase in conjunction with non-radioactive detection (Bond et al., 1990). The ability of LCR to distinguish between normal β^A- and sickel β^S-globin genotypes from 10 μl of whole blood using probes having a mismatch or full complementarity at the point of ligation has recently been reported (Barany 1991), demonstrating the potential of LCR in allele-specific detection.

A schematic of LCR is presented in the accompanying figure. The two pairs of complementary oligonucleotide probes A, A' and B, B' are selected such that they will hybridize in a contiguous fashion on the target DNA sequence. Only the probes B and A' are phosphorylated at their 5' ends. The solution of thermostable DNA ligase, probes, and target DNA

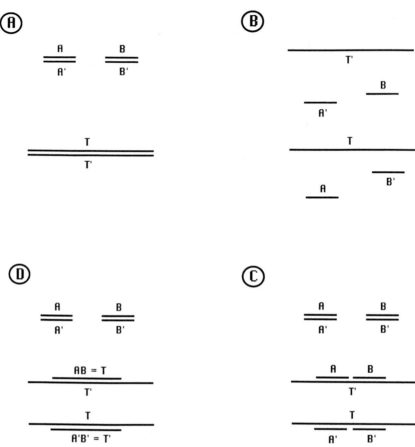

Ligase chain reaction. **A** A large molar excess of two sets of complementary oligonucleotide probes are provided (A plus A', and B plus B') such that A and B can hybridize adjacently on the desired target strand T', and A' and B' can hybridize adjacently on the complementary strand T. **B** The mixture is denatured at high temperature, separating all species into single strands. **C** Upon cooling, the probes hybridize to any target which may be present. Adjacently hybridized probes are joined by the action of DNA ligase. The products of the reaction (A joined to B, or A' joined to B') are functionally equivilant to new target strands (T or T'), effectively doubling the number of targets initially present. The process can be repeated; such repetitions result in exponential growth in the number of target equivalents

are heated and cooled, allowing the target sequence to organize the probes in the correct orientation so that they can be joined by DNA ligase. The resulting ligation product can functionally serve as a DNA target for aligning complementary probes during subsequent rounds of temperature cycling. Repetition of this autocatalytic process results in geometric accumulation of ligated product that can be described by the expression:

$$P_f = T_i (1 + \alpha)^n,$$

in which P_f is the amount of ligated product formed, T_i is the initial amount of target, α is the efficiency of each cycle of amplification and n is the number of cycles.

Oligonucleotide probes ranging in length from 20 to 25 bases with theoretical melting temperatures between 60° and 70°C are optimal for use in LCR. Recently, properties of oligonucleotide probes, thermostable DNA ligase, exogenous DNA, and the effects resulting from varying their relative concentrations in LCR have been discussed (Shimer and Backman, 1992). Our main experience has been detection using isotopic methods and that is what we detail here; however, approaches to nonisotopic detection will be briefly presented at the end of this chapter.

16.4.2 Ligase Chain Reaction Assay

- EPPS, Tris-HCl, Tris-base, boric acid, EDTA, urea, NAD^+ **Reagents**
  (Sigma Chemical Co.)
- Bromophenol blue (Mallinckrodt)
- Xylene cyanole FF, polyacrylamide (19:1), formamide
  (Bio-Rad Laboratories)
- Human placental DNA, bovine serum albumin (Sigma Chemical)
- *T. thermophilus* DNA ligase (Epicentre Technologies)
- Terminal deoxynucleotidyl transferase (Life Technologies/BRL)
- $[\alpha\text{-}^{32}P]\text{-}3'\text{-}dATP$ (Du Pont)

- $10\times$ LCR buffer: 0.4 M EPPS, pH 7.8; 0.8 M K^+ as a combination of **Buffers**
  KOH and KCl, 0.1 M NH_4Cl; 0.1 M $MgCl_2$; 0.1 mg/ml BSA
- TE: 10 mM Tris-HCl, pH 7.8, 1 mM EDTA
- Stop buffer: 9.8 ml deionized formamide, 0.5 mg bromophenol blue,
  0.5 mg xylene cyanol, 0.2 ml 0.5 M EDTA; pH 8.0
- TBE: 80 mM Tris-HCl, pH 8.0; 80 mM boric acid; 1 mM EDTA

- Background DNA solution: 0.1 mg/ml human placental DNA, or its **Solutions**
  equivalent, in TE
- Target DNA solution: Prepared by serially diluting the desired target in
  background DNA solution. If the target DNA is a plasmid it is advised
  that the plasmid be linearized so that the strands can physically separate
  upon denaturation
- Oligonucleotide probe solutions: 10^{12} oligonucleotides per µl in TE
- DNA ligase solution: *T. thermophilus* DNA ligase at 8.0×10^{-6} units/µl
  (Barker et al., 1985) in $1\times$ LCR buffer supplemented with 0.1 mM
  NAD^+

- COY Model 50 TempCycler (Coy Laboratory Products)   **Equipment**

Either probe B or A′ is $[^{32}P]$-labeled by combining 1.0 µl of oligonucleotide **Labeling**
at 10^{13} molecules/µl with 10.0 µl $[\alpha\text{-}^{32}P]\text{-}3'\text{-}dATP$ and 1.0 µl of terminal **reaction**

deoxynucleotidyl transferase (Tu and Cohen, 1980) in a 20 μl volume using the buffer supplied by the manufacturer. The reaction is incubated at 37°C for 1.5 h and the unincorporated $[\alpha\text{-}^{32}P]\text{-}3'\text{-dATP}$ is separated from the oligonucleotide by column chromatography on a 1.0 ml Sephadex G-50 column equilibrated with TE. Peak fractions are pooled, dried, and dissolved in 100 μl of H_2O.

LCR assay Obtaining the most accurate discrimination between samples containing the correct target sequence and the target-independent background reaction is best achieved by preparing a single assay solution consisting of buffer, probes (including the $[^{32}P]$-labeled probe) and NAD^+ that is sufficient to run the desired number of assays. This solution is aliquoted into separate assays followed by the addition of the necessary amounts of background DNA, with and without target DNA, and ligase. Since target-independent ligation results in product formation in the absence of any targets, it is necessary to run controls containing everything but target DNA. The final LCR assay of 20 μl contains 10^{12} each oligonucleotide probe (including the $[^{32}P]$-labeled probe), 0.3 μg background DNA, 1× LCR buffer, 0.5 mM NAD^+, and 1.2×10^{-6} units of DNA ligase.

LCR assays on two zero target control and two target-containing reactions is performed, step by step as follows:

1. Mix together:
   - 7.3 μl 10× LCR buffer
   - 4.3 μl 10 mM NAD^+
   - 4.3 μl oligonucleotide probe A at $10^{12}/\mu l$
   - 4.3 μl oligonucleotide probe A' at $10^{12}/ml$ phosphorylated at 5'-end
   - 3.7 μl oligonucleotide probe B at $10^{12}/\mu l$ phosphorylated at 5'-end
   - 4.3 μl oligonucleotide probe B' at $10^{12}/\mu l$
   - 12.0 μl $[^{32}P]$-labeled probe B at ca. $5 \times 10^{10}/\mu l$
   - 19.8 μl deionized H_2O
2. Pipette 14.0 μl of this mixture into each of four siliconized 0.65 ml Eppendorf tubes.
3. Add 3.0 μl of background DNA plus target DNA into two tubes and an equal amount of background DNA solution with no target DNA into the remaining two tubes.
4. Heat the samples to 100°C for 2 min to denature the target DNA, cool to room temperature, centrifuge to collect the sample in the bottom of the tube, and cool on ice for several minutes.
5. The LCR assay is initiated by adding 3.0 μl of the ligase solution to each tube on ice. While still on ice, overlay the solutions with 10 μl of mineral oil and then briefly centrifuge at room temperature before transferring to the temperature cycling apparatus.
6. Cycle the samples between 90°C and 55°C for the required number of cycles. With the Coy model 50 TempCycler use a ramp time equal to 0.01 with soak times of 30 s at the upper and lower temperatures. The temperature cycling may also be accomplished by transferring the samples between water baths at 90°C and 55°C every minute.

7. After the desired number of cycles the samples are briefly centrifuged at room temperature following the 55 °C segment of the cycle. Aliquots (typically 1.4 μl) are transferred to 2.0 μl of stop buffer.
8. Samples from subsequent cycles (generally at 3 cycles intervals) are collected in the same manner.

After all the aliquots have been collected, they are heated to 95 °C for 2−3 min. The ligated and unligated probes are resolved by electrophoresis at 18 W for 45 min on a 20×40×0.04 cm, 19:1 cross-linked 15% polyacrylamide gel containing 50% (w:v) urea in TBE. Following electrophoresis the gel is autoradiographed to visualize the extent of amplification. The amount of ligated product is readily quantified by liquid scintillation counting (LSC), using the autoradiograph as a template to excise the ligated and unligated oligonucleotides. These gel slices are counted in 4.0 ml of an aqueous-based LSC cocktail. The extent of amplification is determined from the ratio of radioactivity associated with the ligated oligonucleotides to the total radioactivity associated with the ligated and unligated oligonucleotides in each lane. These values are corrected by subtracting separate background for the LSC instrument and a zero ligation control lane from the polyacrylamide gel. No correction for quenching is necessary.

Detecting and quantitating the extent of ligation

16.4.3 Nonradioactive Detection of Amplification Products

− Biotin phosphoramidite (Glen Research)
− ELISA amplification system (Life Technologies/BRL)
− DIG-[11]-dUTP (Boehringer Mannheim)
− Anti-digoxigenin: alkaline phosphatase (Boehringer Mannheim)
− Streptavidin-coated microtiter plates (Omega Specially Instrument)

Additional reagents

Nonradioactive detection of LCR products has been achieved by replacing the A probe with its chemically 5′-biotinylated counterpart and with a B probe that has been enzymatically modified with dUTP-DIG and terminal deoxynucleotidyl transferase (Nickerson et al., 1990). The LCR is performed in the same manner as above except only four 4.0 μl aliquots are removed to 50 μl of 50% formamide in 1× LCR buffer. The samples are loaded onto the streptavidin-coated microtiter plate, incubated for 30 min at room temperature, and the plates are washed and color developed as detailed in Nickerson et al. (1990). LCR performed in this manner is able to discriminate between only 10^5 targets and the background reaction.

Detection reaction

References

Backman KC, Wang C-NJ (1989) Method for detecting a target nucleic acid sequence. Eur Pat Appl, 0 320 308

Barany F (1991) Genetic disease detection and DNA amplification using cloned thermostable ligase. Proc Natl Acad Sci USA 88:189−193

Barker DG, Johnson AL, & Johnson LH (1985) An improved assay for DNA ligase reveals temperature-sensitive activity in cdc9 mutants of Saccharomyces cerevisiae. Mol Gen Genet 200:458−462

Nickerson DA, Kaiser R, Lappin S, Stewart J, Hood L, Landegren U (1990) Automated DNA diagnostics using an ELISA-based oligonucleotide ligation assay. Proc Natl Acad Sci USA 87:8923−8927

Orgel LE (1989) Ligase based amplification method. World Intellectual Property Organization, WO 89/09835

Richards RM, Jones T (1989) Method and reagents for detecting nucleic acid sequences. PCT Int Appl WO 89/12696

Royer GP, Cruickshank KA, Morrison LE (1989) Template-directed photoligation. Eur Pat Appl 0 324 616

Shimer Jr. GH, Backman KC (1992) Ligase chain reaction, in methods in molecular biology volume; modern bacteriology methods. In: Howard JJ, Walker JM (eds) Humana Press, USA, in press

Wallace BR (1989) Method of amplifying and detecting nucleic acid sequences. Eur Pat Appl, 0 336 731

Wu DY, Wallace RB (1989) The ligation amplification reaction (LAR): Amplification of specific DNA sequences using sequential rounds of template-dependent ligation. Genomics 4:560−569

16.5 Isothermal Target Amplification

E. JAMES

16.5.1 Principle and Applications

Nucleic acid sequence-based amplification (NASBA™*) is a rapid and simple nucleic acid amplification methodology. NASBA yields amplification of a nucleic acid target in the order of one billion-fold in less than 2 h at a single temperature and requires no specialized equipment for its use.

Characteristically, NASBA technology directly amplifies specific single-stranded RNA sequences. Using a natural RNA template, a standard NASBA reaction comprises three enzymes, two specifically designed primers, nucleoside triphosphates, and appropriate buffer components.

Primer 1 comprises approximately 45 bases and contains at its 5' end a sequence recognized by T7 RNA polymerase. The remaining 18−22 bases are complementary to the 3' side of the target sequence. The approximately 20-base primer 2 derives from the opposite (5') side of the target sequence. The three enzymes involved in the reaction are T7 RNA polymerase, AMV reverse transcriptase (RT), and RNase H.

To illustrate the typical sequence of events, a single RNA molecule can be followed through the NASBA process.

---

* NASBA™ is a trademark of Cangene Corp.

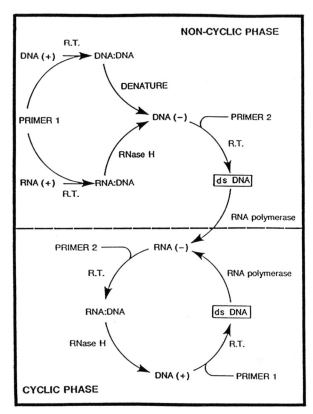

Process for the continuous homogeneous isothermal amplification of RNA or DNA.
[for references see: Compton J (1991) Nature 350:91−92 and Sect. 15.2.2]

In the „noncyclic" phase of the reaction (see figure):

1. Primer 1 anneals to the RNA target sequence.
2. RT, which is present in the reaction mixture, uses the dNTPs, also in
   the mixture, to extend the 3′ end of primer 1, creating a cDNA copy of
   the RNA template and forming an RNA:DNA hybrid.
3. RNase H recognizes this hybrid as substrate and hydrolyzes the RNA
   portion of the hybrid, leaving single-stranded DNA.
4. Primer 2 anneals to the resulting DNA, again forming a substrate suitable for RT extension of the primer.
5. The preceding reaction renders the promoter portion of the nucleic acid
   sequence double-stranded and transcriptionally active.
6. Recognizing the now functional promoter, T7 RNA polymerase generates an RNA transcript using the cDNA strand as template, generating
   many copies from each template molecule.
7. Each „new" RNA molecule becomes available as template for further
   amplification in the „cyclic" phase (see figure) of NASBA. This phase
   is operationally similar to the previous phase, except the RNA template

is now anti-sense to the original. Consequently, primer annealing and extension occur in reverse order. Again, many copies are generated from each RNA template, resulting in exponential amplification.

The NASBA process continues as described above in a homogeneous fashion under isothermal conditions, yielding a 10^9-fold amplification in approximately 1½h. While these steps have been outlined separately for clarity, all the described reactions occur concomitantly and continuously.

The NASBA process produces three types of specific products: double-stranded DNA (dsDNA), single-stranded RNA (ssRNA), and RNA:DNA hybrid, with all three being visible as bands on an agarose gel.

In both purified and heterologous samples, NASBA technology produces target-specific products (unpublished data from Dr. Larry Malek, Cangene Corporation, presented at the 1991 Miami Winter Symposium). For particularly complex samples or an extremely low proportion of target, the use of nested primers offers further enrichment of the final product. In this circumstance, primer 1 of the nested set, P1', anneals to the amplified RNA product of the first reaction and amplifies as before. The amplified product of this set of reactions is the same sense as the original target.

NASBA technology has been documented to have a <0.3% error frequency (based on cloned products). Sensitivity of NASBA has been demonstrated clinically; as few as ten HIV molecules in a 100 µl sample of human blood can be detected. Nucleic acid isolated from 1 ml of either fresh or frozen plasma samples taken from AIDS patients consistently resulted in specific product amplification, with essentially the same sensitivity as model systems (unpublished data from Dr. Peter Lens, Organon Teknika, presented at the 1991 Miami Winter Symposium).

NASBA is applicable in most research situations where other amplification technologies are used and has specific features of particular benefit. Working directly from an RNA template, NASBA technology requires no additional preparation for amplifying RNA samples. The RNA products can be sequenced directly from the reaction with 99% readability (unpublished data from Roy Sooknanan). The DNA products from NASBA reactions clone well also.

The isothermal nature of the NASBA reactions does not promote amplification of dsDNA. Consequently, ssRNA, even in low concentrations, can be amplified in the presence of native genomic DNA in the sample. This feature proves useful when investigating the active expression of a gene; the presence of the mRNA can be distinguished from the gene itself.

Certain investigations, such as *in situ* amplification, may also benefit from the single, low reaction temperature. As a general benefit, NASBA technology needs no specialized equipment.

NASBA also holds great promise in the area of clinical diagnostics. Many diseases of clinical concern are caused either by RNA viruses or by the activation of „silent" genes, resulting in mRNA formation.

The ability of NASBA to detect rare RNA molecules allows detection of both gene activation and the presence of infectious agents in blood, even in the presence of a large uninfected background, and long before an immune response is detectable. This latter feature has particular value in HIV screening, due to the potentially lengthy lag time between infection and antibody formation in AIDS.

Besides disease detection, clinical uses of NASBA also extend to: screening for genetic disorders, diagnosis and monitoring of cancers, and testing for disease susceptibility.

Finally, NASBA may enable rapid detection of viable pathogens in food samples and will also have application in agriculture, aquaculture, veterinary diagnostics, and forensic sciences.

16.5.2 Nucleic Acid Sequence Based Amplification: General Procedure

Stock reagents/ solutions

- 1 M Tris-HCl, pH 8.5 (Sigma Chemical Company)
- 2 M KCl
- 1 M MgCl$_2$ (Sigma Chemical Company)
- 100 mM each ATP, CTP, GTP, UTP (Pharmacia)
- 100 mM each dATP, dCTP, dGTP, dTTP (Pharmacia)
- 250 mM dithiothreitol (DTT)
- 100 pmoles/µl primer stocks
- Dimethylsulfoxide (DMSO) (BDH, Anala'R' Grade)
- Bovine serum albumin (BSA) (Boehringer Mannheim Canada)
- T7 RNA polymerase (Pharmacia)
- AMV reverse transcriptase (Seikagaku)
- RNase H (Pharmacia)
- RNAguard^{TM} (Pharmacia)
- Ultrapure water (Baxter)

Final Reaction Conditions

- 40 mM Tris-HCl, pH 8.5
- 50 mM KCl
- 12 mM MgCl$_2$
- 2 mM NTPs (each)
- 1 mM dNTPs (each)
- 10 mM DTT
- 0.2 µM Primers (each)
- 15% [v/v] DMSO
- 100 µg/ml BSA
- 20 units T7 RNA polymerase
- 8 units AMV reverse transcriptase
- 0.1–0.3 units RNase H
- 12.5 units RNAguard
- Template determined by sample (see below)
- Water to 25 µl

- Add primers as a mixture containing 5 pmoles of each primer, 3.75 μl of 100% DMSO and water to make a final volume of 6.0 μl
- Add enzymes as a mixture containing the three enzymes plus BSA
- NTP and dNTP mixtures are made by mixing equal volumes of each of the four 100 mM stocks.
- Mg^{2+} concentration is critical; tested commercial stocks are recommended.
- Add buffer as a 2.5× concentrate which includes Tris, $MgCl_2$, KCl, dNTPs and NTPs.
- The number of template molecules in the reaction may be determined by the sample used. For control reactions approximately 10^4 molecules is suggested.
- Incubate the final reaction mixture at 40 °C for 90 min.

16.5.3 Special Hints for Application and Troubleshooting

Single-stranded DNA amplification — NASBA technology will amplify nucleic acid templates other than natural single-stranded RNA. Single-stranded DNA molecules with discrete ends, such as those prepared from restriction enzyme digests, enter directly into the „cyclic" (see figure) phase of the reaction. In general, however, DNA must have an additional priming step performed before beginning the standard NASBA amplification reaction previously described. Stringency in this preliminary priming step is important for superior amplification; if this requirement is observed, amplification in the same order of magnitude as with ssRNA results.

Cross-contamination As with other amplification systems, problems that occur in performing NASBA relate to its sensitivity; care must be taken not to cross-contaminate samples or contaminate the work area with reaction products. In addition, when working with RNA templates, additional care should be taken to avoid RNase contamination of reagents and reaction vessels.

16.6 The Potential of rDNA in Identification and Diagnostics

ERKO STACKEBRANDT and WERNER LIESACK

16.6.1 Principle and Applications

More than any other molecule, analyses of ribosomal RNA species, especially the small subunit rRNA, have changed our understanding about the evolution and genealogy of organisms and opened the field of rapid diagnostic probing of pathogens and strains isolated from the environment. A prerequisite was the development of rapid sequencing techniques of rRNA. These allowed analysis of the primary structures of rRNA and rRNA genes from about 2000 strains of pro and eukaryotes in the relatively short period of 12 years. Starting with 5S rRNA analysis (Hori and Osawa 1979) and oligonucleotide cataloging of 16S rRNA (Woese and Fox, 1977) in the late 1970s, cloning of large rDNA species (Olsen et al., 1986) and, above all, reverse transcriptase (RT) sequencing of rRNA (Lane et al., 1985) marked a breakthrough in the rapid elucidation of natural relationships of bacteria. Thus, information about the location of evolutionarily conserved sequences in the large rDNA species 16S/18S and 23S/28S accumulated and led to the design of a set of primers that facilitated elucidation of the primary structure of rDNA. Today, with gene amplification and sequence analysis of double-stranded PCR products (Böttger 1989) even the RT method can be considered as outdated. Conserved stretches located at both termini of the 16S rDNA genes allow amplification of almost the entire gene. Internal primers that were developed for RT sequencing can also be used as amplification and sequencing primers in the generation of partial rDNA amplicons and in their sequence analysis.

The following areas of proven or potential application of 16S rDNA have been explored:

1. Diagnostics of pure cultures:
   − Amplification with taxon-specific primers and identification via fragment length
   − Amplification with conserved primers (e.g., the 8/27−1525/1541 pair) followed by probing with a series of taxon-specific probes (dot blot, Southern), sequence analysis of the complete or taxonomically significant regions of the primary structure, and restriction fragment length polymorphism analysis
2. Identification of communities including unculturable strains
   − Amplification with conserved primers carrying restriction sites, generation of a clone library, sequencing of a representative number of clones, development of specific probes and PCR primers for subsequent analysis of clones, isolated strains, and detection of organisms in situ

The diagnostic potential of rRNAs and their genes is based on (a) the presence of taxon-specific stretches that can be targeted by oligonucleotide probes and PCR primers and (b) the large number (about 10^4) of molecules per cell, allowing identification of individual organisms. The latter approach to identification requires detection by nonradioisotopic labels, which are incorporated into the molecule either directly, e.g., by fluorescent dyes (fluorescein, tetramethylrhodamine, nitrobenzofuran), or via enzymes (alkaline phosphatase, horseradish peroxidase detected by chemiluminescence) linked to capture systems (primary and/or secondary antibodies, streptavidin) that detect reporter groups, e.g., digoxigenin, biotin, sulfone, or N-2-acetylaminofluorene.

This section briefly describes methods to obtain DNA of sufficient purity to perform PCR amplification of rRNA genes. The decision about which lysis techniques to use depends upon whether the sample is a pure culture, mixed culture or environmental in origin. Various 16S rDNA-based techniques that are available for identification of culturable and unculturable strains are also outlined. Information on the amplification of eukaryotic 16S-like coding regions has been published (Sogin 1990; White et al., 1990) but will not be discussed here.

16.6.2 Isolation of Genomic DNA from Pure Cultures

The original protocols for the isolation of bacterial DNA, published by Marmur (1964) and Kirby (1964), have been reviesed by Owen and Pitcher (1985) and Johnson (1985 a, b; 1991). For a broad range of, in particular, Gram-negative bacteria, a „soft lysis" can easily be obtained by incubation of the bacterial suspension with sodium dodecyl sulphate (SDS) at 60°C (Ibrahim et al., 1992).

Most Gram-positive organisms can be lysed by incubation with lysozyme followed by treatment with SDS. In cases where these procedures fail, other gentle enzymatic lysis techniques have to be developed. Mycobacteria, for example, can be lysed by successive treatment with lysozyme and proteinase K (Böddinghaus et al., 1990). Application of these techniques allows isolation of high molecular weight DNA, which not only can be used in PCR-mediated amplification but also in restriction enzyme analyses or the generation of gene libraries. Alternative lysis procedures include „microwave oven" treatment (Bollet et al., 1991) or freeze-thaw lysis (Woods and Cole 1989). The latter procedure includes several cycles of freezing of the sample in a −70°C dry-ethanol bath and thawing in a hot water bath. Moderate temperatures (up to about 75°C) for the thawing process will rlease high molecular weight DNA, harsher treatment (temperatures above 90°C) will lead to some extent to fragmentation of the DNA. PCR-mediated amplification of rDNA can be performed successfully with fragmented genomic DNA. Consequently, whenever ‚soft lysis' techniques fail mechanical lysis procedures can be applied, e.g., French press or mini-bead beater (Weisburg et al., 1991). Ribosomal RNA genes have even been amplified from DNA isolated from autoclaved material (Barry and Gannon 1991).

All described lysis techniques can be performed in 1.5 ml tubes and applied even to lyophilized material (Wisotzky et al., 1990; Weisburg et al., 1991) reducing the need for growing cultures.

In most cases purification of genomic DNA suitable for PCR-mediated amplification can be obtained from lysed samples either by performing several extractions with phenol/chloroform followed by ethanol precipitation or by using commercial kits according to the instructions of the manufacturer.

16.6.3 Isolation of Genomic DNA from Mixed Cultures and Environmental Samples

Cells from aquatic environments are obtained either by centrifugation, by different filtration techniques (Bej et al., 1990; Giovannoni et al., 1990), or by using filter cartridges (Sommerville et al., 1989). Lysis be the same as described above for pure cultures and PCR can be performed either directly after lysis (Bej et al., 1991) or after additional purification steps (Sommerville et al., 1989).

Isolation of DNA from soil or sediment samples is likely to be more laborious. Additional purification steps depend upon the composition and amount of humic acids, clay, and organic compounds and have to be empirically adjusted for each habitat. The reader is referred to the appropriate literature, e.g., Ogram et al., 1987; Steffan et al., 1988; Pillai et al., 1991; Tsai and Olson 1992; Liesack and Stackebrandt 1992).

16.6.4 PCR-Mediated Amplification of 16S rDNA

The protocol described below is suitable for amplification of the complete 16S rDNA (about 1.5 kb) from almost all bacteria using the primer pair targeting *E. coli* positions $9-27$ and $1492-1513$ (see table).

The annealing temperature indicated yields mostly the one specific 16S rDNA product only, but should be optimized whenever unspecific PCR products are detected in analytical gels. The tables list additional primers for amplification of 16S and 23S rDNA. While many of these primers have already been proven to be successful in PCR studies often have only been included in RT sequencing of rRNA or direct sequencing of PCR products. Their suitability as amplification primers still needs to be tested.

16.6.5 Amplification of 16S rDNA Genes

1. Add $10-500$ ng of genomic DNA dissolved in TE buffer (0.01 M Tris HCl, pH 8.0, 1 mM EDTA), 10 µl of 10× amplification buffer (0.5 M KCL, 0.1 M Tris HCl, pH 8.4, 0.015 M $MgCl_2$, 0.1% gelatin), 100 pg of each primer dissolved in TE buffer (pH 8.0), 10 µl of dNTP (20 nmol each) from 2 mM solutions of dATP, dGTP, dCTP, or dTTP (each at pH 7.0) to a 0.5 ml tube; add water to give a final volume of 99.5 µl.
2. Incubate the mixture for 2 min at 98 °C to denature the genomic DNA, centrifuge for a few seconds, add 0.5 µl (2.5 units) of *Taq* polymerase (Boehringer Mannheim) and overlay with about 80 µl light mineral oil.
3. Starting with the annealing step, perform 30 cycles using the thermal profile: (1) annealing: 48 °C for 2 min; (2) primer extension: 72 °C for 2.5 min, and (3) denaturation: 94 °C for 1 min. Primer extension of the last cycle should be for 8 min to fully extend all PCR products.

Summary of primers for the PCR amplifications of genes coding for the small subunit rRNA of broad taxonomic groups

E. coli location	5' Oligomer 3'	Specificity
Foward primers		
8–27	AGAGTTTGATCMTGGCTCAG	Most bacteria[a]
50–68	AACACATGCAAGTCGAACG	Most bacteria[b], not planctomycetes
333–348	TCCAGGCCCTACGGG	Archaeobacteria[b]
342–357	CTACGGGRSGCAGCAG	Most bacteria[b]
515–533	GTGCCAGCMGCCGCGG	Almost universal[a]
906–922	AAACTYAAAKGAATTGACGG	Almost universal[a]
1100–1115	GCAACGAGCGCAACCC	Most bacteria[a]
Reverse primers		
124–107	ACGYGTTACKCACCCGT	Many bacteria[a]
325–309	TCAGGCTCCCTCTCCGG	Many eukaryotes[b]
357–343	CTGCTGCCTCCCGTA	Most prokaryotes[b]
536–519	GWATTACCGCGGCKGCTG	Almost universal[a]
704–685	TCTACGRATTTCACCYCTAC	α-, σ-Proteobacteria, *Fusobacteria*[a]
704–685	TCTACGCATTTCACYGCTAC	β-, γ-Proteobacteria[a]
704–685	TCTRCGCATTYCACCGCTAC	Most gram-positives, cyanobacteria[a]
704–690	TCTACGCATTTCACC	Most bacteria[b]
704–690	TTACAGGATTTCACT	Most Archaeobacteria[b]
704–690	TCCAAGAATTTCACC	Most eukaryotes[b]
803–786	CTACTSGGGTATCTAATC	Most bacteria[c]
926–907	CCGTCAATTCMTTTRAGTTT	Almost universal[a]
1115–1100	AGGGTTGCGCTCGTTG	Almost universal[b]
1242–1224	CATTGTAGCATGCGTGAAG	Many bacteria[c]
1301–1282	CGGTGTGTRCAAGGCCC	Many bacteria[c]
1407–1391	ACGGGCGGTGTGTRC	Almost universal[a]
1513–1492	TACGGYTACCTTGTTACGACTT	Almost universal[a]
1541–1525	AAGGAGGTGWTCCARCC	Most prokaryotes[a]
Forward primers with linkers		
8–27	ccgaattcgtcgacaacAGAGTTTGATCCTGGCTCAG EcoRI SalI	Most bacteria[d]
9–27	gcgggatccGAGTTTGATCCTGGCTCAG BamHI	Most bacteria[e]
Reverse primers with linkers		
1243–1224	ggccgtcgacGCCATTGTAGCACGTGTGYA SalI	Most bacteria[e]
1512–1492	cccgggatccaagcttACGGCTACCTTGTTACGACTT XmaI HindIII BamHI	Many bacteria[d]
1541–1525	cccgggatccaagcttAAGGAGGTGATCCAGCC XmaI HindIII BamHI	Almost universal[d]

Nonstandard abbreviations: M = A or C; R = A or G; Y + C or U; K = G or T; W = A or T.
[a] Lane (1991).
[b] N. Pace, personal communication
[c] Stackebrandt and Charfreitag (1990)
[d] Weisburg et al. (1991)
[e] Liesack and Stackebrandt (1992)

Primers for the sequence analysis of genes coding for the large subunit rRNA that could also be used for PCR amplification (From: Lane, 1991)

E. coli location	5' Oligomer 3'

Forward primers

130	CCGAATGGGVAAGGG
256	AGTAGYGGCGAGCGAA
577	GCGTRCCTTTTGTAKAATG
1104	WGCGTAAYAGCTCAC
1623	AAACCGWCACAGGTRG
1948	GTAGCGAAATTCCTTGTCG
2069	GACGYAAAGACCCCRTG
2253	GGYACAAADTGACCCC
2512	CCTCGATGTCGRCTC
2669	AGTACGAGAGGACCGG
0008	YTGAARGCATCTAA

Reverse primers

115	GGGTTBCCCCATTCGG
242	KTTCGCTCGCCRCTAC
559	CATTMTACAAAAGGYACGC
1091	RGTGAGCTRTTACGC
1608	CYACCTGTGWCGGTTT
1930	CGACAAGGAATTTCGCTAC
2053	CAYGGGGTCTTTRCGTC
2241	ACCGCCCCAGTHAAACT
2498	GAGYCGACATCGAGG
2654	CCGGTCCTCTCGTACT
2747	GYTTAGATGCYTTC

Acknowledgements. Work performed for this contribution was supported by a grant from the Australian Research Council AD 8931593.

References

Barry T, Gennon F (1991) Direct genomic DNA amplification from autoclaved infectious microorganisms using PCR technology. PCR Meth Appl 1:75

Bej AK, Mahbubani MH, Miller R, DiCesare JL, Haff L, Atlas RM (1990) Detection of coliform bacteria in water by polymerase chain reaction and gene probes. Appl Environ Microbiol 56:307–314

Bej AK, Mahbubani MH, DiCesare JL, Atlas RM (1991) Polymerase chain reaction-gene probe detection of microorganisms by using filter-concentrated samples. Appl Environ Microbiol 57:3529–3534

Böddinghaus B, Wolters J, Heikens W, Böttger EC (1990) Phylogenetic analysis and identification of different serovars of Mycobacterium intracellulare at the molecular level. FEMS Microbiol Lett 70:197–204

Böttger EC (1990) Rapid determination of bacterial ribosomal RNA sequences by direct sequencing of enzymatically amplified DNA. FEMS Microbiol Lett 65:171–176

Bollet C, Gevaudan MJ, de Lambarellie X, Zandotti C, de Micco P (1991) A simple method for the isolation of chromosomal DNA from Gram positive or acid-fast bacteria. Nucleic Acids Res 19:1955

Giovannoni SJ, De Long EF, Schmidt TM, Pace NR (1990) Tangential flow filtration and preliminary phylogenetic analysis of marine picoplankton. Appl Environ Microbiol 56:2572−2575

Hori H, Osawa S (1979) Evolutionary change in 5S rRNA secondary structure and a phylogenetic tree of 54 5S rRNA species. Proc Natl Acad Sci USA 76:381−385

Ibrahim A, Liesack W, Stackebrandt E (1992) Polymerase chain reaction − gene probe detection system exclusive to pathogenic strains of Yersinia enterocolitica. J Clin Microbiology 30:1942−1947

Johnson JL (1985a) Determination of DNA base composition. Meth Microbiol 18:1−31

Johnson JL (1985b) DNA reassociation of RNA hybridization of bacterial nucleic acids. Meth Microbiol 18:33−74

Johnson JL (1991) Isolation and purification of nucleic acids. In: Stackebrandt E, Goodfellow M (eds) Nucleic Acid Techniques in Bacterial Systematics, John Wiley & Sons, Chichester, pp 1−20

Kirby KS (1964) Isolation and fractionation of nucleic acids. Progr Nucl Acid Res 3:1−31

Lane DJ (1991) 16S/23S rRNA Sequencing. In: Stackebrandt E, Goodfellow M (eds) Nucleic Acid Techniques in Bacterial Systematics, Wiley & Sons, Chichester, pp 115−176

Liesack W, Stackebrandt E (1992) Occurence of novel types of bacteria as revealed by analysis of the genetic material isolated from an Australian terrestrial environment. J Bacteriol 174:5072−5078

Marmur J (1961) A procedure for the isolation of deoxyribonucleic acid from microorganisms. J Mol Biol 3:208−218

Ogram A, Sayler GS, Barkay T (1987) The extraction and purification of microbial DNA from sediments. J Microbiol Meth 7:57−66

Olsen GJ, Lane DJ, Giovannoni SJ, Pace NR (1986) Microbial ecology and evolution: A ribosomal RNA approach. Ann Rev Microbiol 40:337−365

Owen RJ, Pitcher D (1985) Current methods for estimating DNA base composition and levels of DNA-DNA hybridization. In: Goodfellow M, Minnikin DE (eds) Chemical methods in bacterial systematics, Academic Press, London, p 67−93

Paul JH, Cazares L, Thurmond J (1990) Amplification of the rbcL gene from dissolved and particulate DNA from aquatic environments. Appl Environ Microbiol 56:1963−1966

Pillai SD, Josephson KL, Bailey RL, Gerba CP, Pepper IL (1991) Rapid method for processing soil samples for polymerase chain reaction amplification of specific gene sequences. Appl Environ Microbiol 57:2283−2286

Sogin ML (1990) Amplification of ribosomal RNA genes for molecular evolution studies. In: Innis MA, Gelfand DH, Sninsky JJ, White TJ (eds) PCR Protocols, Academic Press, San Diego, pp 307−314

Sommerville CC, Knight IT, Straube WL, Colwell RR (1989) Simple, rapid method for direct isolation of nucleic acids from aquatic environments. Appl Environ Microbiol 55:548−554

Stackebrandt E, Charfreitag O (1990) Partial 16S rRNA primary structure of five Actinomyces species: phylogenetic implications and development of an Actinomyces israelii-specific oligonucleotide probe. J Gen Microbiol 136:37−43

Steffan RJ, Goksoyr J, Bej AK, Atlas RM (1988) Recovery of DNA from soils and sediments. Appl Environ Microbiol 54:2908−2915

Tsai Y-L, Olson BH (1992) Detection of low numbers of bacterial cells in soils and sediments by polymerase chain reaction. Appl Environ Microbiol 58:754−757

Weisburg WG, Barns SM, Pelletier DA, Lane DA (1991) 16S ribosomal DNA amplification for phylogenetic study. J Bacteriol 173:697−703

White TJ, Bruns T, Lee S, Taylor J (1990) Amplification and direct sequencing of fungal ribosomal RNA genes for phylogenetics. In: Innis MA, Gelfand DH, Sninsky JJ, White TJ (eds) PCR protocols, Academic Press, San Diego, pp 315−322

Wisotzkey JD, Jurtshuk Jr P, Fox GE (1990) PCR amplification of 16S rDNA from lyophilized cell cultures facilitates studies in molecular systematics. Curr Microbiol 21:325–327

Woese CR, Fox GE (1977) Phylogenetic structure of the prokaryotic domain: the primary kingdoms. Proc Natl Acad Sci USA 74:5088–5090

Woods S, Cole ST (1989) A rapid method for the detection of potentially viable Mycobacterium leprae in human biopsies: a novel application of PCR. FEMS Microbiology Letters 65:305–310

17 Signal Amplification Systems: Substrate Cascade

RONALD I. CARR, F. KWAN WONG, and DAMASO SADI

17.1 Principle and Applications

Self and colleagues (Self, 1985; Johansson et al., 1985; 1988; Zaid et al., 1988) described a technique for markedly increasing the sensitivity of alkaline phosphatase dependent enzyme-linked immunosorbent assay (ELISA) procedures. This increase is achieved by having alkaline phosphatase, the primary enzyme, produce an activator for a secondary enzyme-substrate system, within which marked amplification occurs. We have used this system to study antibodies to casein, bovine serum albumin, ovalbumin, cardiolipin (Carr et al., 1987), and DNA (Jones et al., 1989); to quantitate murine immunoglobulin levels in serum and tissue culture supernatants (Carr et al., 1992); and to quantitate immune complexes (Carr et al., 1987). It is very flexible since it can be sparing of limited samples to be screened (e.g., mouse serum or monoclonals), or if the added sensitivity is not needed up to 100-fold less alkaline phosphatase conjugate can be used.

17.2 Amplification of Alkaline Phosphatase Substrates

Reagents
- Nicotinamide adenine dinucleotide phosphate (NADP), MW 765.5 (Sigma)
- p-Iodonitrotetrazolium violet (INT), MW 505.7 (Sigma)
- Diaphorase (Sigma)
- Alcohol dehydrogenase (Sigma)
- Ethanol
- 0.4 N HCl

Buffer preparation
- Diethanolamine buffer:
  5.20 ml of 98% [v/v] stock diethanolamine (BDH) is diluted to 800 ml with distilled water. 0.203 g $MgCl_2$ is added, dissolved, and the pH adjusted to 9.5 with 0.2 N HCl (approximately 10−15 ml). Adjust to a final volume of 1 l. This gives final concentrations of 50 mM diethanolamine, 1.0 mM $MgCl_2$.
- Sodium phosphate buffer:
  Add 90 ml 0.2 M Na_2HPO_4 and 35 ml 0.2 M NaH_2PO_4 to 800 ml distilled

water. Adjust pH to 7.2 with 1 N NaOH or 1 N H_3PO_4. Adjust to a final colume of 1 l. This gives 25 mM sodium phosphate buffer, pH 7.2.

These should be made up just before use and stored in the dark.

Substrate solutions

1. Primary substrate:
   Dissolve NADP at 1.53 mg/10 ml (0.2 mmol/l) of diethanolamine buffer. 40 μl/well, or about 4 ml per 96-well microtiter plate, will be needed.
2. Secondary substrate (Amplifier):
   Add the following to 25 mM sodium phosphate buffer, pH 7.2. 10.56 ml of the buffer will be needed per plate, to which will be added 0.44 ml ethanol just before use (see below), making a total volume of 11 ml/plate, therefore the following amounts should be multiplied by 11 for the amount needed for each plate:
   – Alcohol dehydrogenase (ADH) 75 U/ml
   – Diaphorase 2.1 U/ml
   – Iodonitrotetrazolium violet (INT) 0.278 mg/ml

Note: Add the ethanol just before the secondary substrate will be added to the plates (i.e., no earlier than 1–3 min before use) and swirl to dissolve the INT. This gives much lighter blanks than if the ethanol is added even 10 min before actual use. We now add ethanol for no more than 6 plates at a time, since with a 12-channel pipettor it takes less than 3 min to add the secondary substrate to 6 plates.

Follow whatever protocol is normally used for coating, postcoating, incubating with samples, washing, and antisera and/or alkaline phosphatase-conjugated reagents. After dumping the alkaline phosphatase conjugate, wash normally, then

Amplification

1. Add 40 μl of primary substrate to each well and incubate plates in the dark (we wrap in aluminium foil) for 15 min at room temperature.
2. Without removing the primary substrate, add 110 μl of the secondary substrate, and incubate for 10–15 min at room temperature, again in the dark.
3. Stop color development by addition of 25 μl 0.4 M HCl to each well.
4. Read at 490–495 nm. Do not use double wavelength, which some people do to compensate for differences in well-to-well plastic absorption, etc. (We started out reading at 490 and 405 nm and discovered that there is considerable absorption at 405 nm, which decreased sensitivity substantially.)

17.3 Special Hints for Application and Troubleshooting

A kit containing the reagents for this system sufficient for ten 96-well plates is available from Life Technologies/BRL.

One can also use a one-step method for addition of the amplifier and substrate, which according to the authors could theoretically double the sensitivity (Johannson and Bates, 1988). *Replace the sodium phosphate with 5 mM Tris-HCl in all the procedures above, since phosphate buffer is inhibitory in the one-step system.* 200 μl amplifier (called secondary substrate in the above protocol) is added, followed immediately by 100 μl primary substrate. Unfortunately the background may also increase as well, because of potential contamination of the amplifier enzymes with phosphatases since phosphatases are ubiquitous. The authors state that it may be necessary to screen the reagents, particularly enzymes for this approach. Please see Johansson and Bates (1988) for further details.

DuPont (New England Nuclear, NEN) (DuPont, 1992) has new substrate amplification kits for peroxidase, called the ELAST amplification system for ELISAs and the BLAST amplification system for blotting (peroxidase, alkaline phosphatase) based on the catalyzed reporter deposition technique of Bobrow and colleagues (1989; 1991). Biotinyl tyramide is activated by horseradish peroxidase and reacts with tyrosines, tryptophans, and other electron-rich molecules on the surface of the solid phase. This results in the binding of biotin to such molecules regardless of whether they are used for blocking the plate, or bound in the specific antibody interactions. The trick is that the amount of biotin is a function of the enzyme bound, since it is required to catalyze the activation of the biotinyl tyramide. Thus even though the binding of the biotin is not antigen specific, the amount bound is directly related to the amount of antigen or antibody as in the usual ELISA. Subsequently, streptavidin horseradish peroxidase or even streptavidin alkaline phosphatase can be added, which can further enhance the sensitivity of the assay. Bobrow et al., applying it to membrane immunoassays, found an increase in sensitivity of membrane assays of from 8- to 200-fold depending on the chromagen used and the amplification format. In ELISA assays they found an 8- to 30-fold or more increase in sensitivity in the detection of mouse immunoglobulin G, human immunodeficiency virus (HIV) p24, and herpes simplex virus (HSV), depending on the combinations of enzymes and substrates used (e.g., streptavidin-horseradish peroxidase and *o*-phenylenediamine (OPD), streptavidin-alkaline phosphatase and *p*-NPP, or streptavidin-β-galactosidase and 4-methylumbelliferyl-β-D-galactoside). It should be possible to combine the Bobrow amplification system with the Self amplification system and achieve sensitivity increases of as much as 100-fold or more, although we have no experience with this approach as yet.

Another substrate amplification approach has been recently published by Bystryak and Mekler (1992). They used a photochemical technique for increasing the sensitivity of horseradish peroxidase-dependent ELISAs by as much as 50-fold. The basic method is a standard ELISA using high-purity OPD as the substrate, resulting in the formation of 2,3-diaminophenazine (DAP) in the dark. This is then followed by illuminating the reaction tubes at 400–500 nm for several minutes, which results in a great increase in the amount of DAP formed, allowing the detection of as little as 50pg/ml of human carcinoembryonic antigen (CEA).

References

Bobrow MN, Harris TD, Shaughnessy KJ, Litt GJ (1989) Catalyzed reporter deposition, a novel method of signal amplification. Application to immunoassays. J Immunol Meth 125:279−285

Bobrow MN, Shaughnessy KJ, Litt GJ (1991) Catalyzed reporter deposition, a novel method of signal amplification. II. Application to membrane immunoassays. J Immunol Meth 137:103−112

Bystryak SM, Meklar VM (1992) Photochemical amplification for horseradish peroxidase-mediated immunosorbent assay. Anal Biochem 202:390−393

Carr RI, Mansour M, Sadi D, James H, Jones JV (1987) A substrate amplification system for enzyme-linked immunoassays: Demonstration of its general applicability to ELISA systems for detecting antibodies and immune complexes. J Immunol Meth 98:201−208, and corrigendum J Immunol Meth 103:157

DuPont (1992) BiotechUpdate 7(2):120−124

Helle M, Boeije L, De Groot E, De Vos A, Aarden L (1991) Sensitive ELISA for interleukin-6. Detection of IL-6 in biological fluids: Synovial fluids and sera. J Immunol Meth 138:47−56

Johansson A, Bates DL (1988) Amplification by second enzymes. In: Kemeny DM, Challacombe SJ (eds) ELISA and Other Solid Phase Immunoassays. John Wiley, Chichester, pp 85−106

Jones JV, Mansour M, James H, Sadi D, Carr RI (1989) A substrate amplification system for enzyme-linked immunoassays; II. Demonstration of its applicability for measuring anti-DNA antibodies. J Immunol Meth 118:79−84

Kemeny DM, Richards D, Durnin S, Johansson A (1989) Ultrasensitive enzyme-linked immunosorbent assay (ELISA) for the detection of picogram quantities of IgE. J Immunol Meth 120:251−258

Self CH (1985) Enzyme amplification − a general method applied to provide an immunoassisted assay of placental alkaline phosphatase. J Immunol Meth 76:389−393

Seth R, Zaidi M, Fuller JQ, Self CH (1988) A highly specific and sensitive enzyme-immunometric assay for calcitonin gene-related peptide based on enzyme amplification. J Immunol Meth 111:11−16

IV Application Formats

18 Overview of Applications Formats

CHRISTOPH KESSLER

The detection of nucleic acids by nonradioactively labeled probes can be performed either with immobilized analytes on blots, by in situ approaches or in solution. The table shows an overview of important application formats including cross reference to the formats described in Chaps. 19−22. With blot and in situ formats, the presence or absence of a particular sequence is recorded, whereas detection in solution allows quantitative measurements. For detection of nucleic acids, a number of blot formats have been established: dot blot, slot blot, Southern blot (DNA analytes), northern blot (RNA analytes), southwestern blot (protein-binding DNA sequences), and genomic blot (analysis of whole genomes). DNA sequencing on blots have also been developed (Richterich et al., 1989; Höltke et al., 1992). A variety of formats have been described for detection oc nucleic acids in situ: colony hybridization (bacterial colonies), plaque hybridization (phage plaques), in situ hybridizations with isolated metaphase chromosomes, tissue sections, biopsies, fixed cells, or whole organisms such as *Drosophila* embryos. Both proteins and glycoproteins are most often analyzed in western blots. For review of the alternative formats see Matthews and Kricka, 1988; Wilchek and Bayer, 1988; Kessler, 1991; Kricka, 1992.

For quantitative detection of nucleic acids in diagnostic systems, reaction formats such as sandwich (Ranki et al., 1983), detection of DNA:RNA hybrids with DNA:RNA-specific antibodies (Coutlee et al., 1989a, 1989b, 1989c), hybridization protection (Arnold et al., 1989), energy transfer, or enzyme channeling formats (Heller et al., 1982; Taub, 1986; Cardullo et al., 1988) can be used. Compared with immunological antigen or antibody detection systems, the nucleic acid-based systems have the following advantages (Zwadyk and Cooksey, 1987; Parsons, 1988; Pasternak, 1988; Kohne, 1990):

− High specificity of analyte recognition (base pairing)
− Sensitivity because of the possibility of a prior amplification reaction (analyte amplification)
− Analyte detection before antigen expression (latency).

With the sandwich technique, two probes are used to hybridize with two different regions of the analyte (Ranki et al., 1983). The replacement technique is based on the replacement of a prebound labeled oligonucleotide by the analyte from a binding complex; in this reaction format, the amount of released labeled oligonucleotide is a measure of the input analyte (Collins et al., 1988).

Application formats

Format	Target analyte	Cross-reference	Reference(s)
Blot formats			
Nucleic acids			
Dot blot	DNA	19.2	Langer et al. (1981); Kessler et al. (1990)
Southern blot	DNA	19.2	Langer et al. (1981); Kessler et al. (1990)
Northern blot	RNA	19.2	Höltke and Kessler (1990)
Colony hybridization	Bacterial DNA/RNA	19.3	Seibl et al. (1990)
Plaque hybridization	Phage DNA/RNA	19.3	Seibl et al. (1990)
Fingerprint	Chromosomal DNA	19.4	Zischler et al. (1989)
Restriction mapping	DNA	19.5	Zuber and Schumann (1991)
DNA sequencing	DNA	19.6, 19.7	Martin et al. (1991); Höltke et al. (1992)
DNA sequencing on blots	DNA	19.8	Pohl and Beck (1987); Richterich et al. (1989)
Proteins/glycoproteins			
Western blot	Protein/glycoprotein	20.1	Bayer et al. (1985); Wilchek and Bayer (1987)
Southwestern blot	DNA:protein complex	20.2	Dooley et al. (1988)
In situ formats			
Nucleic acids			
Virus detection in fixed cells	DNA/RNA	21.1	Brigati et al. (1983); Heiles et al. (1988)
Differentiation of viral/chromosomal DNA	DNA	21.2	Herrington et al. (1989a, 1989b)
Virus detection in biopsy	DNA/RNA	21.3	Heino et al. (1989); Hukkanen et al. (1990)
Hybridization on banded chromosomes (ISHB)	DNA	21.4	Ward et al. (1991); Wienberg et al. (1992)
One step labeling (PRINS)	DNA	21.5	Gosden et al. (1991); Koch et al. (1992)
Multiple chromosome mapping (FISH)	DNA	21.6	Lichter et al. (1990)
Mapping polytene chromosomes	DNA	21.7	Langer-Safer et al. (1982)
mRNA/fixed cells/RNA probes	mRNA	21.8	Herget et al. (1988); Zimmermann et al. (1988)
mRNA/fixed/tissue/RNA probes	mRNA	21.9	Morris et al. (1990)
mRNA/fixed tissue/oligonucleotide probes	mRNA	21.10	Baldino et al. (1989)
mRNA/*Drosophilia* embryo/ss or ds DNA probes	mRNA	21.11	Tautz and Pfeifle (1989)
mRNA/*Drosophilia* embryo/PCR probes	mRNA	21.12	Dynlacht et al. (1989); Hortsch et al. (1990); Perkins et al. (1990)
Proteins/glycans			
mRNA/proteins double-labeling	mRNA/protein	21.13	Cohen (1990)
Quantitative formats			
Nucleic acids			
Sandwich	DNA/RNA	22.1	Van Prooijen-Knegt et al. (1982); Stollar and Rashtchian (1987); Rashtchian et al. (1987); Yehle et al. (1987); Coutlee et al. (1989a, 1989b, 1989c)

DNA:RNA-specific antibodies	DNA/RNA	22.2	Ranki et al. (1983); Syvänen et al. (1986a, 1986b); Yehle et al. (1987); Jungell-Nortama et al. (1988); Nichols and Malcolm (1989); Newmann et al. (1989)
Strand displacement	DNA/RNA	22.3	Vary et al. (1986); Ellwood et al. (1986); Vary (1987); Collins et al. (1988)
Hybridization protection	DNA/RNA	–	Arnold et al. (1989)
Energy transfer (FRET)	DNA/RNA	22.4	Cardullo et al. (1988)
Enzyme channeling	DNA/RNA	–	Albarella et al. (1985); Taub (1986); Miller (1987)
Streptavidin-MTP/-beads	DNA/RNA	22.5	Seibl and Eberle (1992)
Proteins/glycans			For review see Linke and Küppers (1986) and Kessler (1991)

The homogeneous hybridization protection assay is based on selective stabilization of acridinium labels in hybrid-bound detector probes (Arnold et al., 1989).

In energy transfer formats, two probes are bound to the analyte. The first probe is labeled with an energy donor and the second is labeled with an energy acceptor (Cardullo et al., 1988). Energy donor and acceptor moiety are located directly together. Thus energy transfer occurs only when both probes are bound to the analyte. In the unbound state no energy transfer is possible. An example of this format uses probes modified with the chemiluminescent peroxidase substrate luminol as energy donor and rhodamine as energy aceptor. The related enzyme channeling format uses the target-directed binding of substrate donor and acceptor probes directly together (Taub, 1986).

In nucleic acid detection systems, unlike immunological detection systems, analyte amplification steps can be integrated into the reaction format (see Part III). A well-established analyte amplification reaction is the PCR in which amplification factors of 10^6-10^7 can be obtained (Saiki et al., 1985; 1988). By combination of these amplification reactions with nonradioactive labeling and detection systems, the quantitative analysis of single molecules is possible with isolated nucleic acids or in whole cells as analytes. This level of sensitivity is unique in bioanalytical indicator systems.

References

Albarella JP, DeRiemer LHA, Carrico RJ (1985) Hybridization assay employing labeled pairs of hybrid binding reagents. Eur Pat Appl 0144914

Arnold LJ, Hammond PW, Weise WA, Nelson NC (1989) Assay formats involving acridinium-ester-labeled DNA probes. Clin Chem 35:1588–1594

Baldino F Jr, Robbins E, Grega D, Meyers SL, Springer JE, Lewis ME (1989) Non-radioactive detection of NGF-receptor mRNA with digoxigenin-UTP labeled RNA probes. Neurosc Abstr 15:864

Bayer EA, Zalis MG, Wilchek M (1985) 3-(N-Maleimido-propionyl)biocytin: a versatile thiol-specific biotinylating reagent. Anal Biochem 149:529–536

Brigati DJ, Myerson D, Leary JJ (1983) Detection of viral genomes in cultured cells and paraffin-embedded tissue sections using biotin-labeled hybridization probes. Virol 126:35–50

Cardullo RA, Agrawal S, Flores C, Zamecnik PC, Wolf DE (1988) Detection of nucleic acid hybridization by nonradioactive resonance energy transfer. Proc Natl Acad Sci USA 85:8790−8794

Cohen SM (1990) Specification of limb development in the *Drosophilia* embryo by positional cues from segmentation genes. Nature 343:173−177

Collins M, Fritsch EF, Ellwood MS, Diamond SE, Williams JI, Brewen JG (1988) A novel diagnostic method based on strand displacement. Mol Cell Probes 2:15−30

Coutlee F, Bobo L, Mayur K, Yolken RH, Viscidi RP (1989a) Immunodetection of DNA with biotinylated RNA probes: a study of reactivity of a monoclonal antibody to DNA-RNA hybrids. Anal Biochem 181:96−105

Coutlee F, Viscidi P, Yolken H (1989b) Comparison of colorimetric, fluorescent, and enzymatic amplification substrate systems in an enzyme immunoassay for detection of DNA-RNA hybrids. J Clin Microbiol 27:1002−1007

Coutlee F, Yolken RH, Viscidi RP (1989c) Nonisotropic detection of RNA in an enzyme immunoassay using a monoclonal antibody against DNA-RNA hybrids. Anal Biochem 181:153−162

Dooley S, Radtke J, Blin N, Unteregger G (1988) Rapid detection of DNA-binding factors using protein-blotting and digoxigenin-dUTP marked probe. Nucleic Acids Res 16:11829

Dynlacht BD, Attardi LD, Admon A, Freeman M, Tijian R (1989) Functional analysis of NTF-1, a developmentally regulated *Drosophila* transcription factor that blinds neuronal cis elements. Genes Dev 3:1677−1688

Ellwood MS, Collins M, Fritsch EF, Williams JI, Diamond SE, Brewen JG (1986) Strand displacement applied to assays with nucleic acid probes. Clin Chem 32:1631−1636

Gosden J, Hanratty D, Starling J, Mitchell A, Porteous D (1991) Oligonucleotide-primed in situ DNA synthesis (PRINS): a method for chromosome mapping, banding, and investigation of sequence organization. Cytogenet Cell Genet 57:100−104

Heiles HBJ, Genersch E, Kessler C, Neumann R, Eggers HJ (1988) In situ hybridization with digoxigenin-labeled DNA of human papillomaviruses (HPV 16/18) in HeLa and SiHa cells. BioTechniques 6:978−981

Heino P, Hukkanen V, Arstila P (1989) Detection of human papilloma virus (HPV) DNA in genital biopsy specimens by in situ hybridization with digoxigenin-labeled probes. J Virol Meth 26:331−338

Herget T, Goldowitz D, Oelemann W, Starzinski-Powitz A (1988) Description of putative ribosomal RNAs with low abundance, developmental regulation, and the identifies sequence. Exp Cell Res 176:141−154

Herrington CS, Burns J, Graham AK, Evans M, McGee JO (1989a) Interphase cytogenetics using biotin and digoxigenin labeled probes. I: Relative sensitivity of both reporter molecules for detection of HPV16 in CaSki cells. J Clin Pathol 42:592−600

Herrington CS, Burns J, Graham AK, Bhatt B, McGee JD (1989b) Interphase cytogenetics using biotin and digoxigenin labeled probes. II: Simultaneous detection of human and papilloma virus nucleic acids in individual nucleic. J Clin Pathol 42:601−606

Höltke H-J, Kessler C (1990) Nonradioactive labeling of RNA transcripts in vitro with the hapten digoxigenin (DIG); hybridization and ELISA-based detection. Nucleic Acids Res 18:5843−5851

Höltke H-J, Sagner G, Kessler C, Schmitz G (1992) Sensitive chemiluminescent detection of digoxigenin-labeled nucleic acids: a fast and simple protocol and its application. BioTechniques 12:104−113

Hortsch M, Patel NH, Bieber AJ, Traquina ZR, Goodman CS (1990) *Drosophila neurotactin,* a surface glycoprotein with homology to serin esterases, is dynamically expressed during embryogenesis. Development 10:1327−1340

Hukkanen V, Heino P, Sears AE, Roizman B (1990) Detection of herpes simplex virus latency-associated RNA in mouse trigeminal ganglia by in situ hybridization using non-radioactive DNA and RNA probes. Meth Mol Cell Biol 2:70−81

Jungell-Nortamo A, Syvänen AC, Luoma P, Söderlund H (1988) Nucleic acid sandwich hybridization: enhanced reaction rate with magnetic microparticles as carriers. Mol Cell Probes 2:281−288

Kessler C, Höltke H-J, Seibl R, Burg J, Mühlegger K (1990) Non-radioactive labeling and detection of nucleic acids: I. A novel DNA labeling and detection system based on digoxigenin:anti-digoxigenin ELISA principle (digoxigenin system). Mol Gen Hoppe-Seyler 371:917−927

Kessler C (1991) The digoxigenin:anti-digoxigenin (DIG) technology − a survey on the concept and realization of a novel bioanalytical indicator system. Mol Cell Probes 5:161−205

Koch J, Mogensen J, Pedersen S, Fischer H, Hindkjær J, Kølvraa S, Bolund L (1992) Fast one step procedure for the detection of nucleic acids in situ by primer induced sequence specific labelling with fluorescein-12-dUTP. Cytogenet Cell Genet 60:1−3

Kohne DE (1990) The use of DNA probes to detect and identify microorganisms. Adv Exp Med Biol 263:11−35

Kricka LJ (1992) Nonisotopic DNA probe techniques. Academic Press, San Diego

Langer PR, Waldrop AA, Ward DC (1981) Enzymatic synthesis of biotin-labeled polynucleotides: novel nucleic acid affinity probes. Proc Natl Acad Sci USA 78:6633−6637

Langer-Safer PR, Levine M, Ward DC (1982) Immunological method for mapping genes on *Drosophila* polytene chromosomes. Proc Natl Acad Sci USA 79:4381−4385

Lichter P, Tang C-JC, Call K, Hermanson G, Evans GA, Housman D, Ward DC (1990) High-resolution mapping of human chromosome 11 by in situ hybridization with cosmid clones. Science 247:64−69

Linke R, Küppers R (1988) Nicht-isotopische Immunoassays − ein Überblick. In: Borsdorf R, Fresenius W, Günzler H, Huber W, Kelker H, Lüderwald I, Tölg G, Wisser H (eds) Analytiker Taschenbuch, Springer Verlag, Berlin/Heidelberg, pp 127−177

Martin C, Bresnick L, Juo R-R, Voyta JC, Bronstein I (1991) Improved chemiluminescent DNA sequencing. BioTechniques 11:110−113

Matthews JA, Kricka LJ (1988) Analytical strategies for the use of DNA probes. Anal Biochem 169:1−25

Miller JA (1987) Polynucleotide hybridization assays employing catalyzed luminescence. US Pat 4670379

Morris RG, Arends MJ, Bishop PE, Sizer K, Duvall E, Bird CC (1990) Sensitivity of digoxigenin and biotin labelled probes for detection of human papillomavirus by in situ hybridization. J Clin Path 43:800−805

Newman CL, Modlin J, Yolken RH, Viscidi RP (1989) Solution hybridization and enzyme immunoassay for biotinylated DNA-RNA hybrids to detect enteroviral RNA in cell culture. Mol Cell Probes 3:375−382

Nicholls PJ, Malcolm ADB (1989) Nucleic acid analysis by sandwich hybridization. J Clin Lab Anal 3:122−135

Parsons G (1988) Development of DNA probe-based commercial assay. J Clin Immunoassay 11:152−160

Pasternak JJ (1988) Microbial DNA diagnostic technology. Biotech Adv 6:683−695

Perkins KK, Admon A, Patel NH, Tijian R (1990) The *Drosophila* fos-related AP-1 protein is a developmentally regulated transcription factor. Genes Dev. 4:822−834

Pohl FM, Beck S (1987) Direct transfer electrophoresis used for DNA sequencing. Meth Enzymol 155:250−259

Ranki M, Palva A, Virtanen M, Laaksonen M, Söderlund H (1983) Sandwich hybridization as convenient method for the detection of nucleic acids in crude samples. Gene 21:77−85

Rashtchian A, Eldredge J, Ottaviani M, Abbott M, Mock G, Lovern D, Klinger J, Parsons G (1987) Immunological capture of nucleic acid hybrids and application to nonradioactive DNA probe assay. Clin Chem 33:1526−1530

Richterich P, Heller C, Wurst H, Pohl FM (1989) DNA sequencing with direct blotting electrophoresis and colorimetric detection. BioTechniques 7:52−59

Saiki RK, Gelfand DH, Stoffel S, Scharf SJ, Higuchi R, Horn GT, Mullis KB, Erlich HA (1988) Primer-directed enzymatic amplification of DNA with a thermostable DNA polymerase. Science 239:487−491

Saiki RK, Scharf S, Faloona F, Mullis KB, Horn GT, Erlich HA, Arnheim N (1985) Enzymatic amplification of beta-globin genomic sequences and restriction site analysis for diagnosis of sickle cell anemia. Science 230:1350−1354

Seibl R, Eberle J (1992) Quantification of reverse transcriptase activity by ELISA. J Virol Meth, in press

Seibl R, Höltke H-J, Rüger R, Meindl A, Zachau H-G, Rasshofer G, Roggendorf M, Wolf H, Arnold N, Wienberg J, Kessler C (1990) Non-radioactive labeling and detection of nucleic acids: III. Applications of the digoxigenin system. Mol Gen Hoppe-Seyler 371:939−951

Stollar BD, Rashtchian A (1987) Immunochemical approaches to gene probe assays. Anal Biochem 161:387−394

Syvänen AC, Laaksonen M, Söderlund H (1986a) Fast quantification of nucleic acid hybrids by affinity-based hybrid collection. Nucleic Acids Res 14:5037−5048

Syvänen AC, Tchen P, Ranki M, Söderlund H (1986b) Time-resolved fluorometry: a sensitive method to quantify DNA-hybrids. Nucleic Acids Res 14:1017−1028

Taub F (1986) An assay for nucleic acid sequences, particularly genetic lesions. PCT Int Appl WO 86/03227

Tautz D, Pfeifle C (1989) A nonradioactive in situ hybridization method for the localization of specific RNAs in *Drosophilia* embryos reveals translational control of the segmentation gene hunchback. Chromosoma 98:81−85

Van Prooijen-Knegt AC, Van Hoek JF, Bauman JG, Van Duijin P, Wool IG, Van der Ploeg M (1982) In situ hybridization of DNA sequences in human metaphase chromosomes visualized by an indirect fluorescent immunocytochemical procedure. Exp Cell Res 141:397−407

Vary CPH (1987) A homogeneous nucleic acid hybridization assay based on strand displacement. Nucleic Acids Res 15:6883−6897

Vary CPH, McMahon FJ, Barbone FP, Diamond SE (1986) Nonisotopic detection methods for strand displacement assays of nucleic acids. Clin Chem 32:1696−1701

Ward DC, Lichter P, Boyle A, Baldini A, Menninger J, Ballard SG (1991) Gene mapping by fluorescent in situ hybridization and digital imaging microscopy. In: Lindsten J, Petterson U (eds) Etiology of Human Diseases at the DNA Level, Raven Press, Boca Raton, Florida, pp 291−303

Wienberg J, Stanyon R, Jauch A, Cremer T (1992) Homologies in human and *Macaca fuscata* chromosomes revealed by in situ suppression hybridization with human chromosome specific DNA libraries. Chromosoma 101:265−270

Wilchek M, Bayer EA (1987) Labeling glycoconjugates with hydrazide reagents. Meth Enzymol 138:429−442

Wilchek M, Bayer EA (1988) The avidin-biotin complex in bioanalytical applications. Anal Biochem 171:1−32

Yehle CO, Patterson WL, Boguslawski SJ, Albarella JP, Yip KF, Carrico RJ (1987) A solution hybridization assay for ribosomal RNA form bacteria using biotinylated DNA probes and enzyme-labeled antibody to DNA:RNA. Mol Cell Probes 1:177−193

Zimmermann K, Herget T, Salbaum JM, Schubert W, Hilbich C, Multhaup G, Kang J, Lemaire H-G, Beyreuther K, Starzinski-Powitz A (1988) Localization of the putative precursor of Alzheimer's disease-specific amyloid at nuclear envelopes of adult human muscle. EMBO J 7:367−372

Zwadyk P, Cooksey RC (1987) Nucleic acid probes in clinical microbiology. CRC Crit Rev Clin Lab Sci 25:71−103

Zischler H, Nanda I, Schäfer R, Schmid M, Epplen JT (1989) Digoxigenated oligonucleotide probes specific for simple repeats in DNA fingerprinting and hybridization in situ. Hum Gen 82:227−233

Zuber V, Schumann W (1991) Tn5cos: a transposon for restriction mapping of large plasmids using phage lambda terminase. Gene 103:69−72

19 Blot Formats: Nucleic Acids

19.1 Factors Influencing Nucleic Acid Hybridization

CHRISTOPH KESSLER

Binding between analyte and modified probe, resulting in stable hybrid molecules, is most prominently mediated by specific interaction between complementary purine and pyrimidine bases forming A:T and G:C base pairs.

The stability of the generated hybrids is the result of the generation of a variable number of hydrogen bonds between complementary bases and of the effect of both electrostatic and hydrophic forces.

G:C base pairs are more stable than A:T pairs because three hydrogen bonds can form between G and C. In contrast, between A and T only two hydrogen bonds are possible. Therefore, G:C-rich sequences are more stable than A:T-rich sequences and thus will have higher T_ms.

Electrostatic forces are caused predominantly by the phosphate molecules of the nucleic acid backbone; thus, double-stranded sequences are stabilized by increasing ionic strength. The presence of a hydrophilic hydroxyl group at the 2′ position of the ribose also stabilizes the double-stranded structure of nucleic acids. Thus, the T_m of DNA/RNA and RNA/RNA hybrids are significantly higher than those of the respective DNA/DNA hybrids. Hydrophobic interactions between the staggered bases also contribute to hybrid stability; this explains the destabilizing effect of organic solvents on hybrid formation. Due to central role of hydrogen bonding in hybrid formation and hybrid stability, the most appropriate sites for base modifications are those not involved in hydrogen bonding: C-5 of uracil and cytosine (Langer et al., 1981; Ruth, 1984; Ruth et al., 1985; Jablonski et al., 1986; Jablonski and Ruth, 1986; Haralambidis et al., 1987; Cook et al., 1988 [C-5 modification of uracil mimics thymine residues, making these compounds useful as thymine analogue]), C-6 of cytidine (Verdlov et al., 1974), and C-8 of guanine and adenine (Reisfeld et al., 1987; Huynh Dinh et al., 1987; Keller et al., 1988).

The N^4 position of cytosine (Gillam and Tener, 1986; Urdea et al., 1987; Gebeyehu et al., 1987; Urdea et al., 1988) and the N^6 position of adenine (Jablonski and Ruth, 1986) or guanine (Tchen et al., 1984) are involved in hydrogen bonding and are therefore less useful (Viscidi et al., 1986; Gebeyehu et al., 1987).

The rate of reassociation between analyte and probe nucleic acid depends on both probe complexity, i.e., probe length as well as length and number of unique sequences, and probe concentration (Britten and Kohne, 1968). Reassociation rates are defined as C_0t values, in which C_0 is the initial DNA molar concentration and t the incubation time in seconds. The reassociation rate is quantified in terms of $C_0t_{1/2}$ values; these values define when half of the nucleic acid molecules have reassociated. At a given concentration the reassociation rate is inversely proportional to probe complexity, i.e., to probe length and the length of the unique sequences within the probe (Davidson and Britten, 1979).

C_0t analyses are performed in solution using equal concentrations of analyte and probe. Thus, in these studies second-order rates are applied for calculating reaction parameters. Since most hybridizations are performed with analytes immobilized on solid supports, such as nitrocellulose or nylon membranes, only first-order kinetics are used for the calculation of association parameters.

With excess probe concentrations — as with most hybridization experiments — the hybridization rate is primarily dependent on probe complexity and probe concentration. Meinkoth and Wahl (1984) described the first-order equations for hybridization with single-stranded probes as follows:

$$t_{1/2} = \frac{\ln 2}{k \times C}$$

in which $k =$ is the rate constant for hybrid formation in mol $\times$ l/number of nucleotides $\times$ seconds and $C =$ is the molar probe concentration.

Within short hybridization periods (<4h) double-stranded probes can be handled similarly; with longer hybridization periods (16h) the actual probe concentration decreases because of probe reassociation: the rate constant k is dependent on probe complexity as well as reaction parameters such as temperature, ionic strength, pH value, and viscosity of the incubation mixture, that is:

$$k = \frac{k_n \times L^{0.5}}{N}$$

in which L is the probe length, N is the length of unique probe sequences and k is the nucleation constant..

The nucleation constant k_n is 3.5×10^5 for standard hybridization conditions, i.e., Na^+ concentrations of $0.4 - 1.0$M, pH values between 5 and 9, and T_m 25°C (Marmur and Doty, 1959).

Under these standard hybridization conditions, the hybridization rate $t_{1/2}$ (as measured in seconds) is calculated as follows:

$$t_{1/2} = \frac{N \times \ln 2}{3.5 \times 10^5 \times L^{0.5} \times C}$$

For a fragment probe of 500 bp, $t_{1/2}$ is calculated for a molar probe concentration of 6×10^{-10}:

$$t_{1/2} = \frac{500 \times 0.693 \,(\text{s})}{3.5 \times 10^5 \times 22 \times 6 \times 10^{-10}} = 7.5 \times 10^3 \,\text{s} = 20\,\text{h}$$

For a 25mer oligonucleotide probe $t_{1/2}$ calculates to 1 h. However, this value has to be considered a rough estimate because experimentally obtained $t_{1/2}$ values fit into the calculated linear relationship in a first approach only (Keller and Manak, 1989).

In addition to probe complexity and probe concentration, the reaction parameters of temperature, salt concentration, base mismatches, and hybridization accelerators also influence the rate of reassociation (Hames and Higgins, 1985).

Inert polymers such as dextran sulphate or polyethylene glycol (PEG) can be used as accelerators of the hybridization reaction with longer fragment probes (Wahl et al., 1979). Nonpolymeric accelerators include phenol (Kohne et al., 1977) and chaotropic salts, e.g., guanidinium isothiocyanate (Thompson and Gillespie, 1987). These chemicals act as water-exclusion reagents and thus lower the energy difference between single- and double-stranded nucleic acids.

The specificity of hybrid formation is determined by the stringency of the hybridization conditions and the stability of the formed hybrid complexes. Like the reassociation rate, the hybrid stability is also directly related to T_m, which is affected by base composition, salt concentration, presence of formamide, fragment length, nature of the nucleic acid within the hybrid (DNA:DNA, RNA:DNA, RNA:RNA), and mismatch formation.

For DNA:DNA hybrids the influence of the first four parameters on T_m is represented by the following equation:

$$T_m = \frac{(81.5\,^\circ\text{C} + 16.6\log M + 0.41\,(\%\,\text{G} + \text{C})) - 500}{n - 0.61\,(\%\,\text{formamide})}$$

in which $M = c_{Na+}\,[M]$ and n is the length of hybridizing sequence.

With respect to the nature of the formed hybrids, the T_m of RNA:DNA hybrids is $10^\circ - 15\,^\circ\text{C}$ higher than that of DNA:DNA hybrids. The T_m of RNA:RNA hybrids is $20^\circ - 25\,^\circ\text{C}$ higher; to lower the T_m of these hybrids, formamide is often added.

Mismatches are less stable than normal base pairing: therefore the presence of mismatches also reduce T_m, which decreases about $1\,^\circ\text{C}$ for every 1% mismatch (Hutton and Wetmur, 1973; Britten et al., 1974).

The specificity of hybrid formation is predominantly influenced by the stringency of the final washing steps after hybridization. Stringency is mostly enhanced during the final washing steps by increasing the temperature to only $5^\circ - 15\,^\circ\text{C}$ below T_m and lowering the salt concentration from $5\times$ SSC ($0.75\,\text{M Na}^+$) to $0.1\times$ SSC ($0.015\,\text{M Na}^+$). This holds true especially for oligonucleotides, for which the washing temperature is usually $5\,^\circ\text{C}$ below T_m. However, the optimal washing temperature has to be determined experimentally and a compromise has to be made with respect to the washing conditions in which the probe binds strongly to the analyte and weakly to unspecific heterologous analyte components.

References

Britten RJ, Graham DE, Neufeld BR (1974) Analysis of repeating DNA sequences by reassociation. Meth Enzymol 29:363–418

Cook AF, Vuocolo E, Brakel CL (1988) Synthesis and hybridization of a series of biotinylated oligonucleotides. Nucleic Acids Res 16:4077–4095

Davidson EH, Britten RJ (1979) Regulation of gene expression: possible role of repetitive sequences. Science 204:1052–1059

Gebeyehu G, Rao PY, SooChan P, Simms DA, Klevan L (1987) Novel biotinylated nucleotide analogs for labeling and colorimetric detection of DNA. Nucleic Acids Res 15:4513–4534

Gillam IC, Tener GM (1986) N^4-(6-aminohexyl)cytidine and -deoxycytidine nucleotides can be used to label DNA. Anal Biochem 157:199–207

Haralambidis J, Chai M, Tregear GW (1987) Preparation of base-modified nucleosides suitable for nonradioactive label attachment and their incorporation into synthetic oligonucleotides. Nucleic Acids Res 15:4857–4876

Hutton JR, Wetmur JG (1973) Length dependence of the kinetic complexity of mouse satellite DNA. Biochem Biophys Res Commun 52:1148–1155

Huynh Dinh T, Sarfati S, Igolen J, Guesdon JL (1987) Markers for detecting nucleic acids are derived from 2'-desoxy adenosine derivatives. Eur Pat Appl 0254646

Jablonski E, Moomaw EW, Tullis RH, Ruth JL (1986) Preparation of oligonucleotide-alkaline phosphatase conjugates and their use as hybridization probes. Nucleic Acids Res 14:6115–6128

Jablonski E, Ruth JL (1986) Synthesis of oligonucleotide-enzyme conjugates and their use as hybridization probes. DNA 5:89

Keller GH, Cumming CU, Huang DP, Manak MM, Ting R (1988) A chemical method for introducing haptens onto DNA probes. Anal Biochem 170:441–450

Keller GH, Manak MM (1989) DNA probes. Stockton Press, New York

Langer PR, Waldrop AA, Ward DC (1981) Enzymatic synthesis of biotin-labeled polynucleotides: novel nucleic acid affinity probes. Proc Natl Acad Sci USA 78:6633–6637

Marmur J, Doty P (1959) Heterogeneity in DNA. I. Dependence on composition of the configurational stability of deoxyribonucleic acids. Nature 183:1427–1428

Meinkoth J, Wahl G (1984) Hybridization of nucleic acids immobilized on solid supports. Anal Biochem 138:267–284

Reisfeld A, Rothenberg JM, Bayer EA, Wilchek M (1987) Nonradioactive hybridization probes prepared by the reaction of biotin hydrazide with DNA. Biochem Biophys Res Commun 142:519–526

Ruth JL, Morgan C, Pasko A (1985) Linker arm nucleotide analogs useful in oligonucleotide synthesis. DNA 4, 93

Tchen P, Fuchs RPP, Sage E, Leng M (1984) Chemically modified nucleic acids as immunodetectable probes in hybridization experiments. Proc Natl Acad Sci USA 81:3466–3470

Thompson J, Gillespie D (1987) Molecular hybridization with RNA probes in concentrated solutions of guanidine thiocyanate. Anal Biochem 163:281–291

Urdea MS, Running JA, Horn T, Clyne J, Ku L, Warner BD (1987) A novel method for the rapid detection of specific nucleotide sequences in crude biological samples without blotting or radioactivity; application to the analysis of hepatitis B virus in human serum. Gene 61:253–264

Urdea MS, Warner BD, Running JA, Stempien M, Clyne J, Horn T (1988) A comparison of nonradioisotopic hybridization assay method using fluorescent, chemiluminescent and enzyme labeled synthetic oligodeoxyribonucleotide probes. Nucleic Acids Res 16:4937–4956

Verdlov ED, Monastyrskaya GS, Guskova LI, Levitan TL, Sheichenko VI, Budowsky EI (1974) Modification of cytidine residues with a bisulfite-O-methylhydroxyl-amine mixture. Biochem Biophys Acta 340:153–165

Viscidi RP, Connelly CJ, Yolken RH (1986) Novel chemical method for the preparation of nucleic acids for nonisotopic hybridization. J Clin Microbiol 23:311–317

Wahl GM, Stern M, Stark GR (1979) Efficient transfer of large DNA fragments from agarose gels to diazobenzyloxymethyl-paper and rapid hybridization by using dextran sulfate. Proc Natl Acad Sci USA 76:3683–3687

19.2 Dot, Southern, and Northern Blots

BARBARA RÜGER AND CORTINA KALETTA

19.2.1 Principle and Applications

Nucleic acid hybridization on solid supports (nitrocellulose or nylon membranes) has become one of the most important techniques in molecular biology. Here, we will specifically deal with applications of the digoxigenin (DIG) system in dot blot, Southern, and northern blot hybridizations.

In principle molecular hybridization is the formation of double-stranded nucleic acid molecules by sequence-specific base pairing of complementary single strands. For dot blot hybridization, DNA or RNA is spotted directly onto a membrane, while for Southern or northern blot hybridization DNA fragments or mRNAs, respectively, are transferred to the membrane after size separation on an agarose gel by capillary, vacuum, pressure or electroblotting and subsequently hybridized with a labeled probe that can detect a specific sequence. How to prepare DNA or RNA agarose gels is described elsewhere (see Sambrook et al., 1989).

Standard labeling of DNA and RNA probes is described in Sect. 3.1. Basically, DIG can be incorporated into nucleic acids by random primed labeling using Klenow enzyme and during PCR with *Taq* DNA polymerase (see Sect. 3.2 as well as 16.1 and Rüger et al., 1991) for DNA fragments. Oligonucleotides can be labeled by terminal transferase and RNA probes can be prepared during in vitro transcription. The hapten DIG can also be introduced by non enzymatic coupling to nucleic acids during a photoreaction. For this purpose, DIG is bound to a photoreactive group via a hydrophilic spacer. It is also possible to introduce DIG at the 5' end of the oligonucleotide during chemical synthesis. Since the corresponding DIG-labeled phosphoamidite for direct incorporation during DNA synthesis is not yet available, the DIG molecule should be chemically added to an amino-linked ester after synthesis of the oligonucleotide.

19.2.2 Preparation of Southern and Northern Blots

Prior to hybridization, DNA fragments have to be transferred to a solid support. Before starting the transfer of a given DNA several practical and

Pretreatment and transfer conditions

Type of nucleic acid	Pretreatment recommended	Depurinization with 0.25 M HCl (~10 min)	UV light[a]	Concentration of transfer buffer
Small DNA fragments (<2–4 kb) only	No	–	–	20× SSC
Medium size DNA-fragments (5–10 kb) only	No	–	–	20× SSC
Large DNA fragments (>10 kb) only	Yes	+*	+*	Pretreatment ≥20× SSC; no pretreatment ≥6× SSC
All sizes of DNA fragments	Optional, depending on expected length of hybridizing fragment	–	Irradiates large fragments only	20× SSC
Super-coiled plasmid DNA	Yes	+	+	10–20× SSC
RNA (all sizes)	No	–	–	20–24× SSC

* may be used alternatively

[a] UV irradiation: For breakage of ds DNA, exposure time has to be established prior to the experiment. It depends on the UV source, fragment length, and gel specifics. The advantage is that only certain parts of a gel need be exposed, i.e., only parts containing large fragments cen be reached by this treatment, leaving smaller size fragments unaffected.

theoretical aspects have to be considered. Depending on the size and structure of the target DNA, pretreatment of the nucleic acid inside the gel may be necessary to obtain optimal transfer. The table lists various pretreatment possibilities and transfer conditions. After denaturation/neutralization, small DNA fragments (up to 10 kb) are transferred readily with the use of 20× SSC as transfer buffer and DNA fragments as small as 300 bp can be efficiently transferred. Reduction of the SSC concentration to 10× −6× SSC would favor the transfer of larger molecules only. Fragments longer than 10 kb and supercoiled plasmids have to be broken at their specific position in the gel. Thus the fragment remains at the position where it has migrated to but is degraded in a defined way to smaller subfragments that can be transferred more easily. In principle there are two ways to achieve this. The first method is to depurinate the DNA by rinsing the gel for approximately 10 min in 0.25 M HCl. The second method is to break DNA fragments by UV light. The latter has the advantage that only the parts of the gel that contain long fragments need be exposed to UV light, leaving smaller fragments unaffected. The disadvantage is that for every transilluminator, irradiation conditions have to be defined empirically so at least one test series is necessary. Normally RNA is single-stranded; folded back structures are prevented by the denaturation step prior to loading of the gel and by size separation on a denaturing gel. Therefore the denaturation and neutralization steps of the gel that have to be performed with double-stranded DNA can be omitted.

Unless alkaline transfer is performed, double-stranded DNA has to be denatured in the gel prior to transfer, because only single-stranded nucleic acids can form hybrids with the labeled probe. This is either accomplished with 1 M KOH (for gels only) or with 0.5 M NaOH; 1.5 M NaCl (for gels and membranes, e.g., colony hybridization or plaque lifts).

After denaturation the gel has to be neutralized prior to transfer. This is achieved with 1 M Tris; 1.5 M NaCl, pH 7.4, for approximately 1 h (or 1 M Tris; 0.5 M NaCl, pH 5.0, for 5−10 min). The high salt concentration is necessary to prevent reannealing of the DNA strands. Especially if a transfer to nitrocellulose membranes is intended, it is important that the actual pH of the gel (not of the transfer buffer!) is controlled after neutralization. It should be below pH 9, otherwise the filters will turn yellow and break during hybridization. For this purpose, one edge of the gel where no DNA has been loaded is lifted, a pH strip pressed onto it, and the ph checked.

The transfer of nucleic acids from the gel to a membrane is usually performed with SSC buffer (0.15 M NaCl; 0.15 M Na-citrate). Other buffers, e.g., SSPE, or alkaline transfer also may be used. The SSC buffer is soaked through the gel with the help of paper towels and thereby the DNA is mobilized and transported to the membrane by capillary diffusion overnight. How to set up a capillary transfer is described elsewhere (see Sambrook et al., 1989). For vacuum, pressure, or electroblot, follow the procedure given by the manufacturer.

After the transfer, the nucleic acid has to be fixed to the membrane either by baking at 120°C for 30 min (Nylon membrane, positively charged, Boehringer Mannheim GmbH) or 2 h at 80°C (most other membranes) or UV cross-linking. The efficiency of the transfer should be controlled by staining the gel after transfer, again with ethidium bromide, and visualizing residual DNA under UV light.

19.2.3 Southern Blot Hybridization

A prehybridization step is necessary to block all sites of the membrane which do not have transferred DNA or RNA bound to them. If this blocking step is omitted, the labeled probe will bind unspecifically all over the membrane. Blocking of the membrane can be achieved with Boehringer Mannheim specific blocking solution when using the DIG system. Some labs use nonfat dry milk powder, but then problems due to DNase and RNase contaminations may occur. Denhardt's solution also can be used with the DIG system, but best results are obtained using the manufacturer's protocol (Boehringer Mannheim). The blocking reagent may be supplemented for specific applications with denatured calf thymus or fish sperm DNA or with total yeast RNA but it is not necessary on a routine basis. When supplementing the hybridization solution with a nonspecific nucleic acid there may be a chance that the specific probe is blocked because of cross homologies between the target DNA and the nonspecific DNA. When, e.g., working with mammalian DNA, yeast RNA is preferred to calf thymus DNA. Preincubation is performed in $5\times$ SSC hybridization buffer (see below) either at 68 °C or, if 50% formamide is used, at the hybridization temperature, e.g., 37 °C.

In the subsequent hybridization step, appropriate conditions have to be chosen, depending on GC content, homology of the probe to the target sequence, and length of the hybrid. First a decision has to be made as to whether stringent or relaxed conditions have to be applied for a certain experiment. Stringent conditions will allow hybrid formation only if the homology between probe and target is about 80% – 100%. Relaxed conditions can be set so that even only 30% homology between probe and target is detected. Hybridization conditions are discussed in more detail in Chap 19.1. In general, stringency is influenced by the following parameters.

1. Temperature: High temperatures increase stringency while low temperature decrease stringency.
2. Formamide: Increases stringency, favors correct base pairing and decreases background. Formamide decreases the melting point of DNA as follows: 1% formamide lowers T_M by 0.72 °C. Therefore, the hybridization temperature can be decreased but the conditions will still be stringent when formamide is included in the hybridization solution.
3. Salt concentration: Low salt increases stringency and high salt decreases stringency.

These parameters are included in equations for the calculation of optimal hybridization conditions: $T_{opt} = T_M - 25 °C$ (about 18% mismatch allowed), in which T_{opt} is the optimal hybridization temperature; and

$$T_M = 16.6 \log \text{mol Na}^+ + 0.41 \text{ (GC in \%)} + 81.5$$

(1.4 °C below T_M means 1% mismatch allowed).

Filters have to be washed after hybridization to release unspecifically bound DNA or sequences only paired at a short stretch of base pairs. Again stringent or relaxed conditions can be applied, depending on the homology of the probe to the target nucleic acid. Stringent washes are performed at 68 °C or, if necessary, the temperature could be increased up to T_M. The salt (SSC) concentration is between 0.5 and $0.1\times$ for stringent washes and can be as high as $6\times$ for relaxed washes. Stringent washing steps are started with a SSC concentration of $2\times$ and then the SSC concentration is stepwise lowered to $0.5-0.1\times$. SDS is included in all washing solutions to reduce unspecific background on the membrane.

With the DIG system hybridization signals have to be immunologically detected. Alkaline phosphatase is coupled to an anti-DIG antibody. After binding of the antibody to the DIG molecules incorporated in the hybrid, a substrate for alkaline phosphatase, either colored or chemiluminescent, is added. The given substrate is then converted by alkaline phosphatase either into a colored precipitate or into a light emitting substance. Chemiluminescence detection is achieved either with X-ray film or with a special Polaroid camera.

- Blocking reagent (Boehringer Mannheim)   **Standard**
- SDS (Boehringer Mannheim)   **reagents**
- NaCl, Na-citrate (Merck)
- *N*-lauroylsarcosine (Sigma)
- Nylon membrane, positively charged (Boehringer Mannheim)
- Maleic acid (Serva)

- 20× SSC: 3 M NaCl; 0.3 M Na-citrate; pH 7.0/20 °C   **Standard**
- Hybridization solution: 5× SSC; 1−5% [w/v] blocking reagent; 0.1% **solutions**
  [w/v] *N*-lauroylsarcosine, sodium salt; 0.02% [w/v] SDS. Blocking reagent is dissolved in maleic acid buffer (100 mM maleic acid; 150 mM
  NaCl, adjusted to pH 7.5 with concentrated or solid NaOH). A 10%
  stock solution of blocking reagent can be prepared by dissolving the solution in a microwave or by stirring on a magnetic heater and subsequently autoclaving. The stock solution, which remains turbid, can
  be stored at 4 °C or at −20 °C for several months. One can also add formamide to 50% [v/v] to the hybridization solution. In this case the concentration of blocking reagent may be increased to 5% [w/v]. Hybridization with 50% [v/v] formamide is performed between 37 °C and 42 °C
  depending on the homology of the probe.
- Solutions for the enzyme-linked immunoassay see Sect. 3.2.

1. Transfer the DNA to be probed to a nitrocellulose or nylon membrane **Standard**
   by dot blot, plaque lift, colony hybridization, or Southern blot.   **hybridi-**
2. Prepare nitrocellulose membranes by presoaking in water and then **zation**
   20× SSC. Nylon membranes can be used without pretreatment.
3. Prehybridize filters in a sealed plastic bag or box with at least 20 ml hybridization solution per 100 cm^2 of filter at 68 °C for at least 1 h. Periodically, distribute the solution.
4. Replace the prehybridization solution with hybridization solution containing freshly denatured DNA. The optimal concentration of
   labeled DNA in the hybridization mixture depends on the amount of
   DNA to be detected on the filter. Usually 10−50 ng of labeled DNA per
   ml hybridization solution is used. About 2.5 ml of hybridization solution per 100 cm^2 of filter − for very small filters slightly more − are
   required.
5. Incubate the filter overnight at 68 °C. Higher DNA concentrations in
   the hybridization solution can be used to reduce the hybridization times
   down to approximately 2 h, but there is always a higher risk of producing strong background, especially in chemiluminescent detection.
6. Wash filters 2 × 5 min at room temperature with at least 50 ml of
   2× SSC; 0.1% [w/v] SDS per 100 cm^2 filter and 2 × 15 min at 68 °C with
   0.1× SSC; 0.1% [w/v] SDS.

7. Filters can than be used directly for detection of hybridized DNA or stored air-dried for later detection (see Sect. 3.1).

In principle all protocols for using radioactive probes can also be used with the DIG system.

19.2.3.1 Special Aspects of Southern Blot Hybridization

A special hybridization protocol, originally described by Church and Gilbert (1984) and adapted for the DIG system, is recommended for use in hybridization of genomic DNA with single copy gene probes (e.g., human, mammalian, or plant) or restriction fragment length polymorphism (RFLP) probes. This results in highly sensitive detection with very low background. Unspecific cross-hybridization with vector sequences is suppressed as well. The main difference compared to other common hybridization protocols is the addition of 7% [w/v] SDS.

Solutions   — Prehybridization solution (final concentrations): 7% [w/v] SDS; 50% [v/v] formamide (deionized); 5× SSC, 2% [w/v] blocking reagent (in maleic acid buffer, see above); 0.1% [w/v] N-lauroylsarcosine; 50 mM Na-phosphate, pH 7.0; yeast RNA 50 µg/ml.

The prehybridization solution is prepared from the following stock solutions which are combined in the following order:

- 250 ml 100% [v/v] formamide (deionized)
- 83 ml 30× SSC
- 25 ml 1 M Na-phosphate, pH 7.0
- 100 ml 10% [w/v] blocking solution (see Sect. 3.2.9)
- 5 ml 10% [w/v] N-lauroylsarcosine
- 25 mg yeast RNA

Pour this mixture into an Erlenmeyer flask containing 35 g SDS (careful: wear respiratory protection). Heat the solution while shaking to dissolve SDS, then fill up to 500 ml with autoclaved water. The hybridization solution can be stored at −20 °C and reused after heating to 65 °C.

Prehybridization   Prehybridize for 1 h at 39 °C in a hybridization oven using 20 ml of buffer per roller tube. If using a plastic bag add 20 ml hybridization buffer per 100 cm² of filter.

Hybridization   The hybridization solution consists of the prehybridization solution and the denatured hybridization probe. Use 6 ml of hybridization solution (per roller tube) or 2.5 ml per 100 cm² filter and 26 ng/ml freshly denatured DIG-labeled DNA probe. Incubate overnight at 39 °C or at the calculated hybridization temperature.

Hybridization washes   Wash 2 × 5 min at room temperature with 2× SSC; 0.1% [w/v] SDS and 2 × 15 min at 65 °C with 0.5× SSC; 0.1% [w/v] SDS. For genomic hybridization the stringent washes should be performed at 65°C instead of 68°C. It is important that the washing solution is preheated!

Immunological detection protocols are provided in Sects. 3.2.11 and 3.2.12. If using chemiluminescent detection, it is necessary to also include Tween-20 in the washing buffer. The antibody is diluted 1 : 10000. The special detection protocol in Sect. 3.2.12 must be followed. **Immuno-logical detection**

Under the described conditions it is possible to detect a single copy gene in 0.25 µg human DNA. This protocol has been successfully applied in human and plant systems and can easily be adapted to accomodate restriction enzyme digests of plasmids, cosmids, and lambda bacteriophages. For RFLP analysis the „Church" buffer is used without formamide at 55°C.

19.2.4 Special Hints for Application and Troubleshooting

- The 7% [w/v] SDS solution will precipitate quickly, particularly if hybridization conditions without formamide are used. Therefore the hybridization solution has to be preheated to about 65°C, especially when sealed bags are used. We recommend using a hybridization oven if possible. **Hybridiza-tion**

- If PCR probes are used, probe concentrations should be evaluated in mock hybridizations. In these, small pieces of membrane (3 × 3 cm) should be shortly prehybridized prior to adding different dilutions of PCR probe in hybridization buffer. This is followed by incubation for 2 h and immunological detection. The probe concentration that gives the lowest background should be used in the main experiment. Commonly, PCR probe concentrations of 6–12 ng/ml are sufficient.

- When roller tubes are used, the actual temperature inside the tube has to be tested in a prerun with, e.g., distilled water. In many ovens the temperature set for the oven is not maintained inside the tube.

- Perform the washes and antibody reaction in trays. Do not use the same roller tube for hybridization and detection reactions. Protein from the antibody detection will bind to the tube and cause background in subsequent hybridizations. For the washes, change trays after the antibody reaction. **Detection reaction**

- Buffer 3, containing $MgCl_2$, may be turbid after autoclaving. It is possible to use buffer 3 without $MgCl_2$.

- Depending on the size of the membrane as little as 500 µl of Lumiphos™ 530* or Lumigen™ PPD* solution can be used for chemiluminescent detection. Start by placing a drop of 500 µl of Lumiphos™ 530 or Lumigen™ PPD on a plastic transparency or seal type bag and place the membrane, DNA side facing the Lumiphos™ 530 or Lumigen™ PPD, on this drop. Spread the Lumiphos™ 530 or Lumigen™ PPD solution by covering membrane with a second transparency sheet and let excess fluid drip off. Incubate for 5–10 min at room temperature and then for about 10 min at 37°C and then start making exposures.

---

* Trademark of Lumigen Inc., Detroit, MI, USA Lumiphos™ and Lumigen™PPD are the subject of U.S. patents 4,962,192 and 4,969,182 granted to Lumigen Inc., Detroit, MI, USA

- If the membrane from Boehringer Mannheim is used baking of the DNA on the filter must be performed for 30 min at 120°C. UV cross-linking can be applied with the same efficiency and is often preferred due to the short reaction time.

Background reduction
- proteinase K treatment of the labeled probe may reduce unspecific hybridization of a probe to the membrane. First, the labeling mixture should be precipitated with ethanol and then resuspended in sterile water. A $\frac{1}{10}$ volume of 20 mg/ml proteinase K solution (Boehringer Mannheim) in 10 mM Tris/HCl, pH 7.5 is then added. Incubate 2 h at 37°C and add the appropriate amount of denatured probe to the hybridization solution.
- When using chemiluminescent detection, problems can arise because background from both sides of the membrane will be detected on the X-ray film. This can be reduced by UV cross-linking of both sides of the membrane.
- Change trays after hybridization and antibody reactions and between antibody reaction and washes (see above).

Increasing signal intensity
- The intensity of bands can be increased about two- to three-fold by adding 10% [w/v] dextran sulfate to the Church buffer with formamide. A disadvantage is that the background may be increased too. The same results can be achieved when 6% [w/v] PEG 8000 is added to the Church buffer with formamide and the background is not as strongly increased as with dextran sulfate.

19.2.5 Northern Blot Hybridization

For RNA:DNA hybridization the Church protocol (see Sect. 19.2.3.1) is highly recommended. For RNA:RNA hybridizations either the standard hybridization protocol (see Sect. 19.2.3) or the Church protocol can be used. In general, RNA:RNA hybridizations give better results than RNA:DNA hybridizations. This is also true for radioactive labeled probes. Methods for denaturation of RNA and electrophoresis on agarose gels are described elsewhere (see Sambrock et al., 1989). The RNA gels can be transferred to the membrane without further pretreatment. All solutions needed for northern blot hybridization should be treated with diethylpyrocarbonate (DEPC) and then sterilized by autoclaving (DEPC is a suspected carcinogen and should be handled with care).

Hybridization
- Follow the standard protocols described above for either Southern blot hybridization or the modified Church procedure. For RNA:DNA hybridization the temperature should be at least 50°C and for RNA:RNA hybridizations at least 68°C, since RNA:DNA hybrids are more stable than DNA:DNA hybrids and RNA:RNA hybrids are even more stable.

  In principle all protocols for using radioactive probes can also be used with the DIG system.

19.2.6 Special Hints for Application and Troubleshooting

- The blocking reagent must be autoclaved. Dissolve the blocking reagent in a microwave oven prior to autoclaving. A 10% stock solution can be stored frozen or at 4 °C for several months (see Southern hybridization protocol). For northern hybridization all required solutions should be autoclaved.
- Instead of DEPC, dimethyldicarbonate, which is less toxic, can be used. Although it decomposed well without autocalving dimethyldicarbonate treated solutions should still be autoclaved prior to use in northern blots.

19.2.7 Dot Blot Hybridization

The same protocols as described for Southern and northern hybridization may be used.

19.2.8 Hybridization with Oligonucleotides in Southern and Northern Blotting

When using oligonucleotides as probes, the best hybridization temperature depends on the length and nucleotide composition of the oligonucleotide. The hybridization is usually carried out at 5°–10°C below the melting temperature (T_M) of a perfect matched hybrid. For oligonucleotides shorter than 18 nucleotides, the T_M can be estimated by adding 2 °C for each A and T and 4 °C for each C and G. A more elaborate equation takes the ionic strength, the G/C content, and the length of the oligonucleotide into account and is suitable for oligonucleotides from 14 to 70 nucleotides complementary to the target sequence: $T_M = 81.5 + 16.6$ $(\log_{10}[Na^+]) + 0.41$ (%G + C) − $(600/N)$, where N is the length of the oligonucleotide and $[Na^+]$ is the concentration of sodium ions in the final stringent wash solution. Hybridization is carried out at temperature 5°–10°C below the T_M value. However, the equations mentioned above provide only a guideline for the optimal hybridization temperature. The hybridization conditions have to be determined empirically with the calculated hybridization temperature as a starting point. In addition a method has been described (Wood et al., 1985) that involves the use of tetramethylammonium chloride. In this salt solution the melting temperature is not dependent on the G/C content but rather is a function of the length of the oligonucleotide only. This method is particularly useful for the hybridization of a pool of degenerated oligonucleotides.

- 5× SSC
- 1% [w/v] blocking reagent in maleic acid buffer (see above protocol for Southern hybridization)

Standard hybridization solutions

- 0.1% [w/v] *N*-lauroylsarcosine
- 0.02% [w/v] SDS

The hybridization is performed as described above for Southern hybridization except using the hybridization buffer given above and the temperature predicted for the oligonucleotide. The optimal concentration of the labeled probe in the hybridization mixture and the time for hybridization depends on the amount of DNA or RNA to be detected on the filter. Usually 1–10 pmol tailed oligonucleotide per ml hybridization solution are used and hybridization is carried out for 1–6 h. The filter should be washed 2 × 5 min at the hybridization temperature with at least 50 ml 2× SSC; 0.1% [w/v] SDS per 100 cm^2 filter and 2 × 5 min at the hybridization temperature with 0.1× SSC; 0.1% [w/v] SDS. To avoid loss of signal, especially with short or degenerated oligonucleotides, the washing steps can also be carried out with 5× SSC; 0.1% [w/v] SDS, first at room temperature then at 2–10 degrees below the T_M predicted for the oligonucleotide.

19.2.8.1 Hybridization in Tetramethylammonium Chloride

Solutions
- TMACl stock solution: 5 mol/l TMACl; dissolve TMACl, heat to 68 °C if required, stir with 10% activated charcoal for 20–30 min; filter sterilize solution; and store in the dark at room temperature. The precise concentration of TMACl is determined by measuring the refractory index of the solution and determination of the exact molarity by the equation: molarity = (refractive index TMACl solution − 1.331)/0.018
- TMAClwash solution: 3 mol/l TMACl; 50 mmol/l Tris-HCl; 2 mmol/l EDTA; 0.1% [w/v] SDS; pH 8.0/25 °C
- 5× SSC

1. Prepare the nylon membrane as described above for Southern hybridization. The hybridization temperature is lowered to between 37 °C and 42 °C. Probes are hybridized less stringently in this procedure; specificity is determined in the TMACl wash. Nitrocellulose membranes are not stable for extended periods of time in TMACl solutions and are not recommended for this reason.
2. Rinse the membrane three times with 5× SSC at 4 °C; incubate membrane 2 × 30 min in 5× SSC at 4 °C.
3. Carefully rinse the membrane twice with TMACl wash solution at room temperature.
4. Incubate the membrane in TMACl wash solution at 68 °C for 20 min for stringent washing of the labeled oligonucleotide. Wash temperatures for oligonucleotides of different length are: for 16 nt, 50 °C; for 20 nt, 55 °C; for 30 nt, 68 °C; and for 50 nt, 75 °C.
5. Repeat step 4. Blot the membrane on chromatography paper to remove excess liquid. Do not allow the filters to dry out if rehybridization is to be performed. The filters are now ready for immunological detection.

References

Church G, Gilbert W (1984) Genomic sequencing. Proc Natl Acad Sci USA 81: 1991–1995

Rüger R, Höltke H-J, Reischl R, Sagner G, Kessler C (1991) Labeling of specific DNA-sequences with digoxigenin during polymerase chain reaction. In: Rolfs A, Schumacher HC, Marx P (eds) PCR Topics, Springer-Verlag, Berlin

Sambrook J, Fritsch T, *Maniatis* (1989) Molecular cloning, A Laboratory Manual, (2nd ed), Cold Spring Harbor Laboratory Press

Southern EM (1975) Detection of specific sequences. J Mol Biol 98:503–517

Thomas PS (1980) Hybridization of denatured RNA and small DNA-fragments transferred to nitrocellulose. Proc Natl Acad Sci USA 77:5201–5205

Wood WI, Gitschier J, Lasky LA, Lawn RM (1985) Base composition-independent hybridization in tetramethylammonium chloride: a method for oligonucleotide screening of highly complex gene libraries. Proc Natl Acad Sci USA 82:1585–1593

19.3 Colony and Plaque Hybridization

Thomas Walter

19.3.1 Principle and Applications

Colony and plaque hybridization have been developed for the rapid screening of bacterial or phage recombinant genomic libraries for specific DNA sequences [1, 2]. Bacterial colonies or phage particles are transferred to a nylon or nitrocellulose membrane. Colonies are lysed and phage particles are disassembled by alkaline treatment. The denatured DNA is then immobilized on the membrane in situ. Positive signals are detected by hybridization with a labeled DNA or RNA probe. Digoxigenin labeled probes have been applied to colony and plaque hybridization with the availability of the related nonradioactive nucleic acid labeling techniques [3, 4].

19.3.2 Methods

– Nylon membrane, positively charged (Boehringer Mannheim).   **Material**
– Nitrocellulose BA85 membrane (Schleicher and Schuell).

– *LB-freeze medium:* 10 g/l Bacto-tryptone; 5 g/l Bacto-yeast extract; **Reagents**
10 g/l NaCl; 25% [v/v] glycerol. For plates add 15 g/l Bacto-agar just **and**
before autoclaving; antibiotics are added as appropriate when the solu- **solutions**
tion is cooled to 55 °C.
– *Denaturing solution:* 0.5 N NaOH; 1.5 M NaCl.
– *Neutralization solution:* 1.5 M NaCl; 0.5 M Tris-HCl, pH 7.4/25 °C.

- *20 × SSC:* 3M NaCl; 0.3M Na-citrate; pH 7.0/25°C.
- *SDS:* 10% [w/v] Sodium-dodecylsulfate, 10% [w/v] (Boehringer Mannheim).
- *Proteinase K in solution:* 20mg/ml (Boehringer Mannheim).
- *PMSF:* 40μg/ml Phenylmethylsulfonylfluoride (Boehringer Mannheim).

Transfer To transfer phage particles or bacterial colonies, after incubation of the plate and colony or plaque formation, chill the plates for at least 1h, place a nylon membrane on the cold agar plate, and leave for 5min (bacteria) to 10min (phages). Label filter asymmetrically with a needle and record the orientation of the filter on the plate. Remove the membrane with care to avoid smearing of the colonies.

Replicas of phage libraries can be directly made this way. The original plates are stored at +4°C. Replicas of bacterial plates can be made by transferring the pattern of colonies to another agar plate. This is easily done with the aid of a piece of sterile velvet, which is placed on a cylinder about the same size as the plate. When freeze medium is used for preparation of the plates, the library can be stored at −70°C after transfer to a membrane.

Processing of the filter Phage particles are bound to the membrane by air drying the filter for 10min. Filters with bacterial colonies or phage particles are subsequently treated identically.

Place 3 sheets of Whatman 3MM paper (approximately the size of the membrane) side by side on plastic foil and saturate with the following solutions:

- first paper: denaturing solution
- second paper: neutralization solution
- third paper: 2× SSC solution

Place the nylon membrane with the colony or phage particles site up for 5min (colony 15min) on the first filter paper and subsequently for 5min on the second filter paper. Finally, leave the membrane for 15min on the third filter paper.

For bacterial colonies, shaking the membrane for 1h in 3 × SSC; 0.1% [w/v] SDS at 68°C is required; afterwards gently rub the top of the membrane with gloved fingertips or place a sheet of wet Whatman 3MM paper on the membrane and rub with a plane tool to transfer cell debris to the paper.

Alternatively, bacterial colonies can be treated with proteinase K at this stage. Therefore, the membrane is incubated for 1h at 37°C with proteinase K solution. For later immunological detection it is necessary to inactivate the proteinase K by incubation for 5min in PMSF, 40μg/ml, at room temperature, followed by two short washes in 2 × SSC.

Caution: PMSF is extremely harmful if swallowed, inhaled or absorbed through the skin. In case of contact, immediately flush eyes or skin with water. PMSF is readily inactivated in aqueous solutions at pH >8.6 and can be discarded after storage for several hours at room temperature.

Let the membrane air dry before performing UV fixation of nucleic acids. Alternatively, DNA can be fixed to the membrane by baking for 15−30 min at 120 °C in a vacuum; the vacuum is only required for baking nitrocellulose membranes.

The membrane can now either be used directly for hybridization or stored for later use.

Any type of DNA, RNA or oligonucleotide probe, as described in Sects. 3.2.3 to 3.2.8, can be used for colony and plaque hybridization. Make sure that the probe does not contain any sequences homologous to the vector used for library construction. Hybridization is performed as described in Sects. 3.2.9 or 3.2.10.

For the detection reaction either the color reaction (NBT/BCIP or fast dyes, Sects. 3.211, 12.1, 12.2) or the chemiluminescent reaction (Sects. 3.2.12, 13.2) can be used.

19.3.3 Special Hints for Application and Troubleshooting

It is not recommended that colonies be grown on nylon membranes when nonradioactive detection is employed, since this leads to false positive signals.

For colony and plaque screening the spoty background, which is sometimes observed with the chemiluminescent detection, can be suppressed just by using a buffer without Mg^{2+} ions (buffer 3). In addition, the antibody vial can be centrifuged briefly, once, and aliquots are taken from the top.

19.3.4 Example

The efficiency of an vitro mutagenesis reaction can be confirmed by colony hybridization with a DIG-labeled oligonucleotide. Competent *E. coli* TG1 cells are transformed with an aliquot of the in vitro mutagenesis reaction mixture of phT3T713M (amber) with a 30mer oligonucleotide that contains the wild type sequence and spans the region with the amber mutation in phT3T713M (amber). After plating and overnight incubation on X-Gal/IPTG indicator plates, revertants are identified by generation of blue colonies. The genotype is investigated by hybridization with the same oligonucleotide that is used for the mutagenesis experiment. The colonies are transferred to nitrocellulose BA85 and lysed following the above protocol. Nucleic acids are fixed by baking at 80 °C for 90 min in a vacuum. After 1 h prehybridization at 58 °C, the solution is replaced by the hybridization solution containing 2 pmol of labeled 30mer oligonucleotide per milliliter and the reaction is continued for 6 h. The hybridization solution can be discarded or stored for reuse at −20 °C. The unbound probe is then

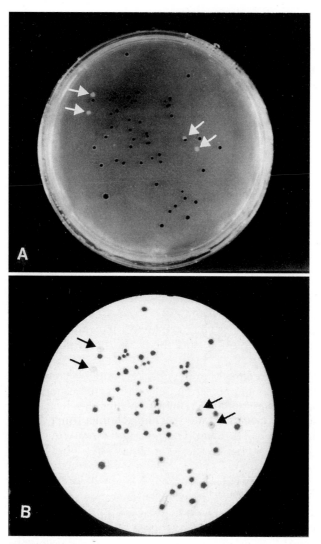

Colony hybridization with a DIG-labeled oligonucleotide and color detection with NBT/ BCIP. Only the colonies with a blue phenotype on the X-Gal/IPTG indicator plate **(A)** give rise to a positive hybridization signal with the specific oligonucleotide **(B).** Mutant white colonies *(arrows)* can be clearly distinguished from specific hybridization signal (Reproduced with permission of Academic Press from [5])

removed by two washes in $2 \times$ SSC for 5 min at room temperature, followed by a stringent wash, twice for 15 min in $0.1 \times$ SSC, at the hybridization temperature of 58°C. The immunological detection reaction is then performed using NBT/BCIP as the substrate for 15 min. Only the colonies which resulted in a blue phenotype on the indicator plate give rise to a positive hybridization signal as expected (Fig.).

References

1. Grunstein M, Hogness DS (1975) Colony hybridization: A method for the isolation of cloned DNAs that contain a specific gene. Proc Natl Acad Sci USA 72:3961–3965
2. Benton WD, Davis RW (1977) Screening λ gt recombinant dones by hybridization to single plaques in situ. Science 196:180–182
3. Voss H, Wirkner U, Jakobi R, Hewitt NA, Schwager C, Zimmermann J, Ansorge W, Pyerins W (1991) Structure of the gene encoding human casein kinase II subunit β. J Biol Chem 266:13706–13711
4. Gibson MA, Sandberg LB, Grosso LE, Cleary EG (1991) Complementary DNA cloning establishes microfibril-associated glycoprotein (MAGP) to be a discrete component of the elastin-associated microfibrils. J Biol Chem 266:7596–7601
5. Schmitz GG, Walter T, Seibl R, Kessler C (1991) Non-radioactive labeling of oligonucleotides in vitro with the hapten digoxigenin by tailing with terminal transferase. Anal Biochem 192:222–231

19.4 Multilocus DNA Fingerprinting Using Nonradioactively Labeled Oligonucleotide Probes Specific for Simple Repeat Elements

Jörg T. Epplen and Judith Máthé

19.4.1 Principle and Applications

Simple DNA sequences can be used as practical tools for probing simple repeat elements. In 1986 we differentiated human genomes using the ^{32}P-labeled synthetic probes $(GATA)_4$ and $(GACA)_4$ and other sequentially simple quadruplet repeat oligonucleotides [1]. Additional probes harboring basic repeat motifs of from two to six nucleotides were subsequently chemically synthesized, ^{32}P-labeled, and hybridized to DNAs from various sources. In parallel, panels of 5' biotinylated and/or digoxigenated oligonucleotides have been chemically synthesized [14] and tested for multilocus DNA fingerprinting in about 300 fungal, plant, and animal species [3, 13]. To date, in every species, at least one of the probes has been found to be informative with respect to genetic individualization [2]. Using the probe $(CAC)_5$ or its complement $(GTG)_5$, the demonstration of individuality is possible in every human being, except for monozygotic twins [9]. Nürnberg et al. [9] have meanwhile also established the somatic stability of $(CAC)_5/(GTG)_5$ fingerprints in human tissues and determined the mutation rate to be less than 0.001 per fragment per gamete. Somatic stability and similar or even reduced mutation rates have been determined for the $(GACA)_4$ fingerprint bands [11], which mainly stem from sequences in the nucleolus organizer regions (NOR) in humans [7].

In comparison to other identification methods, the advantages of multilocus fingerprinting in general include the technical simplicity and reproducibility, the high degree of informativeness, and, at the same time, absolute protection of personal data. In addition to solving identification problems, other potential applications are obvious: The fingerprint patterns of tumors are often changed due to the gain or loss of chromosomes and/or intrachromosomal deletion and amplification events [10]. After bone marrow transplantation complete and mixed chimerism can readily be demonstrated [6]. In certain animal species particular oligonucleotides identify simple repeat accumulations on the sex chromosomes or on other heterochromatic parts of the genome [7, 8]. In forensic applications not only human materials and stains can be identified but also, for example, the genealogy of valuable birds can be determined (see below). Thus the simple methodology of oligonucleotide DNA profiling offers considerable advantages over conventional identification and other DNA fingerprinting techniques over a broad range of applications. For all routine purposes, in which sufficient material is available, nonradioactive fingerprinting is highly recommended.

The mere presence and quantity of simple tandem repeats in the genome can be assessed effectively by a convenient slot blot hybridization method [4]. This, however, does not in itself prove the suitability of a given probe for multilocus fingerprinting. The latter has to be investigated with restriction enzyme-digested and electrophoretically separated DNA from the respective species. Hence, the principle of trial and error has to be adopted. The length of the probes range from 15 to 24 bases thus allowing the calculation of both hybridization and stringent wash temperatures according to the formula T_m (°C) $-5 = [(A+T) \times 2 + (C+G) \times 4] - 5$ (valid for 5' ^{32}P-labeled oligonucleotides). This rule applies for a salt concentration of 1 M NaCl. In practice, all hybridization and washing steps are conveniently performed in a temperature range from room temperature to 65°C, solely depending on the length and the base composition of the probe.

Three different approaches can be chosen to establish nonradioactive fingerprints: (1) In-gel hybridization and signal development (depending on the molecular weight of the antibody conjugate); (2) blot hybridization and signal detection via dye precipitation; (3) blot hybridization and signal detection via 3-(4-methoxyspiro [1,2-dioxetane-3,2'-tricyclo [3.3.1.1^{3,7}] decan]-4-yl) phenyl phosphate (AMPPD) and X-ray film.

In summary nonradioactive oligonucleotide fingerprinting is advantageous compared to related methodologies for a number of reasons: (a) the constant probe quality in a single batch of chemically synthesized oligonucleotide; (b) the economy due to nearly unlimited reusability of the hybridization solution; (c) the fast and technically simple procedure due to short hybridization, limited standardized washing steps, and short signal developing times; (d) the possibility of repeated hybridization to the same DNA sample. Above all, the most significant advantage over other fingerprint probe methods is base-specific hybridization under the appropriate conditions: even a single mismatch can abrogate the hybridization. Therefore, by following the established protocols, almost absolute reproducibility is ensured.

19.4.2 Demonstration of DNA Fingerprints by DIG-Labeled Oligonucleotides

- Restriction enzymes (Boehringer Mannheim, New England Biolabs, **Standard** Pharmacia) **reagents**
- DNA quick preparation kit („Genomix"; Kontron)
- Agarose (Sigma # A 6877)
- Immobilon-P PVDF transfermembrane (Millipore) (PVDF)membrane (Millipore)
- Hybond N (Amersham)
- DIG-labeled oligonucleotide probes (Fresenius)
- Blocking reagent (Boehringer Mannheim)
- Anti-digoxigenin antibody conjugated to alkaline phosphatase (<DIG>:AP) (Boehringer Mannheim)
- Dimethylformamide (DMF; Merck)
- Nitroblue tetrazolium (NBT; Boehringer Mannheim, ENZO, Sigma)
- 5-bromo-4-chloro-3-indolyl phosphate (BCIP; Boehringer Mannheim, ENZO, Sigma)
- AMPPD (Boehringer Mannheim)

For all standard labeling, hybridization, and detection solutions see Sect. **Standard** 3.2. **solutions**

- Oligonucleotide hybridization solution; final concentration of solutions: $5\times$ SSPE, $5\times$ Denhardt's solution, $10\,\mu g/ml$ fragmented and denatured *E. coli* DNA (denatured at $95\,°C$ for 5 min in distilled water), and 0.1% SDS. For 10 ml hybridization solution:
  - 2.5 ml $20\times$ SSPE (for 1 l: 3M NaCl, 0.2M $NaH_2PO_4 \cdot H_2O$, 0.02M $Na_2EDTA \cdot 2H_2O$; adjust the pH to 7.4 with 10M NaOH and fill up to 1 l with distilled H_2O. Autoclaving is not absolutely necessary if used up within several weeks).
  - 0.5 ml $100\times$ Denhardt's solution: 2g polyvinylpyrrolidone, 2g bovine serum albumin, 2g Ficoll 400. Dissolve in 100 ml autoclaved H_2O and store in small aliquots at $-20\,°C$.
  - 0.1 ml 10% SDS (Biorad)
  - 100 μg *E. coli* DNA (Sigma)
- 75 mg/ml NBT in 70% DMF
- 50 mg/ml BCIP in 100% DMF

1. DNA can be prepared from peripheral blood leukocytes of most animals according to standard protocols [12], or preferably using a kit (e.g., Genomix), in 45 min. The nucleated erythrocytes of birds, reptiles, and fishes may be preincubated in an anticoagulant preservative solution [2]. **Restriction digest, Southern blot, and hybridization**
2. A total of $3–10\,\mu g$ DNA per individual is restriction enzyme digested according to the manufacturers' recommendations. In general *Hin*fI, *Mbo*I, *Alu*I, or *Hae*III are preferable for fingerprinting. For certain plant or fungal species *Taq*I or *Mbo*II may be even more informative [13].

3. Electrophoresis is usually performed in 0.7%−0.8% agarose gels to resolve 1.5−30 kb DNA fragments. For hybridization and signal detection restriction enzyme-digested and size-fractionated genomic DNA is either fixed directly in the gel matrix by drying [14] or it is transferred onto immobilizing membranes.

4. After denaturation, DNA is blotted under alkaline conditions [12] onto PVDF or nylon membranes, which yield the optimal signal to background ratios. For filter hybridization, preblocking and pre-hybridization steps are necessary to avoid unspecific binding of the probe to the membrane. The duration of these two incubations is not critical. (Since blocking reagent dissolves only after heating, prepare this solution in advance and allow to cool to room temperature. For in-gel hybridization no pretreatment is required.

5. Hybridization is carried out for approximately 2−3 h with 10 pmoles of digoxigenated oligonucleotide probe per ml hybridization solution (see below). In order to increase signal intensity, the hybridization temperature should be $T_m-10°C$ (see table). During and after hybridization do not allow the membrane to dry out as this could cause severe background problems. The appropriate volume for bags is 0.05−0.1 ml/cm² blot and for rotating cylinders 0.025 ml/cm² gel.

6. After hybridization the gels or membranes are washed two or three times for 20−30 min at room temperature in 6× SSC and for 1 min at the hybridization temperature (stringent wash; see table).

Hybridization and stringent wash temperatures ($T_m - 10°C$)

Temperature	Oligonucleotide
40 °C	$(GTG)_5/(CAC)_5$
38 °C	$(GACA)_4$
30 °C	$(GATA)_4$
38 °C	$(GGAT)_4$

Signal development

Digoxigenated oligonucleotides are detected with a monospecific antibody coupled to alkaline phosphatase. Obviously the side of the membrane where the DNA is bound must be in contact with the antibody solution. If one applies the antibody solution in rolling cylinders be sure not to cover parts of the membrane by overlapping. The antibody solution can also be spotted directly onto the membrane. The phosphatase staining reaction is done at pH 9.5 in the presence of NBT, BCIP, and Mg^{2+} ions. The dye precipitation patterns are documented by photography of the moist filters or the gels. An example of the AMPPD detection procedure is explicitly described in the figure legend. Documentation is by exposure to X-ray films.

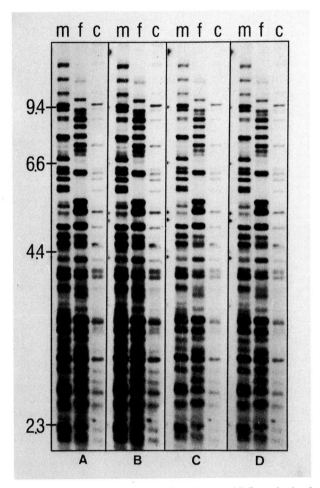

Multilocus fingerprints of a family of falcons (*m*, father; *,f* mother; *c*, child) as obtained with the digoxigenated oligonucleotide probe (GGAT)₄. All offspring bands can be traced to the parents' patterns proving the questioned upbringing in captivity and excluding theft of eggs from natural habitats. DNA has been digested with *Hin*fI, electrophoresed, blotted onto a nylon membrane, and hybridized as described in the text. Signal development: 30 min in 1:10 000 diluted antibody conjugate solution in blocking buffer (room temperature; 2 × 20 min washing solution (room temperature); 2 × 5 min in buffer 2 (room temperature); 5 min in 1:100 diluted AMPPD in buffer 2. The moistened membrane is wrapped airtight with plastic wrap and incubated at 37 °C for 15 min. Before the filter is exposed to X-ray film (Kodak XAR-5 or any equivalent), it should be stored for some time since the resulting intensity of the patterns increases considerably. *A*, 10 min exposure, 2 h after incubation; *B*, 5 min exposure, 6 h after incubation; *C*, 3 min exposure, 6 h after incubation; *D*, 1.5 min exposure, 6 h after incubation. Molecular weight markers are indicated in kilobases on the *left*

19.4.3 Special Hints for Application and Troubleshooting

- Avoid partial restriction enzyme digestions and inhomogeneitis of the electric field.
- Unfortunately, AMPPD signal development does not work directly in the gel.
- Substrate dyes must not precipitate, not even in the hardened agarose.
- Removal of dye precipitates and probe: From PVDF membranes dye precipitate can be removed, to a certain extent, by treatment with DMF. Subsequently the probe can be detached in low salt buffer at an elevated temperature. The membrane is then ready for a second hybridization, which should be started with the preblocking step.

References

1. Ali S, Müller CR, Epplen JT (1986) DNA fingerprinting by oligonucleotides specific for simple repeats. Hum Genet 74:239–243
2. Arctander P (1988) Comparative studies of avian DNA by restriction fragment length polymorphisms analysis: convenient procedures on blood samples from live birds. J Ornitol 129:205–216
3. Epplen JT, Ammer H, Epplen C, Kammerbauer C, Roewer L, Schwaiger W, Steimle V, Zischler H, Albert E, Andreas A, Beyermann B, Meyer W, Buitkamp J, Nanda I, Schmid M, Nürnberg P, Pena SDJ, Pöche H, Sprecher W, Schartl M, Yassouridis A (1991) Oligonucleotide fingerprinting using simple repeat motifs: a convenient, ubiquitously applicable method to detect hypervariability for multiple purposes. In: Burke T, Dolf G, Jeffeys AJ, Wolff R (eds) DNA Fingerprinting: Approaches and Applications, Birkhäuser-Verlag, Basel, 1991, pp 51–69
4. Epplen JT (1988) On simple repeated GATA/GACA sequences: a critical reappraisal. J Hered 79:409–417
5. Epplen JT (1992) The methodology of multilocus DNA fingerprinting using radioactive or nonradioactive oligonucleotide probes specific for simple repeat motifs. In: Chrambach, Dunn, Radola BJ (eds) Adv. Electrophoresis, VCH-Verlag, Weinheim, pp 59–114
6. Mittermüller J, Hartwig R, Epplen JT, Mönch T, Simon L, Kolb HJ (1991) DNA-Fingerprinting zur Überprüfung des Chimärismus nach allogener Knochenmarkstransplantation. In: Radola BJ (ed) Elektrophorese Forum '91, pp 166–171
7. Nanda I, Deubelbeiss D, Guttenbach M, Epplen JT, Schmid M (1990) Heterogeneitis in the distribution of $(GACA)_n$ simple repeats in the karyotypes of primates and mouse. Hum Genet 85:187–194
8. Nanda I, Schmid M, Epplen JT (1991) In situ hybridization of nonradioactive oligonucleotide probes to chromosomes. In: Adolph KW (ed) Advanced Techniques in Chromosome Research, Marcel Dekker, New York, pp 117–134
9. Nürnberg P, Roewer L, Neitzel H, Sperling K, Pöpperl A, Hundrieser J, Pöche H, Epplen C, Zischler H, Epplen JT (1989) DNA fingerprinting with the oligonucleotide probe $(CAC)_5/(GTG)_5$; somatic stability and germline mutations. Hum Genet 84:75–78
10. Nürnberg P, Zischler H, Fuhrmann E, Thiel G, Losanova T, Kinzel D, Nisch G, Witkowski R, Epplen JT (1991) Co-amplification of simple repetitive DNA fingerprint fragments and the EGF receptor gene in human gliomas. Genes Chromosomes Cancer 3:79–88
11. Roewer L, Nürnberg P, Fuhrmann E, Rose M, Prokop O, Epplen JT (1991) Stain analysis using oligonucleotide probes specific for simple repetitive DNA sequences. Forensic Sci Internatl 47:59–70

12. Sambrook J, Fritsch EF, Maniatis T (1989) Molecular cloning: a laboratory manual. Cold Spring Harbor Laboratory Press, Cold Spring Harbor

13. Weising K, Ramser J, Kaemmer D, Kahl G, Epplen JT (1991) Oligonucleotide fingerprinting in plants and fungi. In: Burke T, Jeffreys AJ, Wolff R, Dolf G (eds) DNA-fingerprinting: approaches and applications, Birkhäuser Verlag, Basel, pp 312−331

14. Zischler H, Nanda J, Schäfer R, Schmid M, Epplen JT (1989) Digoxigenated oligonucleotide probes specific for simple repeats in DNA fingerprinting and hybridization in situ. Hum Genet 82:227−233

19.5 Tn5*cos* Restriction Mapping of Large DNA Plasmids

ULRICH ZUBER and WOLFGANG SCHUMANN

19.5.1 Principle and Applications

Restriction mapping of large DNA fragments by standard techniques can be a slow and tedious process. Partial cleavage of linear molecules with appropriate restriction endonuclease, labeling at only one end, electrophoretic separation of the digestion products, and detection of the end-labeled cleavage products by autoradiography can simplify the mapping procedure (Smith and Birnstiel, 1976). The disadvantage of this method is the preparation of fragments labeled only at one end.

To overcome this experimental problem, Rackwitz and coworkers developed a method which allows specific labeling of linear DNA fragments at one end (Rackwitz et al., 1984; 1985). Their *cos* mapping method is based on the property of the lambda *cos* site to yield 12 nucleotides long single-stranded overhangs after treatment with lambda terminase (Terlineraritation). These single-stranded regions are nonpalindromic; therefore, specific oligonucleotides can be annealed to the left and right *cos* sites. These oligonucleotides have been designated ON-L and ON-R.

Rackwitz and coworkers used the *cos* mapping procedure to establish restriction maps for inserts within cosmids and lambda vectors. We expanded the application of the *cos* mapping procedure to virtually any plasmid from gram-negative bacteria (Zuber and Schumann, 1991). The principle of our method consists of the use of a mobile *cos* site, which can integrate into the plasmid of choice by transposition. The *cos* site was thus inserted into Tn5, a mobile element that can transpose at random into any DNA molecule in a wide variety of gram-negative species.

Further applications of our Tn5*cos* mapping system can be envisaged. The Tn5*cos* element can be used for directly establishing a restriction map of a localized region of interest within the bacterial chromosome. Furthermore, the *cos* cassette can be inserted into transposable elements active in other organisms such as gram-positive bacteria, yeast, and other lower and even higher eukaryotes. Restriction maps at predetermined sites can be established by first integrating the *cos* cassette into a restriction fragment and then recombining the *cos* region into the target molecule via homologous recombination.

19.5.2 Reaction Scheme

Tn5*cos* restriction mapping is comprised of seven steps:

1. Genetic labeling of the plasmid with Tn5*cos*
2. DIG labeling of the oligonucleotides ON-L and ON-R using terminal deoxynucleotide transferase
3. Linearization of plasmid: Tn5*cos* with lambda terminase
4. Partial cleavage with restriction enzyme
5. Annealing of DIG-ON-L and DIG-ON-R
6. Pulse field gel electrophoresis (PFGE)
7. Detection of DIG-labeled fragments

The standard reaction scheme for genetic labeling and restriction mapping of plasmids containing Tn5*cos* is shown in the figure.

19.5.3 Tn5*cos* Restriction Mapping

Standard reagents
- ON-L, ON-R (Amersham-Buchler)
- DIG-[11]-dUTP, DIG-[11]-ddUTP (Boehringer Mannheim)
- Terminal deoxynucleotidyl transferase (TdT) (BRL)
- Lambda terminase extract (Amersham-Buchler)
- Nitroblue tetrazolium salt (NBT) (Boehringer Mannheim)
- 5-bromo-4-chloro-3-indolyl phosphate (BCIP) (Boehringer Mannheim)
- Anti-DIG:AP conjugate (Boehringer Mannheim)

Standard solutions
- TE buffer: 10 mM Tris-HCl; 1 mM EDTA; pH 8.0/25 °C
- 5× Tailing buffer: 500 mM potassium cacodylate; pH 7.2/25 °C; 10 mM $CoCl_2$; 1 mM DTT
- EDTA solution: 0.2 M EDTA, pH 8.0/25 °C
- Gel loading solution: 0.25% [w/v] bromophenol blue; 0.25% xylene cyanol FF; 30% [v/v] glycerol
- Buffer A: 150 mM Tris-HCl; 22.5 mM $MgCl_2$; 7.5 mM EDTA; pH 8.0/25 °C
- Buffer B: 6 mM Tris-HCl; 18 mM $MgCl_2$; 30 mM spermidine-HCl; 60 mM putrescine; pH 7.4/25 °C
- TBE buffer: 89 mM Tris-HCl; 89 mM boric acid; 2 mM EDTA; pH 8.0/25 °C
- Buffer I: 100 mM Tris-HCl; 150 mM NaCl; pH 7.5/25 °C
- Buffer II: 100 mM Tris-HCl; 100 mM NaCl; 50 mM $MgCl_2$; pH 9.5/25 °C

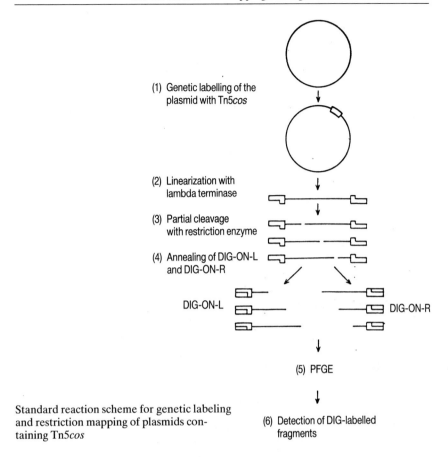

(1) Genetic labelling of the plasmid with Tn5*cos*

(2) Linearization with lambda terminase

(3) Partial cleavage with restriction enzyme

(4) Annealing of DIG-ON-L and DIG-ON-R

DIG-ON-L

DIG-ON-R

(5) PFGE

Standard reaction scheme for genetic labeling and restriction mapping of plasmids containing Tn5*cos*

(6) Detection of DIG-labelled fragments

– NBT solution: 75 mg/ml NBT in dimethylformamide
– BCIP solution: 50 mg each BCIP and toluidinium salt

To genetically label the plasmid of choice, the in vitro constructed transpo- **Genetic** son Tn5*cos* has to be moved into that plasmid. At present, three donor **labeling** replicons are available, the *E. coli* chromosome (strain C600::Tn5*cos*), ColE1::Tn5*cos,* and the suicide vector pSUP202 (pSUP202::Tn5 *cos*). If the labeling can be done in *E. coli,* the plasmid is transferred into C600::Tn5*cos* by conjugation, mobilization, or transformation, in which Tn5*cos* spontaneously transposes into the plasmid. To recover a transposition event the plasmid is crossed out into another *E. coli* strain: transconjugants are selected on kanamycin-containing plates. Alternatively, the plasmid can be isolated and transformed into an *E. coli* strain. If a plasmid in a gram-negative bacterium other than *E. coli* should be genetically marked, the suicide vector pSUP202::Tn5*cos* is mobilized into that strain, and Tn5*cos*-containing bacteria are isolated by one of the methods described for *E. coli*.

DIG labeling of ON-L and ON-R

1. Incubate the following reaction mixture for 20 min at 37°C:
   - 5 µl ON-L and ON-R, each (1 pmol/µl)
   - 1 µl 1 mM DIG-[11]-dUTP or DIG-[11]-ddUTP
   - 2 µl 5× tailing buffer
   - 2 µl TDT (0.6 units/µl)
2. Add:
   - 3 µl EDTA solution to stop the reaction
3. Add:
   - 125 µl 10 mM TE
   - 50 µl 1 M NaCl
   - 75 µl gel loading solution
   - 250 µl
4. Store labeled oligonucleotides at −20°C

Plasmid isolation, lineariza-tion, and partial cleavage

1. Isolate the plasmid by a standard procedure.
2. Treat with lambda terminase:
   - 11 µl plasmid DNA (2−5 µg)
   - 2 µl buffer A
   - 2 µl buffer B
   - 1 µl 75 mM MgATP
   - 2 µl 50 mM DTT, freshly prepared
   - 2 µl lambda terminase extract
3. Incubate for 30 min at 20°C.
4. Stop reaction by incubating for 3 min at 65°C.
5. Cleave partially with appropriate restriction endonuclease(s).
6. Treat reaction mixture once with phenol-chloroform-isoamylalcohol (50:48:2).
7. Precipitate DNA with ethanol, centrifuge, and dry under vacuum.

Hybridiza-tion and product separation

1. Mix in two separate reactions:
   - 5 µl 1.5 µg of Ter-linearized and partially cleaved plasmid DNA
   - 5 µl labeled DIG-ON-L
   - 5 µl 2.5 µg of Ter-linearized and partially cleaved plasmid DNA
   - 5 µl labeled DIG-ON-R
2. Incubate for 3 min at 70°C to disrupt the cohesive termini.
3. Incubate for 30 min at 41°C to anneal the oligonucleotides to the *cos* sites.
4. Load probes onto a 1% agarose gel.
5. Carry out PFGE (Biorad) at 18°C and 16 V/cm for 15−17 h in TBE buffer; the higher ionic strength of the buffer (instead of the recommended 0.5× TBE) is necessary to stabilize the hybridization of the oligonucleotides; the switching interval was increased from 1.0 to 8.0 s.

Detection reaction

1. Vacuum dry the gel for 120 min at room temperature.
2. Shake the gel for 90 min in 20 ml buffer I containing 5 µl of Anti-DIG-AP conjugate.

3. Remove unbound antibodies by washing the gel in 100 ml of buffer I for 30 min.
4. Seal the gel in a plastic bag containing:
   10 ml   buffer II
   45 µl   NBT solution
   35 µl   BCIP solution
5. Stain for 9−10 h.
6. Soak gel in TE to stop the staining reaction.
7. Dry gel for 30 min at 60 °C.
8. Read the result.

References

Rackwitz HR, Zehetner G, Frischauf A, Lehrach H (1984) Rapid restriction mapping of DNA cloned in lambda phage vectors. Gene 30:195−200
Rackwitz HR, Zehetner G, Murialdo H, Delius H, Chai JH, Poustka A, Frischauf A, Lehrach H (1985) Analysis of cosmids using linearization by phage lambda terminase. Gene 40:259−266
Smith HO, Birnstiel ML (1976) A simple method for DNA restriction site mapping. Nucleic Acids Res 3:2387−2400
Zuber U, Schumann W (1991) Tn5*cos:* a useful tool for restriction mapping of large plasmids. Gene 103:69−72

19.6 DIG DNA Sequencing with Chemiluminescent or Dye Substrates

GREGOR SAGNER

19.6.1 Principle and Applications

DNA sequencing based on the enzymatic dideoxy sequencing technique (Sanger et al., 1977) mainly employs two different procedures. Large-scale sequencing projects are often performed with automated sequencers using fluorescent detection (Ansorge et al., 1987; Connell et al., 1987), whereas routine sequencing in the majority of laboratories uses radioactive labeling. In recent years the technique of nonradioactive labeling and enzyme-coupled detection using colorimetric and chemiluminescent substrates has been shown to be applicable to DNA sequencing (Beck, 1987; Beck et al., 1989).

A 5′-labeled oligonucleotide is used as a primer in an extension reaction for DNA sequencing or as a probe for hybridization to a target DNA. After electrophoresis the sequencing reaction products are transferred to a nylon membrane. This can be performed either with a direct blotting electrophoresis (DBE) device (see Sect. 19.7) or with a simple and efficient blotting procedure from standard sequencing gels. The latter method uses equipment available in any biochemical laboratory (sequencing electrophoresis device, gel dryer, plastic bags, shaker, heat sealer) and is compatible with the widely used radioactive sequencing methods. A hard copy of the sequence data can be stored either on X-ray film using chemiluminescent detection or on nylon membrane using colorimetric detection.

The nonradioactive sequencing system described here relies on the interaction between the hapten digoxigenin (DIG) for 5′ labeling of the sequencing primer and anti-digoxigenin antibodies coupled to alkaline phosphatase for detection of the sequencing reaction products. The method can be used with both the colorimetric substrates BCIP/NBT and the chemiluminescent substrate Lumigen™PPD (see Sects. 3.2.11 and 3.2.12) and yields clear and background-free sequence data. The resolution of the sequence ladder is analogous to radioactive sequencing; up to 300 nucleotides can routinely be read from a single sequence ladder. The detection time of the sequence ladder is reduced to 60–180 min using the chemiluminescent substrate Lumigen™PPD at 10 μl/ml or to 20–60 min at 100 μl/ml concentration compared to 16–40 h using radioactive sequencing techniques. Thus the additional time required for the membrane processing steps is compensated. The transfer of the sequence ladder from the standard polyacrylamide sequencing gel to the membrane is performed within 15 min using a gel drying apparatus. No special laboratory equipment is required.

As reprobing of blots after luminescent detection is easy and fast (Höltke et al., 1992) the DIG chemiluminescent detection chemistry can ideally be combined with the multiplex DNA sequencing technique (Church and Kieffer-Higgins, 1988; Tizard et al., 1990).

A flow diagram of the reaction scheme for DIG sequencing and detection is given in the figure.

19.6.2 DIG DNA Sequencing and Detection

Primer DIG-labeling reagents
- DIG-*N*-hydroxy-succinimide ester (Boehringer Mannheim)
- Aminolink II™ (Applied Biosystems)

Sequencing reaction reagents
- *Taq* DNA Polymerase
- T7 DNA Polymerase
- DIG-labeled M13/pUC sequencing/reverse sequencing primer (Boehringer Mannheim)
- dNTPs, ddNTPs (Boehringer Mannheim)

DNA transfer reagents
- Nylon membrane (Boehringer Mannheim)
- Whatman 3 MM paper

Sequence detection reagents
- Blocking reagent (Boehringer Mannheim)
- Anti-DIG:alkaline phosphatase, F_{ab} fragments (Boehringer Mannheim)

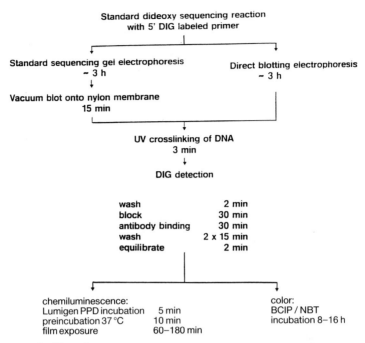

DIG DNA sequencing/detection

- Colorimetric detection: DIG Nucleic Acid Detection Kit (Boehringer Mannheim) or BCIP/NBT solutions (Boehringer Mannheim)
- Chemiluminescent detection: DIG Luminescent Detection Kit for Nucleic Acids (Boehringer Mannheim) or Lumigen PPD™ stock solution (Boehringer Mannheim)

Sequencing solutions

- Denaturation buffer: 2M NaOH; 2mM EDTA; pH 8.0
- Neutralization buffer: 2M ammonium acetate; pH 4.5
- 5× reaction buffer for *Taq* DNA polymerase: 250mM Tris-HCl; 50mM $MgCl_2$; pH 9.0
- *Taq* DNA polymerae termination mixes:
  - ddA termination mix: 25 µM each dATP, dGTP (or 7-deaza-dGTP), dCTP, dTTP; 850 µM ddATP; 950 µM $MgCl_2$; pH 7.5
  - ddC termination mix: 25 µM each dATP, dGTP (or 7-deaza-dGTP), dCTP, dTTP; 400 µM $MgCl_2$; pH 7.5
  - ddG termination mix: 25 µM each dATP, dGTP (or 7-deaza-dGTP), dCTP, dTTP; 75 µM ddGTP; 175 µM $MgCl_2$; pH 7.5
  - ddT termination mix: 25 µM each dATP, dGTP (or 7-deaza-dGTP), dCTP, dTTP; 1275 µM ddTP; 1370 µM $MgCl_2$; pH 7.5
- 5× reaction buffer for T7 DNA polymerase: 200mM Tris-HCl; 100mM $MgCl_2$; 250mM NaCl; pH 7.5

- T7 DNA polymerase termination mixes:
  - ddA termination mix: 80 µM each dATP, dGTP (or 7-deaza-dGTP), dCTP, dTTP; 8 µM ddATP; 50 mM NaCl
  - ddC termination mix: 80 µM each dATP, dGTP (or 7-deaza-dGTP), dCTP, dTTP; 8 µM ddCTP; 50 mM CaCl
  - ddG termination mix: 80 µM each dATP, dGTP (or 7-deaza-dGTP), dCTP, dTTP; 8 µM ddGTP; 50 mM NaCl
  - ddT termination mix: 80 µM each dATP, dGTP (or 7-deaza-dGTP), dCTP, dTTP; 8 µM ddTTP; 50 mM NaCl
- formamide buffer: 98% formamide; 0.2% bromophenol blue; 0.2% xylene cyanol; 10 mM EDTA; pH 8.0

Detection solutions
- buffer 1: 0.1 M maleic acid, 0.15 M NaCl, pH 7.5
- blocking stock solution: 10% blocking reagent in buffer 1
- washing buffer: buffer 1 + 0.3% Tween 20
- buffer 2: blocking stock solution diluted 1 : 10 in buffer 1
- buffer 3: 0.1 M Tris-HCl, 0.1 M NaCl, 50 mM $MgCl_2$, pH 9.5
- color-substrate solution (freshly prepared): 45 µl
- NBT solution + 35 µl BCIP solution in 10 ml buffer 3
- chemiluminescent substrate solution (freshly prepared): Lumigen™PPD stock solution diluted 1 : 1000 in buffer 3 = 23.5 µM Lumigen™PPD

Primer digoxigenylation
DIG-labeled oligonucleotide sequencing primers are prepared by standard chemistry on an automatic DNA synthesizer and reacted in a final synthesis cycle with a 5′ terminal amino function. After cleavage of the protection group by concentrated ammonia, the oligonucleotide is reacted with the activated N-hydroxysuccinimide ester of digoxigenin (DIG-NHS). As the labeling reaction is not quantitative, separation of the labeled oligonucleotide from the unlabeled compound has to be performed either by reverse phase HPLC or by electrophoresis on a denaturing polyacrylamide gel. The detailed protocol is stated in the product description of DIG-NHS (Boehringer Mannheim).

Sequencing reactions
Only minor modifications of the standard sequencing protocols for dideoxy sequencing with T7 and *Taq* DNA polymerase are necessary for the DIG system. As DIG-labeled primers are used in our technique, the labeling step in both standard T7/*Taq* sequencing protocols can be omitted.

The detailed procedure for nonradioactive sequencing using either *Taq* DNA polymerase or T7 DNA polymerase is as follows:

Standard reaction for single-stranded DNA sequencing
I Primer annealing
1. Mix the folowing components in a sterile microcentrifuge tube:
   - Single-stranded template DNA 0.5 pmol
   - 1 pmol primer
   - 4 µl corresponding reaction buffer (for *Taq* Pol or T7 Pol)
     Add redist. H_2O to a final volume of 10 µl.

2. Centrifuge briefly.
3. Incubate for 10 min at 55 °C.
4. Allow the reaction mixture to cool slowly at room temperature. Subject the mixture to a brief centrifugation.
   While the primer annealing mixture is cooling, prepare the four corresponding extension/termination mixtures (for *Taq* DNA pol or T7 DNA pol), dispense 2 µl of the A, C, G and T mix into reaction vials which have been correspondingly labeled by letter or color.
5. Add 1 µl *Taq* DNA polymerase (3 U/µl) or 2 µl T7 DNA polymerase (1 U/µl) to the primer annealing mixtures.
6. Add sterile redist. H_2O to a final volume of 20 µl

II Extension/termination reaction
1. At room temperature, add 4 µl of the annealing reaction to each of the four marked (A, C, G and T) termination mixtures; centrifuge briefly.
2. Incubate for 3 min at 70 °C (for *Taq* DNA pol) or for 5 min at 37 °C (for T7 DNA pol).
3. Centrifuge the tubes briefly.
4. Add 2 µl of formamide buffer to stop the reaction.
5. The sequencing reactions may be stored at −20 °C before use.

I Denaturation of plasmid DNA
Standard assay for double-stranded sequencing

1. Dilute an aliquot containing 1 pmol plasmid DNA (for *Taq* DNA pol sequencing) or 1.5 pmol plasmid DNA (for T7 DNA pol sequencing) to a final volume of 18 µl with redist water.
2. Add 2 µl denaturation buffer, mix thoroughly, and incubate for 5 min at room temperature.
3. Add 2 µl neutralization buffer for neutralization.
4. Immediately add 100 µl of pre-chilled ethanol (95%, −20 °C), mix, and incubate for at least 5 min at −70 °C.
5. Centrifuge for 15 min at 4 °C in a microcentrifuge (approx. $10000 g$). Decant the supernatant taking care not to lose the DNA pellet.
6. Gently add 1 ml ethanol (70%) to the DNA precipitate, briefly vortex, and centrifuge again for 15 min. Carefully discard the supernatant.
7. Dry the plasmid DNA briefly under a vacuum (the DNA can be stored for a few days at −20 °C).

II Primer annealing
1. Add to the dried, denatured plasmid DNA 1 pmol primer, 4 µl corresponding reaction buffer (for *Taq* DNA pol or T7 DNA pol). Add sterile redist H_2O to a final volume of 10 µl.
2. All following steps are described for single-stranded sequencing beginning with step 2 of the corresponding primer annealing reaction.

Immediately before gel electrophoresis, denature the reaction products for 3 min at 95 °C, place on ice, and subject to a brief centrifugation. Dispense 3 µl of each termination reaction to the lanes of a sequencing gel.

Denaturation

19.6.2

Standard sequencing gel electrophoresis and blotting

Transfer the sequencing products from the polyacrylamide sequencing gel to a membrane after standard acrylamide gel electrophoresis by an efficient and convenient vacuum blot procedure:

1. Sequencing gel electrophoresis:
   - Treat both glass plates with Repel-Silane (E. Merck, Darmstadt) before pouring the gel.
   - Perform electrophoresis under standard conditions.
2. After electrophoresis, carefully remove one of the glass plates.
3. Cut a nylon membrane to the size of the gel.
4. Press the membrane close to the gel avoiding air bubbles.
5. Pull off the gel from the glass plate together with the membrane.
6. Blotting: Use a commercially available gel dryer of appropriate size.
   - Cover the porous gel support with one layer of dry Whatman filter paper.
   - Place the membrane/gel sandwich on top (gel above the membrane).
   - Apply vacuum without heating.
   - Vacuum-blot for 15 min.
7. After blotting, the gel usually sticks to the nylon membrane. Do not try to remove it before performing the washing step for DIG detection.
8. Bind the DNA to the membrane by UV crosslinking for 3 min (crosslinking is not inhibited by the bound gel).

DIG detection

Detection of the DIG-labeled sequencing products on the membrane is performed with minor modifications according to the protocols given in Sect. 3.2 using the chemiluminescent substrate Lumigen™PPD or the color substrates BCIP/NBT.

Chemiluminescent detection

The following modifications of the protocol in Sect. 3.2.12 are recommended:

Perform the washing step in an appropriate tray; the gel is thereby detached from the membrane and can easily be removed from the washing buffer. All subsequent steps are performed in a sealed plastic bag. This reduces the volumes of the solutions used to approx. $10 \, ml/100 \, cm^2$ membrane size. Repeated cutting and resealing of the plastic bag in the course of the detection steps makes frequent changing of the bag unnecessary (exception: use of a new plastic bag for membrane exposure in the chemiluminescent detection protocol). All steps are performed with thorough shaking or mixing to achieve an uniform membrane incubation in the small buffer volumes.

Alternatively, the detection reactions can be performed in a cylindrical roller of the appropriate size. Don't let the membrane overlap during incubation steps.

As the format of blotting membranes for sequencing gels is relatively large, for chemiluminescent detection it is recommended that Lumigen™ PPD is used at a concentration of $10 \, \mu g/ml$ (1 : 1000 dilution of Lumigen™ PPD stock solution) instead of the $100 \, \mu g/ml$ (1 : 100 dilution) recommended for hybridization procedures; this extends the film exposure times to acceptable short 60−180 min. For sequencing applications the diluted Lumigen™ PPD solution should not be reused. The Lumigen™PPD concentration of 10 µg/ml keeps the costs for the DIG detection low. When shorter exposure times are desirable the "standard" Lumigen™PPD concentration of 100 µg/ml can also be used for sequencing applications.

Colorimetric detection with BCIP/NBT is performed as described in Sect. 3.2.11. The membrane is processed in a sealed hybridization bag analogous to chemiluminescent detection, except that the colorimetric substrate reaction step is performed without shaking. Colorimetric detection is usually complete after 8−16 h. **Colorimetric detection**

19.6.3 Special Hints for Application and Troubleshooting

All pitfalls stated in Sect. 3.2.13 may be relevant for DIG DNA sequencing. Try to perform the procedures exactly as described in the protocols. Some additional problems may be correlated with DIG DNA sequencing:

- Keep the membrane in close contact to the sequencing gel during the transfer. Before pulling the gel/membranes sandwich off the glass plate, press the membrane close to the gel avoiding air bubbles. **Uneven transfer, poor resolution**

- The membrane may not be incubated uniformly during the processing steps. When detection in hybridization bags is performed make sure the membrane is thoroughly agitated and air bubbles are removed. **Uneven background**
- If using a cylindrical roller don't let the membrane overlap during incubation steps.

- Check the digoxigenylation of the custom made sequencing primer used by comparing it to the DIG M13/pUC sequencing primers provided by Boehringer Mannheim. 1 fmole of 5′ DIG-labeled primer should be detectable in a spot assay. **Low sensitivity**

References

Ansorge W, Sproat B, Stegemann J, Schwager C, Zenke M (1987) Automated DNA Sequencing: Ultrasensitive detection of fluorescent bands during electrophoresis. Nucleic Acids Res:4593−4602
Beck S (1987) Colorimetric-detected DNA sequencing. Anal Biochem 164:514−520
Beck S, O'Keefe T, Coull J, Köster H (1989) Chemiluminescent detection of DNA: application for DNA sequencing and hybridization. Nucleic Acids Res 17:5115−5123
Church GM, Kieffer-Higgins S (1988) Multiplex DNA sequencing. Science 240:185−188
Connell C, Fung S, Heiner C, Bridgeham J, Chakerian V, Heron E, Jones B, Menchen S, Mordan W, Raff M, Recknor M, Smith L, Springer J, Woo S, Hunkapiller M (1987) Automated DNA sequence analysis. BioTechniques 5:342−438

Höltke H-J, Sagner G, Kessler C, Schmitz G (1992) Sensitive chemiluminescent detection of digoxigenin-labeled nucleic acids: a fast and simple protocol and its applications. Bio Techniques 12:104−113

Sanger F, Nicklen S, Coulson AR (1977) DNA sequencing with chain-terminating inhibitors. Proc Natl Acad Sci USA 74:5463−5467

Tizard R, Cate R, Ramachandran KL, Wysk M, Voyta J, Murphy O, Bronstein J (1990) Imaging of DNA sequences with chemiluminescence. Proc Natl Acad Sci (USA) 87:4514−4518

19.7 DNA Sequencing: Chemiluminescent Detection with the 1,2-Dioxetane CSPD

CHRIS S. MARTIN and IRENA BRONSTEIN

19.7.1 Principle and Applications

DNA sequencing methods, including both Maxam-Gilbert and Sanger dideoxy procedures, have traditionally involved the detection of DNA fragments labeled with radioactive isotopes. More recently, other techniques for imaging DNA sequence ladders with nonisotopic labels have become available. These include colorimetric BCIP and NBT) (Richterich et al., 1989), fluorescent (Prober et al., 1988), and chemiluminescent (Beck et al., 1989; Tizard et al., 1990; Creasey et al., 1991; Martin et al., 1991; Richterich and Church 1992), methods.

Here we describe the chemiluminescent detection of DNA sequence ladders with the novel chemiluminescent substrate disodium 3-(4-methoxyspiro[1,2-dioxetane-3,2'-(5'-chloro)tricyclo [3.3.1.1.^{3,7}]decan]-4-yl)phenyl phosphate (CSPD) for the enzyme alkaline phosphatase (Bronstein et al., 1991; Martin et al., 1991). Enzymatic dephosphorylation of CSPD results in a strongly electron-releasing destabilized dioxetane anion, which further fragments with the emission of light (see Sect. 13.1.2). The emission maximum in aqueous solution is at 477 nm. The dephosphorylated dioxetane anion has a half-life of approximatelx 40 min on nylon membrane.

The procedure we developed for DNA sequencing with chemiluminescence is based on standard dideoxy methods. The DNA fragments resulting from the DNA sequencing reactions are labeled through the incorporation of biotinylated primers and detected by binding a streptavidin alkaline phosphatase conjugate which is subsequently detected with CSPD chemiluminescent substrate. All the steps, including the DNA sequencing reactions, the gel electrophoresis, the DNA transfer from gel to a nylon membrane, and the chemiluminescent detection can be performed in 7−8h. The quality of the DNA sequence data is comparable to results obtained with radioactive isotopes.

The following procedure was initially developed with the 1,2-dioxetane substrate AMPPD; however, CSPD exhibits dramatically improved performance compared to AMPPD when used for DNA sequencing and is the preferred substrate for this application. The presence of a chlorine atom appended to the adamantyl group in CSPD provides control of chemiluminescence kinetics, aggregation, and hydrophobic/hydrophilic balance. This results in a shorter time lapse before reaching steady state light emission when detecting nucleic acid fragments on nylon membrane. Also, the more hydrophobic nature of the chloro-appended adamantyl group decreases the rate of diffusion of dephosphorylated anion and limits the loss of imaged band resolution associated with the use of AMPPD. Hence, high quality DNA sequence data can be imaged over longer periods of time.

19.7.2 DNA Sequencing and Detection with CSPD

- DNA sequencing reaction kit (Tropix SEQ-Light or equivalent)
- Biotinylated DNA sequencing primer (Tropix or other)

DNA sequencing reagents

- Whatman 3MM paper
- Tropilon-45 nylon membrane (Tropix)

DNA transfer reagents

The reagents (except for the bags) are available together from Tropix as the SEQ-Light DNA sequencing detection kit:
- I-Block casein blocking reagent (Tropix)
- Avidx streptavidin alkaline phosphatase conjugate (Tropix)
- Diethanolamine (99%) (Tropix)
- CSPD chemiluminescent substrate (100× concentrate) (Tropix)
- 40 × 53 cm heat sealable bags (Tropix)

Chemiluminescent detection reagents

- Plastic wrap (SaranWrap best, Dow Brands Inc.)
- XAR-5 X-ray film (Kodak)

Film exposure reagents

- 10× TBE (0.9 M Tris borate, 20 mM EDTA)
- 10× Phosphate buffered saline (PBS) (0.58 M Na_2HPO_4, 0.17 M NaH_2PO_4, 0.69 M NaCl, pH 7.2)
- Blocking buffer (0.2% I-Block reagent, 0.5% SDS, 1× PBS)
- Conjugate solution (0.2% I-Block reagent, 1× PBS, 0.5% SDS, $\frac{1}{5000}$ dilution of AvidX-AP alkaline phosphatase conjugate)
- Wash buffer (0.5% SDS, 1× PBS)
- Assay buffer (0.1 M diethanolamine, 1 mM $MgCl_2$, pH 10.0)
- CSPD detection solution ($\frac{1}{100}$ dilution of CSPD in assay buffer)

Standards buffers

- UV light source (hand-held lamp, germicidal bulb, or UV transilluminator)
- Plastic bag sealer

Standard equipment

19.7.2

- Film cassette with clamps
- Optional transfer apparatus for electroblotting (OWL Scientific Plastics)

DNA sequencing reactions

The DNA sequencing reactions can be performed using the procotol and reagents in the Tropix SEQ-Light DNA sequencing reaction kit. However, most DNA sequencing reaction procedures can be easily modified to permit chemiluminescent detection by eliminating the labeling step to incorporate radioactive nucleotides and by substituting biotinylated primers. Biotinylated primers behave normally in DNA sequencing reactions and should be used at the same concentration as nonbiotinylated primers.

DNA sequencing gel electrophoresis

Separate the DNA sequencing reactions on a standard DNA sequencing gel with 7.67 M urea as a denaturant. Make sure that one of the glass plates is siliconized to allow easy separation of the plates before proceeding to the DNA transfer step.

DNA transfer to membrane

The DNA fragments in a sequencing gel can be transferred to the nylon membrane support by either electrotransfer or capillary transfer method. Fixing the gels or removing the urea is not necessary.

Capillary transfer: A simple capillary transfer method, which does not require an external buffer, can be used to transfer 10%−20% of the DNA to the membrane. This quantity of DNA can be easily detected with chemiluminescence. However, in this case, it is important to use an adequate quantity of high quality DNA template in the DNA sequencing reactions.

The procedure for the transfer method is as follows:

1. Disassemble the gel apparatus and separate the glass plates.
2. Remove the gel from the glass plate with a dry piece of Whatman filter paper. Place the paper with gel attached back on the plate with the gel side up.
3. Cut a section of membrane which will cover the region of the gel to be blotted and wet it thoroughly with TBE.
4. Gently place the wet membrane on the gel. Remove any air bubbles by rigorously rolling a pipette or rod over the membrane.
5. Place two or three pieces of dry Whatman paper on top of the membrane, the other glass plate, and an approximately 2 kg weight on top. Allow the transfer to proceed for 1 h and subsequently perform the UV cross-linking step.

Electrotransfer: DNA transfer using an electric field requires a large format electroblotting apparatus. A vertical blotting apparatus (VEP-1) and a semi-dry horizontal unit (HEP-3) are available from OWL Scientific Plastics. Detailed instructions for the use of these devices is provided by the manufacturer.

Direct transfer electrophoresis: A direct transfer electrophoresis apparatus is available from Betagen, Inc., Hoefer Scientific Instruments and GATC GmbH. These devices transfer DNA fragments to a membrane while the gel is running. Following electrophoresis, the user can proceed directly to the DNA cross-linking to membrane step.

Following the transfer step, DNA must be immobilized on neutral nylon membrane by UV irradiation. Hand-held UV lamps, UV transilluminators, and germicidal UV sources can be used to immobilize the DNA. Irradiation times of 1–10 min are usually adequate. Following UV cross-linking, the user can proceed to the detection protocol; alternatively, the membrane can be stored between two pieces of plastic wrap at 4°C. **UV cross-linking DNA to membrane**

Note: If genomic or multiplex DNA sequencing, which requires probe hybridization, is performed, UV irradiation conditions are more critical and the optimum cross-linking time should be accurately determined.

The following protocol has been developed for the detection of immobilized biotinylated or hybridized biotinylated DNA on neutral nylon membranes. All recommendations of volumes of reagents in the following protocol apply to a single membrane ($1000-1200\,cm^2$) processed in a large hybridization bag or optionally in a tray. Solution changes when using hybridization bags are performed by cutting one corner of the bag, draining the previous solution, adding the new solution, and then resealing the bag. A funnel can be used to facilitate the addition of buffers to hybridization bags and to avoid the introduction of air. All procedures should be performed at room temperature with moderate shaking (150–200 rpm). **Chemiluminescent detection**

1. Place the membrane in the plastic hybridization bag. This can be accomplished by cutting all but one edge of the bag, peeling the bag open, placing the membrane down flat, and then resealing the bag on all sides.
2. Add 500 ml of blocking buffer and incubate for 10 min.
3. Incubate for 20 min in 500 ml of conjugate solution.
5. Wash 1 × 5 min in 500 ml of blocking buffer.
6. Wash 3 × 5 min in 500 ml of wash buffer.
7. Wash 2 × 1 min in 500 ml of assay buffer.
8. Add 50 ml CSPD detection solution and incubate for 5 min.
9. Drain excess substrate from the bag, smooth out any wrinkles, and reseal it. Alternatively, the membrane may be removed and placed between two pieces of plastic wrap. Extreme care should be taken to avoid any wrinkles in the hybridization bag or plastic wrap during this step. Do not blot the membrane or allow it to dry at any time. The membrane must remain wet.

19.7.3

Film exposure and development Plastic wrapped membranes are imaged by direct contact with standard Kodak XAR5 X-ray or other photographic films. We recommend an initial 30 min exposure to assess the intensity of the signal. Additional shorter or longer exposures can be performed, if necessary. Due to the kinetics of light emission on nylon membranes, incubation of the plastic wrapped membrane at room temperature from 1 h to overnight will permit shorter film exposures.

19.7.3 Special Hints for Application and Troubleshooting

It is important that only ultrapure water and other reagents which are free of alkaline phosphatase contamination be used.

High background
- For best results, the assay buffers should be prepared daily.
- Insure that the membrane is well agitated in the plastic bag, i.e., increase the shaker speed, increase the volume of liquid, remove any air bubbles.
- Splotchy images may result from bacterial contamination of the membrane. Determine that all buffers are free of contamination prior to use and that the membrane, blotting paper, and hybridization bags are clean and fingerprint-free. Avoid cross-contamination of the wash solutions by the conjugate solution, i.e., cover solutions during storage, wash the funnel used for liquid transfers, clean the outside of the bag after each solution addition.
- Increase the incubation time in the blocking buffer or increase the number of wash steps after the incubation with the enzyme conjugate.

Poor band resolution
- Membrane may not be in good contact with the gel during the transfer. Press membrane onto the gel with a pipette or similar rod. Add a heavier weight during the capillary transfer procedure.
- Film may not be in good contact with the membrane. Reseal the membrane with plastic wrap and avoid wrinkles. Use exposure cassette with clamps.

References

Beck S, O'Keeffe T, Coull JM, Koster H (1989) Chemiluminescent detection of DNA: application for DNA sequencing and hybridization. Nucleic Acids Res 17:5115–5123

Bronstein I, Juo R-R, Voyta JC, Edwards B (1991) Novel chemiluminescent adamantyl 1,2-dioxetane enzyme substrates. In: Stanley P, Kricka LJ (eds) Bioluminescence and chemiluminescence: Current Status, John Wiley, Chichester, pp 73–82

Creasey A, D'Angio Jr. L, Dunne TS, Kissinger C, O'Keeffe T, Perry-O'Keefe H, Moran LS, Roskey M, Schildkraut I, Sears LE, Slatko B (1991) Application of a novel chemiluminescence-based DNA detection method to single-vector and multiplex DNA sequencing. BioTechniques 11:102–109

Martin C, Bresnick L, Juo R-R, Voyta JC, Bronstein I (1991) Improved chemiluminescent DNA sequencing. Bio Technique 11:110–113

Prober JM, Trainor GL, Dam RJ, Hobbs FW, Robertson CW, Zagursky RJ, Cocuzza AJ, Jensen MA, Baumeister K (1987) A system for rapid DNA sequencing with fluorescent chain-terminating dideoxynucleotides. Science 238:336−341

Richterich P, Heller C, Wurst H, Pohl FM (1989) DNA sequencing with direct blotting electrophoresis and colorimetric detection. Bio Techniques 5:52−59

Richterich P, Church GM (1992) DNA sequencing with direct transfer electrophoresis and nonradioactive detection. Meth Enzymol, Recombinant DNA, Volume H, in press

Tizard R, Cate RL, Ramachandran KL, Wysk M, Voyta JC, Murphy OJ, Bronstein I (1990) Imaging of DNA sequences with chemiluminescence. Proc Natl Acad Sci USA 87:4514−4518

19.8 Direct Blotting Electrophoresis for DNA Sequencing

Thomas M. Pohl

19.8.1 Principle and Applications

For nonradioactive sequencing of DNA, two very different methods have been developed over the last few years:
(1) Transferring the band pattern from the separation gel to a membrane surface; this allows very sensitive detection methods to subsequently be employed (Richterich et al., 1989; Beck, 1987). (2) Labeling the DNA fragments with a fluorescent chromophore and observing the band pattern in the gel directly by excitation with a laser light (Smith et al., 1986).

Different methods for transfering separated bands to solid supports have been proposed, e.g., capillary transfer (Southern blot), vacuum blotting, or electrophoretic transfer. These methods involve dismantling of the gel and handling of delicate and thin polyacrylamide gels.

In order to simplify these procedures we have developed direct blotting or „direct transfer electrophoresis" (Beck and Pohl, 1984; Pohl and Beck, 1987) in which separation of bands and transfer to a membrane are done at the same time. When the bands reach the end of the gel they are eluted and come in contact with a membrane, which moves with constant speed across the bottom of the gel. DNA fragments bind to the membrane and are thus removed from the gel. In this way the temporal order of eluting from the gel is transformed into a spatial separation. There are a number of advantages, e.g., equal spacing between bands over a very large range, and the convenience of using short gels. In addition, the elution of all bands is possible within a single run, and a very high resolution can be achieved by using very thin slab gels, the distance between bands can be controlled by the speed of the conveyor belt, and the gel may be used again.

Here we provide protocols for performing DNA sequencing runs with direct bloting electrophoresis. These have been tested extensively in our lab using biotin − or digoxigenin − as labeled probes. Over the last few months more than half a million nucleotides as raw sequence data were generated with this method by shot-gun sequencing of two cosmids for the EEC project of sequencing *Saccharomyces cerviciae*. On average, more than 400 nucleotides per clone were read with about 2700−4000 band read per 15 × 40 cm nylon membrane (Pohl and Feger, 1992).

19.8.2 DNA Sequencing by Direct Blotting Electrophoresis

Standard reagents
- Wash ethanol (Roth GmbH)
- Silan-solution (GATC GmbH)
- TEMED (GATC GmbH)
- Ammonium persulfate (APS) (GATC GmbH)
- Urea, acrylamide, bisacrylamide (GATC GmbH)
- Nylon membrane (GATC GmbH)

Standard solutions
- 30% [w/v] Acrylamide and 0.8% [w/v] bisacrylamide
- Urea diluent: 42% [w/v] urea; 9.35% [v/v] 10× TBE, pH 8.8; 36.65% [v/v] H_2O
- Silan-solution: 50 μl silan; 30 μl 100% [v/v] acetic acid; 0.3 ml H_2O: adjust volume to 10 ml with 100% [v/v] ethanol
- 10× TBE: 15.45% [w/v] Tris base; 2.62% [w/v] boric acid; 0.9% [w/v] NaEDTA; 81.03% [w/v] H_2O
- 10% APS: 10% [w/v] APS; 90% [v/v] H_2O

Standard equipment
- Direct blotter GATC 1500 (GATC GmbH)
- Glass plates (30 cm × 20 cm × 4 mm), spacers, precomb, shark stooth comb (0.125 mm thick) (GATC GmbH)
- Power supply 3000 V (Pharmacia Biosystems)

Glass plate preparation
1. Matched plates, spacer, and precomb are washed with ethanol.
2. Pipette 600 μl silan solution onto each plate, wipe with soft tissue, and let react for 5−15 min.
3. Wipe plates with wash ethanol-wetted tissue paper to remove excess silan solution.
4. Insert a 1.5 cm wide and 18 cm long parafilm strip in the corner of the pouring table and remove the paper wrapper from the parafilm.
5. Clean the plate without ears with an anti-dust tissue and put it on the pouring table; position the spacer and clean the plate with the buffer chamber. Carefully lay the plate onto the other plate (but do not touch the front edge).
6. Check the spacers: they should not stick out at the front edge, but rather at the buffer chamber. Fix them with adhesive tape to the lower plate, otherwise they will move upon pouring the gel.

1. Dilute 1.5 ml 30% acrylamide solution (30:0.8) with 10.5 ml urea **acryl-** diluent. **amide**
2. Degas the solution for 5 min (the gel should be room temperature). **gel**
3. Add 12 μl TEMED. Mix and add 50 μl 10% APS solution. Mix again. **prepara-**
4. Take up the solution into a 20 ml syringe and start immediately to pour **tion** the gel (see below).

Pouring should not take more than 5 min.

1. Take off the upper glass plate sideways, by holding it at the buffer **Pouring the** chamber end. **gel**
2. Put the front edge (approximately 1 cm away from the end) at an angle of 30°−45° on the slot site of the lower plate.
3. Pour the gel solution from one side to the other without getting air bubbles in between (approximetaly 2−3 ml).
4. Start to move the upper plate to the front and lower the angle slowly to 0°, always keeping gel solution at the front edge of the upper plate.
5. Move slowly to the end of the gel and watch the upper edge of the lower plate taking care that no air bubbles develop in the gel (if some occur, move backwards a little bit).
6. Press both plates to the front (the front edge should be parallel and aligned).
7. Clamp the plate with four clamps at the front side and in the middle. Add some gel solution to the slot and insert the precomb.
8. Clamp the middle of the precomb and then on both sides of the buffer chamber.
9. Press the plates with an elastic band towards the front edge of the pouring table.
10. After 30 min of polymerization rinse the precomb with water or TBE buffer to remove excess gel solution.
11. Let gel sit for at least 1−2 h.

1. Degas 0.711× TBE buffer for 10−15 min and pour it into the lower buf- **Setup and** fer chamber. Insert the membrane (normally 42 cm long and 15 cm **prerun** wide), securing it with a piece of adhesive tape onto the moving belt.
2. Remove the elastic band and the clamps from the gel.
3. Wash away the urea and polymerized gel under running water.
4. If some polymerization has occurred on the lower edge of the plates, remove it with a wet tissue or with a spatula, but do not touch the front of the gel.
5. Put the gel assembly into the direct blotter, starting with one corner, and then lower it carefully down so that no air bubbles get under the gel.
6. Fix the aluminium plate with two clamps at the left and right sides of the buffer chamber and fix the gel in the direct blotter device.
7. Add some buffer to the comb and pull it out carefully.
8. Add 500 ml of 1× TBE buffer into the upper buffer chamber and clean the slot with a 10 ml syring to remove any remaining polyacrylamide.
9. Start the prerun with 1800 V (approximately 14−16 mA) for 30 min.

19.8.2

19.8.2

Sequence run

1. Wash the slot with buffer to remove the urea and insert the sharkstooth comb carefully.
2. Load the sample and run for 15 min at 400 V to remove the salt front of the sequencing reactions.
3. Start the run with 2100 V and start the stepper motor to move the membrane with a speed of about 18−19 cm/h.
4. It takes 45 min to move the membrane under the gel; at the same time the bromphenol blue front will reach the edge of the gel.
5. Stop the membrane under the gel and wait until the bromphenol blue elutes onto the membrane.
6. Start the membrane with a constant speed of between 18 and 20 cm/h (run takes about 2.5 h).
7. If any white spaces appear at the bromphenol blue line, air bubbles are under the gel!
8. After the membrane has passed the gel, stop the direct blotting apparatus and move the membrane out of the buffer.

Membrane development

1. The membrane is removed from the blotter and the DNA is fixed by baking and UV cross-linking.
2. Depending on the labeling reaction (biotin, digoxigenin, etc.), colorimetric or chemiluminescent enzymatic detection follows.
3. The amount of DNA detected is comparable to that detected by radioactive labeling but in a shorter amount of time.
4. If colorimetric detection is used, the membrane itself is saved as a hard copy.

Clean-up

1. Take out the membrane and remove gel and the lower buffer chamber.
2. Add water into lower buffer chamber and wash the moving belt by moving it with the fast gear from the motor controlling unit two to three times through the water.
3. Remove the water and dry the belt with soft tissue: move it dry backwards to the starting point.

References

Beck S (1987) Colorimetric-defected DNA sequencing. Anal Biochem 164:514−520
Beck S, Pohl FM (1984) DNA sequencing with direct blotting electrophoresis. EMBO J 3:2905−2909
Pohl FM, Beck S (1987) Direct transfer electrophoresis used for DNA sequencing. Meth Enzymol 155:250−259
Pohl FM, Feger GM (1992) Boehringer Mannheim Biochemica Colloquium 2:17−18
Richterich P, Heller C, Wurst H, Pohl FM (1989) DNA sequencing with direct blotting electrophoresis and colorimetric detection. BioTechniques 7:52−59
Smith LM, Sanders JZ, Kaiser RJ, Hughes P, Dodd C, Connell CR, Heiner C, Kent SBH, Hood LE (1986) Fluorescence detection in automated DNA sequence analysis. Nature 321:674−679

20 Blot Formats: Proteins and Glycoproteins

20.1 Detection of Proteins and Glycoproteins on Western Blots

ANTON HASELBECK and WOLFGANG HÖSEL

20.1.1 Detection of Proteins

The analysis of proteins on blots after electrophoretic transfer from gels rather than analyzing the gels themselves is becoming increasingly popular due to several advantages, e.g., application of immunological techniques („classical western blotting"); membranes do not shrink and are easy to handle; and several types of membranes with different advantages are available. However, general protein staining techniques of blots are by far not as well established as the very sensitive silver staining method of gels. Staining techniques using gold probes are approaching this sensitivity and are very useful [1]. By adapting the immunological digoxigenin (DIG)/ anti-digoxigenin:alkaline phosphate (<DIG>:AP) system for the general staining of proteins on blots, we developed a second very sensitive system which should be even more useful due to its greater flexibility and convenient combination with other immunological detection methods of blots.

The method is applicable to different type of membranes used for protein and peptide blotting (nitrocellulose; PVDF) and the various types of electrophoreses (e.g., SDS-PAGE; IEF; 2-D electrophoresis). In addition, the method can be used to selectively label and detect − NH2, -SH and S-S groups of proteins and peptides as well as the combination of all groups. Protein staining can be conveniently combined with other DIG-based staining methods (e.g., all glycoprotein detection methods described later) and specific immunological detection methods, thus allowing direct comparison of parallel and twin blots. The sensitivity of DIG-based protein labeling can be further increased by using chemiluminescent substrates of AP (e.g., AMPPD) in combination with nylon membranes (e.g., Zeta Probe), thus allowing, for example, investigation of the protein content of single cells as described previously [2].

20.1.2 Detection of Glycoproteins

The various DIG-based glycan labeling methods described in Sect. 3.3 allow a whole array of different investigations of glycoproteins on blots, including:

- General detection of the presence of carbohydrates in proteins (glycoproteins; yes or no?).
- Selective detection of sialic acids and terminal galactose units.
- Investigations of glycan structural features based on the specific binding of lectins (for details on the lectin specificity see [3] and the references cited therein). The structural conclusions obtained can be corroborated by employing competition experiments with low molecular weight carbohydrates in combination with the lectins.
- The combination of the various DIG-based glycan detection methods with the different types of exo- and endoglycosidases of well known carbohydrate specificity available [4–6] considerably increases the number of possible structural investigations.
- It is well known that glycoproteins with a high carbohydrate content (e.g., >50%) do not produce good protein staining. However, they can be detected very well by the general glycan staining procedure described. Thus a combination of general protein and glycan staining of separated protein mixtures on blots (e.g., from animal or plant origins) complement each other nicely and will therefore render a much more complete picture of substances present than either staining procedure alone.
- The various detection methods can be applied to the investigation of blotted glycopeptides obtained after protease digestion and separation on reversed phase HPLC [see 7]. Thus it is possible to analyze single glycosylation sites of glycoproteins on blots. They can also be used for the detection of glycoconjugates in serum or other body fluids, especially after isoelectric focusing and transfer onto a membrane [8].

References

1. Gillespie PG, Hudspeth AJ (1991) Chemiluminescence detection of proteins from single cells. Proc Natl Acad Sci USA 88:2563–2567
2. Haselbeck A, Schickaneder E, vd Eltz H, Hösel W (1990) Structural characterization of glycoprotein carbohydrate chains by using digoxigenin-labeled lectins on blots. Anal Biochem 191:25–30
3. Kobata A (1979) Use of endo glycosidases and exo glycosidases for structural studies of glyco conjugates. Anal Biochem 100:1–14
4. Mäder M, Retzlaff K, Felgenhauer K (1991) A cationic glycoprotein pattern in human serum and cerebrospinal fluid with pathological implications. Eur J Clin Chem Clin Biochem 29:481–485
5. Maley F, Trimble RB, Tarentino AL, Plummer Jr TH (1989) Characterization of glycoproteins and their associated oligosaccharides through the use of endoglycosidases. Anal Biochem 180:195–204

6. Moeremans M, Daneels G, De Mey J (1985) Sensitive colloidal metal (gold or silver) staining of protein blots on nitrocellulose membranes. Anal Biochem 145:315−321
7. Tarentino AL, Trimble RB, Plummer Jr TH (1989) Enzymatic approaches for studying synthesis and processing of glycoproteins. Meth Cell Biol 32:111−139
8. Weitzhandler M, Hardy M (1990) A sensitive blotting assay for the detection of glycopeptides in peptide maps. J Chromatogr 510:225−232

20.2 Southwestern Analysis Using DIG

STEVEN DOOLEY

20.2.1 Principle and Applications

The existence of site-specific DNA-protein interactions in eukaryotes has been demonstrated by several approaches. A limitation of most of these techniques is that, despite identifying the site of binding within the DNA, they do not characterize the specific proteins involved. Information about their nature can be obtained only after extensive purification procedures. Here, I wish to describe a procedure, which uses protein blotting and requires no prior purification steps, that has the potential to directly identify the polypeptides that interact with a target DNA sequence. A further advantage of this procedure is the use of DIG-[11]-dUTP instead of ^{32}P-dNTP for labeling the target DNA; thus overcoming radioactive handling and long exposure times.

20.2.2 DIG Southwestern Analysis

− DIG oligonucleotide tailing kit (Boehringer Mannheim)	**Labeling reagents**
− EDTA (Serva)	
− Push columns (Stratagene)	
− Double-stranded oligonucleotide binding site	
− NaCl (Sigma)	
− Trizma base (Sigma)	
− HCl (Fluka)	
− pBR328 [control DNA, DIG-labeled] (Boehringer Mannheim)	**Control reagents**
− Rainbow protein molecular weight markers (Amersham)	**PAGE reagents**
− Acrylamide-bisacrylamide mixture (Roth)	
− Ammonium persulfate (APS, Serva)	
− N,N,N',N'-Tetramethylethyldiamine (TEMED, Sigma)	
− SDS (Serva)	

- Isopropanol (Fluka)
- Bromphenolblue (Serva)
- 2-Mercaptoethanol (Sigma)

Western blot reagents
- Glycine (Sigma)
- Methanol (Roth)
- Whatman 3MM paper (Schleicher & Schüll)

Southwestern reagents
- poly-[dI-dC] (p(dIdC) (Boehringer Mannheim)
- High molecular weight (HMW) *E. coli* DNA

Detection reagents
- DIG nucleic acid detection kit (Boehringer Mannheim)
- $MgCl_2$ (Merck)
- Nitrocellulose membrane (BA 85, Schleicher & Schüll)

Labeling solutions
- Terminal transferase buffer, 5 × (1M Potassium cacodylate; 0.125M Tris-HCl; pH 6.6, 25°C; 1.25 mg/ml bovine serum albumin, BSA)
- $CoCl_2$ solution (25 mM)
- DIG-[11]-dUTP (1 mM)
- dATP (10 mM)
- Terminal transferase (50 U/μl)
- Glycogen solution (20 mg/ml)
- EDTA (0.2 M; pH 8.0, 25°C)
- 1× STE buffer (20 mM Tris-HCl, pH 7.5, 25°C; 100 mM NaCl; 10 mM EDTA)

PAGE solutions
- 10× SDS probe buffer: 125 mM Tris-HCl; pH 6.8, 25°C; 1% SDS; 0.05% bromphenolblue; 1.3 M 2-mercaptoethanol
- Acrylamide stock solution: 25% acrylamide; 0.6% bisacrylamide
- 4× Electrophoresis buffer: 1.5 M Tris-HCl; pH 8.8, 25°C
- 10% SDS
- Stacking gel solution: 5% acrylamide; 0.13% bisacrylamide; 125 mM Tris-HCl; pH 6.8, 25°C
- 10% APS
- Gel running buffer: 50 mM Tris; 133 mM glycine; 0.1% SDS

Western blotting solutions
- 1× Transfer buffer: 48 mM Tris; 39 mM glycine; 20% methanol

Southwestern solutions
- 1× Buffer E: 25 mM NaCl; 25 mM Hepes; 5 mM $MgCl_2$; 0.5 mM DTT; pH 7.9, 25°C (add DTT fresh before using the solution)
- 1× Buffer F: buffer E including 5% blocking reagent
- 1× Buffer G: buffer E including 0.25% blocking reagent

Detection solutions
- DIG detection kit (Boehringer Mannheim)
- Buffer I: 100 mM Tris-HCl, pH 7.5, 25°C; 150 mM NaCl
- Buffer II: 1% blocking reagent in buffer I

– Buffer III: 100 mM Tris-HCl; pH 9.5, 25°C; 100 mM NaCl; 50 mM $MgCl_2$

1. Pipette the following into a microfuge tube on ice: **Labeling**
   4 µl   5× reaction buffer   **reaction**
   4 µl $CoCl_2$ solution
   20 pm double-stranded oligonucleotide containing binding site of
   interest
   1 µl   DIG-[11]-dUTP
   1 µl   dATP solution
   1 µl   terminal transferase (50 U)
   Make up to:
   20 µl   with sterile redistilled water
2. Mix well and incubate for 15 min at 37°C.
3. Mix 1 µl glycogen solution to 200 µl EDTA solution (0.2 M; pH 8.0) and
   add 2 µl of this mixture to stop the reaction.
4. Purify labeled oligonucleotide from unincorporated nucleotides with
   the push column system (Stratagene) exactly according to the manufacturer's recommendations.
5. After purification check the labeling efficiency by comparison with the
   labeled control plasmid via direct detection.

For further details see Ausubel et al., 1987.   **PAGE**

1. Assemble the glass plate sandwich using two clean glass plates and two
   0.75 mm spacers.
2. Lock the sandwich to the casting stand (Protean II, Biorad equipment).
3. Prepare 50 ml separating gel solution. For a 12.5% gel:
   25 ml   acrylamide stock solution
   12.5 ml 4× gel buffer
   500 µl 0% SDS
   Degas the solution using desiccator for 10 min, add:
   250 µl 10% APS
   30 µl   TEMED
   Stir gently to mix.
4. Using a Pasteur pipette, apply the separating gel solution to the
   sandwich along an edge of one of the spacers until a defined height,
   leaving 1 cm stacking gel distance to the comb.
5. Using another Pasteur pipette, slowly cover the top of the gel with a
   layer (1 cm thick) of isopropanol.
6. Allow the gel to polymerize for 1 h at room temperature.
7. Pour off the layer of isopropanol and rinse with water.
8. Prepare the stacking gel, solution pipetting together:
   10 ml   stacking gel solution
   130 µl 10% APS
   10 µl   TEMED
   Stir gently to mix.

9. Using a Pasteur pipette, slowly allow the stacking gel solution to trickle into the sandwich.
10. Insert a 0.75 mm comb into the layer of stacking gel solution.
11. Allow the stacking gel to polymerize 30–45 min at room temperature.
12. Dilute 100 µg nuclear protein extract with ⅒ volume 86% glycerol and ⅒ volume 10× SDS probe buffer. Heat 5 min at 100°C before loading the gel. Use 20 µl of Rainbow protein molecular weight markers (Amersham).
13. Connect the power supply to the „Protean cell" and run at 10 mA of constant current for a slab gel 0.75 mm thick in 1× electrophoresis buffer until the bromphenolblue tracking dye enters the separating gel. Then increase the current to 15 mA and run the gel until the bromphenolblue tracking dye has reached the bottom of the separating gel.

Western blotting
1. Carefully slide one of the spacers halfway from the edge of the sandwich along its entire length. Use the exposed spacer as a lever to pry the glass plate, exposing the gel.
2. Carefully remove the gel from the lower plate. Cut a small triangle at one corner of the gel so the lane orientation is not lost during the following procedure.
3. Equilibrate the gel for 30 min in 1 × blotting buffer.
4. Transfer the proteins from the gel to a nitrocellulose sheet using Biorad semidry „Trans Blot" electrophoretic transfer cell exactly as recommended (any other equipment for western blotting can be used according to manufacturer's recommendations).

Southwestern interaction
For further details see (Bowen et al., 1980; Miskimins et al., 1985; Dooley et al., 1988; Freyaldenhoven et al., 1989).
1. Wash the filter for 30 min in buffer E to renature the proteins.
2. Block the filter by incubation for 30 min in buffer F.
3. After a short wash with 50 ml buffer G, the filter is exposed to labeled probe DNA in 10 ml buffer G, complemented with 5 µg/ml p(dIdC) or E. coli DNA for unspecific binding competition, for 2 h at room temperature.
4. The filter is then washed 3 × 5 min in 100 ml buffer G.
5. The filter is dried and prepared for detection.

Detection reaction
This procedure was done exactly according to the Boehringer Mannheim DIG DNA Detection Kit protocol.

20.2.3 Special Hints for Application and Troubleshooting

Low protein transfer
– SDS can be used in blotting buffer to increase transfer efficiency.

Unspecific binding
– Concentration of p(dIdC) or E. coli DNA in the binding assay can be increased until 100 µg/ml to compete away unspecific bands.

20.2.3

– It is possible that the binding activity observed in the gel shift experi- **No**
ment is a function of protein dimerization. If the monomers have diffe- **binding**
rent positions in the gel after electrophoresis they cannot dimerize to
receive binding activity.

References

Ausubel FM, Brent R, Kingston RE, Moore DD, Seidmann JG, Smith JA, Struhl K
(1987) Current protocols in molecular biology

Bowen B, Steinberg J, Laemmli UK, Weintraub H (1980) The detection of DNA-binding proteins by protein blotting. Nucleic Acids Res 8:1−20

Dooley S, Radtke J, Blin N, Unteregger G (1988) Rapid detection of DNA-binding factors using protein blotting and digoxigenin-dUTP marked probes. Nucleic Acids Res
16:11893

Freyaldenhoven MP, Royer-Pokora B, Napierski I, Royer HD (1989) DNA binding
proteins present in guanidinium isothiocyanate lysates of cells are suitable for specific
binding site blotting. Nucleic Acids Res 17:8891

Miskimins WK, Roberts MP, McClelland A, Ruddle FH (1985) Use of a protein blotting
procedure and a specific DNA probe to identify nuclear proteins that recognize the
promoter region of the transferrin receptor gene. Proc Natl Acad Sci (USA)
82:6741−6744

21 In Situ Formats

21.1 Virus Detection in Fixed Cells

ELKE GENERSCH, B. J. HEILES and R. NEUMANN

21.1.1 Principle and Applications

Soon after the method of in situ hybridization (ISH) had been published (Pardue and Gall, 1969), reports appeared showing that it could also be used with great success in the study of virus-infected systems (Orth et al., 1970; Geukens and May, 1974). Now it was possible to study the biology of viruses and the mechanisms of viral infections in detail and to both improve diagnosis and form the basis of prognosis of viral diseases.

In the case of hepatitis B virus (HBV), ISH has been used to verify the postulated asymmetric replication model (via an RNA intermediate and free minus-strand DNA) in human hepatocytes (Blum et al., 1984).

ISH has also been used to elucidate the mechanisms of virus latency and persistence in vivo. For instance it was possible to detect adenoviral sequences in human tonsils and in cell lines established therefrom in the absence of infectious virus (Neumann et al., 1987), indicating that adenoviruses can persist in human tissues. The neurons of normal trigeminal ganglia have been identified as the site of persistence of varicella zoster virus (VZV) and herpes simplex virus (HSV) using ISH (Vafai et al., 1988).

If it is possible to detect small copy numbers of viral genomes, as has been done in the study of viral persistence, the ISH can also be used to localize sites of active viral infections, especially when ISH is applied together with the demonstration of late viral antigens and cytopathy. Therefore ISH can be a very useful tool in the diagnosis of viral diseases.

Furthermore ISH has been of great value in investigating the link between viral infection and carcinogenesis.

Studies have been performed in which sequences of human papilloma viruses (HPV 16/18) were hybridized to cervical smears to determine the contribution of genital HPV 16/18 infection to the development of cervical cancer (Neumann et al., 1989; Heiles et al., 1988). In spite of great efforts, it is still uncertain whether the demonstration of persistence of these viruses in cells of the genital tract is of prognostic value.

Although causal inferences between viral infections and tumorgenesis are difficult to verify, ISH on the chromosomal level may allow for the establishment of models explaining the mechanism(s) of viral carcinogenesis by demonstrating the sites of integration of viral sequences into the host genome with respect to transformation events.

21.1.2 Detection of Viral Sequences in Fixed Cells and Tissues by In Situ Hybridization

– Proteinase K (Merck, Darmstadt)   **Standard**
– DNase I (Worthington, Freehold, USA)   **reagents**
– RNase A (Worthington, Freehold, USA)

– Prehybridization buffer: 6× SSC, 45% formamide, 5× Denhardt's sol- **Standard**
  ution, 100 µg/ml denatured salmon sperm DNA   **buffers**
– Hybridization buffer: 6× SSC, 45% formamide, 5× Denhardt's solution, 10% dextran sulphate, labeled DNA
– Washing buffers after hybridization: (1) 6× SSC, 45% formamide;
  (2) 2× SSC; (3) 0.2× SSC
– 1× Denhardt's solution: 0.02% BSA, 0.02% PVP-360, 0.02% Ficoll
  400

– Quadriperm cell culture dishes (Heraeus)   **Standard**
– Heating plate (BRL)   **equipment**
– Fixogum (Marabu)
– Incubation oven (Heraeus)

The following is a detailed protocol for a selected example

1. Cells were cultured on clean, UV irradiated slides in cell culture dishes (Quadriperm) in appropriate culture medium. In the case of HeLa cells for the detection of HPV sequences, MEM/5% NCS (Eagle 1959) was used.
2. When the cells were semiconfluent, the slides were removed from the dishes and washed several times in PBS (phosphate-buffered saline).
3. The cells were fixed to the glass surface with 100% ethanol.
4. Rehydration of the cells was performed with a decreasing alcoholgradient: 70%, 50%, 30% ethanol, 5 min each.
5. The slides were then washed twice for 10 min in PBS containing 5 mM $MgCl_2$. This step was followed by incubating the slides in 0.2 N HCl for 20 min to denature the proteins and to disturb membrane structures. The slides were then washed twice in 2× SSC, 5 mM EDTA for 30 min at 50 °C. At this stage a proteinase K treatment (1 µg/ml in PBS, 15 min at 37 °C) may be performed to further open cell membranes. The slides must be subsequently washed with 0.2% glycine in PBS for 10 min to inhibit the enzyme activity. This step was not necessary for HeLa cells.

6. The cells were then fixed with 4% paraformaldehyde for 10−45 min and washed twice with PBS/5 mM $MgCl_2$ for 15 min.

7. Prehybridization was performed for at least 15 min at 42 °C in 50 μl prehybridization solution per slide.

8. After carefully removing the prehybridization solution, hybridization was carried out under siliconized cover slips using 5 ng labeled DNA in 20 μl hybridization solution. The cover slips were fixes into place with Fixogum.

9. Denaturation of the cellular and the probe DNA was carried out simultaneously by placing the slides onto a heated plate for 3−5 min (temperature approximately 90 °C on the slide). The temperature was monitored using a thermopile.

10. The slides were placed on ice and when cooled down to 4 °C incubated overnight in a damp chamber at 42 °C. The hybridization buffer and temperature were adopted to the detection of HPV sequences and therefore have to be varied for sequences with a different GC content.

11. After removal of the cover slips, the slides were washed as follows: 2 × 15 min at 42 °C with 6× SSC, 45% formamide; 2 × 5 min at room temperature with 2× SSC; and 2 × 15 min at 50 °C with 0.2 × SSC.

12. Color detection was performed according to the standard procedures for the detection of digoxigenin- or biotin-labeled probes.

21.1.3 Special Hints for Application and Troubleshooting

− The pretreatment of the cells is a critical point. While the cell morphology must be maintained, the probe must nonetheless be able to reach the target DNA or RNA in the following hybridization step.

− Therefore it is necessary to test empirically the HCl and proteinase K incubation conditions.

− The critical step in the hybridization procedure is the denaturation and renaturation of the cellular and probe DNAs. The DNAs have to be totally denatured and cooled down as fast as possible to prevent renaturation before hybridization.

− The washing procedure after hybridization has to be carried out intensively to minimize background, but take care not to loosen cells.

− Before using nonradioactive-labeled probes in ISH, be sure that the cellular proteins will not react unspecifically with the antibody used for during the color reaction.

− ISH on tissue sections often carries with it the difficulty that the sections will tend to float off during the hybridization and washing procedures. Different pretreatments of the slides and fixation methods must be tested. In our hands, the method of Tourtellotte (Tourtellotte et al., 1987) proved to be the most suitable one.

References

Blum HE, Haase AT, Harris JD, Walker D, Vyas G (1984) Asymmetric replication of hepatitis B virus DNA in human liver: demonstration of cytoplasmic minus-strand DNA by blot analysis and in situ hybridization. Virology 139:87−96

Eagle H (1959) Amino acid metabolism in mammalian cell cultures. Science 130:432−439

Geukens M, May E (1981) Ultrastructural localization of SV 40 viral DNA in cells, during lytic infection, by in situ molecular hybridization. Exp Cell Res 87:175−185

Heiles BJ, Genersch E, Kessler C, Neumann R, Eggers HJ (1988) In situ hybridization with digoxigenin labeled DNA of human papilloma viruses (HPV 16/18) in HeLa and SiHa cells. BioTechniques 6:978−981

Neumann R, Genersch E, Eggers HJ (1987) Detection of adenovirus nucleic sequences in human tonsils in the absence of infectious virus. Virus Res 7:93−97

Neumann R, Eggers HJ, Zippel HH, Remy B, Nelles G, Heiles BJ, Molitor E, Schulz KD (1989) Beitrag zur klinischen Relevanz des Nukleinsäurenachweises der humanen Papillomaviren (HPV) in Abstrichzellen der Cervix uteri. Geburtsh u Frauenheilk 49:11−16

Orth G, Jeanteur P, Croissant O (1970) Evidence for and localization of vegetative viral DNA replication by autographic detection of RNA-DNA hybrids in sections of tumours induced by Shope papilloma virus. Proc Natl Acad Sci USA 68:1876−1881

Pardue ML, Gall JG (1969) Chromosomal localization of mouse satellite DNA. Science 168:1356−1358

Tourtellotte WW, Verity AN, Schmid P, Martinez S, Shapshak P (1987) Covalent binding of formalin fixed paraffin embedded brain tissue sections to glass slides suitable for in situ hybridization. J Virol Meth 15:87−95

Vafai A, Murray RS, Welish M, Devlin M, Gilden DH (1988) Expression of varicella-zoster virus and herpes simplex virus in normal human trigeminal ganglia. Proc Natl Acad Sci USA 85:2362−2366

21.2 Differentiation of Viral and Chromosomal Nucleic Acids in Individual Nuclei

C. Simon Herrington

21.2.1 Principle and Applications

Multiple in situ nucleic acid detection can be achieved in two ways: by sequential hybridization with probes labeled with the same reporter and by simultaneous hybridization using probes labeled with different reporters. Any reporter molecules which can be used individually can be combined to allow dual nucleic acid detection, but we have found digoxigenin and biotin the most generally useful as they are safe and sensitive. We find that nick-translated probes give more consistent and sensitive results than those labeled by random priming and the protocols described here therefore utilize nick-translated probes (Herrington et al., 1989a). The maximum sensitivity achieved to date with high signal resolution is one to two copies of human papilloma virus (HPV) (Herrington et al., 1992a) in cultured cells and 2.5–12 copies of HPV in archival biopsies (Herrington et al., 1991), and digoxigenin-labeled HPV probes have been used to investigate the role of HPV infection in cervical neoplasia both in clinical biopsy material (Herrington et al., 1990a, b; Cooper et al., 1991a, b; Cooper et al., 1992a, b) and cervical smears (Herrington et al., 1992a, b; Troncone et al., 1992).

For a successful hybridization reaction to take place, the cell/tissue and its nucleic acid content must be fixed such that morphology is preserved but the nucleus is sufficiently permeable for labeled probe to reach its target. We find that probes with a median size of 200–400 base pairs are suitable for cells/tissues fixed in aldehyde and subjected to varying degrees of proteolysis. Aldehyde fixation increases sensitivity, where this is desirable (Herrington et al., 1989a). Unmasking of nucleic acids is achieved by use of either proteinase K or pepsin HCl. The former is a more rigorous treatment and is of particular use in the investigation of HPV infection. The latter is more gentle and is more appropriate for more friable tissues.

Denaturation is achieved by heat and hybridization occurs when the reaction temperature falls below the melting temperature (T_m) of the duplex formed between probe and target. Stringency conditions, which determine the degree to which the probe cross hybridizes with closely related sequences, can be varied according to individual requirements (Herrington et al., 1990; Herrington and McGee, 1992c). For the detection of HPV and genomic sequences in archival cervical biopsies, the addition of exogenous nucleic acid is unnecessary. However, this requirement should be established by experiment for other tissues and probes. Sheared human DNA is required for detection of viral sequences and sheared herring or salmon sperm DNA for human sequences in cytological material (Herrington et al., 1992b; Herrington and McGee, 1992d).

The principles of discriminative detection are similar to those of immunocytochemistry, with the additional steps being the degree of unmasking required and the hybridization reaction itself. Adequate contrast between substrate products is essential in multiple labeling techniques, particularly when studying nucleic acids and proteins in the same cellular compartment, i.e., nuclear, cytoplasmic. We find that, for ordinary light microscopy, red and blue/black products give good contrast (Herrington et al., 1989b).

Here, methods developed for the detection of HPV and repetitive genomic DNA sequences in cultured cells and routinely processed surgical biopsies are described. These methods are particularly useful when the relationship between DNA sequences in individual nuclei is of interest, such as in the interphase cytogenetic study of tumors (Herrington et al., 1990c) or when clinical material is in short supply.

The methods described below have been applied to the discrimination of morphologically dissimilar signals in archival biopsies, e.g., the Y chromosome and HPV sequences in male condylomata acuminata (Herrington et al., 1989b). The discrimination of morphologically similar signals, e.g., two HPV signals can also be achieved both in archival biopsies (Herrington et al., 1990a) and routine cervical smears (Herrington et al., 1992a; 1992b; Troncone et al., 1992). The sequential hybridization method has been used to analyze the relationship between chromosomes in fine-needle aspirates from breast tumors (Herrington and McGee, unpublished observations).

21.2.2 Non-Fluorescent In Situ Hybridization

- 4 Spot multiwell slides (Hendley)
- Aminopropyltriethoxysilane (Sigma)
- Decon 90 (BDH)
- Paraformaldehyde (BDH)
- Methanol and glacial acetic acid (BDH)

Slide preparation and cell/tissue fixation reagents

- Proteinase K (Boehringer Mannheim)
- Pepsin (Sigma, P7000)
- Microtiter or Terasaki plates (Gibco/Nunc).

Unmasking reagents

- Tris-HCl (Boehringer Mannheim)
- EDTA (Sigma)
- Sodium pyrophosphate
- Polyvinylpyrolidone (MW 40000) (Sigma)
- Ficoll (MW 400000) (Sigma)
- Formamide (Sigma)
- Dextran sulphate (Sigma)
- NaCl, Na citrate (Sigma)
- Herring sperm and human placental DNA (Sigma)

Hybridiza-tion reagents

- The following antibodies and conjugates were obtained from Dakopatts: monoclonal anti-biotin; biotinylated rabbit anti-mouse (F(ab')$_2$ fragment); rabbit anti-mouse immunoglobulin; avidin-peroxidase
- Monoclonal anti-digoxin (Sigma)
- Anti-digoxigenin alkaline phosphatase conjugate (Boehringer Mannheim)
- Dimethylformamide (DMF) (Sigma)
- Nitroblue tetrazolium (NBT) (Sigma)
- 5-bromo-4-chloro-3-indolyl phosphate (BCIP) (Sigma)
- Amino-9-ethylcarbazole (AEC) (Sigma)
- Diaminobenzidine (DAB) (Polysciences)

Detection reagents

- Aminopropyltriethoxysilane solution: mix 12 mls of aminopropyl-triethoxysilane with 588 mls of acetone immediately prior to use.
- Methanol/acetic acid (MAA) (3:1 [v/v]): this should be mixed fresh and cooled to −20°C prior to use.
- Phosphate-buffered saline (PBS): 10 mM phosphate; 150 mM NaCl; pH 7.4.

Slide preparation and cell/tissue fixation solutions

- PBS glycine: dissolve 0.2 g glycine (Sigma, UK) in 100 mls PBS to give a 0.2% solution.
- Paraformaldehyde (4%, w/v): boil 100 ml PBS containing of paraformaldehyde in a fume hood. Cool on ice prior to use. The final pH of this solution should be 7.2−7.4 without adjustment.

Unmasking solutions
- Pepsin solution (0.1%, w/v): dissolve 0.1 g pepsin (Sigma) in 96 ml distilled water prewarmed to 37°C and add 4 ml 5 M HCl slowly.
- Proteinase K solution: dissolve proteinase K to 500 μg/ml for biopsies and 1 μg/ml for cells in PBS.

Hybridization solutions
- Tris-EDTA (TE) buffer: 10 mM Tris HCl, 1 mM EDTA, pH 8.0.
- TE-PPF buffer: 500 mM Tris-HCl, pH 8.0, containing 1% (w/v) sodium pyrophosphate, 2% (w/v) polyvinylpyrolidone (MW 40 000), 2% Ficoll (MW 400 000), and 50 mM EDTA. Dissolve the reagents by heating to 65°C. Once dissolved, hold at this temperature for 15 mins. The buffer can be stored at room temperature.
- Human DNA (for viral sequence): extract DNA from peripheral blood lymphocytes or use human placental DNA and dissolve in water to a concentration of 10 mg/ml. Shear by autoclaving for 20 min in either a pressure cooker or a commercial autoclave.
- Herring sperm DNA (for genomic sequence): dissolve in water to 10 mg/ml and shear as described for human DNA.
- Standard saline citrate (SSC, 1×): 150 mM NaCl, 15 mM sodium citrate.
- Hybridization mixture: add 1 ml of 50% (w/v) dextran sulphate in distilled H_2O and 1 ml of 20× SSC to 5 ml deionized formamide. Adjust the mixture to pH 7.0 using 5 M HCl and store at 4°C. Under these conditions, it lasts up to 1 year.

Detection solutions
- Tris-buffered saline (TBS); 50 mM Tris-HCl, 100 mM NaCl, pH 7.2.
- TBT buffer (Tris-BSA-Triton): TBS containing 3% (w/v) bovine serum albumin (fraction V) and 0.05% (v/v) Triton X-100.
- Alkaline phosphatase substrate buffer: 50 mM Tris-HCl, 100 mM NaCl, 1 mM $MgCl_2$, pH 9.5.
- 20 mM Acetate buffer, pH 5.0−5.2.
- NBT/BCIP alkaline phosphatase substrate: prewarm 30 ml of alkaline phosphatase substrate buffer (see above) to 37°C. Dissolve 10 mg of NBT (Sigma) in 200 μl DMF and add to 1 ml of prewarmed substrate buffer. Add this mixture dropwise to the remaining substrate buffer. Dissolve 5 mg of BCIP in 200 μl of DMF; add this slowly to the mixture and store in 4 ml aliquots at −20°C.
- AEC/H_2O_2 peroxidase substrate: this is prepared fresh daily by dissolving 2 mg AEC in 1.2 ml dimethylsulfoxide in a glass tube. This mixture is added to 10 ml 20 mM acetate buffer, pH 5.0−5.2. Immediately prior to use, 1 μl 30% (v/v) H_2O_2 is added. The final mixture may require filtration prior to use.

- DAB/H_2O_2 substrate: dissolve DAB to 0.5 mg/ml in distilled water and add 10 µl 30% (v/v) H_2O_2 per ml of solution immediately prior to use.
- Glycerol jelly: dissolve 10 g of gelatin in 60 ml of distilled water on a hot stirrer. Add 70 ml of glycerol and 0.25 g of phenol and mix thoroughly. Glycerol/gelatin can be stored at room temperature (solid) or at 42°C (liquid).

- Incubation ovens (e.g., from Fisons/Gallenkamp) at 37°C, 42°C, 75°C and 95°C **Standard equipment**
- Standard water baths (e.g., Grant/BDH)
- Pipettmen (over the range 1 µl − 1 ml) and appropriate tips (e.g., Gilson/Anachem)
- Staining tanks and tray

1. Place the multiwell slides in a slide rack and immerse in 2% (v/v) Decon 90 in distilled water at 60°C for 30 min. **Preparation of slides from archival material**
2. Rinse thoroughly in distilled water, then acetone, and air dry.
3. Immerse in 2% (v/v) aminopropyltriethoxysilane solution for 30 min.
4. Rinse in acetone, wash in distilled water, and air dry at 37°C. Slides prepared in this way can be stored indefinitely at room temperature.
5. Cut 5 µm sections from routine paraffin-embedded blocks onto slides prepared as above.
6. Bake the sections either overnight at 60°C or for 45 min at 75°C. The sections can be stored at room temperature at this stage.
7. Dewax the sections by heating them to 75°C for 15 min then immersing them in two changes of xylene for 5 min each.
8. Remove the xylene by washing in two changes of 99% ethanol (industrial grade) for 5 min each at room temperature and wash in distilled water.

Proteinase K: **Unmasking of nucleic acids**
1. Spot the proteinase K solution on to the slides (100 µl per spot), place them in Terasaki plates, and float in a water bath at 37°C for 15 min.
2. Wash in distilled water and air dry at 75°C.

Pepsin HCl:
1. Incubate sections in pepsin solution in a Koplin jar for 15 min at 37°C.
2. Wash in distilled water and air dry at 75°C.

1. Cells in suspension, e.g., from fine-needle aspirates, should be fixed in methanol/acetic acid (3:1, v/v) for 15 min at room temperature; they can then be stored at −20°C. When required, the cells are either spotted or cytospun on to coated slides and air-dried. (Cells prepared in this way can be processed without further fixation by direct application of probe.) Adherent cultured cells can be grown directly onto coated slides, washed in PBS, then fixed in methanol/acetic acid as above. Routinely collected cervical smears should be fixed in methanol/acetic acid as above. **Preparation of cytological material**

2. Fix all preparations in 4% paraformaldehyde for 15 min at room temperature.
3. Rinse in 0.2% (w/v) glycine in PBS (5 min), PBS (5 min), and air dry.
4. Unmask nucleic acids using 1 µg/ml proteinase K in PBS for 15 min at 37 °C.
5. Postfix the cells in fresh 4% paraformaldehyde, then rinse in 0.2% (w/v) glycine in PBS (5 min), PBS (5 min), and air dry.

Probe preparation/ hybridization

This and the following methods apply to slides prepared by either of the two methods described above.

1. To each 35 µl of hybridization mixture, add 1 µl human or herring sperm DNA (as required) and 1 µl of each labeled probe required. We use probes at a final concentration of 1−2 µg/ml but this should be determined by experiment. Probes may be added individually or in combination and may be labeled with one or more reporter molecules.
2. Add TE buffer (for archival biopsies) or TE-PPF buffer (for cytological material) to a final volume of 50 µl and vortex briefly.
3. Apply the resultant mixture to biopsies or smears and cover with the appropriate glass coverslip. For cells in suspension and small archival biopsies, we use 4 spot multiwell slides, which require approximately 5.5 µl of probe mix per well and a 14 mm round coverslip. Routine cervical smears and larger biopsies require 50 µl per slide and a 22 × 50 mm coverslip.
4. Place two slides in each microtiter plate with a small volume of water to prevent drying and denature the DNA by heating for 15 min at 95 °C in a hot air oven. Preparations fixed only in MAA should be denatured at 75 °C for 6−7 min.
5. Transfer the plates to a hot air oven at 42 °C and incubate for 2 h.

Stringency washing and blocking

1. Wash the slides in two changes of 4× SSC at room temperature for 5 min each.
2. Wash in the appropriate stringency washing solution, e.g., 50% formamide/0.1× SSC, if required for discriminating closely homologous sequences. Adjust all washing solutions to pH 7.0 with 5 M HCl. The temperature of the solution should be monitored directly using a mercury thermometer. Washing should be carried out for 30 min.
3. Wash in 4× SSC at room temperature for 5 min.
4. Incubate for 10−15 min in blocking solution TBT at room temperature.

Simultaneous double probe detection

All incubations in antibody/avidin/enzyme conjugates are carried out at room temperature for 30 min unless otherwise stated. The substrate reactions are carried out at room temperature and signal development monitored by light microscopy. The substrate incubation times are therefore determined empirically for each experiment.

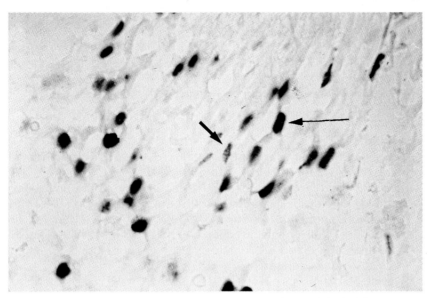

A cervical biopsy showing CIN 3 with HPV-associated morphological features was simultaneously hybridized with biotinylated HPV 33 and digoxigenin-labeled HPV 16. Detection was performed using the amplified dual probe detection system, with the HPV 33 probe developed as red (seen here as light gray, *small arrow*) and the HPV 16 probe as blue/black (seen here as black, *large arrow*). The signals localize to different cells demonstrating coinfection of the lesion with both viral types

Single step dual detection:

1. Pipette 1 µl of avidin peroxidase and 6 µl alkaline phosphatase conjugated anti-digoxigenin into each 600 µl TBT containing 5% (w/v) nonfat milk and incubate the slides in this solution for 30 min at room temperature.
2. Wash in two changes of TBS for 5 min each at room temperature.
3. Develop the red (biotin) signal using AEC/H_2O_2 substrate for 15–30 min at room temperature and wash in TBS for 5 min twice.
4. Develop the blue/black (digoxigenin) signal using NBT/BCIP for 30–60 min at room temperature, wash thoroughly in distilled water and air dry.
5. Mount in aqueous mountant, e.g., glycerol gelatin.

Amplified dual probe detection (see figure):

1. Incubate the preparations in monoclonal anti-biotin diluted 1:50 in TBT for 30 min at room temperature and wash in TBS twice.
2. Incubate in biotinylated rabbit anti-mouse (F(ab)$_2$ fragment) diluted 1:200 in TBT for 30 min at room temperature.
3. Follow steps 1–5 of the single step dual detection protocol.

Sequential probe detection This protocol was developed for the discriminative detection of chromosome-specific repeat probes in cytological specimens obtained by fine-needle aspiration. No cross-reaction between the detection systems occurs, presumably due to inactivation of the first peroxidase enzyme by the denaturation step. The cells are fixed in MAA (3 : 1, v/v) for 15 min and air dried prior to probe application (see above).

1. Denature probe and target for 6−7 min at 75 °C.
2. Hybridize for 2 h at 42 °C.
3. Carry out stringency washes to give required level of stringency (determined empirically) and block with TBT.
4. Detect the presence of probe hybrids using the appropriate monoclonal antibody (monoclonal anti-digoxin is used at a dilution of 1 : 10 000 and monoclonal antibiotin at 1 : 50 in TBT) followed by biotinylated rabbit anti-mouse (F(ab)$_2$ fragment) diluted 1 : 200 in TBT and avidin-peroxidase diluted 1 : 75 in TBT containing 5% (w/v) nonfat milk for 30 min each at room temperature.
5. Develop the signal using DAB/H$_2$O$_2$ substrate, wash in distilled water, and air dry at 42 °C.
6. Apply the second probe and repeat steps 1−4.
7. Develop the second signal using AEC/H$_2$O$_2$ substrate, wash in distilled water, air dry, counterstain in hematoxylin, and mount in glycerol/gelatin.

21.2.3 Special Hints for Application and Troubleshooting

Section dehiscence The major problems encountered using the above procedures are: section dehiscence; weak or absent signal; lack of reproducibility of signal; and high background staining.

− The adherence of sections to glass slides is dependent on the tissue under investigation and the adhesive properties of the glass and adhesive compound used. If dehiscence is a problem, then the concentration of proteolytic enzyme should be reduced (for proteinase K) or the enzyme changed (e.g., from proteinase K to pepsin).

Signal absence − Absence of signal in positive control material can occur for many reasons, from inadequate probe labeling to accidental omission of an antibody incubation step. In practice, we find that the usual reason for a weak signal is suboptimal unmasking which is often due to spontaneous enzyme inactivation on storage or variation of activity between lots. Our approach to investigation of weak signals is to check the incorporation of label and probe size and, if adequate, repeat the experiment using a different batch of unmasking enzyme. Adequate controls must be included in each experiment and interpreted appropriately to exclude technical variation.

- High background staining may be due to nonspecific probe binding, **High** nonspecific antibody/avidin binding, or nonspecific substrate deposi- **back-** tion. Nonspecific antibody/avidin binding can be reduced by using **ground** $F(ab)_2$ fragments and preincubation in bovine serum albumin or, if this is ineffective, human or animal antisera. Nonspecific avidin binding is reduced by either using modified avidin (Dakopatts) or streptavidin both of which are neutrally charged at pH 7, or by incubation in nonfat milk. Nonspecific substrate deposition may be due to endogenous enzyme activity, spontaneous substrate conversion, or deposition on noncellular components of the specimen.

Acknowledgements. The support of the Cancer Research Campaign (CRC) (UK) and the help of Professor J O'D McGee are gratefully acknowledged.

References

Cooper K, Herrington CS, Graham AK, Evans MF, McGee JO'D (1991a) In situ HPV genotyping of cervical intraepithelial neoplasia in South African and British patients: evidence for putative HPV integration in vivo. J Clin Pathol 44:400−405

Cooper K, Herrington CS, Graham AK, Evans MF, McGee JO'D (1991b) In situ evidence for HPV 16, 18, 33 integration in cervical squamous cell cancer in Britain and South Africa. J Clin Pathol 44:406−409

Cooper K, Herrington CS, Stickland JE, Evans MF, McGee JO'D (1992a) Episomal and integrated HPV in cervical neoplasia demonstrated by nonisotopic in situ hybridization. J Clin Pathol (in press)

Cooper K, Herrington CS, Lo ES, Evans MF, McGee JO'D (1992b) HPV 16 and 18 integration in cervical adenocarcinoma. J Clin Pathol (in press)

Herrington CS, Burns J, Graham AK, Evans MF, McGee JO'D (1989a) Interphase cytogenetics using biotin and digoxigenin labeled probes I: relative sensitivity of both reporters for detection of HPV 16 in CaSki cells. J Clin Pathol 42:592−600

Herrington CS, Burns J, Graham AK, Bhatt B, McGee JO'D (1989b) Interphase cytogenetics using biotin and digoxigenin labeled probes II: simultaneous detection of two nucleic acid species in individual nuclei. J Clin Pathol 42:601−606

Herrington CS, Burns J, Graham AK, McGee JO'D (1990a) Discrimination of closely homologous HPV types by in situ hybridization: definition and derivation iof Tm's. Histochem J 22:545−554

Herrington CS, Flannery DMJ, McGee JO'D (1990b) Single and simultaneous nucleic acid detection in archival human biopsies: application of non-isotopic in situ hybridization and the polymerase chain reaction to the analysis of human and viral genes. In: Polak JM & McGee JO'D (eds) In situ hybridization: principles and practice, Oxford University Press, Oxford, pp 187−215

Herrington CS, McGee JO'D (1990c) Interphase cytogenetics. Neurochem Res 4:467−474

Herrington CS, Graham AK, McGee JO'D (1991) Interphase cytogenetics using biotin and digoxigenin labeled probes: III. Increased sensitivity and flexibility for detecting HPV in cervical biopsy specimens and cell lines. J Clin Pathol 44:33−38

Herrington CS, de Angelis M, Evans MF, Troncone G, McGee JO'D (1992a) High risk HPV detection in routine cervical smears: a strategy for screening. J Clin Pathol 45:385−390

Herrington CS, Troncone G, McGee JO'D (1992b) Screening for high and low risk HPV types in single routine cervical smears by nonisotopic in situ hybridization (NISH). Cytopathology 3:71−78

Herrington CS, McGee JO'D (1992c) Principles and basic methodology of DNA/RNA detection by in situ hybridization. In: Herrington CS & McGee JO'D (eds) Diagnostic molecular pathology: a practical approach Vol 1, Oxford University Press (in press)

Herrington CS, McGee JO'D (1992d) In situ hybridization in diagnostic cytopathology. In: Herrington CS & McGee JO'D (eds) Diagnostic molecular pathology: A practical approach Vol 1, Oxford University Press (in press)

Troncone G, Herrington CS, Cooper K, de Angelis ML, McGee JO'D (1992) HPV detection in matched cervical smears and biopsies by nonisotopic in situ hybridization. J Clin Pathol 45:308–313

21.3 Virus Detection in Biopsy Specimens

PIRKKO HEINO and VEIJO HUKKANEN

21.3.1 Principle and Applications

In situ hybridization (ISH) is the method of choice for localization of viral nucleotide sequences in infected tissue and individual cells. In situ detection of viral mRNA enables the study of viral gene expression in the infected cells. ISH methods can also be combined and correlated with immunocytochemical detection of viral and cellular proteins.

The use of nonisotopic ISH methods has become common because most of the nonradioactive methods are now as sensitive as autoradiography but they do not have the disadvantages of the isotopic methods, such as the long assay time and the requirement for special laboratory facilities. Nonisotopic in situ procedures conventionally include the use of biotinylated probes, detected immunologically by anti-biotin antibodies and fluorescein- or enzyme-labeled second antibodies or by use of a biotin-streptavidin-polyalkaline phosphatase complex (Pinkel et al., 1986; Lewis et al., 1987; Allan et al., 1989).

Biotinylated probes for ISH have been used to demonstrate papillomavirus nucleotides in biopsy specimens (Beckmann et al., 1985; Crum et al., 1986; Burns et al., 1987; Syrjänen et al., 1988). The method has also been applied to detection of, e.g., adenovirus (Unger et al., 1986), cytomegalovirus (Unger et al., 1986), Epstein-Barr virus (Murphy et al., 1990), hepatitis B virus (Negro et al., 1985; Choi 1990), herpes simplex virus (Burns et al., 1986), and measles virus (Cosby et al., 1989; McQuaid et al., 1990) in biopsy specimens.

The background staining due to the presence of endogenous biotin in tissue can be problematic when biotinylated probes are used. Digoxigenin (DIG) labeling has proved to have a sensitivity equal to or better than that of biotin or radioactive labeling systems (Heino et al., 1989; Hukkanen et al., 1990; Morris et al., 1990; Permeen et al., 1990). The localization of the DIG-labeled hybrids at the cellular level was better than with autoradiography, because the autoradiographic grains tend to scatter around the compartment of the cell in which the hybridization occurs (Heino et al., 1989; Hukkanen et al., 1990).

ISH with DIG-labeled DNA probes have been used to detect human papillomavirus (Heino et al., 1989; Furuta et al., 1990; Konno et al., 1990; Morris et al., 1990), Epstein-Barr virus (Permeen et al., 1990), cytomegalovirus (Musiani et al., 1990) and herpes simplex virus (Hukkanen et al., 1990) in biopsy specimens. DIG-labeled RNA probes have been used to detect transcription from specific genes of herpes simplex virus in tissue sections as a marker for cells harboring the latent virus (Hukkanen et al., 1990).

21.3.2 Reaction Scheme

The standard procedure for detection of viral nucleic acids in biopsy specimens consists of the following reactions:

1. Labeling of the DIG-DNA probes by a random primed reaction or by transcription of DIG-labeled RNA probes from linearized plasmid templates
2. Sectioning and pretreatment of the tissue specimens
3. Hybridization of the pretreated tissue specimens with DIG-labeled DNA or RNA probes
4. Washing the specimens free of the unbound probes
5. Detection of the DIG-labeled hybrids using an anti-digoxigenin antibody:alkaline phosphatase conjugate and the enzyme-catalyzed color reaction.

21.3.3 In Situ Hybridization of Tissue Specimens with DIG-[11]-dUTP-Labeled DNA and with DIG-[11]-UTP-Labeled RNA Probes

Detection of HPV DNA in genital biopsy specimens and of the latency-associated RNA of herpes simplex virus (HSV) in mouse trigeminal ganglia are given as sample ISH protocol (see figure).

Labeling reagents

The labeling and detection reagents can be obtained from Boehringer Mannheim as the DIG DNA Labeling and Detection Kit and the DIG RNA Labeling Kit (SP6/T7)

Prehybridization reagents

- Proteinase K (Merck)
- Tris-HCl ((Sigma)
- CaCl$_2$ (Merck)
- RNase A, ribonuclease A (Sigma)
- NaCl, Na-citrate (Merck)
- Acetic anhydride (Merck)
- Triethanolamine-HCl (Sigma)

21.3.3

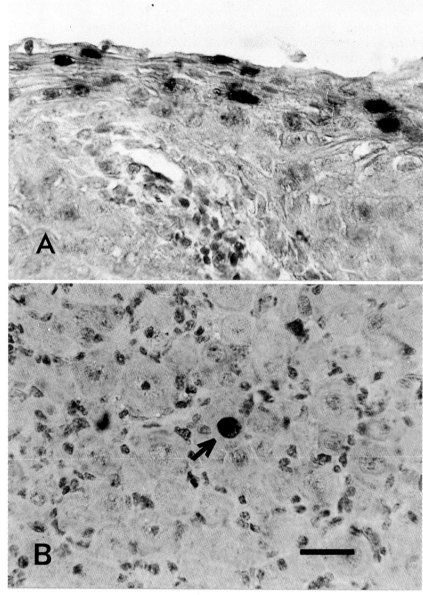

In situ detection of human papilloma virus type 18 DNA in a genital biopsy specimen, using a digoxigenin-labeled DNA probe **(A)** and detection of the latency-associated RNA of herpes simplex virus (HSV) in trigeminal ganglion of a mouse latently infected with HSV type 1 by use of a DIG-labeled single-stranded RNA probe **(B)**. The *arrow* in **B** indicates the location of the neuronal nucleus harboring the latency-associated RNA. The scale bar represents 32 μm in **A** and 25 μm in **B**

- Ficoll (type 400) (Sigma)
- BSA, bovine serum albumin (Sigma)
- PVP, polyvinylpyrrolidone (Sigma)
- Formamide (Carlo Erba)
- Salmon sperm DNA (Sigma)

- Formamide (Fluka)
- Dextran sulfate (Pharmacia)
- NaCl (Merck)
- DTT, dithiothreitol (Sigma)
- Ficoll (type 400) (Sigma)
- BSA, bovine serum albumin (Sigma)
- PVP, polyvinylpyrrolidone (Sigma)
- Tris-HCl (Sigma)
- EDTA (Boehringer Mannheim)
- Salmon sperm DNA (Sigma)
- tRNA (type XXI, from *E. coli*) (Sigma)

Hybridization reagents

- DIG DNA Labeling and Detection Kit (Boehringer Mannheim)
- DIG RNA Labeling Kit (Boehringer Mannheim)
- Tris-HCl (Sigma)
- EDTA (Boehringer Mannheim)

Detection reagents

- Proteinase K: 1 or 40 μg/ml in 10 mM Tris-HCl, 2 mM $CaCl_2$, pH 7.4
- Ribonuclease A: 100 μg/ml in 2× SSC (1× SSC: 0.015 M NaCl, 0.15 M sodium citrate; pH 7.0/25 °C)
- Acetylation solution: 0.25% acetic anhydride in 0.1 M triethanolamine-HCl buffer, pH 8.0
- Prehybridization buffer: 50% deionized formamide, 2× SSC, 1× Denhardt's solution (0.02% Ficoll, 0.02% bovine serum albumin, 0.02% polyvinylpyrrolidone) and 300 μg/ml of denatured salmon sperm DNA (the 10 mg/ml DNA stock solution is boiled for 5 min before each use, chilled in an ice bath, and added to the solution through a 24–25 gauge needle)

Prehybridization solutions

- Hybridization buffer for human papilloma virus: 50% formamide, 1× Denhardt's solution, 10% dextran sulfate, 0.6 M sodium chloride, 10 mM DTT, 10 mM Tris-HCl, pH 7.0, 0.5 mM EDTA, 0.5 mg/ml of salmon sperm DNA, 250 μg/ml of tRNA
- Hybridization buffer for herpes simplex virus: 50% formamide, 10% dextran sulfate, 0.3 M NaCl, 5 mM Tris-HCl, pH 7.4, 1 mM EDTA, 0.02% Ficoll, 0.02% polyvinylpyrrolidone, 1 mg/ml BSA, 5 mM DTT, and 0.5 mg/ml denatured mouse brain nucleic acids

Hybridization solutions

The probes are first dissolved in the buffer components without formamide and dextran sulfate. The dextran sulfate is dissolved separately in deionized formamide at a concentration of 200 mg/ml in a 55 °C waterbath by intermittent vortexing for 30 min. The dextran sulfate/formamide is then added to the probes dissolved in the other buffer components.

- Herpes simplex virus RNA washing buffer: 45% formamide, 2× SSC, 10 mM Tris-HCl, pH 7.4, and 1 mM EDTA (4 liters are needed).

Detection solutions
- Buffer 1 (100 mM Tris-HCl, 150 mM NaCl, pH 7.5)
- Blocking buffer: the blocking reagent from the DIG DNA Labeling and Detection Kit (Boehringer Mannheim) is dissolved as 1% solution in buffer 1 for 1 h at 65 °C a with intermittent shaking.
- Anti-DIG antibody:alkaline phosphatase conjugate: dilute into buffer 1 at a concentration of 750 mU/ml.
- Buffer 3: 100 mM Tris-HCl, 100 mM NaCl, 50 mM $MgCl_2$, pH 9.5.
- Color solution: 4.5 ml NBT solution and 3.5 ml BCIP solution of the kit in 1 ml of buffer 3.
- TE buffer: 10 mM Tris-HCl, 1 mM EDTA, pH 8.0.

1. Fix the genital biopsy specimens with 10% formalin and the mouse trigeminal ganglia in MOCA fixative 73% ethanol, 24.5% glacial acetic acid, 0.74% formaldehyde, and 1.76% water). Embed the specimens in paraffin. Section the specimens at a thickness of 4−5 μm and mount them on organo-siliconized activated slides (Maples, 1985). Incubate the sections on slides at 60 °C for at least 30 min, deparaffinate (see below) or store at 20 °C in a dry box until use.
2. For deparaffination of the specimens incubate them in xylene 2 × 5 min and dip in two serial 99% ethanol solutions, 8 times each. Fix the sections of the mouse trigeminal ganglia at this stage again for 3−5 h at 20 °C in MOCA fixative. Dehydrate specimens for 5 min each in two changes of 99% ethanol and one change of 95% ethanol solution. Allow the slides to dry at room temperature for a period up to 18 h.
3. Rehydrate the slides by serial incubations in 95%, 70%, and 50% ethanol, 5 min each. Soak in distilled water and incubate in 0.2 N HCl at 20 °C for 20 min. Dip twice in distilled water and incubate in a pre-warmed, predigested (30 min at 37 °C) solution of proteinase K (40 μg/ml for genital biopsy specimens and 1 μg/ml for mouse trigeminal ganglia; 15 min at 37 °C). Wash slides twice in distilled water.

Control sections for the HSV procedure can be pretreated with RNase which is made free of DNase (incubate RNase A solution at 37 °C for 2 h).

The slides are now acetylated with acetylation solution by vigorous shaking and subsequent incubation for 10 min at 20 °C. Wash the slides in distilled water and dehydrate them in 50%, 70%, and 95% ethanol, for 5 min each. Let them dry.

For HSV: Prehybridization takes place in a 45 °C incubator for 2 h in a glass dish containing prehybridization buffer.

1. Prepare the probes either by random primed labeling of the DNA frag- **In situ** ment (DIG DNA probes) or by transcription of a linearized plasmid **hybridi-** DNA template (DIG RNA probes) as described in Chap. 3. Concen- **zation** trate the probes by ethanol precipitation and dissolve them in a 20 μl TE buffer. Store at −20°C.

2. For genital biopsy specimens: Dilute the probes at a concentration of 0.2 μg/ml into the hybridization solution. Cover the tissue specimens with 15−50 μl of the probe solution (depending on the sizes of the sections), apply a coverslip, and seal it using rubber cement. Denature the HPV DNA probe and DNA in the specimen by heating the slides in a 93°−95°C oven for 8−12 min. Incubate the slides at 42°C for 16−20 h in a well-humidified chamber.

3. For mouse ganglion specimens: Prepare the probe cocktails in hybridization solution at concentrations of 0.1 μg/ml (DNA probes) or 0.5 μg/ml (RNA probes). You can use 20 μl of final hybridization mixture for a group of 3−4 mouse ganglion sections. The final probe mixtures are heated at 95°C for 10 min (DNA probes) or 3 min (RNA probes) and chilled on ice. The prehybridization solution is wiped away from the reverse side of the slide and from the specimen side so that only the sections remain covered by the prehybridization mixture. Apply 20 μl probe mixture to each group of sections and cover the area with a 15 ml Falcon conical centrifuge tube lid. The lid should not have contact with the solution or the sections. Attach the lid to the slide by dipping the lid rims in rubber cement. Place the slides in a tight, well-humidified box and incubate the box at 45°C for 18−20 h in a well-humidified incubator (e.g., a waterbath with low level of water).

Genital tissue: **Washing**

1. Dip the slides after hybridization into a tube which contains 0.5× SSC/ **the sections** 1 mM EDTA/1 mM DTT. Wash the slides at 25°C (unless otherwise indicated) in 0.5× SSC/1 mM EDTA/1 mM DTT (2 × 5 min). All washings should be carried out with gentle agitation.

2. Wash the slides in 0.5× SSC/1 mM EDTA (2 × 5 min), in 50% formamide/0.15 M NaCl/5 mM Tris-HCl, pH 7.4/0.5 mM EDTA for 10 min, in 0.5× SSC (4 × 5 min at 55°C), and in 0.5× SSC (1 × 5 min).

Mouse trigeminal ganglia:

1. Remove the probe by dipping the slide into a tube which contains 2× SSC. Place the slides into a dish containing 2× SSC buffer. Change the buffer twice at 10 min intervals.

2. If DIG RNA probes were used, the sections can be washed by digestion with RNase A (40−50 μg/ml) and RNase T1 (10 U/ml; Boehringer Mannheim) in 10 mM Tris-HCl, 0.3 M NaCl, pH 7.5, for 40 min at 37°C. Transfer the slides to 2× SSC after the incubation. You may need to optimize the RNase A concentration for your experiment. Digoxigenin-labeled RNA probes may be more sensitive to RNase than radiolabeled probes.

3. Wash the sections, hybridized with either DNA or RNA probes, for 72 h at 25 °C using 2 liters of HSV RNA washing buffer with one buffer change at 24 h, and gentle magnetic stirring.

Detection of DIG-labeled nucleic acids in tissue sections

1. Transfer the slides into 2× SSC buffer.
2. Incubate them for 1 min in buffer 1.
3. The blocking reagent from the DIG DNA Labeling and Detection Kit (Boehringer Mannheim) should be dissolved in buffer 1 for 1 h at 65 °C, with intermittent shaking. Incubate the slides in 1% blocking reagent for 30 min at 20 °C.
4. Dip slides in buffer 1.
5. Dilute the anti-digoxigenin antibody: alkaline phosphatase conjugate into buffer 1 at a concentration of 750 mU/ml. Reserve 70 μl for each group of sections. Wipe the slides one by one as was done for the hybridization and pipette the conjugate onto the sections. **Note:** the conjugate dries out easier than the hybridization mixtures did. Incubate the slides in a humidified plastic chamber at 20 °C for 30 min.
6. Dip the slides in a large tube containing buffer 1 and wash the slides twice at 20 °C for 15 min in buffer 1.
7. Incubate the sections for 2 min in buffer 3.
8. Wipe the slides as above and apply 70 μl of color solution on each section. Incubate in a dark, humidified chamber at 20 °C for 5−12 h. Do not move the slides while the color precipitate forms.
9. Wash the slides in TE buffer. You can place them briefly under the microscope now in order to follow the color development. Do not let the slides dry out.
10. Stain the slides for 30−60 s in Gill's hematoxylin 1 (Sigma GHS 1−80), wash with tap water (and stain for 20 s with aqueous Eosin Y solution). The coverslips can be mounted with an aqueous mountant such as Gurr's Aquamount (BDH, Poole, England).
11. The positive hybridization result is a dark brown/purple precipitate.

21.3.4 Special Hints for Application and Troubleshooting

- Endogenous alkaline phosphatase activity can be inhibited by immersing the specimens in acetic acid before hybridization (Morris et al., 1990)
- The background can be decreased by pretreatment of the sections with DNase or with RNase (Furuta et al., 1990).

References

Allan GM, Todd D, Smyth JA, Mackie DP, Burns J, McNulty MS (1989) In situ hybridization: an optimised detection protocol for a biotinylated DNA probe renders it more sensitive than a comparable ^{35}S-labeled probe. J Virol Meth 24:181−190

Beckmann AM, Myerson D, Daling JR, Kiviat NB, Fenoglio CM, McDougall JK (1985) Detection and localization of human papillomavirus DNA in human genital condylomas by in situ hybridization with biotinylated probes. J Med Virol 16:265−273

Burns J, Redfern DRM, Esiri MM, McGee JO'D (1986) Human and viral gene detection in routine paraffin embedded tissue by in situ hybridization with biotinylated probes: viral localisation in herpes encephalitis. J Clin Pathol 39:1066−1073

Burns J, Graham AK, Frank C, Fleming KA, Evans MF, McGee JO'D (1987) Detection of low copy human papilloma virus DNA and mRNA in routine paraffin sections of cervix by nonisotopic in situ hybridization. J Clin Pathol 40:858−864

Choi YJ (1990) In situ hybridization using a biotinylated DNA probe on formalin-fixed liver biopsies with hepatitis B virus infections: In situ hybridization superior to immunohistochemistry. Mod Pathol 3:343−347

Cosby SL, McQuaid S, Taylor MJ, Bailey M, Rima BK, Martin SJ, Allen IV (1989) Examination of eight cases of multiple sclerosis and 56 neurological and non-neurological controls for genomic sequences of measles virus, canine distemper virus, simian virus 5 and rubella virus. J Gen Virol 70:2027−2036

Crum CP, Nagai N, Levine RU, Silverstein S (1986) In situ hybridization analysis of HPV 16 DNA sequences in early cervical neoplasia. Am J Pathol 123:174−182

Furuta Y, Shinohara T, Sano K, Meguro M, Nagashima K (1990) In situ hybridization with digoxigenin-labeled DNA probes for detection of viral genomes. J Clin Pathol 43:806−809

Heino P, Hukkanen V, Arstila P (1989) Detection of human papilloma virus (HPV) DNA in genital biopsy specimens by in situ hybridization with digoxigenin-labeled probes. J Virol Meth 26:331−338

Hukkanen V, Heino P, Sears AE, Roizman B (1990) Detection of herpes simplex virus latency-associated RNA in mouse trigeminal ganglia by in situ hybridization using nonradioactive digoxigenin-labeled DNA and RNA probes. Meth Mol Cell Biol 2:70−81

Konno R, Shikano K, Horiguchi M, Endo A, Chiba H, Yaegashi N, Sato S, Yajima H, Tase T, Yajima A (1990) Detection of human papillomavirus DNA in genital condylomata in women and their male partners by using in situ hybridization with digoxigenin labeled probes. Tohoku J Exp Med 160:383−390

Lewis FA, Griffiths S, Dunnicliff R, Wells M, Dudding N, Bird CC (1987) Sensitive in situ hybridization technique using biotin-streptavidin polyalkaline phosphatase complex. J Clin Pathol 40:163−166

Maples JA (1985) A method for the covalent attachment of cells to glass slides for use in immunohistochemical assays. Am J Clin Pathol 83:356−363

McQuaid S, Isserte S, Allan GM, Taylor MJ, Allen IV, Cosby SL (1990) Use of immunocytochemistry and biotinylated in situ hybridization for detecting measles virus in central nervous system tissue. J Clin Pathol 43:329−333

Morris RG, Arends MJ, Bishop PE, Sizer K, Duvall E, Bird CC (1990) Sensitivity of digoxigenin and biotin labeled probes for detection of human papillomavirus by in situ hybridization. J Clin Pathol 43:800−805

Murphy JK, Young LS, Bevan IS, Lewis FA, Dockey D, Ironside JW, O'Brien CJ, Wells M (1990) Demonstration of Epstein-Barr virus in primary brain lymphoma by in situ DNA hybridization in paraffin wax embedded tissue. J Clin Pathol 43:220−223

Musiani G, Gentilomi G, Zerbini M, Gibellini D, Gallinella G, Pileri S, Baglioni P, La Placa M (1990) In situ detection of cytomegalovirus DNA in biopsies of AIDS patients using a hybrido-immunocytochemical assay. Histochem 94:21−25

Negro F, Berninger M, Chiaberge E, Gugliotti P, Bussolati G, Actis GC, Rizzetto M, Bonino F (1985) Detection of HBV-DNA by in situ hybridization using a biotin-labeled probe. J Med Virol 15:373−382

Permeen AMY, Sam CK, Pathmanathan R, Prasad U, Wolf H (1990) Detection of Epstein-Barr virus DNA in nasopharyngeal carcinoma using a nonradioactive digoxigenin-labeled probe. J Virol Meth 27:261−268

Pinkel D, Straume T, Gray JW (1986) Cytogenetic analysis using quantitative, high-sen-
sitivity, fluorescence hybridization. Proc Natl Acad Sci 83:2934−2938

Syrjänen S, Partanen P, Mäntyjärvi R, Syrjänen K (1988) Sensitivity of in situ hybridization techniques using biotin- and ^{35}S-labeled human papillomavirus (HPV) DNA
probes. J Virol Meth 19:225−238

Unger ER, Budgeon LR, Myerson D, Brigati DJ (1986) Viral diagnosis by in situ hybridization. Description of a rapid simplified colorimetric method. Am J Surg Pathol
10:1−8

21.4 Fluorescent In Situ Hybridization on Banded Chromosomes

N. Arnold, M. Bhatt, T. Ried, J. Wienberg, and D. C. Ward

21.4.1 Principles and Techniques

In situ hybridization of nonradioactive DNA probes to metaphase
chromosomes is becoming an increasingly important tool in clinical and
tumor cytogenetics and for gene mapping studies (for review see Lichter
et al., 1991; Raap et al., 1990; Ward et al., 1991). The positive identification of chromosomes is the prerequisite to ordering DNA probes on the 24
different human chromosomes. Several probe sets are currently available
in the nonisotopic format for this purpose, the most commonly used being
chromosome specific centromeric repeat clones (Moyzis et al., 1987; Willard and Waye, 1987) and „composite probe" sets for chromosome „painting" (Collins et al., 1991; Dilla et al., 1990). More recently, region-specific
reference clones have become available, including cosmid or YAC clones,
and microlibraries derived from microdissected chromosome bands
(Boyle et al., 1990; Landegent et al., 1987; Lichter et al., 1990a; Lengauer
et al., 1992; Meltzer et al., 1992; Slim et al., 1991; Tkachuk et al., 1990;
Trask et al., 1991; Trautmann et al., 1991). These chromosome identification procedures can be readily combined with fluorescence in situ hybridization of other DNA probes. However, they are hampered by the lack of
precision compared to classical chromosome banding analysis. In addition, in situ hybridization on banded chromosomes would make it possible
to immediately assign and compare mapping positions with respect to the
well characterized framework of conventional banding patterns.

Classical cytogenetic banding techniques such as Giemsa (G) or reverse
(R) banding provide landmarks on which in situ hybridization signals can
be mapped with high resolution. A limited number of these banding
techniques can be used in combination with fluorescence in situ hybridiza-

tion techniques. Several protocols have been published that allow a clear-cut identification of hybridization signals even on high resolution chromosome banding patterns (Ambros et al., 1987; Baldini and Ward 1991; Cherif et al., 1990; Fan et al., 1990; Klever et al., 1991; Kuwano et al., 1991; Lawrence et al., 1990; Lemieux et al., 1992; Smit et al., 1990; Takashi et al., 1990; Tucker et al., 1988). In this manual we will review the protocols for a G-banding procedure, BrdU replication banding, and an in situ hybridization banding (ISHB) technique, the latter two giving patterns that resemble R banding.

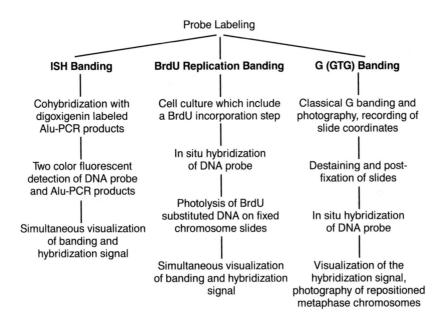

21.4.2 Detailed Standard Procedure

Reagents for Cell Culture and Chromosome Preparation   **Standard reagents**

Reagents	**Source and Catalog No.**
RPMI 1640	Boehringer Mannheim, cat. no. 209945
Fetal calf serum (FCS)	Boehringer Mannheim, cat. no. 210471
Phytohemagglutinin (PHA)	Wellcome, cat. no. HA15
Methotrexate (MTX)	Sigma, cat. no. M8407
5-Bromo-2'-deoxyuridine (BrdU)	Serva, cat. no. 15240
5-Fluoro-2'-deoxyuridine (FUdR)	Serva, cat. no. 21555
Penicillin/Streptomycin	Flow, cat. no. 16-700-49
Colcemid	Life Technologies/BRL, cat. no. 120-5210-AD

21.4.2

Reagents for Chromosome Banding

Hoechst 33258	Serva, cat. no. 15090
Bacto-Trypsin	Difco, cat. no. 0153-59
Giemsa Stain	Merck, cat. no. 9204
Acid free formaldehyd (37%)	Merck, cat. no. 3999
Formamide	Sigma, cat. no. F-7503

Reagents for Alu-PCR

Taq DNA polymerase	Perkin-Elmer/Cetus, cat. no. 182415
dNTPs	Boehringer Mannheim, cat. no. 1051440, 1051458, 1051466, 1051482
Human placental DNA	Sigma, cat. no. D-7011

For reagents for in situ hybridization, see Chap. 21.6 of this manual.

Solutions for BrdU-replication banding

- PHA, store at 4°C
- Colcemid, store at 4°C
- Methotrexate 10^{-3} M, stock solution: Dissolve 1 mg MTX in 2.2 ml ddH$_2$O, filtrate, and store at −20°C in the dark. For 10^{-5} M working solution make a 100-fold dilution and aliquots, store at −20°C
- Hypotonic solution: Dissolve 5.6 g KCl (0.075 M) in 1 liter dH$_2$O (use up to 1 month).
- Hoechst 33258 solution: Prepare a stock solution with 0.5 mg/ml in ddH$_2$O which should be kept frozen in dark vials. The solution can be used up to a few months. Working solution has a concentration of 2.5 µg/ml.
- BrdU solution (1 mg/ml): Dissolve 1 mg of FUdR in 10 ml of dH$_2$O and add 0.5 ml of this solution to 9.5 ml of distilled water to which 10 mg of BrdU will be added. Filter BrdU through a 0.45 µm filter. Store frozen in the dark. Solution can be used for at least 2−3 weeks.
- Blood Culture set-up (for each T75 flask):

RPMI 1640	40.0 ml
FCS	10.0 ml
Penicillin/Streptomycin	0.5 ml
PHA	2.0 ml
Blood	5.0 ml

- Alu-PCR reaction mixture (final reaction volume 100 µl)

2.5 mM	MgCl$_2$
10 mM	Tris-HCl, pH 8.5/25°C
50 mM	KCl
0.001%	Gelatin
300 µM	each of the four dNTPs
2.5 U	*Taq* DNA polymerase
1 µM	primer (No. 517; Nelson et al., 1989)
100 ng	needle-sheared human genomic DNA

- Primer sequence
  CGACCTCGAGATCT(C/T) (G/A)GCTCACTGCAA
- TE Buffer
  10 mM   Tris-HCl, pH 8.0/25 °C
   1 mM   EDTA

- 10× PBS pH 7.0 (11):
  Dissolve 80g of NaCl, 2g of KCl, 14.4g of Na_2HPO_4 and 2.4g of KH$_2$PO$_4$ in 800 ml of distilled dH_2O. Adjust the pH to 7.4 with HCl. Add dH_2O to 1 liter. Dispense the solution into aliquots and sterilize them by autoclaving. Store at room temperature.
- 20× SSC pH 7.0 (11):
  Dissolve 175.3 g of NaCl and 88.2 g of sodium citrate in 800 ml of dH_2O. Adjust the pH to 7.0 with a few drops of a 10 N solution of NaOH. Adjust the volume to 1 liter with dH_2O. Dispense into aliquots. Sterilize by autoclaving.
- Sørensen buffer pH 7.0:
  For solution A: dissolve 9.078 g KH_2PO_4 in 1 liter dH_2O.
  For solution B: dissolve 11.876 g $Na_2HPO_4 \times 2 H_2O$ in 1 liter dH_2O.
  Mix 38.8 ml solution A with 61.2 ml solution B and, if necessary, adjust pH 7.0.
- Formaldehyde (3.7% in PBS):
  Mix 10 ml 37% formaldehyde solution with 10 ml 10× PBS, adjust the volume to 100 ml.
- 70% Formamide/2× SSC:
  Mix 70 ml formamide with 3 ml 20× SSC, adjust the volume to 100 ml and pH to 7.0.

Solutions for G banding

21.4.3 General Protocol

We present three protocols for chromosome banding in combination with in situ hybridization which are used in our laboratories routinely. We selected these protocols since they are simple to use and give reproducible results. Nevertheless, other protocols or variations of those presented here may work as well.

When using "replication R banding" by the BrdU incorporation technique (Vogel et al., 1986; Manuelidis and Borden, 1988; Takashi et al., 1990), a special cell culture set-up is necessary. Methotrexate is used to synchronize the cells and the thymidine analog BrdU is added to the culture in late S phase of the cell cycle. FUdR is added to inhibit thymidine synthesis to increase BrdU incorporation. Preparation of the slides and in situ hybridization is according to standard procedures (see protocol in Raap et al., chap 21.6 this book). After in situ hybridization, the slides are stained with Hoechst 33258 and exposed to UV light, which induces photolysis of BrdU substituted chromatin. Then, standard detection of the in situ hybridization signal with fluorochrome-coupled antibodies or avidin is followed.

ISH banding with Alu sequences results in R banding, on which the probe can be mapped simultaneously. The Alu DNA repeat family, which is represented about 300 000 to 900 000 times in the human genome, is not distributed randomly over the chromosomes. In situ hybridization of cloned Alu sequences showed that they are concentrated in the R (reverse) bands (Manuelidis and Ward, 1984; Korenberg and Rykowsky, 1988; Moyzis et al., 1989). Accordingly, they can be used for chromosome identification during in situ hybridization experiments (Lichter et al., 1990b). An improvement in this technique was introduced when using it for in situ hybridization of PCR products from a single Alu primer (No. 517; Nelson et al., 1989) and human genomic DNA for Alu amplification instead of using cloned Alu sequences (Baldini and Ward, 1991). A higher contrast of the banding pattern was achieved, which may be due to the fact that the PCR products may represent divergent Alu repeats and therefore hybridize to a larger number of target sequences on metaphase chromosomes.

Various protocols have been published to combine G banding before or after in situ hybridization. We use routine GTG banding before in situ hybridization, which gives us the most reproducible results. Metaphases are photographed and the coordinates carefully documented to facilitate the repositioning of the metaphase after in situ hybridization. Slides are destained and postfixed with formaldehyde. Postfixation is essential to preserve chromosome morphology in subsequent in situ hybridization experiments (Klever et al., 1991).

All three techniques give reproducible banding patterns. The contrast of the banding patterns with the ISH (and BrdU incorporation) technique can be drastically improved when using digital imaging procedures.

However, all techniques also exhibit some drawbacks, which should be mentioned. Since a special cell culture set-up and pretreatment of the slides is needed, the workload with the BrdU incorporation technique is increased. R banding by ISH banding with Alu sequences is restricted to chromosome preparation of humans and primates. However, dispersed repetitive sequences from other species are also known that are nonrandomly distributed throughout the genome (e.g., mouse LINEs, Boyle et al., 1990). Both R banding techniques have the advantage that the hybridization signal can be analyzed directly on banded chromosomes by just changing the filter set. When using the BrdU incorporation technique, more bands are discernable, in particular with respect to chromosomes 3, 7, 9, 11, 12, and 19. In figure 1, we present a direct comparison between both R banding patterns on the same metaphase plate.

When using chromosome G banding previous to in situ hybridization, the banding pattern is of high quality (Yunis, 1976). However, since metaphase chromosomes have to be relocated on the microscope for inspection of the hybridization signal, the workload is increased significantly. Since only a portion of all metaphases on a slide exhibit optimal hybridization signals, this technique should be restricted to experiments where the success of in situ hybridization is predictable, e.g., cosmid clones or chromosome specific DNA libraries. In figure 2, we give the G banding pattern and the hybridization signal of a human chromosome 8 specific plasmid DNA library hybridized to chromosomes from a primate species *(Macaca fuscata)* (Wienberg et al., 1992).

21.4.4 Chromosome Banding Protocols

General hint for storing slides

- Prepare slides as normal (optimal humidity = 50%−60%)
- Use acetic acid (70%−100%) on slides to remove cytoplasm.
- Soak slides in 70%, 90%, and 100% ethanol for 5 min each, air dry.
- Store slides at 4°C (or at −70°C for extended periods) in a black box and seal with parafilm.
- Age fresh or freshly stored slides at least 3 days to one week before use for in situ hybridization.

Lymphocyte culture for the BrdU incorporation technique (for 10 ml peripheral blood)

1. Culture for 72 h at 37°C (shake flasks once a day).
2. Add MTX (10^{-5} M) 500−550 µl/flask (final concentration is 10^{-7} M), shake gently.
3. Incubate for 17 h at 37°C.
4. Transfer to 50 ml tubes (29 ml to each tube).
5. Centrifuge at 900 rpm for 8 min.
6. Remove supernatant, leaving 1−2 ml of medium to resuspend pellet.
7. Resuspend in 25 ml of FCS-free RPMI 1640 per tube.
8. Repeat steps 5−7.
9. Repeat steps 5−6.
10. Resuspend in 25 ml of RPMI 1640 with 20% FCS per tube.
11. Add BrdU solution (final concentration 30 µg/ml), mix, transfer tubes to two T75 culture flasks.
12. Incubate for 5 h at 37°C.
13. Add colcemid (10 µg/ml) 300 µl/flask, and mix well by gently shaking or pipeting (final concentration 0.06 µg/ml).
14. Transfer to 4 × 50 ml tubes immediately.
15. Put back into 37°C incubator (total time of colcemid treatment is 5−10 min).

Lymphocyte preparation

16. Start centrifugation as in step 5.
17. Remove supernatant as in step 6.
18. Add 5 ml of prewarmed (37°C) 0.075 M KCl/each tube drop by drop, mix by tapping.
19. Continue adding KCl slowly to total volume of 40 ml/tube.
20. Put in 37°C water bath for 16−18 min total.
21. Add 1 ml of freshly prepared fixative (3:1 methanol and glacial acetic acid) and mix gently.
22. Centrifuge as in step 5.

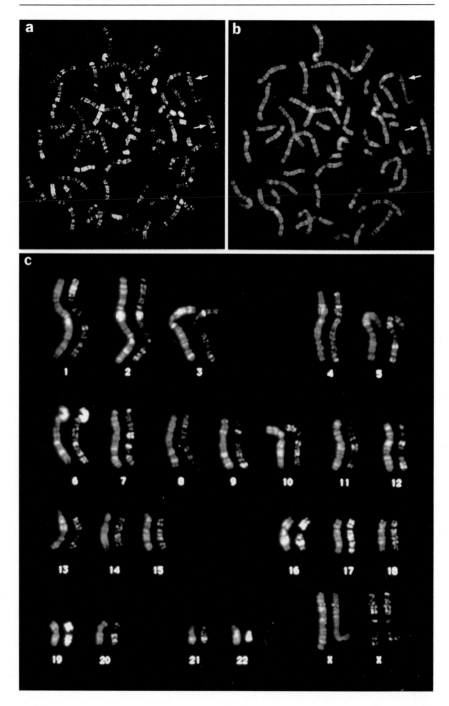

23. Remove supernatant as much as possible and tap well to mix pellet.
24. Add freshly prepared fixative drop by drop up to 5 ml (pipette well but *gently* to make sure no clumps are present).
25. Add more fixative (up to 25 ml/tube).
26. Leave at room temperature for 20–30 min.
27. Centrifuge as step 5.
28. Discard supernatant and tap well to mix pellet.
29. Add 25 ml of freshly prepared fixative slowly to each tube, pipetting to prevent clumping.
30. Centrifuge as step 5.
31. Repeat steps 28–30 at least three more times until the pellet becomes white.

In situ hybridization with the BrdU incorporation technique

– Perform routine day 1 procedure for in situ hybridization (i.e., probe labeling, denaturation of slides and probe, in situ hybridization).
– On day 2: follow standard washing protocol up to and including blocking step.
– FP (fluorescence-photolysis):
Incubate slides in the dark with 2.5 µg/ml Hoechst 33258 at room temperature for 15 min in a coplin jar. Briefly wash with water.
Expose slides (in a dish with 2× SSC) to the light of a UV mercury lamp at a distance of 10–20 cm for 1 h (15 W lamp).
Incubate slides in 2× SSC at 60 °C for 1 h.
Detection steps, amplification of the in situ hybridization signals, counterstain with DAPI or PI and antifade are as in routine protocol [see protocol in Raap et al. (1990), Sect. 21.6 and figure 1].

←——————————————————————————

Figure 1. Metaphase chromosomes of peripheral blood lymphocytes from a normal female donor. Cells were prepared following methotrexate synchronization and BrdU incorporation as described in the protocol. **a** shows co-hybridization of digoxigenin labeled Alu-PCR products (Baldini and Ward, 1991) and a biotinylated cosmid clone specific for the Duchenne muscular dystrophy gene (Ried et al., 1990). The biotinylated probe was detected with avidin FITC and the digoxigenin labeled Alu-PCR products were visualized using anti-digoxigenin rhodamine resulting in an R banding pattern. Each of the X chromosomes reveals a hybridization signal on both sister chromatids on chromosomal band position Xp21 *(arrows)*. **b** Same metaphase spread as in **a** stained with DAPI, which also display an R banding pattern. The same biotinylated clone is shown to hybridize on Xp21. Note that with the BrdU incorporation technique the active and inactive X chromosomes can easily be distinguished *(arrows)*. **c** Karyotype composite of chromosomes banded by Alu-PCR products *(right)* and the same chromosomes with BrdU incorporation visualized with DAPI stain *(left)*. It is noticeable that more bands are discernable using the BrdU method. Metaphase plates were imaged with a cooled CCD camera. Merging of the pictures and montage of the karyotype was performed with the programs Gene-Join (Ried et al., 1992) and Adobe Photoshop. Photographs were taken directly from the screen using a Kodak 100 HC color slide film

In situ hybridization banding PCR conditions for the generation of Alu sequences from human genomic DNA were optimized for the thermocycler used (Ericomp). After initial denaturation at 95°C for 3 min, 35 cycles of PCR were carried out with denaturation at 94°C for 1 min, annealing at 55°C for 2 min and extension at 72°C for 4 min (last cycle 7 min). We want to stress that PCR conditions might vary slightly depending on the thermocycler used. The products were ethanol precipitated in the presence of 2M ammonium acetate and resuspended in TE. PCR products were labeled by standard nick translation with biotin-[11]-dUTP (see Chap. 4.1) or DIG-[11]-dUTP (see Chap. 3.2). DNase concentration was adjusted to give a probe size of about 200−500 bp.

Some 5−10 ng/µl labeled PCR products were added to the standard hybridization mixture if no suppression hybridization was necessary.

If a genomic probe required a preannealing step with human competitor DNA of the Cot 1 fraction, the probe DNA was resuspended in 5 µl hybridization solution, denatured separately, and allowed to preanneal. Immediately before placing the probe on the slides, the denatured Alu-PCR products (also resuspended in 5 µl hybridization solution) were combined with the probe DNA. The probe DNA and the Alu-PCR products were detected with different fluorochromes. In our laboratories, we routinely use avidin-FITC for probe detection and anti-DIG: rhodamine for visualizing Alu banding.

G banding previous to in situ hybridization Prepare in coplin jars for banding:
- 0.5 ml Bacto-Trypsin (resolved in 10 ml ddH$_2$O) in 100 ml 1× PBS
- 2 Coplin jars with 1× PBS
- 5 ml Giemsa stain in 100 ml Sørensen buffer (pH 7.0)

Prepare in coplin jars destaining and postfixation:
- Xylene
- Xylene/ethanol; 1:1
- 2 Coplin jars with methanol/acetic acid (3:1)
- Ethanol: 100%, 90%, 70% each
- 1× PBS
- Formaldehyde (4% in 1× PBS)
- 2 Coplin jars 1× PBS
- 70% formamide/2× SSC

Staining G banding was performed according to standard procedures. The time for trypsin treatment and Giemsa staining may vary between different cells (lymphocytes, fibroblasts etc.) and according to the age of the solutions. Before staining a series of slides, a single slide should always be used first to adopt the appropriate timings.

1. Prewarm trypsin solution to 37°C. Incubate slides for 5−20 s and wash briefly in 1× PBS.
2. Stain 10 min in freshly prepared Giemsa, and remove excess staining solution under floating tap water.

21.4.4

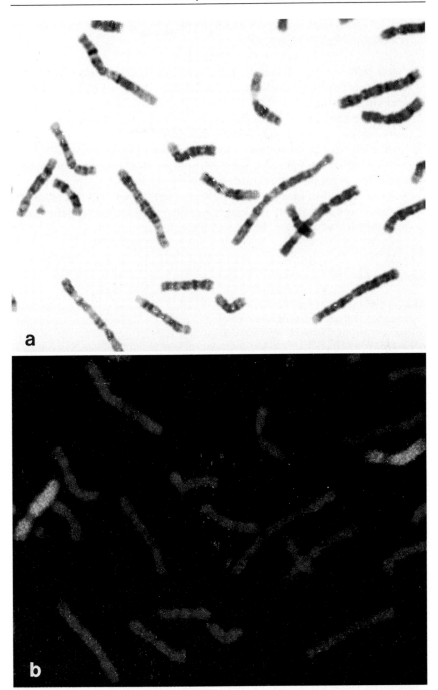

Figure 2. G banding of chromosomes previous to in situ hybridization. **a** G-banded metaphase chromosomes, **b** the same metaphase plate after destaining, postfixation, and subsequent in situ hybridization. The probe was a human chromosome 8 specific plasmid library hybridized to chromosomes of the old world monkey Macaca fuscata (Wienberg et al., 1992). The hybridization signals are found on macaque chromosomes that are homologous to human chromosome 8

3. Cover slides with a coverslip and check for appropriate banding and staining under the microscope.

Post-fixation
1. After photography, the slides are rinsed in xylene, xylene/ethanol and two times methanol/acetic acid, 5 min each, to remove immersion oil and Giemsa stain.
2. Slides are rehydrated in an alcohol series and rinsed twice in 1× PBS, 5 min each.
3. Specimens are postfixed with formaldehyde, 15 min, washed 2× for 5 min each in 1× PBS and finally stored in 70% formamide/2× SSC overnight at room temperature. At this point, the slides are ready for in situ hybridization (see protocol in Raap et al., chap 21.6 this manual). If slides are to be stored for longer periods, they should be dehydrated in an alcohol series (see the second figure).

References

Ambros PF, Karlic HI (1987) Chromosome insertion of human papillomavirus 18 sequences in HeLa cells detected by nonisotopic in situ hybridization and reflection contrast microscopy. Hum Genet 77:251–254

Baldini A, Ward DC (1991) In situ hybridization of human chromosomes with Alu-PCR products: a simultaneous karyotype for gene mapping studies. Genomics 9:770–774

Boyle AL, Ballard SG, Ward DC (1990) Differential distribution of long and short interspersed elements in the mouse genome: Chromosome karyotyping by fluorescence in situ hybridization. Proc Natl Acad Sci USA 87:7757–7761

Cherif D, Julier C, Delattre O, Derre J, Lathrop GM, Berger R (1990) Simultaneous localization of cosmids and chromosome R-banding by fluorescence microscopy: Application to regional mapping of human chromosome 11. Proc Natl Acad Sci USA 87:6639–6643

Collins C, Kuo WL, Segraves R, Pinkel D, Fuscoe J, Gray JW (1991) Construction and characterization of plasmid libraries enriched in sequences from single human chromosomes. Genomics 11:997–1006

Dilla van MA, Deaven LL (1990) Construction of gene libraries for each human chromosomes. Cytometry 11:208–218

Fan YS, Davis LM, Shows TB (1990) Mapping small DNA sequences by fluorescence in situ hybridization directly on banded metaphase chromosomes. Proc Natl Acad Sci USA 87:6223–6227

Klever M, Grond-Ginsbach C, Scherthan H, Schroeder-Kurth T (1991) Chromosomal in situ suppression hybridization after Giemsa banding. Hum Genet 86:484–486

Korenberg JR, Rykowsky MC (1988) Human genome organization: Alu, LINES and molecular structure of metaphase chromosome bands. Cell 53:391–400

Kuwano A, Ledbetter SA, Dobyns WB, Emanuel BS, Ledbetter DH (1991) Detection of deletions and cryptic translocations in Miller-Dieker syndrome by in situ hybridization. Am J Hum Genet 49:707–714

Landegent JE, Jansen in de Wal N, Dirks RW, Baas F, van der Ploeg M (1987) Use of whole cosmid cloned genomic sequences for chromosomal localization by non-radioactive in situ hybridization. Hum Genet 77:366–370

Lawrence JB, Singer RH, McNeil JA (1990) Interphase and metaphase resolution of different distances within the human dystrophin gene. Science 249:928–932

Lemieux N, Dutrillaux B, Viegas-Peqiugnot E (1992) A simple method for simultaneous R- or G-banding and fluorescence in situ hybridization of small single-copy genes. Cytogenet Cell Genet 59:311–312

Lengauer C, Green ED, Cremer T (1992) In situ hybridization of YAC clones after Alu-PCR amplification. Genomics 13:826—828

Lichter P, Jauch A, Cremer T, Ward DC (1990a) Detection of Down syndrome by in situ hybridization with chromosome 21 specific DNA probes. In: Patterson D (ed) Molecular Genetics of Chromosome 21 and Down Syndrom. Liss, New York, pp 69—78

Lichter P, Tang CC, Call K, Hermanson G, Evans GA, Housman D, Ward DC (1990b) High resolution mapping of chromosome 11 by in situ hybridization with cosmid clones. Science 247:64—69

Lichter P, Boyle AL, Cremer T, Ward DC (1991) Analysis of genes and chromosomes by non-isotopic in situ hybridization. Genet Anal Techn Appl 8:24—35

Manuelidis L, Ward DC (1984) Chromosomal and nuclear distribution of the HindIII 1.9-kb human DNA repeat segment. Chromosoma 91:28—38

Manuelidis L, Borden J (1988) Reproducible compartimentalization of individual chromosome domaine in human CNS cells revealed by in situ hybridization and three dimensional reconstruction. Chromosoma 96:397—410

Meltzer PS, Guan X-Y, Burgess A, Trent JM (1992) Rapid generation of region specific probes by chromosome microdissection and their application. Nature Genet 1:24—28

Moyzis RK, Albright KL, Bartholdi MF, Cram LS, Deaven LL, Hildebrand CE, Joste NE, Longmire JL, Meine J, Schwarzacher-Robinson T (1987) Human chromosome specific repetitive DNA sequences: Novel markers for genetic analysis. Chromosoma 95:375—386

Moyzis RK, Torney DC, Meyne J, Buckingham JW, Wu JR, Burks C, Sirotkin KM, Good WB (1989) The distribution of interspersed repetitive DNA sequence in the human genome. Genomics 4:273—289

Nelson DL, Ledbetter SA, Corbo L, Victoria MF, Ramirez-Solis R, Webster TD, Ledbetter DH, Caskey CT (1989) Alu polymerase chain reaction: A method for rapid isolation of human-specific sequences from complex DNA sources. Proc Natl Acad Sci USA 86:6686—6690

Raap AK, Nederlof PM, Dirks JW, Wiegant JCAG, Van der Ploeg M (1990) Use of haptenized nucleic acid probes in fluorescent in situ hybridization. In: Harris N, Williams EG (eds) In Situ Hybridization: Application to Developmental Biology and Medicine. Cambridge University Press, Cambridge, pp 33—41

Ried T, Mahler V, Vogt P, Blonden C, van Ommen GJB, Cremer T, Cremer M (1990) Direct carrier detection by in situ suppression hybridization with cosmid clones for the Duchenne/Becker muscular dystrophy locus. Hum Genet 85:581—586

Ried T, Baldini A, Rand TC, Ward DC (1992) Simultaneous visualization of seven different DNA probes by in situ hybridization using combinatorial fluorescence and digital imaging microscopy. Proc Natl Acad Sci USA 89:1388—1392

Slim R, Weissenbach J, Nguyen VC, Danglot G, Bernheim A (1991) Relative order determination of four Yp cosmids on metaphase and interphase chromosomes by two-color competitive in situ hybridization. Hum Genet 88:21—26

Smit VTHBM, Wessels JW, Mollevanger P, Schrier PI, Raap AK, Beverstock GC, Cornelisse CJ (1990) Combined GTG-banding and nonradioactive in situ hybridization improves characterization of complex karyotypes. Cytogenet Cell Genet 54:20—23

Takahashi E, Hori T, O'Connell P, Leppert M, White R (1990) R-banding and nonisotopic in situ hybridization: precise localization of the human type II collagen gene (COL2A1). Hum Genet 86:14—16

Tkachuk DC, Westbrook CA, Andreeff M, Donlon TA, Cleary ML, Suryanarayan K, Homge M, Redner A, Gray J, Pinkel D (1990) Detection of bcr-abl fusion in chronic myelogeneous leukemia by in situ hybridization. Science 250:559—562

Trask BJ, Massa H, Kenwrick S, Gitschier J (1991) Mapping of human chromosome Xq28 by two-color fluorescence in situ hybridization of DNA sequences to interphase cell nuclei. Am J Hum Genet 48:1—15

Trautmann U, Leuteritz G, Senger G, Claussen U, Ballhausen WG (1991) Detection of APC region-specific signals by nonisotopic chromosomal in situ suppression (CISS)-hybridization using a microdissection library as a probe. Hum Genet 87:495—497

Tucker JD, Christensen ML, Carrano AV (1988) Simultaneous identification and banding of human chromosome material in somatic cell hybrids. Cytogenet Cell Genet
48:103–106

Vogel W, Autenrieth M, Speit G (1986) Detection of bromodeoxyuridine-incorporation
in mammalian chromosomes by a bromodeoxyuridine-antibody. I. Demonstration of
replication patterns. Hum Genet 72:129–132

Ward DC, Lichter P, Boyle A, Baldini A, Menninger J, Ballard SG (1991) Gene mapping by fluorescent in situ hybridization and digital imaging microscopy. In: Lindsten
J, Petterson U (eds) Etiology of human diseases at the DNA level. Raven, New York,
pp 291–303

Willard HF, Waye JS (1987) Hierachical order in chromosome-specific human alpha
satellite DNA. Trends in Genet 3:192–198

Wienberg J, Stanyon R, Jauch A, Cremer T (1992) Homologies in human and Macaca
fuscata chromosomes revealed by in situ suppression hybridization with human
chromosome specific DNA libraries. Chromosoma 101:265–270

Yunis JJ (1976) High resolution mapping of human chromosomes. Science
191:1268–1270

21.5 Probe Labeling and Hybridization in One Step

J. KOCH

21.5.1 Principle and Applications

Primed in situ labeling (PRINS) is a technique based on the sequence
specific annealing of unlabeled DNA probes to target sequences in situ.
This induces site specific synthesis of labeled DNA catalyzed by a
polymerase utilizing the probe as primer and the target as template.
PRINS is a remarkably simple, fast and sensitive technique, which has
already proven useful in the analysis of minute substructures in repeated
DNA and the organization of DNA sequences in the centromere, in
studies on gene expression, and in the study of chromosome aberrations
occurring in clinical genetics (Gosden et al., 1991; Hindkjær et al., 1991;
Koch et al., 1989; Mitchell et al. 1992; Moens and Pearlman 1990a, 1990b,
1991; Winterø et al., 1992). Furthermore the technique holds promise for
the localization of cloned cDNA sequences, for the investigation of mis-
spliced or mutated mRNA sequences, and for the investigation of single
copy sequences in interphase nuclei.

Originally the technique was designed to give as much signal with oligonucleotide probes as is obtained with cloned probes, thus making it possible to exploit the high discriminatory power of the oligonucleotide probes
under conditions of optimal sensitivity (Koch et al., 1989). This goal was
achieved and it appeared that a number of additional advantages were also
obtained. One of these relates to the fact that the probe is unlabeled, which
means that nonspecific sticking of the probe to the specimen does not
result in increased background staining. It is therefore possible (and

advantageous) to use very high concentrations of probe. As a consequence, PRINS is much faster than in situ hybridization. In fact, with the newly introduced fluorochrome labeled nucleotides it is possible to detect repeated sequences in chromosomes in a one-step reaction of a few minutes (Gosden and Hanratty 1992; Koch et al., 1992). Another advantage of the technique is a better preservation of structure in chromosomes, cells, and tissues. This probably to some extent is due to the speed of the reaction, but may also be related to some stabilizing effect of the enzyme buffer and the exclusion of formamide from most of the protocols.

Since the original invention of the technique we have developed approaches for the study of DNA as well as RNA and for the study of cells fixed to a solid support as well as cells in suspension (Hindkjær et al., 1991; Koch et al., 1989, 1991, 1992; Mogensen et al., 1991). However, the technique has been most intensly applied to the study of repeated DNA sequences in methanol/acetic acid fixed spreads of metaphase chromosomes and nuclei. Therefore protocols given here are for this application. Protocols for other applications are available from the author upon request. The reaction is depicted in the accompanying schema.

Prepare chromosome spreads

add probe (primer) and enzyme under denaturing conditions

transfer to annealing

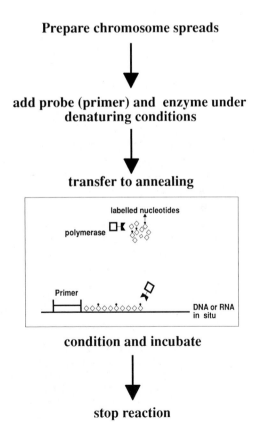

condition and incubate

Reaction scheme of PRINS

stop reaction

21.5.2

21.5.2 Procedure

Reagents − Oligonucleotides (synthesized on an Applied Biosystem DNA synthesizer and used without any special purifications).
− DIG-[11]-dUTP, fluorescein-[12]-dUTP, hydroxycoumarin-[6]-dUTP, dATP, dCTP, dGTP, dTTP, ddATP, ddCTP, ddGTP and ddTTP (Boehringer Mannheim).
− Bio-[11]-dUTP (Sigma).
− Rhodamine-dUTP (Amersham).
− Klenow and *Taq* DNA polymerase (Boehringer Mannheim).
− Amplitaq (Perkin Elmer Cetus).
− T4 DNA ligase (Boehringer Mannheim).
− Restriction enzyme *Dde*I (Boehringer Mannheim).
− FITC-avidin DCS and biotinylated anti-avidin antibody (Vector Laboratories).
− Anti-DIG fluorescein, Fab fragment (Boehringer Mannheim).
− Antifade (Sigma).
− All other chemicals are standard reagents of analytical quality.

Equipment − Two thermostate controlled heating devices (one for denaturation and one for annealing/chain elongation) − suitable instrumentation includes incubators and metal containers in a waterbath and heating blocks.
− A standard fluorescence microscope − if a confocal scanning laser microscope (CLSM, Leica) is available this can be used to increase sensitivity and store images.

Solutions *PRINS and DNA with short oligonucleotide probes*
− Reaction mixture: 2−400 pmol oligonucleotide, 100 μM each of dATP, dCTP and dGTP plus labeled dUTP (either 50 μM of either bio-[11]-dUTP or fluorochrome labeled dUTP (choice of blue (hydroxycoumarin), green (fluorescein), or red (rhodamine) label) or 35 μM DIG-[11]-dUTP supplemented with 65 μM dTTP[1] in 25 μM nick translation (NT) buffer. (NT buffer: 50 mM Tris-HCl, pH 7.2; 10 mM $MgSO_4$; 100 μM dithiothreitol; 50 μg/ml bovine serum albumin). The reaction mixture is most easily prepared from 10× stocks of NT buffer and mixed nucleotides. The volume is adjusted with TE buffer or ddH_2O.
− Stop buffer: 100 ml 50 mM EDTA, pH 8.0; 50−500 mM NaCl.
− ddNTP blocking solution: 50 μM each of ddATP, ddCTP, ddGTP and ddTTP in 25 μl NT buffer.

---

[1] Originally we worked with bio-[11]-dUTP and most of the work has been performed with this nucleotide analogue. However, it appears that DIG-[11]-dUTP gives far more sensitivity. If signals are very strong it is possible to use nucleotide analogues directly labeled with fluochromes. This, of course, simplifies the reaction, but be aware of the lowered signal intensity. Of the directly labeled nucleotides fluorescein-[12]-dUTP is the most sensitive.

PRINS of DNA with denatured double-stranded probes (or long oligonuc-leotide probes)

– Reaction mixture: 100 ng-12 μg PCR product or cloned DNA cut into short pieces (50–200 bases) with a suitable restriction enzyme such as *Dde*I or 100 ng-2 μg oligonucleotide in 50 μl *Taq* buffer (according to supplier of *Taq* DNA polymerase) containing dATP, dCTP, and dGTP to a final concentration of 100 μM each plus labeled dUTP (either 50 μM of either bio-[11]-dUTP or fluorochrome labeled dUTP (choice of blue (hydroxycoumarin), green (fluorescein), or red (rhodamine) label) or 35 μM DIG-[11]-dUTP supplemented with 65 μM dTTP (as above). The reaction mixture is most easily prepared from 10× stocks of *Taq* buffer and mixed nucleotides. The volume is adjusted with TE buffer or ddH$_2$O.

– Stop buffer is as above.

– ddNTP blocking solution: 50 μM each of ddATP, ddCTP, ddGTP and ddTTP in 50 μl *Taq* buffer.

Visualization of biotin in situ

– BN buffer: 100 mM NaHCO$_3$, pH 8.0; 0.01 % Nonidet P-40.

– blocking solution: 5 % powdered nonfat milk in BN buffer.

PRINS of DNA with short oligonucleotide probes (probes that anneal at or below 55 °C) (Koch et al. 1989)   **Detailed protocol**

– Denature chromosomal DNA by incubating in 100 ml 70 % deionized formamide, 2 × SSC at 70 °C for 2 min.

– Dehydrate immediately through an ice cold ethanol series [70 %−90 %−99 % ethanol (v/v)] and air dry.
  (At this stage, the slide can be stored dry until further use).

– Preincubate denatured and dehydrated metaphase spreads at the anne-aling temperature for 20−30 min (excess time can be shortened).

– Coverslips and reaction mixture are likewise preheated.

– After the preincubation 1 U Klenow polymerase is added to the reac-tion mixture which is rapidly transferred to the slide and spread with a coverslip.

– Polymerization is performed for 15−20 min at the annealing tempera-ture, strong reactions being visible within seconds, while weak reac-tions may require incubation for 1−3 h. Terminate the reaction by washing in stop buffer at 65 °C for 1−5 min.

– Transfer the slide to BN buffer.

PRINS of DNA with denatured double-stranded probes (or long oligonuc-leotide probes) (Koch et al., 1991)

This procedure also works with oligonucleotides that anneal below 55 °C, but at reduced efficiency due to the temperature requirements of the *Taq* DNA polymerase. Prior to use in PRINS cloned probes are digested with the restriction enzyme *Dde*I (2.5−3.0 U/μg DNA for 2−3 h at 37 °C). A sample of the digested DNA is analyzed by electrophoresis in a neutral agarose gel to check that the digestion is complete.

- Preheat the slide and the coverslip to 94°C for 10 min (as above).
- Preheat the reaction mixture to 94°C for 5 min (as above).
- Add 1 U of *Taq* DNA polymerase.
- Apply the mixture on the slide, spread with a preheated coverslip and transfer to a moisture chamber at 55−78°C (depending on the probe).
- After 15−30 min of incubation (as above), terminate the reaction by washing for 5 min at the annealing temperature in stop buffer.
- Transfer the slide to BN buffer.

Visualization of biotin in situ
- Slides equilibrated in BN buffer are incubated for 5 min at room temperature (RT) with 50 μl blocking solution under a plastic coverslip (cut blocking from a chateque), and then under the same coverslip with FITC avidin diluted in blocking solution (5 ng/μl) for 10 min at RT followed by wash in BN buffer for 3 × 5 min.
- Insufficient signals can be amplified as follows: Incubation with biotinylated anti-avidin antibody diluted in blocking solution (5 ng/μl) for 10 min at RT followed by wash in BN buffer for 3 × 5 min and incubation with FITC avidin as above.

Dig-[11]-dUTP is visualized with fluorochrome labeled anti-digoxigenin antibody as recommended by the supplier. Incubation times may be adjusted from 30 min to 24 h depending on whether the signals are expected to be strong or weak.

For *microscopy*, stained slides are mounted in antifade solution (1 μg/μl p-phenylenediamine dihydrochloride in glycerol/PBS (9/1)) with the pH adjusted to between 8 and 11.

If counterstaining is desired, one of the following can be used. Blue counterstaining: stain in Hoechst either prior to mounting or by including the stain in the antifade solution (3 parts antifade and 1 part Hoechst), green counterstaining (Q banding): stain in quinacrine mustard prior to mounting and red counterstaining: include propidium iodide (1 μg/ml) in the antifade solution.

Results can be evaluated either by standard fluorescence microscopy or by confocal laser scanning microscopy (CLSM).

21.5.3 Special Hints for Application and Troubleshooting

Some series of slides work best if postfixed (after spreading) in 3% paraformaldehyde for 2 min at RT, followed by dehydration through an ethanol series. This treatment improves the preservation of the chromatin, especially in older preparations.

Nicks in the chromosomal DNA may cause endogenous labeling of the chromosomes. With most slides this is not detectable. However, with a few (especially older) slides, it may be useful to take this phenomenon into account. One way of dealing with the problem is by closing the nicks with

.T4 DNA ligase (1 U/slide in 50 μl ligase buffer, 1 h at RT). Another way is to incubate for 15 min with ddNTP blocking solution and 1 U of the appropriate DNA polymerase, which will make the 3′ ends in the nicks unable to sustain chain elongation. Both treatments are followed by incubation in stop buffer and dehydration through an ethanol series.

One problem associated with the use of denatured double-stranded probes as primers for a PRINS reaction is that they may reassociate in solution. This will cause endogenous labeling of the probe DNA if sticky ends are generated. Such probe labeling may lead to background staining and may also reduce the specific staining through competition for reagents. From studies with incorporation of radiolabeled nucleotides and subsequent precipitation with trichloroacetic acid (TCA), we have found that cloned DNA digested with *Dde*I produces an amount of endogenous labeling that corresponds to <10% of the labeling obtained when the same probe is labeled by nick translation. Making the probe blunt-ended with Klenow polymerase and closing nicks with T_4-DNA ligase reduce this figure to 1%. This reduction of an already low figure is not significant enough to warrant this extra treatment when repeated DNA is detected. Still, it may help when attempting to detect single copy sequences.

Typical Examples

The two illustrations depict typical examples. In the first, there is staining of centrometric alpha satellite DNA by PRINS with 100 ng PCR product specific for chromosome 11, labeling with bio-[11]-dUTP in 30 min chain elongation at 70 °C, visualization with one layer of FITC-avidin, and counterstaining with propidium iodide. In the second there is staining of telomeric repeat sequences by PRINS with 2 μg 42 base oligonucleotide ((TAACCC)$_7$), labeling with DIG-[11]-dUTP in 1 h chain elongation at

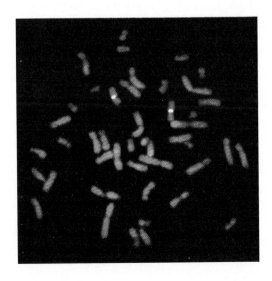

21.5.3

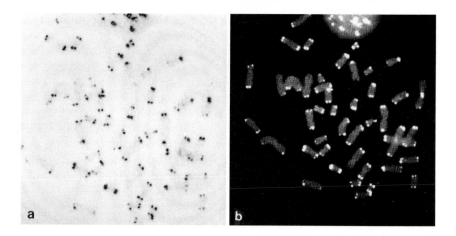

55 °C, and staining with anti-digoxigenin fluorescein. Parts a and b show the appearence without counterstaining and if counterstained with pro-pidium iodide. Note that not all chromosome ends are stained. In each spread there are metaphases where all ends are labeled; however, on the average only about 90% of ends are stained. Note also that counterstaining may hide signals.

References

Gosden J, Hanratty D, Starling J, Mitchell A, Porteous D (1991) Oligonucleotide-primed in situ DNA synthesis (PRINS): a method for chromosome mapping, banding, and investigation of sequence organization. Cytogenet Cell Genet 57:100–104

Gosden J, Hanratty D (1992) Comparison of sensitivity of three haptens, in the PRINS (oligonucleotide PRimed IN Situ synthesis) reaction. Cytogenet Cell Genet (in press)

Hindkjær J, Koch J, Mogensen J, Pedersen S, Fischer H, Nygard M, Junker S, Greger-sen N, Kølvraa S, Bolund L (1991) In situ labeling of nucleic acids for gene mapping, diagnostics and functional cytogenetics. Biotech Forum Europe 12:752–756

Koch J, Kølvraa S, Corneliussen M, Gregersen N, Petersen KB, Bolund L (1988) Treat-ment of genomic DNA with T4 DNA ligase improves Southern blot analysis. Nucleic Acids Res 16:10387

Koch J, Kølvraa S, Gregersen N, Bolund L (1989) Oligonucleotide-priming methods for the chromosome-specific labeling of alpha satellite DNA in situ. Chromosoma 98:259–265

Koch J, Hindkjær J, Mogensen J, Kølvraa S, Bolund L (1991) An improved method for chromosom-specific labeling of alpha satellite DNA in situ using denatured double stranded DNA probes as primers in a PRimed IN Situ labeling (PRINS) procedure. GATA 8:171–178

Koch J, Mogensen J, Pedersen S, Fischer H, Hindkjær J, Kølvraa S, Bolund L (1992) Fast one step procedure for the detection of nucleic acids in situ by primer induced sequence specific labeling with fluorescein-12-dUTP. Cytogenet Cell Genet 60:1–3

Mitchell A, Jeppesen P (1992) The organization of repetitive DNA sequence on human chromosomes with respect to the kinetochore analyzed using a combination of oligonucleotide primers and CREST anticentromere serum. Chromosoma (in press)

Moens PB, Pearlman RE (1990) In situ DNA sequence mapping with surface-spread mouse pachytene chromosome. Cytogenet Cell Genet 53:219–220

Moens PB, Pearlman RE (1990) Telomere and centromere DNA are associated with the cores of meiotic chromosomes. Chromosoma 100:8−14

Moens PB, Pearlman RE (1991) Visualization of DNA sequences in meiotic chromosomes. Methods in Cell Biology 35:101−108

Mogensen J, Kølvraa S, Hindkjær J, Petersen S, Koch J, Nygard M, Jensen T, Gregersen N, Junker S, Bolund L (1991) Nonradioactive detection of mRNA subspecies in situ by PRimed IN Situ labeling (PRINS). Eptl Cell Res 196:92−98

Winterø AK, Fredholm M, Thomsen PD (1992) Variable $(dGdT)_n \cdot (dC\text{-}dA)_n$ sequences in the porcine genome. Genomics 12:281−288

21.6 Multiple Fluorescence In Situ Hybridization for Molecular Cytogenetics

ANTON K. RAAP, JOOP WIEGANT, and PETER LICHTER

21.6.1 Principles and Applications

Nucleic acid probes hybridized in situ to nucleic acid targets of chromosome or cell preparations can be detected nonradioactively in several ways. (for reviews, see Lichter et al., 1991; Mc Neil et al., 1991; Raap et al., 1990a, b). Nonradioactive detection has a number of advantages over radioactive detection using microautoradiography, e.g., a higher spatial resolution of the hybridization signals. A distinct advantage of nonradioactive in situ hybridization lies in its potential to detect and differentiate multiple targets simultaneously within the same specimen. In order to achieve simultaneous visualization, for each in situ hybridization probe, different reporter molecules are selected to identify the sites of hybridization. Each kind of reporter molecule yields different colors, thereby permitting multiple in situ hybridization (Hopman et al., 1986; Nederlof et al., 1989, 1990; Wiegant et al., 1991a). For multiprobe/multicolor detection, fluorescence microscopy is the method of choice, as it permits sensitive detection of various fluorochromes with good spectral separation (Ploem and Tanke, 1987).

The principle of in situ hybridization is largely identical to that of other solid phase hybridization techniques described in this book. The main difference lies in the fact that hybrids are detected in morphologically preserved structures such as fixed chromosomes, isolated interphase nuclei, cells, or tissue sections. This special feature can be a source of difficulties as the hybridization and the detection reactions have to take place in a fixed matrix of high density. For instance, the high molecular weight reagents may bind nonspecifically to the fixed chromatin and the reagents may not have access to the nucleic acid targets, necessitating pretreatments which potentially lead to loss of morphological information. Quality control of the fixation and pretreatment steps cannot be performed independently, which is in contrast to the quality control of DNA labeling, immunological detection reagents, and microscope equipment.

For molecular cytogenetics, however, protocols have emerged in recent years that give a good compromise between detection sensitivity and preservation of morphology. These protocols can be readily applied to DNA mapping (see, e.g., Albertson, 1987; Brandiff et al., 1991; Cherif et al., 1990; Kievits et al., 1990; Landegent et al., 1985, 1987; Lawrence et al., 1990; Lichter et al., 1990; Trask et al., 1989, 1991; Wiegant et al., 1991b) and the detection of structural and numerical chromosome aberrations in meta- and interphase cells (see, e.g., Arnoldus et al., 1990; 1991; Cremer et al., 1988; Dauwerse et al., 1990; Hopman et al., 1991; Lichter et al., 1988b; Pinkel et al., 1988; Ried et al., 1990, 1991; Tkatchuk et al., 1991).

21.6.2 Reaction Scheme

A general in situ hybridization protocol for analyzing cytogenetic preparations consists of a number of steps. The standard scheme is shown in the figure:

Labeling of DNA probes by nick translation with biotin-, digoxigenin-, or fluorescein-dUTP

↓

Preparation of hybridization solution

↓

Pretreatment of specimen

↓

Denaturation of target and probe DNA

↓

In situ hybridization

↓

Posthybridization washings

↓

Immunocytochemistry

↓

Fluorescence microscopy

21.6.3 Multiple Fluorescence In Situ Hybridization

We will describe in detail: (a) probe labeling using nick translation, (b) preparation of hybridization solutions, (c) pretreatment, (d) in situ hybridization, and (e) immunological detection procedures for multiple fluorescence in situ hybridization using fluorescein-12-, digoxigenin-11-, and biotin-11-dUTP as nucleic acid labels. A distinction will be made between low-complexity probes recognizing highly repetitive sequences e.g., the cloned α satellite DNAs (Willard and Way, 1987) and high-complexity genomic probes such as DNAs cloned in plasmid, phage, cosmid, or yeast artificial chromosome vectors, or probe sets such as DNA libraries from chromosomes or chromosome subregions (Cremer et al., 1988; Fuscoe et al., 1989; Lengauer et al. 1991; Lichter et al., 1988; Pinkel et al., 1988). The high complexity probes need a preannealing step with unlabeled total human DNA or Cot-1 DNA to eliminate the labeled, dispersely occurring repeat sequences (mainly *Alu* repeats) present in such probes from participation in the situ hybridization (Landegent et al., 1987; Lichter et al., 1988; Pinkel et al., 1988). For small unique plasmid probes, the protocol for high-complexity probes should be used, with the exception that preannealing of the probe with competitor DNA can be ommitted.

It should be stressed that the multiplicity of in situ hybridization (i.e., the number of simultaneously detectable probes) is currently not limited by the number of available hapten or fluorochrome DNA modifications but by the number of fluorochromes that can be selected with high efficiency in the fluorescence microscope. However, by double or triple labeling probes, they can, after in situ hybridization, be identified on the basis of their color composition (Nederlof et al., 1990; Ried et al., 1992; Wiegant et al., submitted). Probes may either be labeled separately with biotin-, fluorescein-, and/or DIG-dUTP in the nick translation reaction and mixed in the in situ hybridization solution or the same probe may be labeled with two or three modified nucleotides in the same reaction. It is also possible to identify probes on the basis of color intensity ratio, giving yet another way to expand the multiplicity of in situ hybridization (Nederlof et al., 1990; Nederlof et al., 1992).

It should be emphasized that in situ hybridization is a multistep procedure and that the quality of the end result is determined by the weakest link in the chain of procedural steps. In this respect, the role of fluorescence microscopy should not be underestimated. [See Ploem and Tanke (1987) and James and Tanke (1990) for practical information on fluorescence microscopy.]

Cytogenetic specimens that can be used with the procedures are the standard meta- and prophase chromosomes from any source, e.g., bone marrow, blood, amnion fluid, or tumor cell cultures. For interphase cytogenetics of cells isolated from (solid) tumors, reliable protocols have been established (see, e.g., Arnoldus et al., 1991b; Hopman et al., 1991). Hybridization to tissue sections is so far limited to highly efficient probes, e.g., the alphoid repetitive DNAs (see, e.g., Arnoldus et al., 1991b; Emmerich et al., 1989; Hopman et al. 1991b).

Nick translation, pretreatment, and in situ hybridization reagents

− DNA polymerase I (Promega, New England Biolabs)
− DNase I (Boehringer Mannheim)
− BSA fraction V (Boehringer Mannheim)
− dATP, dCTP, dGTP, dTTP (Boehringer Mannheim)
− Bio-[11]-dUTP (Sigma)
− Fluorescein-[12]-dUTP (Boehringer Mannheim)

- DIG-[11]-dUTP (Boehringer Mannheim)
- Salmon sperm DNA (Sigma)
- Carrier yeast RNA (Sigma)
- Competitor DNA: human placental DNA, Cot-1 fraction of human DNA (Sigma, Life Technologies/BRL)
- Dextran sulfate (Pharmacia)
- RNase A (Boehringer Mannheim)
- Pepsin (Serva)
- Formamide (Baker, Merck, Aldrich, Serva)
- Amberlite MB1, ion exchanger (Serva)
- Rubber cement (Simson)
- Acid-free formaldehyde (37%) (Merck)

Immuno-cyto-chemistry reagents
- Blocking reagent (Boehringer Mannheim)
- Nonfat dry milk (Carnation)
- Tween-20 (Sigma)
- Avidin.D, avidin.D:FITC, avidin.D:TRITC, avidin.D:AMCA (Vector Laboratories)
- Goat anti-avidin.D:biotin (Vector Laboratories)
- Sheep anti-DIG (Boehringer Mannheim)
- Mouse anti-DIG (monoclonal) (Sigma)
- Sheep anti-mouse:DIG (Boehringer Mannheim)
- Sheep anti-DIG:FITC (Boehringer Mannheim)
- Sheep anti-DIG:TRITC (Boehringer Mannheim)
- Donkey anti-sheep:FITC (Chemicon, Sigma)
- Rabbit anti-mouse:FITC (Sigma)
- Rabbit anti-mouse:TRITC (Sigma)
- Goat anti-rabbit:FITC (Sigma)
- Goat anti-rabbit:TRITC (Sigma)
- Goat-anti-mouse:FITC (Sigma)
- Goat anti-mouse:TRITC (Sigma)
- Rabbit anti-goat:TRITC (Sigma)
- Rabbit anti-FITC (Dako[patts])
- Sheep anti-mouse:FITC (Boehringer Mannheim)
- Goat anti-mouse:TRITC (Sigma)
- Rabbit anti-sheep:FITC (Sigma)
- DABCO (1,4-diazabicyclo [2,2,2]-octane) (Sigma)
- PI (propidium iodide) (Sigma)
- DAPI (4,6-diamidino-2-phenylindol · 2HCl) (Serva)

Note: Some companies supply antibodies in the form of solutions without specifying the antibody concentration. The concentrations and dilutions given by us are guidelines.

Standard solutions
- 10× Nick translation buffer: 0.5M Tris-HCl, pH 7.8; 50mM $MgCl_2$; 0.5mg/ml BSA (nuclease free); store at −20°C.
- Nucleotide mix: 0.5mM dATP; 0.5mM dCTP; 0.5mM dGTP; 0.1mM dTTP; store at −20°C.

- RNase A: 10 mg/ml in 2× SSC.
- Pepsin: 10 mg/100 ml 0.01 N HCl; prepare fresh from 10% stock.
- 1% Formaldehyde in 1× PBS containing 50 mM $MgCl_2$.
- 10× PBS pH 7.0: 1.4 M NaCl, 0.1 M sodium phosphate.
- 20× SSC: 3.0 M NaCl, 0.3 M Na-citrate, pH 7.0.
- Deionized formamide for hybridization.
- Commercial formamide for washings with formamide/SSC solutions.
- DABCO-PI/DABCO-DAPI embedding media:
  1. Dissolve 2 g DABCO (an antifading reagent) in 90 ml glycerol for 15−30 min at 60 °C.
  2. Add 10 ml 1.0 M Tris-HCl pH 7.5.
  3. Adjust the pH to 8.0 with a few drops of 5 M HCl.
  4. Cool to room temperature.
  5. Add 100 µl 20% thimerosal (in H_2O).
  6. Add either 50 µl PI (stock solution: 1 mg/ml PI in H_2O) or 15 µl DAPI solution (stock solution: 1 mg/ml DAPI in H_2O).
  7. Store at 4 °C.

1. Prepare the labeling solution on ice as follows:

   Probe labeling by nick translation

− filtered distilled water	26 µl
− nick translation buffer (10×)	5 µl
− DTT (0.1 M)	5 µl
− nucleotide mix (0.5 mM dATP; dGTP; dCTP; 0.1 mM dTTP)	4 µl
− biotin-11-dUTP (1 mM)	2 µl
− probe DNA (1 µg/µl)	1 µl
− DNA polymerase I (10 units/µl)	2 µl
− DNase I (1 : 1000 from a 1 mg/ml stock solution)	5 µl

Note that for other probe DNA concentrations, the probe DNA and water volumes can be balanced against each other.

For labeling with digoxigenin or fluorescein, replace the biotin-11-dUTP by 2 µl digoxigenin-11-dUTP (1 mM) or fluorescein-12-dUTP (1 mM).
2. Incubate for 2 h at 15 °C.
3. To check fragment length of the probe, the following can optionally be done:
   - Put the reaction at 0 °C.
   - Take a 6 µl aliquot from the reaction.
   - Denature at 100 °C for 5 min, chill on ice for 3 min.
   - Run sample on an agarose minigel along with a DNA size marker.
   The denatured probe should range between 100 and 500 nucleotides in length.
   If necessary, add more DNase, incubate longer, and check fragment size again.
4. Add 1 µl 0.5 M EDTA, pH 8, to stop the reaction.
5. Put on ice until further use.

6. Purify the labeled probe by gel filtration using a Pasteur pipet column (150 mm) and Sephadex G50 (fine) equilibrated with TE buffer and previously flushed with 50 μg of salmon sperm DNA and 50 μg of yeast RNA. The labeled DNA elutes in the 2nd 600 μl.

7. Add 50 μg salmon sperm DNA and 50 μg yeast RNA and mix well; high complexity probes can be stored at −20°C at this stage.

Preparation of hybridization solutions

Low complexity probes recognizing repetitive targets:

1. Add 0.1 volumes 3 M sodium acetate (pH 5.5) to the probe solution and mix gently.

2. Add 2.5 volumes 100% ethanol (−20°C). Mix well and place for 30 min on ice.

3. Spin down for 30 min at 4°C in an Eppendorff centrifuge.

4. Remove the supernatant dry the pellet, and dissolve the pellet in 60% deionized formamide/2× SSC/50 mM sodium phosphate, pH 7, to a final labeled probe DNA stock solution of 10 ng/μl. Store at 4°C or −20°C.

5. Typical concentrations for in situ hybridization are 1−2.5 ng/μl. For multiple in situ hybridization, probes can simply be mixed.

High complexity probes, including preannealing to unlabeled competitor DNA:

1. Take the required amount of DNA(s) from the purified stock DNA in TE (final probe DNA concentrations, e.g., 2 ng/μl per cosmid and 10 ng/μl per Bluescript chromosome library) (Collins et al., 1991).

2. Add 250−500× excess human placental DNA or 50× excess human Cot-1 DNA as competitor DNA.

3. Add 0.1 volumes 3 M sodium acetate (pH 5.5) and mix gently.

4. Add 2.5 volumes 100% ethanol (−20°C). Mix well and place for 30 min on ice.

5. Spin down for 30 min at 4°C.

6. Remove supernatant and dissolve the pellet in 10 μl 50% deionized formamide/2× SSC/50 mM sodium phosphate/10% dextran sulfate (prewarmed to 37°C).

7. Dissolve the DNA for 15 min in a 37°C waterbath.

8. Denature the DNA for 5 min at 75°C and quench on ice for 1 min.

9. Anneal the DNA for 2 h at 37°C.

To illustrate the possible range of variations in the protocols, an alternative successful procedure is given. Differences with the above protocol are in the concentrations of carrier and competitor DNA, the temperature changes after probe denaturation, and the preannealing time.

1. Combine probe DNA(s) with typically 1−3 μg human competitor DNA (total human DNA or its Cot-1 fraction) and 7−9 μg salmon sperm DNA. The total DNA concentration in the hybridization mixture is adjusted to 1 mg/ml by the salmon DNA. When DNA probes are particular rich in *Alu* DNA, the amount of competitor DNA should be increased.

2. Precipitate probe mixture and resuspend in hybridization cocktail as described above.
3. Denature DNA for 5 min at 75°C.
4. Transfer tube immediately to 37°C to allow preannealing for 5−20 min.

These pretreatments are optional and may be tried if suboptimal results are obtained: **RNase and pepsin pretreatment of specimen**
1. Locate the area of interest on the slide using phase contrast microscopy.
2. Scribe outlines of target area for hybridization on the back of the slide with a diamond-tipped scribe.
3. Wash the slide for 5 min in 2× SSC at 37°C.
4. Apply 120 µl RNase A (100 µl/ml 2× SSC) to the slide and cover with a 24 × 60 mm^2 glass coverslip.
5. Incubate for 1 h at 37°C in a moist chamber.
6. Wash the slide 3 × 5 min in 2× SSC.
7. Wash the slide for 5 min in prewarmed 1× PBS at 37°C.
8. Place the slide for 10 min in a 10 mg pepsin/100 ml 0.01 M HCl solution in a 37°C waterbath.
9. Wash the slide for 5 min in 1× PBS at room temperature.
10. Place the slide for 10 min in a 1% acid-free formaldehyde/1× PBS/ 50 mM MgCl$_2$, solution at room temperature.
11. Wash the slide for 5 min in 1× PBS at room temperature.
12. Dehydrate the slide through an ethanol series (70%, 90%, 100% ethanol; 5 min each).
13. Air dry.

Low complexity probes for repetitive targets: **Denaturation and in situ hybridization**
1. Apply 5 µl of the diluted probe(mixture) to the target area and cover with an 18 × 18 mm^2 coverslip; sealing with rubber cement is optional.
2. Denature the slide and the probe DNA simultaneously for 2−3 min at 80°C.
3. Hybridize overnight at 37°C in a moist chamber (moistening medium: 60% formamide/2× SSC, pH 7).

High complexity probes:
1. Apply 120 µl 70% deionized formamide/2× SSC/50 mM sodium phosphate, pH 7, solution on the slide.
2. Cover the solution with a 24 × 60 mm^2 coverslip.
3. Denature the DNA 2−3 min at 80°C.
4. Remove the coverslip and place the slide directly in 70% ethanol (−20°C); wash 2 × 5 min.
5. Dehydrate the slide through an ethanol series (90% and 100% ethanol; 5 min each).
6. Place the slide on a 37°C plate and leave it there to dry.
7. Apply 10 µl of the preannealed probe(mixture) on the predenatured slide per target area and cover with a 18 × 18 mm^2 coverslip.

8. Seal the edges of the coverslip with rubber cement.
9. Hybridize overnight at 37°C in a moist chamber (moistening medium: 50% formamide/2× SSC, pH 7).

Post hy-
bridization
washes
Low complexity probes for repetitive targets:
1. Place the slide in prewarmed 60% formamide/2× SSC, pH 7, at 37°C and remove the coverslip by gently shaking.
2. Wash the slide at 37°C in three changes of 60% formamide/2× SSC, pH 7.
3. Wash 2 × 5 min with 2× SSC at room temperature.
4. Wash the slide for 5 min in the appropriate immunology buffer.

High complexity probes:
1. Prepare a 50% formamide/2× SSC, pH 7, solution and bring to 45°C.
2. Prepare a 0.1× SSC solution and bring to 60°C.
3. Remove the rubber cement on the coverslip with forceps.
4. Place the slide in prewarmed 50% formamide/2× SSC, pH 7, at 45°C and remove the coverslip by gently shaking.
5. Wash the slide 3 × 5 min in prewarmed 50% formamide/2× SSC, pH 7, at 45°C.
6. Wash the slide additionally in three changes of 0.1× SSC at 60°C.
7. Wash the slide for 5 min in the appropriate immunology buffer.

Immuno-
cyto-
chemistry
For applications demanding high detection sensitivity, immunological amplification of signals is often a requisite. Avidin and multiple antibodies from various species are used for this purpose. Cross-reactivity should be avoided. The reagents used in the protocol do not lead to perceivable cross-reaction at the immunological level. It is a general immunocytochemical rule that the ionic strength of the buffers used in detection be as high as possible to prevent nonspecific binding as much as possible. For detection of biotinylated probes, this point has been made by Lawrence et al. (1988) and is useful for the detection of small unique probes using amplification with alternate layers of biotinylated goat-anti-avidin and avidin (Pinkel et al., 1986). However, for multiple indirect detection using multiple immunocytochemical layers, the avidity of the reagents for detecting fluorescein and/or digoxigenin may be such that the antibodies do not withstand high salt concentrations, implying that combining multiple antibodies in high salt buffers does not necessarily give optimal results. The protocols given below for three color indirect fluorescence in situ hybridization uses a low salt buffer and mixed immunocytochemical reagents as much as possible. High salt buffers (e.g., 4× SSC/0.05% Tween-20), however, can be used and are recommended for unicolor detection of small (1−5 kb) unique targets with biotinylated probes.

Practical hints:
− Dilute immunocytochemical reagents in 0.1 M Tris-HCl/0.15 M NaCl/ 0.05% Tween-20 (TNT), pH 7.5, containing 0.5% blocking reagent

(Boehringer Mannheim). Alternatively, 5% nonfat dry milk may be used. Centrifuge the solution; if not dissolvable and use the supernatant.

- Use freshly diluted antibody solutions which are centrifuged for at least 1 min at 12 000 g before use.
- Each time apply 120 μl of the antibody solution to the slide and cover with a 24 × 60 mm^2 glass coverslip.
- Incubate at 37 °C; place in a humidified atmosphere using the buffer as humidifier for 30 min.
- Wash 3 × 5 min with TNT between incubations at room temperature.
- If immunocytochemistry is to FITC only, embed in DABCO/PI or DABCO/PI/DAPI (0.5 μg/ml PI and 75 ng/ml DAPI).
- If immunocytochemistry is developed to TRITC or FITC and TRITC, embed in DABCO/DAPI (75 ng/ml).
- If immunocytochemistry is developed to FITC, TRITC, and AMCA, embed in DABCO only.

Three immunological detection protocols for triple fluorescence in situ hybridization will be given: one for low complexity probes to repetitive targets; two for high complexity probes:

- Low complexity probes for repetitive targets:
  1. Avidin.D:AMCA (50 μg/ml) mixed with sheep anti-digoxigenin: TRITC (2 μg/ml)

- High complexity probes (cosmids and chromosome libraries):
  1. Avidin.D:AMCA (50 μg/ml)
  2. Goat anti-avidin.D:biotin (5 μg/ml) mixed with rabbit anti-FITC (1 : 250) and mouse anti-DIG (0.5 μg/ml)
  3. Avidin.D:AMCA (50 μg/ml) mixed with goat anti-rabbit:FITC (1 : 500) and sheep anti-mouse:DIG (2 μg/ml)
  4. Sheep anti-DIG:TRITC (2 μg/ml)

For better visualization of multiple chromosome libraries, the following is suggested to further amplify the signals after step 3:

  4. Goat anti-avidin.D:biotin (2.5 μg/ml) mixed with rabbit anti-FITC (1 : 500) and mouse anti-DIG (0.25 μg/ml)
  5. Avidin.D:AMCA (25 μg/ml) mixed with goat anti-rabbit:FITC (1 : 1000) and sheep anti-mouse:DIG (1 μg/ml)
  6. Sheep anti-DIG:TRITC (2 μg/ml)

21.6.4 Special Hints for Application and Troubleshooting

- Run simple positive procedural control in situ hybridizations (e.g., **Controls** total human DNA to cell hybrids, or alphoid DNAs to human metaphase chromosomes) to help locate the weak links in your procedures.

- Check the DNA labeling by performing a colorimetric filter spot test **No or weak** (either direct or via a hybridization). Approximately 0.1−1.0 pg of **signals** probe should be visible using alkaline phosphatase/BCIP/NBT detection (see Sect. 3).

- Check the fragment size of the labeled DNA by electrophoresis. Fragment sizes should range from approximately 200 to 600 base pairs.
- Check and, if necessary adjust the pH of the weakly buffered formamide/SSC solutions.
- Increase the intensity of the pepsin treatment.
- Increase the denaturation time.
- Check the microscope for optimal installment.
- Adapt the stringency of hybridization.
- Check and/or increase the probe concentration.

Poor morphology
- Decrease the intensity of the pepsin treatment.
- Decrease the denaturation time.

References

Albertson DG (1985) Mapping muscle protein genes by in situ hybridization using biotin-labeled probes. EMBO J 4:2493–2498

Arnoldus EPJ, Wiegant J, Noordermeer IA, Wessels JW, Beverstock GC, Grosveld GC, Van der Ploeg M, Raap AK (1990) Detection of the Philadelphia chromosome in interphase nuclei. Cytogenet Cell Genet 54:108–111

Arnoldus EPJ, Noordermeer IA, Peters ACB, Voormolen JHC, Bots GTAM, Raap AK, Van der Ploeg M (1991a) Interphase cytogenetics of brain tumors. Genes, Chromosomes and Cancer 3:101–107

Arnoldus EPJ, Dreef EJ, Noordermeer IA, Verheggen MM, Thierry RF, Peters ACB, Cornelisse CJ, van der Ploeg M, Raap AK (1991) Feasibility of in situ hybridization with chromosome-specific DNA probes to paraffin wax embedded tissues. J Clin Pathol 44:900–904

Brandiff B, Gordon L, Trask B (1991) A new system for high-resolution DNA sequence mapping in interphase pronuclei. Genomics 10:75–82

Cherif D, Julier D, Delattre O, Derre J, Lathrop GM, Berger R (1990) Simultaneous localization of cosmids and chromosome R-banding by fluorescence microscopy: application to regional mapping of chromosome 11. Proc Natl Acad Sci USA 87:6639–6643

Collins C, Kuo WL, Segraves R, Fuscoe J, Pinkel D, Gray JW (1991) Construction characterization of plasmid libraries enriched in sequence from single human chromosomes. Genomics 11:997–1000

Cremer T, Lichter P, Borden J, Ward DC, Manuelidis L (1988) Detection of chromosome aberrations in metaphase and interphase tumor cells by in situ hybridization using chromosome specific library probes. Hum Genet 80:235–246

Dauwerse JG, Kievits T, Beverstock GC, Van der Keur D, Smit E, Wessels HW, Hagemeijer A, Pearson PL, Van Ommen GJB, Breuning MH (1990) Rapid detection of chromosome 16 inversion in acute nonlymphocytic leukemia, subtype M4: regional localization of the breakpoint in 16p. Cytogenet Cell Genet 53:126–128

Emmerich P, Jauch A, Hofman MC, Cremer T, Walt H (1989) Interphase cytogenetics in paraffin embedded sections from testicular germ cell tumor xenografts and in corresponding cell cultures. Lab Invest 61:235–240

Fuscoe JC, Collins CC, Pinkel D, Gray JW (1989) An efficient method for selecting unique sequence clones from DNA libraries and its application to fluorescent staining of human chromosome 21 using in situ hybridization. Genomics 5:100–109

Hopman AHN, Wiegant J, Raap AK, Landegent JE, van der Ploeg M, van Duijn P (1986) Bicolour detection of two target DNAs by nonradioactive in situ hybridization. Histochemistry 85:1−4

Hopman AHN, Moesker O, Smeets AWGB, Pauwels RPE, Vooijs GP, Ramaekers FCS (1991) Numerical chromosome 1, 7, 9 and 11 aberrations in bladder cancer detected by in situ hybridization. Cancer Res 51:644−651

Hopman AHN, Van Hooren E, Van de Kaa CA, Vooijs GP, Ramaekers FCS (1991) Detection of numerical chromosome aberrations using in situ hybridization in paraffin sections of routinely bladder cancers. Modern Pathol 4:503−513

James J, Tanke HJ (1991) Fluorescence microscopy; in: Biomedical light microscopy, Kluwer Academic Publ., Dordrecht, the Netherlands; Chapter 3, pp50−66

Kievits T, Dauwerse JG, Wiegant J, Devilee P, Breuning MH, Cornelisse CJ, Van Ommen GJB, Pearson PL (1990) Rapid subchromosomal localization of cosmids by nonradioactive in situ hybridization. Cytogenet Cell Genet 53:134−136

Landegent JE, Jansen in de Wal N, Ommen GJB, Baas F, De Vijlder JJM, Van Duijn P, Van der Ploeg M (1985) Chromosomal localization of a unique gene by nonautoradiographic in situ hybridization. Nature 317:175−177

Landegent JE, Jansen in de Wal, Dirks RW, Baas F, van der Ploeg M (1987) Use of whole cosmid cloned genomic sequence for chromosomal localization by nonradioactive in situ hybridization. Hum Genet 77:366−370

Lawrence JB, Villnave CA, Singer RH (1988) Interphase chromatin and chromosome gene mapping by fluorescence detection of in situ hybridization reveals the presence and orientation of two closely linked copies of EBV in a human lymphoblastoid cell line. Cell 52:51−61

Lawrence JB, Singer RH, McNeil JA (1990) Interphase and metaphase resolution of different distances within the human dystrophin gene. Science 249:928−931

Lengauer C, Eckelt A, Weith A, Endlich N, Ponelies N, Lichter P, Greulich KO, Cremer T (1991) Painting of defined chromosomal regions by in situ suppression hybridization of libraries from laser-microdissected chromosomes. Cytogenet Cell Genet 56:27−30

Lichter P, Cremer T, Borden J, Manuelidis L, Ward DC (1988a) Delineation of individual human chromosomes in metaphase and interphase cells by in situ suppression hybridization using recombinant DNA libraries. Hum Genet 80:224−234

Lichter P, Cremer T, Tang CC, Watkins PC, Manuelidis L, Ward DC (1988b) Rapid detection of human chromosome 21 aberrations by in situ hybridization. Proc Natl Acad Sci USA 85:9664−9668

Lichter P, Tang CC, Call K, Hermanson G, Evans G, Housman D, Ward DC (1990) High resolution mapping of human chromosome 11 by in situ hybridization with cosmid probes. Science 247:64−69

Lichter P, Boyle AL, Cremer T, Ward DC (1991) Analysis of genes and chromosomes by nonisotopic in situ hybridization. Genet Anal Techn Appl 8:24−35

McNeil JA, Johnson CV, Carter KC, Singer RH, Lawrence JB (1991) Localizing DNA and RNA within nuclei and chromosomes by fluorescence in situ hybridization. Genet Anal Techn Appl 8:41−58

Nederlof PM, Robinson D, Abuknesha R, Wiegant J, Hopman AHN, Tanke HJ, Raap AK (1989) Three colour fluorescence in situ hybridization for the simultaneous detection of multiple nucleic acid sequences. Cytometry 10:20−27

Nederlof PM, van der Flier S, Wiegant J, Raap AK, Tanke HJ, Ploem JS, Van der Ploeg M (1990) Multiple fluorescence in situ hybridization. Cytometry 11:126−131

Nederlof PM, van der Flier S, Vrolijk J, Tanke HJ, Raap AK (1992) Quantification of in situ hybridization signals by fluorescence digital imaging microscopy. II. Fluorescence ratio measurements of double labeled probes. Cytometry (in press)

Pinkel D, Landegent J, Collins C, Fuscoe J, Segraves R, Lucas J, Gray J (1988) Fluorescence in situ hybridization with human chromosome specific libraries: detection of trisomy 21 and translocation of chromosome 4. Proc Natl Acad Sci 85:9138−9142

Pinkel D, Straume T, Gray JW (1986) Cytogenetic analysis using quantitative high sensitivity fluorescence hybridization. Proc Natl Acad Sci USA 83:2934−2938

Ploem JS, Tanke HJ (1987) Introduction to fluorescence microscopy. In: RMS Micro-scopy Handbooks Series No. 10, Oxford Science Publications

Raap AK, Dirks RW, Jiwa NM, Nederlof PM, Van der Ploeg M (1990a) In situ hybridi-zation with hapten-modified DNA probes. In: Racz P, Haase AT, Gluckman JC (eds) Modern Pathology of AIDS and Other Retroviral Infections, Karger, Basel, pp 17–28

Raap AK, Nederlof PM, Dirks RW, Wiegant JCAG, Van der Ploeg M (1990b) Use of haptenized nucleic acid probes in fluorescent in situ hybridization. In: Harris N, Williams DG (eds) In Situ Hybridization: Application to Developmental Biology and Medicine, Cambridge University Press, Cambridge, pp 33–41

Ried T, Mahler V, Vogt P, Blonden L, van Ommen GJB, Cremer T, Cremer M (1990) Carrier detection by in situ suppression hybridization with cosmid clones of the Duchenne/Becker muscular dystrophy (DMD/BMD)-locus. Hum Genet 85:581–586

Ried T, Baldini A, Rand T, Ward DC (1992) Simultaneous visualization of seven differ-ent DNA probes by in situ hybridization using combinatorial fluorescence and digitial imaging microscopy. Proc Natl Acad Sci USA, in press

Ried T, Lengauer C, Cremer T, Wiegant J, Raap AK, van der Ploeg M, Groitl P, Lipp M (1991) Specific metaphase and interphase detection of the breakpoint region in 8q24 of Burkitt lymphoma cells by triple colour fluorescence in situ hybridization. Genes Chromosomes and Cancer 4:1–6

Tkatchuk D, Westbrook C, Andreef M, Donlon TA, Cleary ML, Suryanarayan K, Homge M, Redner A, Gray JW, Pinkel D (1990) Detection of BCR-ABL fusion in chronic myeologeneous leukemia by two colour fluorescence in situ hybridization Science 220:559–562

Trask B, Pinkel D, Van den Engh G (1989) The proximity of DNA sequences in interph-ase nuclei is correlated to genomic distance and permits ordering of cosmids spanning 250 kilobase pairs. Genomics 5:710–717

Trask BJ, Massa H, Kenwrick S, Gitschier J (1991) Mapping of human chromosome Xq28 by 2-colour fluorescence in situ hybridization of DNA sequences to interphase cell nuclei. Am J Hum Genet 48:1–15

Wiegant J, Ried Th, Van der Ploeg M, Nederlof PM, Tanke HJ, Raap AK (1991a) In situ hybridization with fluoresceinated DNA. Nucleic Acids Res 19:3237–3241

Wiegant J, Galjart N, Raap AK, d'Azzo A (1991b) The gene encoding human protective protein is on chromosome 20. Genomics 10:345–349

Wiegant J, Wiesmeijer CC, Hoovers J, Schuuring E, d'Azzo A, Vrolijk J, Tanke HJ, Raap AK (1992) Sensitive and multiple in situ hybridization with rhodamine-, fluores-cein- and coumarin-labeled DNAs, submitted

Willard HF, Waye JS (1987) Hierarchical order in chromosome-specific human alpha satellite DNA. Trends Genet 3:192–198

21.7 Mapping of Polytene Chromosomes

ERWIN R. SCHMIDT

21.7.1 Principle and Applications

Polytene chromosomes consist of up to several thousands of chromatids and are therefore especially suitable for direct mapping with the help of in situ hybridization. With the introduction of nonradioactive labeling and detection methods, e.g., fluorescence in situ hybridization (FISH) (Langer-Safer et al., 1982), in situ hybridization procedure has become easy to perform and the results can be obtained within a day. Furthermore, the method described here (Schmidt et al., 1988) is a simplified version which additionally allows for the hybridization of more than one DNA probe simultaneously. This double or multihybridization results in very precise mapping of two neighboring DNA probes, provided that these probes are differentially labeled and therefore can be detected by different colors. In this way, it has been possible to simultaneously localize two DNA sequences which were only approximately 30 kb apart from each other.

21.7.2 Mapping of Polytene Chromosomes by Multiple In Situ Hybridization

1. Salivary glands are excised from 4th instar larvae and fixed 1−2 min in 40% (v/v) acetic acid. **Preparation of polytene chromosomes**
2. The glands are transferred with a drop of acetic acid (40% v/v) to a glass slide and squashed under a cover slip according to standard procedures.
3. After squashing, freezing, and removing of the cover slip, the chromosome preparations are transferred into 100% propanol-2 and stored at −20 °C.

Prior to the in situ hybridization the DNA in the chromosomes has to be made single stranded (denatured). There are several possibilities: heat treatment, heat treatment in high concentration of formamide, exonuclease treatment (Schmidt, 1989), treatment with HCl or NaOH (Singh et al., 1977). In principle, all methods work, but the preservation of the chromosomes is different depending on the type and length of treatment. Usually, a short treatment with 0.1 N NaOH is the easiest and cheapest way. Heat stabilization before denaturation (Bonner and Pardue, 1976) helps to prevent damage to the chromosomes. **Denaturation**

1. Rehydrate the chromosomes in a decreasing series of propanol-2: 100%, 70%, 50%, 30% (v/v), 0.1× SSC, 2× SSC.
2. Incubate the preparation at least 30 min in 2× SSC at 80 °C (heat stabilization). Cool to room temperature.
3. Wash in 0.1× SSC.
4. Incubate the preparation in 0.1 N NaOH for 1 min with constant agitation.
5. Wash slides in 0.1× SSC, 30 s.
6. If required, the RNA can be digested with RNase, but this step is not necessary for routine localizations of DNA sequences.
7. Wash in 2× SSC, 30 s.
8. Dehydrate preparations through a series of increasing concentrations of propanol-2: 30%, 50%, 70%, 100% (v/v), 2 min each.
9. Dry the slides in the air (approximately 5 min).

21.7.2

The preparation is ready for hybridization. If the slides are not to be used immediately, they can be stored for several years in 100% propanol-2 at −20°C without any significant loss of hybridization efficiency.

Labeling of the probes Good results are obtained using the random primed oligolabeling procedure (Feinberg and Vogelstein, 1983). The commercially available digoxigenin DNA labeling kit (Boehringer Mannheim) is based on this method and gives routinely good labeling with a broad range of different DNA concentrations. The reaction is not very sensitive to contaminants in the DNA preparation.

1. Dissolve 0.5−1 μg linearized DNA in 15 μl A. dest (destilled water).
2. Denature in a boiling water bath for 10 min. Chill on ice.
3. Use DIG DNA labeling kit (Boehringer Mannheim) as recommended by the manufacturer (total reaction volume of 20 μl).
4. Mix, centrifuge, and incubate either 1−2 h at 37°C or leave the reaction overnight at room temperature.
5. Stop the reaction by heating 10 min in a boiling water bath; this also denatures the digoxigenated DNA. Chill on ice.
6. Add 30 μl of dH$_2$O; 49 μl of 10× SSC; 1 μl of 10% [w/v] SDS (SDS significantly reduces background) to give a volume of 100 μl, which is sufficient for the hybridization of approximately 20 slides. If more preparations are to be hybridized, the volume can be increased. The concentration of the hybridizing DNA is thus lower, which can have an effect on the intensity of the hybridization signal.

An alternatively used hybridization buffer: It might be necessary to use a more complex hybridization buffer to obtain a stronger hybridization signal or to reduce the background. A recommendable mixture is 2× SSC, 50% [v/v] formamide, 10% [w/v] dextran sulfate, 10× Denhardt's solution, 1 μg/ml carrier DNA, 0.1% [w/v] SDS. Due to the formamide, this mixture requires a different hybridization temperature.

Experience over the years has shown that it is absolutely unnecessary to remove unincorporated dNTPs. If background problems arise, then these are most probably due to other reasons.

Hybridization and detection
1. Apply 5 μl of DIG-labeled DNA in hybridization mixture onto the spot of chromosomes (air dried) and cover with coverslip (18 × 18 mm^2).
2. Seal coverslip with rubber cement.
3. Incubate slides at the appropriate temperature between 50°C and 65°C (depending on the probe, AT content, sequence homology, etc.) for 4−6 h or overnight.
4. Remove rubber cement, wash off the coverslip in 2× SSC at room temperature for 2−5 min (longer washing at higher temperature, i.e., hybridization temperature, may reduce background, if this turns out to be a problem).
5. Wash in PBS for 2 min.

6. Remove excess buffer by wiping with soft paper around the chromo-somes, but do not let the preparation dry completely (produces background and unspecific binding)!

7. Incubate with 5 μl of fluorescent dye-labeled anti-digoxigenin anti-body (Boehringer Mannheim) in a 1 : 10 dilution with PBS-BSA (1 mg BSA/ml PBS) for 30 min at room temperature under a coverslip.

8. Wash 5 min in PBS.

9. Embed the chromosome preparation in glycerol-para-phenylene diamine antifading mixture (1 mg p-phenylenediamine in 1 ml of 50% [v/v] phosphate-buffered glycerol: 1 mM Na-phosphate pH 8.0, 15 mM NaCl).

10. The result can be seen in the fluorescence microscope using the appropriate combination of filters.

21.7.3 Special Hints for Application and Troubleshooting

— Instead of antibodies labeled with fluorescent dyes all other types of labeled antibodies (enzyme-conjugated, gold-labeled followed by silver enhancement, mouse monoclonal anti-digoxigenin antibodies followed by labeled secondary antibodies, etc.) may be used. **Alternative labeling**

— For special purposes it is sometimes helpful to hybridize two or more different probes simultaneously with one and the same preparation. This works excellent with the method described here. The only pre-requisite is different labeling of the different probes to be localized, for example, probe 1 is digoxigenin (DIG) labeled and detected with TRITC-labeled anti-DIG antibody; probe 2 is labeled by the incorpora-tion of FITC-dUTP (Boehringer Mannheim), which can be detected directly without any treatment after the hybridization. Thus one probe gives a red and the other a green signal. Many other combinations are possible (biotin, digoxigenin, FITC-dUTP, resorufin-dUTP, etc.). For the double hybridization the different probes are mixed prior to hy-bridization, and the detection is achieved using the appropriate mixture of antibodies. With a special filter device (filter no. 513 803, Leica, Bensheim) it is even possible to observe red and green signals without changing the filter. With this technique, we have been able to localize two probes (separately on the chromosomes) which are only 30 kb apart from each other. **Multiple hybridiza-tion**

When the experiment fails completely, you should first check that this is not a result of a simple mistake, i.e., no probe in the hybridization mixture, wrong antibody, chromosomes not denatured etc. Beside of these trivial reasons there are a number of problems which can cause failure or unsatisfactory results. Some of these problems are listed below.

— According to my experience, there are very few if any probes which cannot be localized by fluorescence in situ hybridization to polytene **Weak/no signal** chromosomes. Very short single copy probes (<300 bp) may be difficult to localize because of the inherently weak signal which is obtained.

- Probe is not efficientyl labeled − check incorporation of digoxigenin; some probes have very high GC-content, which may limit the incorporation of DIG-dUTP.
- The wrong hybridization temperature was used; this is sometimes a problem if heterologous probes are to be hybridized.
- Some chromosome preparations do not hybridize very well. The reason is not known. This can be tested by using a probe which has been shown already to give good hybridization results (positive control). The hybridization efficiency of some chromosomal preparations can be enhanced by mild digestion with trypsin, but the quality of the chromosomes suffers significantly.

Enhancement of the hybridization can be achieved by using 10% dextran sulfate in the hybridization buffer.

- In some chromosomal preparations the denaturation of the DNA in the chromosomes seems to be incomplete after alkaline treatment. In this case it is worth to try denaturation by heat treatment with 70% (v/v) formamide, 2× SSC at 70°C for 2−5 min.
- In some cases we have been able to improve the hybridization signal by incubating the chromosomes with the hybridization mixture at a denaturing temperature (>90°C) and then gradually cooling down to the required optimal renaturation temperature over a period of 1−2 h.

Background − Over the chromosomes: unspecific binding of DNA or antibody to the chromosomes, sometimes a „beautiful banding pattern" („pseudo-hybridization") appears. The possible reasons include: hybridization temperature was too low; preparations dried out during the hybridization procedure; no SDS in the hybridization micture. Solution: increase the SDS concentration or use Denhardt's solution in the hybridization mixture, including unlabeled, single-stranded carrier DNA.
- Over the areas without chromosomes: sticking of the DNA or the antibody to cells, cytoplasm, or any other material left over from the tissue used for preparation of the chromosomes. This background can be suppressed by preincubation with Denhardt's solution supplemented with carrier DNA (10 µg/ml) (a mixture of single- and double-stranded DNA with no homology to the DNA in the chromosomes).

References

Bonner JJ, Pardue ML (1976) Ecdyson-stimulated RNA synthesis in imaginal discs of Drosophilia melanogaster. Assay by in situ hybridization. Chromosoma 58:87−99

Feinberg AP, Vogelstein B (1983) A technique for radiolabeling DNA restriction endonuclease fragments to high specific activity. Anal Biochem 132:6−13

Langer-Safer PR, Levine M, Ward DC (1982) Immunological method for mapping genes on Drosophila polytene chromosomes. Proc Natl Acad Sci USA 79:4381−4385

Schmidt ER (1988) Exonuclease digestion of chromosomes for in situ hybridization. Nucleic Acids Res 16:10381

Schmidt ER, Keyl HG, Hankeln T (1988) In situ localization of two hemoglobin gene clusters in the chromosomes of 13 species of Chironomus. Chromosoma 96:353−359

Singh L, Purdom JF, Jones KW (1977) Effect of denaturing agents on the detectability of specific DNA sequences of various base compositions by in situ hybridization. Chromosoma 60:377−389

21.8 Detection of mRNA in Fixed Cells with DIG-Labeled RNA Probes

ANNA STARZINSKI-POWITZ and KATRIN ZIMMERMANN

21.8.1 Principle and Applications

We describe here the in situ hybridization of fixed tissue culture cells with DIG-labeled RNA probes to detect mRNA molecules. This method is sensitive enough to detect low abundant mRNA species such as the glucocorticoid receptor (GR) mRNA. GR mRNA is hardly detectable in polyA$^+$ RNA from human myogenic cells when tested in conventional Northern blots but can be readily shown to be expressed by polymerase chain reaction and in situ hybridization (see figure).

The advantage of in vitro transcribed anti-sense RNA probes is the ready availability of sense RNA probes from the same sequence as negative control. Although DIG-labeled RNA probes are easier to handle and hybridization signals faster to detect when compared to radioactively labeled probes, there are certain critical steps in the procedure described below. One, for example, is the preparation and evaluation of the probe. Each of the RNA probe preparations should be tested in dot blots for the efficiency of the labeling reaction. Only those probes capable of detecting between 1 and 5 pg of insert DNA are useful for in situ hybridization.

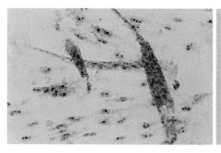

Human myotubes differentiated in culture: in situ hybridization with DIG-labeled RNA probes. **a** DIG-labeled GR anti-sense RNA; **b** DIG-labeled GR sense RNA

The disadvantages of using DIG-labeled probes are: (a) the subcellular localization of hybridization signals is currently not easy and (b) quantitation of the DIG signal is difficult (in contrast to quantitation of radioactive hybridization signals).

21.8.2 DIG-Labeled RNA Probes in the Detection of mRNA in Fixed Cells

Standard reagents
- Bind silane (Pharmacia/LKB)
- Calf skin collagen type III (acid-soluble) (Sigma)
- Paraformaldehyde
- Glycine
- EDTA
- Ethanol
- NaCl
- Na-citrate
- Proteinase K (Sigma)
- Fluoromount G (Southern Biotechnology Associates, Inc.)
- DIG RNA Labeling Kit (Boehringer Mannheim)
- DIG Nucleic Acid Detection Kit (Boehringer Mannheim)
- pSPT18 vector DNA (Boehringer Mannheim)

Standard buffers and solutions
- Bind silane solution: 100 ml ethanol, 3 ml 10% acetic acid, 300 µl bind silane.
- Collagen solutions: For the stock solution, dissolve 50 mg of collagen in 8 ml 100 mM acetic acid at 37 °C. Store at 4 °C. Working solution: 1:500 dilution in cold cell culture medium without FCS.
- PBS: 40 g NaCl, 1 g KCl, 7.21 g $Na_2HPO_4 \cdot H_2O$, 1 g KH_2PO_4 in a final volume of 5 l.
- 4% Paraformaldehyde solution: 4 g paraformaldehyde, 96 ml Ca^{2+}, Mg^{2+}-free PBS, 4 ml 0.1 N NaOH, dissolve at 55 °C; sterilize by filtration. Store at 4 °C. Discard after 1 week!
- Proteinase K stock solution: 5 mg/ml in 10 mM Tris-HCl, pH 7.5. Working solution: 1–10 µg/ml in 0.05 M EDTA, 0.1 M Tris-HCl pH 7.5; 0.02 M $CaCl_2$.
- Ethanol: H_2O solutions with the ethanol concentrations indicated.

Hybridization solutions
- 20× SSC: 3 M NaCl; 0.3 M Na-citrate, pH 7.0; 25 °C.
- 100× Denhardt's solution: 2% Ficoll, 2% polyvinylpyrrolidone, 2% bovine serum albumin.
- Prehybridization solution: 5× SSC, 0.04 M EDTA, 5× Denhardt's solution, 50% formamide (deionized), 250 µg/ml yeast t-RNA (sterile-filtered), 250 µl/ml sonicated and denatured salmon sperm DNA.
- Hybridization solution: does *not* contain salmon sperm DNA but is otherwise identical to the prehybridization solution.

These are prepared according to the manufacturer's protocol in the DIG **Detection** detection kit. **buffers** **21.8.2**

- Buffer 1: 0.1 M Tris-HCl pH 7.9, 0.15 M NaCl.
- Buffer 2: 0.5% (w/v) blocking reagent, 2% fetal cell serum in buffer 1.
- Buffer 3: 0.1 M Tris-HCl pH 9.5, 0.1 M NaCl, 0.05 M $MgCl_2$.
  Staining solution: 337.5 µg/ml nitroblue tetrazolium salt (NBT), 175 µg/ml 5-bromo-4-chloro-3-indolyl-phosphate (BCIP) in buffer 3.
- Buffer 4: 0.1 M Tris-HCl pH 8, 0.01 M EDTA.

There is no special equipment required. Racks for the coverslip and the **Standard** humid chamber are „home-made" (plastic box with grid and liquid in the **equipment** bottom which has the same salt concentration as the solution on the coverslip).

1. Wash coverslips in ethanol and incubate in bind silane solution for **Pretreat-** about 30 min. **ment and**
2. Wash coverslips in ethanol at least three times for about 10 min and dry. **collagen-**
3. Pack coverslips in aluminium foil and sterilize at 180°C for 2−3 h. **coating of**
   coverslips
4. Incubate coverslips in tissue culture dishes with collagen solution at 37°C for at least 4 h.
5. Remove collagen solution, seed cells as usual, and keep them under normal culture conditions.

1. Remove culture medium and wash cells in PBS at room temperature for **Fixation of** 5−10 min. **cells on**
2. Incubate cells in 4% paraformaldehyde at room temperature for **coverslips** 15 min.
3. Wash cells in PBS at room temperature for 10 min.
4. Incubate cells in ethanol/glacial acetic acid (95:5) at −20°C for 5 min and air dry for 5−10 min.
5. Put coverslips in petri dish (cell side up) and shock-freeze over (not in!) liquid nitrogen for 5−15 min.
6. Store coverslip at −20°C.

1. Rehydrate cells at room temperature by incubation in: 100%, 90%, **Pretreat-** 70%, 50% and 30% ethanol:H_2O, 2 min each and in PBS containing **ment of** 5 mM $MgCl_2$, 2 × 10 min. **cells for in**
2. If necessary, treat cells with proteinase K solution (1−10 µg/ml pro- **situ hybridi-** teinase) at 37°C for 15−30 min. **zation**
3. Wash at room temperature with H_2O for 5 min and with 0.2% (w/v) glycine in PBS for 10 min.

Note: Treatment of cells with proteinase K is a critical step to retain cell morphology and to abtain reliable results. Thus, both the concentration of proteinase K and the incubation time should be tested carefully for each particular cell type. For some cells, treatment with proteinase K might not be required.

21.8.2

Probe of prepara- tion DIG-labeled RNA probes were transcribed in vitro from bacterial pro- moters using plasmid pSPT18 as the cloning vector and a DIG RNA Label- ing Kit (both from Boehringer Mannheim).

In situ hy- bridization

1. Incubate cells with prehybridization solution (about $40 \mu l/cm^2$ coverslip) in a humid chamber at $42°-50°C$ (depending on stringency) for 3h. Cover the solution with a small piece of parafilm to avoid evap- oration.
2. Remove prehybridization solution and wash coverslip in increasing ethanol concentration 70%, 90%, 100% at room temperature, 2min each.
3. Air dry coverslips briefly and put them back into the humid chamber.
4. Add DIG-labeled RNA probe ($5-10 ng/\mu l$ in hybridization solution; $25 \mu l/4-5 cm^2$).
5. Cover hybridization samples carefully with siliconized coverslips.
6. Incubate in humid chamber at $42°-50°C$ overnight.

Note: Stringency of hybridization and of washing (see also below) depends on length of the probe and the degree of homology between probe and RNA to be tested. Concentration of probe should be chosen according to the abundance of the mRNA to be detected.

Washing

1. Remove upper coverslip carefully and wash the hybridized samples as follows: $2 \times 15 min$ in $2 \times$ SSC at $42°C$; $1 \times 15 min$ in $0.2 \times$ SSC at $42°C$; $1 \times 15 min$ in $0.1 \times$ SSC at $42°C$; $1 \times 15 min$ in $0.1 \times$ SSC at $42°-65°C$ (depending on stringency and background).

Detection

1. Wash coverslips 1min in buffer 1.
2. Incubate coverslips in buffer 2 at room temperature for $30-60 min$.
3. Wash coverslips briefly in buffer 1.
4. Incubate samples with alkaline phosphatase-labeled anti-DIG anti- body (diluted in buffer 1 according to the manufacturer's manual) at room temperature for 30min.
5. Wash $2-3$ times with buffer 1 for 15min each.
6. Equilibrate samples in buffer 3 at room temperature for 2min.
7. Incubate samples with staining solution in the dark at room tempera- ture overnight (or, if appropriate for shorter periods).
8. Wash samples in buffer 4 at room temperature for 6min.
9. Embed samples with fluoromount G or other embedding reagents. If necessary, air dry samples before embedding.

References

Herget T, Goldowitz D, Oelemann W, Starzinski-Powitz A (1988) Description of puta-
    tive ribosomal RNAs with low abundance, developmental regulation, and the iden-
    tifier sequence. Exp Cell Res 176:141−154
Pardue ML (1985) In situ hybridization. In: Hames BD, Higgins SJ (eds) Nucleic Acid
    Hybridization − A Practical Approach, IRL Press, Oxford, England, Chapter 8,
    pp 179−202
Zimmermann K, Herget T, Salbaum JM, Schubert W, Hilbich C, Multhaup G, Kang J,
    Lemaire H-G, Beyreuther K, Starzinski-Powitz A (1988) Localization of the putative
    precursor of Alzheimer's disease-specific amyloid at nuclear envelopes of adult human
    muscle. EMBO J 7:367−372

21.9 Detection of mRNA in Fixed Tissues Using RNA Probes

KENNETH J. HILLAN

21.9.1 Principle and Applications

The detection of specific nucleic acid sequences in tissue sections by in situ hybridization provides a means by which to study gene expression at a histological level. The ideal detection method should be rapid and provide a sensitive and specific signal without compromising tissue morphology. We have found single stranded riboprobes, internally labeled with DIG-[11]-dUTP, to provide such a system.

Single-stranded RNA probes are prepared from specifically designed plasmids that contain transcription initiation sites for T7, T3, or SP6 RNA polymerase, adjacent to the polylinker region in which the specific cDNA of interest has been cloned. Both antisense (positive) and sense probes (negative control) can be transcribed. The resultant single-stranded RNA probes are free from the problems of double-stranded DNA probes, which reanneal in solution during hybridization (Cox et al., 1984). The use of digoxigenin avoids the difficulty of using biotinylated probes in tissues such as liver and kidney, which are rich in endogenous biotin (Wood and Warnke, 1981; Morris et al., 1990). We have confined our studies to tissue fixed in either 10% buffered formalin or Bouin's. Formalin fixation provides superior morphological resolution while detection of low copy mRNA's appears to be marginally superior in Bouin's fixed tissue. Glutaraldehyde fixed tissue is also suitable.

21.9.2 Standard Procedure

Unless otherwise stated all reagents were purchased from Boehringer Mannheim Biochemicals.

Buffers All solutions are made using autoclaved deionized water treated with a 0.5% solution of DEPC to inactivate RNases.

- Alkaline hydrolysis buffer (10×): 400 mM NaHCO$_3$, 600 mM Na$_2$CO$_3$, pH 10.2.
- Buffer 1 (1 liter): 100 mM Tris-HCl, 150 mM NaCl, pH to 7.5 (can be prepared as a 10× concentrate).
- Buffer 3 (1 liter): 100 mM Tris, 100 mM NaCl, 50 mM MgCl$_2$; pH to 9.5.
- Standard saline citrate (SSC): 0.15 M NaCl, 0.15 M Na-citrate.
- 4% Paraformaldehyde: Add 10 g paraformaldehyde to 250 ml PBS and heat to 70°C until dissolved. When clear chill to 4°C. Always prepare freshly.
- Hybridization buffer: see table.

Hybridization buffer	Final	10 ml	50 ml
2 M Tris pH 7	1 mM	5 µl	25 µl
100× Denhardt's	12.5×	1.25 ml	6.25 ml
20× SSC	2×	1 ml	5 ml
Formamide	50%	5 ml	25 ml
20% SDS	0.5%	0.25 ml	1.25 ml
50% Dextran sulphate	10%	0.5 ml	2.5 ml
Salmon sperm DNA (10 mg/ml)	0.25 mg/ml	0.25 ml	1.25 ml
DEPC/H$_2$O		1.75 ml	8.75 ml

Plasmid vectors Many commercially available plasmids, including the Gemini (Promega) and Bluescript (Stratagene) vectors are suitable. The cDNA insert is ligated into the polylinker region and sense or antisense probes transcribed, after linearization of the plasmid downstream of the insert, with an appropriate restriction enzyme.

Linearization of probe
1. Linearize 5 µg of plasmid DNA with appropriate restriction enzyme in a 20 µl reaction mix for 2 h at 37°C.
2. Add 30 µl of 3 M Na Acetate and 250 µl of dH$_2$O.
3. Phenol/chloroform extract.
4. Ethanol precipitate for at least 2 h at −20°C.
5. Wash and resuspend plasmid in 5 µl of DEPC/dH$_2$O.
6. Check linearization by running 1 µl on a 1% agarose gel.

1. Add into a tube: **Trans-**
   | | | **crip-** |
   5× transcription buffer   4 µl (Life Technologies/BRL)   **tion**
   Dithiothretiol (DTT) (100 mmol)   2 µl   **reaction**
   RNA polymerase (30 U)   2 µl (Life Technologies/BRL)
   ATP, GTP, CTP (each 2.5 mmol)   4 µl (Stratagene)
   UTP (10 mmol)   2 µl (Stratagene)
   DIG-[11]-dUTP (10 mmol)   1 µl
   RNase inhibitor (500 U)   1 µl
   Linearized plasmid DNA (1 µg/µl)   2 µl
   DEPC/H$_2$O   2 µl

   Transcribe for 4 h at 37°C
2. Terminate transcription by addition of 1 µl DNase for 10 min at 37°C.
3. Add 250 µl of DEPC/dH$_2$O and 30 µl of 3 M Na-acetate.
4. Precipitate with two volumes 100% ethanol for 2 h at −20°C.
5. Wash pellet in 100% ethanol and air dry.
6. Resuspend probe in 100 µl DEPC/dH$_2$O by incubating at 37°C for 30 min.

We have found the ideal probe length for hybridization to be between 50   **Alkaline**
and 150 bases. Longer transcripts can be hydrolyzed to this length prior to   **hydrolysis**
use. For probe hydrolisis: **of trans-**
cribed

$$L_o - L_i / (0.11 \times L_i \times L_o) = \text{time in alkaline hydrolysis buffer}$$ **probe**

L_i is the initial probe length; L_o is the required length (kb)

1. Incubate probe at 60°C for required length of time.
2. Add Na-acetate to 0.3 M, 10 µl 5% acetic acid,
   and 1 µl *E. coli* rRNA as carrier.
3. Run hydrolyzed probe through Sephadex G-50 columns (Pharmacia) and collect 200 µl fractions.
4. Blot fractions to identify those containing labeled probe and pool.
5. Add Na-acetate to 0.3 M and ethanol precipitate pooled fractions in two volumes of 100% ethanol as before; spin and wash pellet.
6. Take probe up in 100 µl DEPC/dH$_2$O (approximately 50 ng/µl). Probe is ready for use.

Glass slides (prepare in batches of 200−300):   **Coating**
glass slides
1. Wash slides in 1% Decon overnight.   **and**
2. Rinse thoroughly in several changes of H$_2$O.   **coverslips**
3. Rinse in acetone for 5 min.
4. Dip in 2% 3-aminopropyltriethoxysilane (APES, Sigma) in acetone for 5 min (fume hood).
5. Rinse in several changes of dH$_2$O; dry and store.

Coverslips:
1. Coat coverslips with 5% dimethyldichlorosilane (Sigma) in chloroform for 30 min (toxic and corrosive).

2. Rinse in dH_2O and lay out individually to dry.
3. Rinse coverslip in methanol prior to use.

In situ hybridization All procedures are carried out at room temperature unless otherwise stated:

1. Dewax sections in three changes of xylene and hydrate through alcohols to PBS.
2. Incubate in $0.2\,N$ HCl for 15 min; rinse briefly in PBS.
3. Permeabilize with 0.3% Triton X-100 in PBS for 15 min; rinse in PBS.
4. Digest sections with proteinase K (Sigma) at 37 °C in a humidified chamber; start at 20 µg/µl in PBS for 45 min; rinse in PBS.
5. Fix in freshly prepared 4% paraformaldehyde in PBS for 5 min at 4 °C; rinse in PBS.
6. Prehybridize for 1 h at 37 °C in 2× SSC containing 50% deionized formamide.
7. Dilute probe 1 : 20 to 1 : 40 in hybridization buffer: Put 15 µl of this onto each coverslip and place the section carefully on top of the coverslip. Hybridize overnight in a sealed humidified chamber at 42° – 50 °C.
8. Remove coverslips by immersion in 4× SSC.
9. Wash in 2× SSC twice, 30 min each.
10. RNase A digest (100 µg/ml) in 2× SSC at 37 °C (100 µl per section) to digest any single-stranded probe (optional).
11. Rinse in 0.1× SSC twice, 30 min each at 45 °C.
12. Immerse in buffer 1 for 5 min.
13. Apply alkaline phosphatase conjugated anti-digoxigenin antiserum, diluted 1 : 2000 in buffer 1 containing 20% normal swine serum, for 2 h.
14. Wash in buffer 1, 2 × 15 min; wash in buffer 3, 1 × 5 min.
15. Incubate sections in NBT/BCIP substrate containing levamisole (see buffers).
16. Incubate until developed (between 2 – 12 h) in a humidified chamber protected from light.

Substrate solution
– Nitroblue tetrazolium chloride (NBT) (Sigma): stock solution consists of 100 mg NBT in 1.333 ml 70% dimethylformamide (Sigma).
– 5-bromo-4-chloro-3-indolyl phosphate (BCIP) (Sigma): stock solution consists of 50 mg in 1 ml 100% dimethylformamide.

To make up substrate add 44 µl NBT and 33 µl BCIP to 10 ml buffer 3. Add 2.3 mg of levamisole (Sigma).

21.9.3 Special Hints for Application and Troubleshooting

Poor labeling
– Repurify the plasmid DNA by phenol/chloroform extraction and ethanol precipitation.
– Incubate the labeling reaction overnight.

- Alter proteinase K digestion time/concentration.   **High**
- Try another proteinase, e.g., proteinase VIII (Sigma)   **background**
- Increase hybridization temperature or wash temperature.
- We have found that washing with formamide invariably increases
  background staining.

- Ensure that the transcribed probe is working. Synthesise unlabeled **No signal**
  sense probe and bind this to nitrocellulose membrane. Hybridize with
  the labeled antisense probe.
- Is there a problem with the tissue fixative or means of fixation? Always
  run a positive control probe with each hybridization run.
- Check probe on a denaturing gel.

References

Cox KH, DeLeon DV, Angerer RC (1984) Detection of mRNAs in sea urchin embryos
    by in situ hybridization using asymmetric RNA probes. Develop Biol 101:485−502
Morris RG, Arends MJ, Bishop PE, Sizer K, Duvall E, Bird CC (1990) Sensitivity of
    digoxigenin and biotin labeled probes for detection of human papillomavirus by in situ
    hybridization. J Clin Path 43:800−805
Wood GS, Warnke R (1981) Suppression of endogenous avidin binding activity in tissues
    and its relevance to biotin-avidin detection systems. J Histochem Cytochem 29:1196

21.10 Detection of mRNA In Situ with DIG-Labeled Synthetic Oligonucleotide Probes

FRANK BALDINO JR., ELAINE ROBBINS, and MICHAEL E. LEWIS

21.10.1 Principle and Applications

The use of in situ hybridization (ISH) histochemistry as a tool to study gene
expression has expanded significantly over the past few years. While this
technology was once limited to the domain of the molecular biologist, the
advent of simplified molecular and histochemical methodologies (Lewis
et al., 1985, 1988; Baldino et al., 1989) has literally brought this advanced
technology to the laboratory of any competent biologist.

The ability to obtain a very high degree of cellular resolution and
improvements in the sensitivity of hybrid detection has accounted, in large
part, for the widespread use of this technology in a variety of fields. ISH
has proved to be particularly valuable in the study of the CNS, where the
phenotypic heterogeneity within neuronal populations is immense.

Fundamentally, ISH is designed to detect the expression of specific genes in any properly prepared tissue section. A number of laboratories routinely use this technology to localize relatively rare mRNAs in the cytoplasm of individual neurons. The localization of primary transcripts to individual cells has enabled the identification of cells responsible for the synthesis of a given protein, e.g., an enzyme or neuropeptide precursor. Moreover, this technology has provided new insights into the regulation of genes at the level of single neurons in the CNS (Baldino et al., in press).

ISH is normally performed with synthetic oligonucleotide, cDNA or cRNA probes. While each type of probe offers different advantages, the utility of any RNA or DNA probe is dependent on the specific application (Lewis and Baldino, 1990). In addition, investigators have the choice of using radioactive or nonradioactive probes.

In recent years several nonradioactive markers have been developed to detect specific nucleotide sequences under a variety of hybridization conditions. These nonradioactive markers are particularly useful for ISH studies where they have overcome several limitations inherent to the use of radiolabeled probes. The prolonged exposure times required with low energy radionuclides (e.g., ^{35}S and ^{3}H) on emulsion-coated slides have been cumbersome and difficult to reproduce, as the time of exposure is a function of the degree of hybridization. Also it is difficult, if not impossible, to detect multiple mRNA species within individual cells using standard autoradiography. Another significant problem is the nonspecific background associated with the use of radiolabeled probes, which often precludes the resolution of cells with rare mRNAs. Thus, with nonradioactive markers, the degree of cellular resolution, often the primary rationale for using ISH, is significantly improved over that normally achieved with autoradiography.

Recently, we developed a method which uses terminal deoxynucleotidyl transferase to incorporate DIG-[11]-dUTP (Boehringer Mannheim) into a synthetic oligonucleotide probe (Baldino and Lewis, 1989; Lewis et al., 1990). This DIG-[11]-dUTP-labeled probe is subsequently detected with an alkaline phosphatase-conjugated IgG which is highly specific for the DIG molecule. Unlike the direct conjugation of a single enzyme molecule to the probe, this new methodology offers the advantage of adding multiple labels to the probe sequence, thus amplifying the hybridization signal. In addition, IgG detection also provides the advantage of using "bridging" antibody techniques to further amplify the signal. Thus, a certain degree of amplification is an intrinsic feature of the DIG detection system. We have recently developed the DIG-[11]-dUTP detection system for in situ hybridization histochemistry with synthetic oligonucleotides (see protocol below). Here, we provide a detailed guide to using this high-resolution methodology to perform nonradioactive in situ hybridization histochemistry in neural tissue.

21.10.2 DIG-Labeled Synthetic Oligonucleotide Probes to Detect mRNA In Situ

Labeling Reaction

Tailing of 35 picomoles of an oligonucleotide probe at the 3' end is accomplished using the DIG DNA Tailing Kit (Boehringer Mannheim)

1. To a sterile 1.5 ml Eppendorf tube, add:

Tailing buffer	4 μl
CoCl$_2$ solution	6 μl
DIG-[11]-dUTP	2.5 μl
Oligonucleotide	1 μl
dATP	2 μl of 1:50 dilution made in sterile H$_2$O
Sterile H$_2$O	1 μl
*TdT (50 Units)	1 μl

2. The reaction is incubated at 37 °C for 5 mins. The tailed oligonucleotide is purified from the labeling reaction by ethanol precipitation:
   - Adjust volume to 100 μl with sterile H$_2$O
   - Add 1:10 (10 μl) 4.5 M Na-acetate pH 6.0
   - Add 3 times volume ice cold 100% ethanol (300 μl)
   - Add 1 μl 20 mg/ml glycogen
   - Mix by inversion and tapping the tube
   - Centrifuge briefly (1–2 s)
   - Let chill in dry ice mixed with ethanol for 30 min
   - Centrifuge for 30 min at 12,000 g (minimum speed)
   - Pour off supernatant, keeping the tube inverted.
   - Allow pellet to dry and resuspend pellet in 20 μl sterile H$_2$O

Tissue preparation and prehybridization washes

Fresh frozen sections (12 μm) are sectioned on a cryostat, thaw-mounted on twice-coated slides (porcine gelatin) and stored at −80 °C. Immediately before the experiment the sections are quickly brought to room temperature under a stream of cool air and fixed in:

1. 3% paraformaldehyde in 0.1 M PBS (plus 0.02% DEPC) for 5 min
2. Rinse 3 times 5 min in 0.1 M PBS
3. Rinse 1 time 10 min in 2 times SSC
4. Optional: slides are placed in a humid chamber and approximately 300 μl of hybridization buffer (see below) is pipetted onto each slide. Slides are permitted to incubate at room temperature for 1 h.

Hybridization

1. The DIG-labeled probe is diuluted with hybridization buffer to a volume of 1 ml. The required concentration of probe will depend upon mRNA abundance and must be determined experimentally.

---

* terminal deoxynucleotidyl transferase

Hybridization buffer is prepared fresh by adding:
5 mls deionized formamide
2 mls 20× SSC
0.2 ml Denhardt's
0.5 ml 10 mg/ml salmon sperm DNA (denatured immediately before use by heating to 95°C for 10 min)
0.25 ml 10 mg/ml yeast tRNA
2 ml dextran sulfate

2. Excess prehybridization buffer is removed from the slides by a quick dip in 2× SSC and the glass surrounding the tissue is carefully dried with a tissue. The probe is then added allowing 30 µl per section, after which a parafilm coverslip is applied. The slides are incubated at 37°C overnight (this can be increased to 42°C for oligonucleotides greater than 36-mer).

Posthybridization The slides are washed (with gentle shaking) as follows:
 − 2× SSC for 1 h at room temperature
 − 1× SSC for 1 h at room temperature
 − 0.5× SSC for 0.5 h at 37°C
 − 0.5× SSC for 0.5 h at room temperature

Immunological detection
1. Wash slides for 1 min in buffer 1 at room temperature.
2. Incubate sections with 2% normal sheep serum plus 0.3% Triton X-100 in buffer 1 for 30 min at room temperature.
3. Dilute antibody conjugate (1:500) with buffer 1 containing 1% normal sheep serum and 0.3% Triton X-100. Pipette diluted anti-digoxigenin antibody conjugate onto sections and incubate at room temperature for 3−5 h in a humid chamber. Do not allow sections to dry out.
4. Wash slides for 10 min in buffer 1 with shaking.
5. Wash slides for 10 min in buffer 2 with shaking.
6. Incubate slides with approximately 500 µl/slide of color solution. Slides are placed in light-tight boxes on wetted filter paper backing. Slides can be checked periodically for color development (0.5−24 h).
7. Reaction can be stopped in buffer 3.
8. Dehydrate sections as follows: 1 min in 70% ethanol, 1 min in 80% ethanol, 1 min in 95% ethanol, 1 min in 100% ethanol, 3 min in xylene. Slides can then be coverslipped in Permount (66% Permount/ 3% xylene) mounting medium.

Solutions Buffer 1: 100 mM Tris-HCl; 150 mM NaCl; pH 7.5
Buffer 2: 100 mM Tris-HCl; 100 mM NaCl; 50 mM $MgCl_2$; pH 9.5
Buffer 3: 10 mM Tris-HCl; 1 mM EDTA; pH 8.0.

Color solution:

45 µl NBT solution; 35 µl BCIP solution; and 2.4 mg levamisole are added to 10 ml buffer 2 immediately before use.

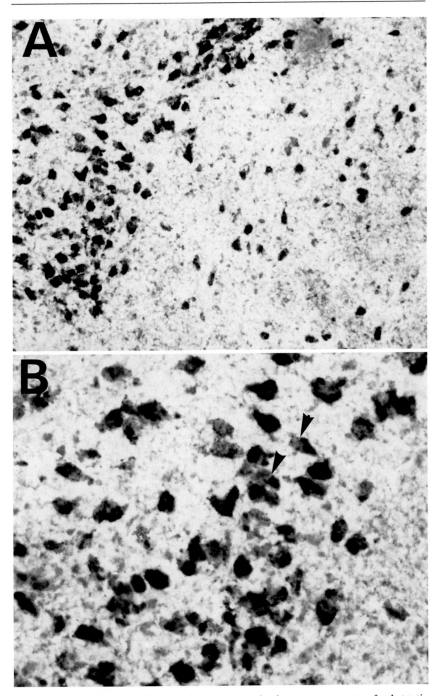

Tyrosine hydroxylase mRNA-containing neurons in the pars compacta of substantia nigra in rat brain. **A** labeled perikarya in the upper part of the photomicrograph ($\times 400$). Note the low density of labeled cells on the border with the pars reticulata in the lower part of the photomicrograph; this area contains principally nondopaminergic (tyrosine hydroxylase-negative) neurons. **B** higher power ($\times 800$) view of labeled pars compacta neurons. *Arrow*, nucleus of positively labeled perikaryon

21.10.3 Results and Concluding Remarks

The degree of cellular resolution obtained with probes labeled with DIG-[11]-dUTP is far superior to that seen with radiolabeled probes. Individual cell profiles were easily defined and multiple cells could be distinguished within deeper layers of the tissue section. The figure shows an example of the resolution achieved with a DIG-[11]-dUTP-labeled oligonucleotide (48 bases) which has been shown to be specific for tyrosine hydroxylase mRNA (Young et al., 1986). The hybridization signal was characterized by dense particulate labeling within the cytoplasm and a clear unlabeled nucleus (arrows). Little, if any, hybridization signal extended beyond the soma with this probe. The nonspecific (i.e., background) labeling was also substantially less than that observed with radiolabeled probes (Baldino et al., 1988). Moreover, hybridization and detection of these probes was obtained within 24 h compared to the 8–10 weeks required to detect single cells profiles with ^{35}S- or ^{3}H-labeled probes.

High-resolution in situ hybridization histochemistry can be readily performed with alkaline phosphatase: DIG-[11]-dUTP-labeled oligonucleotide probes. The ease of performance, safety, and rapidity of detection render this methodology a useful alternative to procedures employing radiolabeled probes for the detection of mRNA in the CNS or any organ system where cellular resolution is essential.

A further advantage of this particular nonradioactive method is that, since several substrates are presently available for alkaline phosphatase, it should be possible to detect multiple mRNAs within a single cell. It is also possible to detect two different species of mRNA by combining an enzyme-labeled probe with a radiolabeled probe (Young, 1989).

One interesting feature inherent to the use of the DIG-[11]-dUTP system is the advantage of amplification by the enzymatic addition of multiple DIG-[11]-dUTPs, which, when coupled with standard double- or triple-bridging techniques with alternate species antisera, should permit the detection of rare mRNA species. To further enhance sensitivity, it is possible to internally label cRNA probes with DIG-[11]-dUTP for in situ hybridization studies (Springer et al., 1991; Robbins et al., 1991); however, this method requires familiarity with microbiological techniques which are not required for the use of synthetic oligonucleotide probes.

References

Baldino F Jr, Chesselet M-F, Lewis ME (1989) High resolution in situ hybridization histochemistry. In: Conn PM (ed) Methods in enzymology: hormone action, Part K, Neuroendocrine peptides Vol 168, Academic Press 761–777

Baldino F Jr, Deutch A Y, Roth RH, Lewis ME (1988) In situ hybridization histochemistry of tyrosine hydroxylase messenger RNA in rat brain. Ann NY Acad Sci 537:484–487

Baldino F Jr, Lewis ME (1989) Nonradioactive in situ hybridization histochemistry with digoxigenin-dUTP labeled oligonucleotides. In: Conn PM (ed) Methods in Neuroscience, Academic Press 282–292

Baldino F Jr, Roberts-Lewis JM, Lewis ME (1992) In situ hybridization histochemistry as a tool for the study of brain function. In: Osbourne NN (ed) Current aspects of the neurosciences Vol 4, Macmillan Publishers, in press

Lewis ME, Baldino F Jr (1990) Probes in situ hybridization histochemistry. In: Chesselet MF (ed) In situ hybridization histochemistry. CRC Press 1—21

Lewis ME, Krause RG, Roberts-Lewis JM (1988) Recent developments in the use of synthetic oligonucleotides for in situ hybridization histochemistry. Synapse 2:308—316

Lewis ME, Robbins E, Grega D, Baldino F Jr (1990) Nonradioactive detection of vasopressin and somatostatin mRNA with digoxigenin-labeled oligonucleotide probes. Ann N Y Acad Sci 579:246—253

Lewis ME, Sherman TG, Watson SJ (1985) In situ hybridization histochemistry with synthetic oligonucleotides. Peptides 6 (suppl 2) 75—87

Robbins E, Baldino F Jr, Roberts-Lewis JM, Meyer S, Grega DS, Lewis ME (1991) Quantitative nonradioactive in situ hybridization of preproenkephalin mRNA with digoxigenin-labeled cRNA probes. Anat Rec 231:559—562

Springer JE, Robbins E, Gwag BJ, Lewis ME, Baldino F Jr (1991) Nonradioactive detection of nerve growth factor receptor mRNA in the rat brain using in situ hybridization histochemistry. J Histochem Cytochem 39:231—234

Young WS III (1989) Simultaneous use of digoxigenin- and radiolabeled oligodeoxyribonucleotide probes for hybridization histochemistry. Neuropeptides 13:271—275

Young WS III, Bonner TI, Brann MR (1986) Mesencephalic dopamine neurons regulate the expression of neuropeptide mRNA in the rat forebrain. Proc Natl Acad Sci 83:9827—9831

21.11 Whole Mount In Situ Hybridization for the Detection of mRNA in *Drosophila* Embryos

DIETHARD TAUTZ

21.11.1 Principle and Applications

In situ hybridization for the detection of RNA in tissues has traditionally been performed on sectioned material. This was necessary since the probes were usually labeled with ^{3}H or ^{35}S, which required the use of a photographic film emulsion covering the sections. The development of highly sensitive nonradioactively labeled probes allows in situ hybridizations to be performed directly in tissues, such as whole *Drosophila* embryos (Tautz and Pfeifle, 1989). This „whole mount" in situ hybridization procedure is highly sensitive and the resolution of details is unparalleled. Complex expression patterns in particular, can only be analyzed in whole embryos, since the reconstruction of a three-dimensional pattern from sections can be very cumbersome. The method is also applicable to other small embryos (Hemmati-Brivanlou et al., 1990; Sommer and Tautz, 1991) and animals (Kurtz et al., 1991) and should therefore be of general usefulness.

The protocol for *Drosophila* includes special procedures to remove the embryonic membranes. These procedures are not necessary if other tissues are used. In these cases it is possible to start directly with the formadehyde fixation (Hemmati-Brivanlou et al., 1990; Kurz et al., 1991).

21.11.2 Whole Mount In Situ Hybridization

Reagents
- Sodium hypochlorite solution (5%; Klorix Colgate-Palmolive)
- Formaldehyde (37%; Merck)
- Triton X-100 (Merck)
- HEPES (Biomol)
- Magnesium sulfate (Merck)
- Magnesium chloride (Merck)
- Tris (Merck)
- EGTA (Sigma)
- Heptane (Merck)
- Methanol (Merck)
- Sodium phosphate (Merck)
- Sodium chloride (Merck)
- Sodium citrate (Merck)
- Tween 20 (Sigma)
- Proteinase K (Type XXVIII; Sigma)
- Glycine (Merck)
- Formamide (BRL)
- Dimethylformamide (Merck)
- Heparin (Sodium salt grade II; Sigma)
- Salmon sperm DNA (Sodium salt, Type III; Sigma)
- Anti-DIG antibody conjugate (Boehringer Mannheim)
- NBT (nitroblue tetrazolium salt; Sigma)
- BCIP (5-bromo-4-chloro-3-indolyl phosphate; Sigma)

Buffer solutions
- Fixation solution: 0.1 M Hepes, pH 6.9; 2 mM $MgSO_4$; 1 mM EGTA
- PBT: 130 mM NaCl; 10 mM sodium phosphate, pH 7.2; 0.1% [v/v] Tween 20
- Hybridization solution: 750 mM NaCl; 75 mM Na-citrate; 50% [v/v] formamide; 0.1% [v/v] Tween 20; 50 µg/ml heparin
- Staining solution: 100 mM NaCl; 50 mM $MgCl_2$; 100 mM Tris-HCl, pH 9.5
- BCIP solution: 50 mg/ml BCIP in 100% [v/v] dimethylformamide
- NBT solution: 75 mg/ml NBT in 70% [v/v] dimethylformamide
- Euparal (Roth)

Fixation of embryos
1. Collect the embryos on an apple juice agar plate and transfer into a little basket made from polyethylene tubing and stainless steel mesh (Wieschaus et al., 1986).
2. Wash embryos with water and dechorionate in a solution of 50% [v/v] commercial bleach (Klorix, about 5% [w/v] sodium hypochlorite) for about 2–3 min. Control this step under the binocular microscope. Dechorionated embryos float to the surface of the solution.

3. Wash with 0.1% [v/v] Triton X-100 and transfer the embryos into a glass scintillation vial containing 4 ml fixation-buffer.
4. Add 0.5 ml 37% [v/v] formaldehyde solution and 5 ml heptane.
5. Shake the vial vigorously for 15−20 min.
6. Remove the lower phase as far as possible (the embryos should swim at the interphase). Add 10 ml methanol and shake vigorously for 10 s. This step causes the vitellin membranes to burst and the devitellinized embryos will sink to the bottom.
7. Transfer the embryos into an Eppendorf tube and wash them with methanol. The embryos may be stored at this stage for several weeks in the refrigerator.

Unless otherwise indicated, all the following steps are done in Eppendorf **Pretreat-** tubes in a volume of 1 ml at room temperature and on a rotating shaker. **ment** Avoid potential sources of RNAse contamination, though it is usually not necessary to take any particular precautionary steps.

1. Wash embryos 3 × 5 min in PBT.
2. Incubate the embryos for 2−5 min in a solution of 50 µg/ml proteinase K in PBT. The exact length of this incubation step should be optimized for each new batch of proteinase K. Too short digestion times result in a loss of signal intensity, too long digestion times may cause the embryos to burst during the subsequent steps.
3. Stop the proteinase K digestion by incubating for 2 min in 2 mg/ml glycine.
4. Wash 2 × 5 min with PBT and refix the embryos with 4% formaldehyde for 20 min.
5. Wash 5 × 5 min in PBT.

1. Wash the embryos in hybridization solution diluted 1 : 1 with PBT for **Hybridi-** 10 min. Then wash 10 min in hybridization solution. This stepwise trans- **zation** fer into the hybridization solution is not strictly required, but embryos which are slightly overdigested with proteinase K would burst, if they were brought directly into the formamide containing solution. The same considerations apply for the washing after the hybridization (see below).
2. Prehybridize for 20−60 min in a waterbath at 45 °C.
3. Remove most of the liquid, leaving about 2 mm of solution above the surface of the settled embryos. This corresponds usually to a hybridization volume of about 100 µl.
4. The probe may be labeled either according to the standard random priming protocol (Sect. 3.2) or according to the PCR protocol (Sect. 21.12). It is possible to use the probe directly after the labeling step without further purification. For one hybridization, we usually use 2 µl of a 20 µl labeling reaction.
5. Add 2 µl of the probe to 1 µl of solution of 10 mg/ml sonicated salmon sperm DNA (this has to be scaled up appropriately if more then one

hybridization is carried out). Denature at 100°C for 3 min and add directly to the embryos in hybridization solution.

6. Mix and incubate at 45°C overnight. Agitation is usually not necessary, but is also of no harm.

Washing The following washing protocol is very extensive and may be necessary if high background is encountered. Fewer steps may be sufficient for many applications. All steps are done at room temperature.

1. Wash 10 min in hybridization solution. Proceed with washes in serial dilutions (e.g., 4:1, 3:2, 2:3, 1:4) of hybridization solution in PBT for 10 min each.
2. Wash 20 min in PBT.

Detection 1. The anti-DIG antibody conjugate should be freshly preabsorbed for 1 h against fixed embryos in order to remove any unspecifically binding materia. The final working dilution of the antibody conjugate is 1:2000. The preabsorption step should be adjusted accordingly. If, for example, ten reactions are processed in parallel, use about 200 µl embryos in 1 ml PBT with an antibody conjugate dilution of 1:200. This solution is then further diluted 1:10 in the next step.

2. Incubate the embryos for 1 h in 500 µl diluted and preabsorbed anti-DIG antibody complex.
3. Wash 3 × 20 min in PBT.
4. Wash 3 × 5 min in staining buffer.
5. Transfer embryos in a small dish with 1 ml staining buffer containing 4.5 µl NBT solution and 3.5 µl BCIP solution.
6. Let the color develop in the dark with occasional inspection under the binocular microscope. Color develops usually within 1 h, but the reaction may also be left overnight.
7. Dehydrate the embryos in an alcohol series (70%, 90%, and 100% [v/v]) and mount in Euparal.

21.11.3 Special Hints for Application and Troubleshooting

Low signal – Check whether the probe is correctly labeled (see Sect. 3.2 and Sect. 21.12)
– Insufficient proteinase K digestion. Depending on the tissue, it may be necessary to perform rather extensive proteinase K digestions. Set up a series of digestion conditions and test these.
– Insufficient devitellinization. Check under the binocular microscope whether the vitellin membranes are fully removed.
– RNAse contamination in one of the solutions. Treat the solutions with diethylpyrocarbonate (Sigma) before use, in particular the PBT solution before the addition of the Tween 20.

- Insufficient washing or insufficient preabsorption of the anti-digoxige- **High** nin complex. Each step may be done for longer times. **background**
- Include levamisole in the staining solution. Levamisole acts as a potent inhibitor for endogenous lysosomal phosphatase. These are, however, usually not a problem in early *Drosophila* embryos.
- Use higher detergent concentration in the PBT. Tween 20 may also be replaced by SDS.
- Include a xylene treatment step after the fixation.

References

Hemmati-Brivanlou A, Frank D, Bolce ME, Brown BD, Sive HL, Harland RM (1990) Localization of specific mRNAs in *Xenopus* embryos by whole-mount in situ hybridization. Development 110:325−330

Kurz E, Holstein T, Petri B, Engel J, David C (1991) Mini-collagens in hydra nematocytes. J Cell Bid 115:1159−1169

Sommer R, Tautz D (1991) Segmentation gene expression in the house fly *Musca domestica*. Development 113:419−430

Tautz D, Pfeifle C (1989) A nonradioactive in situ hybridization method for the localization of specific RNAs in *Drosophila* embryos reveals translational control of the segmentation gene hunchback. Chromosoma 98:81−85

Wieschans E, Nüsslein-Vollhard C (1986) Looking at embryos. In: Roberts DB (ed) *Drosophila* − A Practical Approach. IRL Press, Oxford, pp 199−228

21.12 DIG-Labeled Single-Stranded DNA Probes for In Situ Hybridization

NIPAM H. PATEL and COREY S. GOODMAN

21.12.1 Principle and Applications

We have developed a simple polymerase chain reaction (PCR) procedure for the generation of digoxigenin-labeled single-stranded DNA probes. The protocol takes advantage of the thermostable properties of *Taq* DNA polymerase to repetitively generate single-stranded DNA but does not utilize the chain reaction properties usually associated with PCR because only a single oligonucleotide primer is used in any one reaction. These single-stranded probes are useful for in situ hybridization or any other application in which digoxigenin-labeled nucleic acid probes are used.

The PCR procedure has several advantages over the standard random oligonucleotide priming method. First, due to the repetitive rounds of DNA synthesis, this protocol yields a greater quantity of probe from an equal amount of starting material. A 25 µl reaction containing 200 ng of template DNA can be used to generate almost 1 µg of probe. This also means that the ratio of labeled DNA to unlabeled starting material (which will compete for hybridization) is much higher. The second advantage is that these single-stranded probes are more sensitive than those produced by random oligonucleotide priming. Since the probe strands will not hybridize to themselves, the hybridization of probe to the target sequences is enhanced. This advantage is not very noticeable when the target sequences are abundant, but it becomes an important factor when the transcripts of interest are expressed at low levels per cell. Nighorn et al. (1991) report that *dunce* transcripts, which are estimated to be no more than 5 parts per million of the poly(A)$^+$ RNA fraction, could be detected with single-stranded digoxigenin probes but not with random oligonucleotide primed digoxigenin probes nor with ^{35}S-labeled RNA probes. Finally, random oligonucleotide priming is relatively inefficient at labeling small fragments of DNA (less than 150 bp). The PCR protocol, however, is quite suitable for producing probes from such small fragments.

One potential disadvantage is that it is less straightforward to control the probe size when using the PCR protocol than when using the random oligonucleotide priming procedure. The "PCR" reaction will generate full length single-strands for inserts up to 2–3 kb. For optimal tissue penetration and hybridization, probe fragments should be 50–200 bp in length. To fragment the probe into smaller pieces, we simply boil the DNA for an extended period of time. We have not made any quantitative estimates of the rate of fragmentation, nor have we investigated alternative ways to fragment the probes. Empirically, however, the labeling and fragmentation technique described below works extremely well for in situ hybridization. Digoxigenin-labeled single-stranded probes made by this procedure have been used on whole mount *Drosophila* embryos to localize transcripts from a wide variety of genes including *NTF*-1 (Dynlacht et al., 1989), *dFRA* (Perkins et al., 1990), and *neurotactin* (Hortsch et al., 1990). It has also proven sensitive enough to give excellent single cell resolution for the neuronal expression of *even-skipped* (Patel, Fanning, and Goodman, unpublished data). Probes made by this method have also been used to localize *engrailed* transcripts in whole mount grasshopper embryos (Patel and Goodman, unpublished data). In all cases, the procedure for embryo preparation and probe detection essentially followed the protocol of Tautz and Pfeifle (1989) also summarized in Chap. 21.11.

21.12.2 DIG-Labeled Single-Stranded DNA Probes

Reagents
- DIG-[11]-dUTP (Boehringer Mannheim)
- Glycogen
- Tris-HCl
- Ultrapure dATP, dGTP, dCTP, dTTP (Pharmacia)
- *Taq* DNA polymerase (Perkin Elmer Cetus)
- KCl
- Gelatin

- MgcL$_2$
- NaCl
- Mineral oil
- SK and KS oligonucleotide primers, Bluescript (Stratagene)

- 10× Reaction buffer: 0.5 M KCl; 0.1M Tris-HCl, pH 8.3; 15mM **Solutions** MgCl$_2$; 0.01% [w/v] gelatin
- dNTP mix: 1.0 mM dATP; 1.0mM dCTP; 1.0mM dGTP; 0.65mM dTTP, 0.35mM DIG-[11]-dUTP
- SK oligonucleotide primer: 30ng/µl in dH$_2$O
- KS oligonucleotide primer: 30ng/µl in dH$_2$O

1. Linearize plasmid DNA containing the insert of interest as you would to make an RNA runoff probe. In general, you will want to set up two separate reactions in order to produce probes complementary to each strand. For example, if have a 1.0kb *Eco*RI fragment cloned into the *Eco*RI site of Bluescript, you might linearize the plasmid by cutting one aliquot with *Kpn*I and another aliquot with *Sac*II (assuming there are no *Kpn*I or *Sac*II sites in the insert). After the restriction digestion is complete, phenol extract and precipitate the DNA. Resuspend the DNA in TE at a concentration of 100−200ng/µl. It is also possible to simply heat inactivate the restriction enzyme and then use the DNA digest for the PCR reaction.

    The KpnI-cut DNA can be used with the SK primer to create one single-stranded probe and the SacII-cut DNA can be used with the KS primer to create the complementary single-stranded probe. If you already know which DNA strand is the sense strand, then you will be able to predict which probe (anti-sense probe) will hybridize to the mRNA and you can use the other probe (sense probe) as a control. Alternatively, the in situ results with each probe should reveal which strand is the sense strand.

    We have written the protocol using Bluescript and KS/SK primers, but any plasmid can be utilized with any appropriate combination of primers (T3/T7, M13-forward/M13-reverse, etc.). In addition, primers complementary to sequence within the insert can be used to make probes specific for products generated by alternative mRNA splicing.

2. Set up the individual PCR reactions as follows (25 µl volume per reaction:
    - 8.5 µl   dH$_2$O
    - 2.5 µl   10× reaction buffer
    - 5.0 µl   dNTP mix
    - 5.0 µl   KS (or SK) primer
             (depending on the enzyme used to linearize the plasmid)
    - 2.0 µl   plasmid DNA (100−200ng/µl;
             linearized on one side of the insert)

Add 40 µl mineral oil and centrifuge for a few seconds. Boil 5 min, then add:

  − 2.0 µl   *Taq* DNA polymerase (1.25 U total)

3. Mix the contents and then centrifuge for 2 min. Incubate in a thermal cycler for 30−35 cycles of the following conditions:

  − 95 °C for 45 s
  − 50−55 °C for 30 s (50 °C for a 17-nucleotide primer, 55 °C for a 21-nucleotide primer)
  − 72 °C for 60 s (extend to 90 s if the insert is over 2 kb long)

4. After the cycles are completed, add 75 µl dH$_2$O, then centrifuge.

5. Remove 90−95 µl of the reaction from beneath the oil and transfer to a new tube.

6. Do one ethanol precipitation as follows: Add NaCl to 0.1 M; 10 µg of glycogen can be added as a carrier. Precipitate by adding three volumes 100% [v/v] EtOH. Mix well, transfer to −70 °C for 30 min, then centrifuge for 10 min. Wash with 70% [v/v] ethanol. Air dry the pellet.

7. For tissue in situ hybridization, resuspend the pellet in 300 µl of hybridization buffer (see Sect. 21.11 for hybridization buffer recipe). As described previously, the probe needs to be fragmented for optimal hybridization. To fragment the single-stranded DNA, boil the probe in the hybridization buffer for 40−80 min. The probe can be stored at −20 °C until ready for use.

8. For in situ hybridization to *Drosophila* embryos, follow the embryo preparation and prehybridization protocol outlined in Sect. 21.11. Just before use, boil the probe for an additional 5 min. We usually combine 300 µl of probe in hybridization buffer with 50 µl of embryos. After the hybridization is complete, the probe can be recovered from the embryos and reused many times. Store the probe at −20 °C and boil it for 5 min before each use.

9. Subsequent washing and detection of the hybridized probe in *Drosophila* embryos should be done according to the protocol outlined in Sect. 21.11.

21.12.3 Special Hints for Application and Troubleshooting

Probe synthesis
− For any one insert, two different single-stranded probes can be produced. One will be the anti-sense strand and will reveal the distribution of the mRNA of interest. The other will be the sense strand and will serve as a control for background hybridization. These single-stranded DNA probes are more sensitive than double-stranded DNA probes but probably less sensitive than single-stranded RNA probes. DNA probes, however, are simpler to produce and handle than RNA probes.

− The *Taq* DNA polymerase will generate single strands up to several kb in length, but for large inserts (over 3 kb in length the probes will not be fully representative of the entire clone, as the *Taq* DNA polymerase will not reach the far end of the molecule on the extension step.

- If you have background problems, try diluting the probe in a larger volume of hybridization buffer. The signal to noise ratio can be *dramatically* improved by dilution of the probe. In fact, you may want to routinely resuspend your probe in 0.5−1.0 ml of hybridization buffer from the outset. For especially abundant transcripts, the initial 300 µl of probe can be diluted tenfold and still give excellent results.
- Background may be reduced and signal improved by adjusting the boiling time of the probe. Underboiled probes tend to create high background, and overboiled probes tend to have poor signal.
- The proteinase K treatment step during the preparation of the embryos is also critical to achieving a good signal to noise ratio.
- GC-rich sequences can also result in excessive background. We have found that the best results can often be obtained by using relatively. AT-rich regions from 5′ or 3′ untranslated regions. Insert size can also be an important parameter. For rare transcripts, the best results are obtained by starting with a cDNA fragment somewhere between 1.0 and 2.0 kb in size.

- This labeling technique can also be used to create biotinylated probes using bio-[16]-dUTP (Boehringer Mannheim) and detected using streptavidin-alkaline phosphatase (Clontech). Biotinylated probes, however, are several fold less sensitive than DIG-labeled probes.

References

Dynlacht BD, Attardi LD, Admon A, Freeman M, Tijian R (1989) Functional analysis of NTF-1, a developmentally regulated Drosophila transcription factor that binds neuronal cis elements. Genes Dev 3:1677−1688

Hortsch M, Patel NH, Bieber AJ, Traquina ZR, Goodman CS (1990) Drosophila neurotactin, a surface glycoprotein with homology to serine esterases, is dynamically expressed during embryogenesis. Development 110:1327−1340

Nighorn A, Healy MJ, Davis RL (1991) The cyclic AMP phosphodiesterase encoded by the Drosophila dunce gene is concentrated in the mushroom body neuropil. Neuron 6:455−467

Perkins KK, Admon A, Patel NH, Tijian R (1990) The Drosophila fos-related AP-1 protein is a developmentally regulated transcription factor. Genes Dev 4:822−834

Tautz D, Pfeifle C (1989) A nonradioactive in situ hybridization method for the localization of specific RNAs in Drosophila embryos reveals translational control of the segmentation gene hunchback. Chromosoma 98:81−85

21.13 Double Labeling of mRNA and Proteins in *Drosophila* Embryos

BARBARA COHEN and STEPHEN M. COHEN

21.13.1 Principle and Applications

A variety of new methods to visualize spatially restricted domains of gene expression have become available in recent years. Application of these methods to visualize messenger RNA by in situ hybridization and proteins by antibody reactions have provided tremendous insights into the biology of early embryonic development in a variety of systems. Many of these methods have been developed for application to the *Drosophila* embryo, but are gaining wider use in vertebrate embryo systems including the mouse, the frog *(Xenopus)*, and the zebra fish.

Often it is useful to compare the expression patterns of two or more genes simultaneously in the same animal. This guarantees that circumstances such as age, genetic background, or conditions of a staining reaction are exactly the same and therefore allows for precise comparison

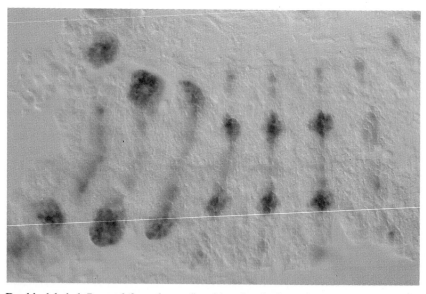

Doubly labeled *Drosophila* embryo, first histochemically stained to detect β-galactosidase activity. The blue stripes identify cells expressing the *wingless* gene. The embryos were subsequently processed for in situ hybridization using a digoxigenin-labeled probe to detect the mRNA product of the *Distal-less* gene (purple spots). The stripes of cells expressing *wingless* bisect the patches of cells expressing *Distal-less*. The double labeling procedure allows the expression patterns of two genes to be compared at cellular resolution of the spatial domains in which the different genes are expressed. Another often crucial question that can be addressed by double labeling is the relative timing of expression of different gene products in a given cell. These considerations are particularly important in understanding gene hierarchies in development. In order for a gene to directly activate another one it should be expressed in the same cells prior to its target.

The techniques described here combine the method of whole mount RNA in situ hybridization with an enzymatic reaction to detect β-galactosidase activity or with a procedure for antibody staining based on the biotin-streptavidin system. We pay special attention to applications involving β-galactosidase since the *lac* Z gene is widely used as a reporter gene. In particular, the application of enhancer detector screening to visualized patterns of expression of endogenous genes using a β-agalctosidase reporter (O'Kane and Gehring 1987) has made *lac* Z the marker of choice in many experiments.

At this point one major weakness of the described techniques should be mentioned: All of the detections are based on enzymatic reactions that may be nonlinear. Therefore precise quantification of the signal is not possible. If only antibody staining is performed this problem can be avoided using fluorescent secondary antibodies. Double labels with different fluorochromes provides unambiguous detection of overlapping staining patterns. Unfortunately, the whole mount in situ hybridization technique has not yet been satisfactory developed for use with fluorescent labels.

21.13.2 Choice of Signal Detection System

A number of options are available in the choice of colorimetric reactions to visualize transcripts and proteins. To facilitate double labeling, the conditions for both in situ hybridization and antibody staining can be manipulated in order to obtain different color reaction products.

The in situ hybridization product is detected using alkaline phosphatase-conjugated antibody and subsequent enzyme-catalyzed color reaction with various substrates available in kits from several suppliers. All kits are based on either reduction of tetrazolium salts or the production of colored diazo compounds. The color of the in situ hybridization reaction depends upon the substrate used for the alkaline phosphatase reaction. Available substrates can produce brown/blue, red or black signals. The NBT/BCIP substrate loses some intensity and changes color from brown to blue when dehydrated and mounted in a nonaqueous medium. However, it is considerably more sensitive than the other commercially available substrates and the slight loss of intensity usually not a problem. In some cases the actual signal to noise ratio seems to be improved after dehydration. For use in combination with an X-gal reaction however it is preferable to mount the double-labeled preparations in the aqueous tris/ glycerol medium to achieve more distinctive colors (brown vs light blue as opposed to two different blues).

Antibody staining can be done with alkaline phosphate or peroxidase detection systems. We find that the Vectastain system produces very clean and intense labeling. This system is based on use of biotin-coupled antibodies, streptavidin and biotin-conjugated enzymes (either horseradish peroxidase or alkaline phosphatase). Biotin is a small, water soluble vitamin that can be readily conjugated to antibodies without influencing biological activity (Guesdon et al., 1979). Streptavidin is a tetrameric protein isolated from the bacterium *Streptomyces avidinii*. It has four sites to bind biotin molecules with high affinity. The specificity and strength of this bond coupled with the signal amplification that occurs allow sensitive immunodetection.

The color of the HRP-antibody staining can be influenced by adding $NiCl_2$ or $CoCl_2$ to 0.04% in the final HRP detection solution (Lawrence et al., 1987; Kellerman et al., 1990). This changes the color of the precipitate from light brown to dark gray/purple. In addition, secondary antibodies coupled to alkaline phosphatase are also available, providing all the possibilities described above. Due to space limitations, the detailed protocol only deals with the HRP detection system for antibody staining. Given the detailed instructions provided by the suppliers of the different staining kits, it should be easy to invent and perform various other possible combinations.

X-gal (5-bromo-4-chloro-3-indoxyl-β-D-galactoside) is a substrate for β-galactosidase. It is used with potassium ferricyanide as a catalyst for the oxidation of the indoxyl substrate leading to a blue signal. Potassium ferrocyanide is added in an equimolar amount to prevent overoxidation to a colorless dehydroindigo.

Application of fluorescent probes to whole mount in situ hybridization would be very useful. Although fluorescent antibodies to DIG are commercially available, these methods have not proven to be useful in labeling of whole mounts. The problematic step seems to be that hybridization of the embryos in 50% formamide results in strong background auto-fluorescence when illuminated at standard wavelengths. If the limitations in fixation and hybridization protocols can be solved we will gain a valuable set of tools.

21.13.3 Double Labeling of mRNA and Proteins

Dechorionation reagents
- Commercial bleach
- Triton X 100

Fixation and devitellinization reagents
- EGTA
- Pipes buffer
- Formaldehyde EM grade 10% (Polysciences)
- Heptane
- Methanol

- Diethylpyrocarbonate (Sigma)
- Formamide (Boehringer Mannheim)
- Herring sperm DNA (Sigma)
- Heparin (176 U/mg, Sigma)
- Anti-DIG antibody (Boehringer Mannheim)
- Nitroblue Tetrazolium (NBT) (Boehringer Mannheim)
- 5-bromo-4-chloro-3-indolyl phosphate (BCIP) (Boehringer Mannheim)
- Alkaline phosphatase substrate kits (Vector labs)
- Hepes
- Dithiothreitol (DTT)
- Random hexanucleotide primer (Boehringer Mannheim)
- dATP, dTTP, dCTP, dGTP
- DIG-dUTP (Boehringer Mannheim)
- Klenow DNA polymerase (Boehringer Mannheim)
- Proteinase K (Boehringer Mannheim)
- Glycine

- X-gal (Boehringer Mannheim)
- $K_3[Fe^{III}(CN)_6]$ (Sigma)
- $K_4[Fe^{II}(CN)_6]$ (Sigma)

- Tween 20 (Fisher)
- Bovine serum albumin (Merck)
- Primary antibody (rabbit, rat, mouse)
- Vectastain Kit, contains secondary antibody, normal serum, ABC reagents (Vector labs)
- Diaminobenzidine (Sigma)
- H_2O_2 (30% stabilized)

- Methylsalycilate (Sigma)
- Canada balsam (BDH)
- Ethanol
- Glycerol
- Tris-HCl

- Fixation buffer: 180 mM Pipes, pH 6.9; 4 mM $MgSO_4$; 2 mM EGTA
- Fixation solution: (per sample) 2 ml fixation buffer; 1.4 ml 10% [w/v] formaldehyde; 5 ml heptane
- PBS: 140 mM NaCl; 7 mM Na_2HPO_4; 3 mM KH_2PO_4; pH 7
- PBT: PBS with 0.1% [v/v] Tween 20
- PBX: PBS with 0.3% [v/v] Triton-X 100
- DEPBT: PBS treated with 0.1% [v/v] diethylpyrocarbonate, autoclaved, and 0.1% [v/v] Tween 20 added
- 20× SSC: 3 M NaCl; 0.3 M Na-citrate; pH 7.2
- Proteinase K stock solution: 4 mg/ml, stored at $-20\,^{\circ}$C in small aliquots
- Methanol/EGTA solution: 95% [v/v] methanol; 5% [v/v] 0.5 M EGTA; pH 8

- Hybridization solution: 50% [v/v] formamide; 5× SSC; 100 µg/ml denatured herring sperm DNA; 50 µg/ml heparin; 0.1% [v/v] Tween 20
- 10× synthesis buffer: 500 mM Hepes pH 6.6; 50 mM $MgCl_2$; 20 mM DTT
- DIG-nucleotide mix: 1 mM dATP; 1 mM dCTP; 1 mM dGTP; 0.65 mM dTTP; 0.35 mM DIG-[11]-dUTP
- X-gal staining solution: 3.1 mM $K_4[Fe^{II}(CN)_6]$; 3.1 mM $K_3[Fe^{III}(CN)_6]$; 0.3% [v/v] Triton-X100; 0.2% [w/v] X-gal in PBS
- BBT: PBT plus 0.1% BSA
- BBT 250: BBT with an additional 250 mM NaCl
- ABC mix: see instructions in the Vectastain kit
- HRP reaction solution: 1 ml PBT with 0.5 mg diaminobenzidine and 0.0015% [v/v] H_2O_2; prepared fresh for each reaction
- Tris/glycerol mounting medium: 70% [v/v] glycerol; 30% [v/v] 100 mM Tris, pH 7.5
- GMM: 1.5 g/ml Canada balsam in methylsalycilate

21.13.3.1 Whole Mount In Situ Hybridization

This protocol is adapted with only minor modifications from the original technique developed by Tautz and Pfeifle (1989). Hülskamp and Tautz (1991) also described the use of substrates other than the NBT/BCIP system.

Collection and fixation of the embryos

1. Collect the embryos with a paintbrush and transfer them into a basket (microcentrifuge tube cut to about 8 mm from top and melted onto fine wire mesh).
2. Rinse with distilled H_2O.
3. Dechorionate embryos by immersing the basket in commercial bleach for approximately 2 min.
4. Rinse thoroughly with 0.1% [v/v] Triton-X 100 and then with distilled H_2O.
5. Remove excess liquid by blotting the basket on a paper towel.
6. Transfer into the fixation solution in a scintillation vial by turning the basket upside down and rinsing the embryos off the mesh bottom of the basket with heptane.
7. Shake vigorously for 20 min at room temperature.

Devitellinization

1. Remove the scintillation vial from the shaker and let phases separate (the embryos float at the interface).
2. Remove as much of the aqueous (lower) phase as possible with a Pasteur pipette.
3. Add 10 ml methanol and shake vigorously (either by hand or using a vortex mixer; forcing the embryos across the phase boundary between heptane and methanol removes the vitelline membrane).
4. Let phases separate.
5. Remove most of the heptane and add another 5 ml of methanol and shake well.
6. Devitellinized embryos settle to the bottom.

1. Transfer embryos into microcentrifuge tube (use blue pipet tip with cut off tip). All subsequent washes are peformed in 1 ml volumes on a rotating shaker at low speed unless otherwise specified.
2. Rinse embryos 3× with methanol/EGTA (ME). At this point embryos can be stored at −20 °C.
3. Rehydrate through sequential washes of a freshly made 7:3, 1:1 and 3:7 mixes of ME and 4% [w/v] formaldehyde in DEPBT (5 min each).
4. Postfix for 20 min in 4% [w/v] formaldehyde in DEPBT.
5. Wash 3 × 5 min in DEPBT.

1. Remove proteinase K aliquot from freezer and add determined amount (see Sect. 21.11.2) to 1 ml of DEPBT immediately when thawed and incubate for 5 min with mixing.
2. Stop reaction by washing embryos for 2 min in 2% [w/v] glycine in DEPBT.

1. Wash 2 × 5 min with DEPBT.
2. Fix for 20 min in 4% [w/v] formaldehyde in DEPBT.
3. Wash 5 × 5 min in DEPBT.
4. Wash 10 min in 1:1 DEPBT : hybridization solution.
5. Prehybridize for 1 h at 45 °C (without shaking).

The probe should be prepared ahead of time:
1. Boil 100−300 ng template DNA in 13 µl H_2O for 10 min to denature.
2. Cool rapidly in ice water bath.
3. Spin 10 s in microfuge and put back on ice.
4. Add 3 µl random primer, 2 µl 10× synthesis buffer, 1 µl DIG-nucleotide mix, 1 µl Klenow (2 units) and mix gently.
5. Incubate at 37 °C overnight.
6. Add 80 µl hybridization solution and store at −20 °C (for most applications it is not necessary to purify the probe DNA by precipitation).

1. Denature probe by boiling for 10 min, cool quickly in ice water, spin 10 sec in microfuge, and mix well.
2. Remove excess hybridization solution from prehybridized embryos (final volume should be around 100 µl).
3. Add 1%−5% of probe to embryos (optimal amount varies with abundance of the transcript and must be empirically determined for each probe).
4. Mix well and incubate at 45 °C overnight (without mixing).

1. Add 1 ml of prewarmed (45 °C) hybridization solution and let embryos settle to rinse away excess probe.
2. Wash 20 min with hybridization solution at 45 °C (without mixing).
3. Wash 20 min with each of the following mixes of hybridization solution: DEPBT 4:1; 3:2; 2:3; 1:4 at room temperature with gentle mixing.
4. Wash 4 × 10 min with DEPBT.

Anti-body incuba-tion Incubate with 1 : 2000 dilution of anti-DIG antibody for 1 h at room temperature in 500 µl volume (antibody should be preabsorbed overnight against fixed embryos at a maximal dilution of 1 : 300 in DEPBT at 4°C; preabsorbed antibody can be stored for months at 4°C).

Detection reaction 1. Wash embryos 4 × 10 min with DEPBT and 2 × 5 min in alkaline phosphatase buffer (the buffer will vary with choice of substrate).
2. Transfer embryos in 1 ml buffer into staining dish (either a watch glass or a small weighing dish).

The actual staining reaction will vary with choice of substrate, each kit will give details. The signal usually becomes visible within the first 30 min, but the reaction can go on for several hours and only needs to be stopped when the background is getting too high. Monitor the staining reaction under the microscope occasionally and stop by washing 3 × 10 min in DEPBT when the desired signal intensity is reached.

Mounting – Option 1: equilibrate into 70% [v/v] glycerol; 100 mM Tris-HCl, pH 7.5 (approximately 2 h at room temperature or overnight at 4°C), transfer embryos onto a slide in 75 µl of solution and cover with 20 × 40 mm coverslip.
– Option 2: dehydrate embryos for 10 min each in 70% and 90% [v/v] ethanol and transfer (with as little ethanol as possible) onto a slide with approximately 500 µl GMM; let ethanol evaporate, spread embryos, and cover.

The cellular morphology of the embryos will look better using option 1, particularly under interference contrast optics. The dehydration procedure in option 2 will change to color of the staining reaction from brown to blue and will slightly reduce the signal intensity. The embryos clear after incubation in GMM, so that internal structures will be more readily visible.

21.13.3.2 X-Gal Reaction Followed by In Situ Hybridization

Application of this technique has been published by Phillips et al. (1990) and Cohen et al. (1991).

Collection and fixation of embryos – Collect, dechorionate, and fix the embryos as described above (Sect. 21.13.3.1)

X-gal staining 1. Stop shaking and let phases separate.
2. Transfer embryos with a cut blue pipet tip into a microcentrifuge tube (the embryos will be sticky and have to be handled slowly and carefully).
3. Remove supernatant (embryos will stick to the tube).
4. Add 1 ml PBX and gently vortex to expose embryos to the detergent.
5. If embryos do not sink, spin for 2 s in the microfuge at low speed.
6. Wash 3 × 5 min with PBX; if necessary continue to vortex and spin (all washes are done in 1 ml solution with mixing unless otherwise specified).

7. Exchange PBX with 500 µl of the X-gal staining solution (this solution should be preincubated for 5 min at 37°C) and incubate at room temperature until the X-gal staining reaction is sufficiently intense. The reaction can be monitored by transfering a sample of embryos with some staining solution into a watch glass or a small weighing boat.
8. Stop reaction by washing embryos 3 × 2 min with PBX.
9. If the X-gal staining reaction is strong, incubation times of less that 1 h may suffice, whereas weaker staining can take up to 24 h. In practice, the minimal time should be used for this reaction, since prolonged incubation will reduce the intensity of the in situ hybridization signals in the subsequent reactions, presumably due to RNA degradation.

1. Transfer embryos with as little PBX as possible into an empty scintillation vial. **Devitellinization**
2. Add 5 ml heptane and then 10 ml methanol (in this order!).
3. Immediately shake vigorously by hand or vortex for 10 s.
4. Remove most of the heptane and add another 5 ml methanol and shake again (most embryos will devitellinize and sink to the bottom).
5. Transfer embryos into microcentrifuge tubes.
6. Proceed with the methanol: EGTA washes after the fixation and follow the entire in situ hybridization protocol without any changes.

21.13.3.3 In Situ Hybridization Followed by Antibody Staining

Application has been described by Cohen (1990) using a slightly more complicated procedure which has been simplified and improved; see also Ingham et al. (1991).

1. Follow the entire in situ hybridization protocol (see above) until the detection reaction is completed.
2. Wash embryos 3 × 5 min in DEPBT.

1. Dehydrate embryos by washing them for 5 min each in 30%, 50%, 2 × 70% ethanol (all washing steps are performed at room temperature with mixing unless otherwise specified). **Dehydration and rehydration**
2. Store embryos overnight at 4°C in 70% ethanol (without rotation).
3. Rehydrate embryos by washing for 5 min each in 50% and 30% ethanol.

1. Wash embryos 3 × 10 min in BBT 250. **Incubation with primary antibody**
2. Add appropriate dilution of antibody into 500 µl BBT 250 and incubate at 4°C overnight (most antibodies should be preabsorbed against fixed embryos in advance. Optimal dilutions will vary and need to be individually determined).

1. Wash 4 × 10 min in BBT 250. **Incubation with secondary antibody**
2. Wash 2 × 20 min in 500 µl BBT 250 with 2% normal serum (from the species in which the secondary antibody was raised).
3. Incubate for 2 h at room temperature in 500 µl BBT-250 containing 2% normal serum and appropriate dilution of secondary antibody (anti-

bodies from the Vectastain Kit are preabsorbed overnight against fixed embryos at 1:10 and used at 1:500).

ABC incu-
bation
1. Wash 6 × 10 min in PBT.
2. Incubate for 1 h in 500 μl ABC mix (components and concentration of reagents in PBT depend on the kit used).

HRP stain-
ing reaction
1. Wash 5 × 10 min in PBT.
2. Transfer embryos in 1 ml PBT into watch glass or small weighing boat.
3. Add 0.5 mg diaminobenzidine (DAB) and 0.0015% H_2O_2 by adding 50 μl of a 10 mg/ml stock solution of DAB (the stock may be frozen and reused repeatedly) and 5 μl of a 1/100 dilution of a 30% stock of H_2O_2.
4. Mix the solution and monitor the developing reaction under the microscope. Be prepared to stop the reaction quickly, as the staining often comes up within seconds. The components are only reactive for approximately 20 min; prolonged incubations will only increase background.
5. Stop by washing 3 × 5 min with PBT. Follow appropriate procedures to dispose of DAB.

Mounting Dehydrate embryos for 10 min each in 70%, 90%, and 100% ethanol. Transfer, with as little ethanol as possible, onto a slide with 500 μl GMM. Let ethanol evaporate, spread embryos, and cover.

21.13.4 Special Hints for Application and Troubleshooting

Optimal
double label
combina-
tions
If more than one method is available for a given gene product the quality of the data can be improved by considering the following criteria:

– In general combining X-gal staining with in situ hybridization will give the most photogenic results. The morphology of the preparation is in general better if the in situ hybridization is combined with the X-gal reaction rather than an antibody staining. If the samples are mounted in glycerol, the colors of the reaction products are quite distinct (brown and blue). If the samples are dehydrated both reaction products are blue, but the colors are still reasonably distinct.

– The time taken for the X-gal reaction should be held to a minimum. Prolonged incubations will lead to reduced intensity in the in situ hybridization reaction, presumably due to loss of RNA. In practical terms, the real time limit depends on the abundance of the RNA to be examined. If the RNA is abundant, longer preincubation times can be tolerated.

– If the abundance of the RNA is low to begin with, the quality of the staining may improve if the in situ hybridization is done first and followed by an antibody staining to β-galactosidase. In our hands, the enzymatic X-gal assay is less sensitive than using antibody staining to detect β-galactosidase. The quality of the anti-β-galactosidase is critically important. We have obtained best results with affinity purified antibody from Cappel.

— If the supply of sample is limiting, it may be preferable to avoid X-gal staining, since this reaction must be performed on fresh preparations. For antibody staining and in situ hybridization, fixed embryos can be stored; therefore, the final sample size can be increased by combining various small samples collected over time.

— A limited range of choice of color is available for the reactions of the antibody staining and in situ hybridization. Several different substrates are available for alkaline phosphatase. We find the NBT/BCIP combination to be most sensitive. The other options may be useful for double labeling if intensity is not a concern. The color of the X-gal reaction cannot be manipulated.

— Detecting synchronous expression of two genes in the same cell is easier if one of the gene products is predominantly nuclear and the other one cytoplasmic. The mRNA will always be localized predominantly in the cytoplasm. Therefore it will be ideal to look at either nuclear X-gal activity or a nuclear localized protein in a region of potential overlap.

Sticky embryos

The X-gal staining is performed without prior removal of the vitelline membrane. These embryos stick to every surface and to each other. To avoid problems all steps following the fixation are done in the presence of higher levels of detergent. However, the transfer of the embryos from the scintillation vial to the microcentrifuge tube immediately following the fixation is somewhat tricky because the embryos tend to stick to the pipet tip or transfer pipet. In most cases this can be avoided when the solution containing the embryos is released very slowly into the microcentrifuge tube. In addition, a tip with embryos sticking to it can be rinsed with the PBX and after pipetting up and down several times the embryos will come off.

Proteinase K treatment

The activity of the proteinase K varies to a certain extent with every batch and even more between suppliers. It is advisable to dissolve the proteinase K at the final concentration (4 mg/ml) and freeze it without delay in small aliquots (20–50 μl). Do not freeze or store the aliquots after thawing. It is necessary to titer the proteinase by following the entire in situ hybridization protocol. Overdigestion will result in breakage of the embryos, underdigestion will lead to increased background. In addition the optimal extent of the treatment depend on the age of the embryos: younger embryos can stand and will need longer incubations than older embryos. Average values are 7 μl and 5 μl of 4 mg/ml stock in 1 ml DEPBT for young and old embryos, respectively.

21.13.4

References

Cohen B, Wimmer E, Cohen SM (1991) Early development of the leg and wing primordia in the *Drosophila* embryo. Mech Devel 33:229–240

Cohen SM (1990) Specification of limb development in the *Drosophila* embryo by positional cues from segmentation genes. Nature 343:173–177

Guesdon J, Ternynck T, Avrameas S (1979) The use of avidin biotin interaction in immunoenzymatic techniques. J Histochem Cytochem 27(8):1131–1139

Hülskamp M, Tautz D (1991) Gap genes and gradients – the logic behind the gaps. Bio Essays 14, No 6:261–268

Ingham P, Taylor AM, Nakano Y (1991) Role of the *Drosophila* patched gene in positional signaling. Nature 353:184–187

Kellermann KA, Mattson DM & Duncan I (1990) Mutations affecting the stability of the fushi tarazu protein of *Drosophila*. Genes & Dev 4:1936–1950

Lawrence PA & Johnston P (1989) Pattern formation in the *Drosophila* embryo: allocation of cells to parasegments by even-skipped and fushi tarazu. Development 105:761–767

O'Kane CJ, Gehring WJ (1987) Detection in situ of genomic regulatory elements in *Drosophila*. Proc Natl Acad Sci USA 84:9123–9127

Phillips RG, Roberts IJH, Ingham PW, Whittle JRS (1990) The *Drosophila* segment polarity gene patched is involved in a position-signaling mechanism in imaginal discs. Development 110:105–114

Tautz D & Pfeifle C (1989) A nonradioactive in situ hybridization method for the localization of specific RNAs in *Drosophila* embryos reveals translational control of the segmentation gene hunchback. Chromosoma 98:81–85

22 Quantitative Formats

22.1 Affinity-Based Collection of Sandwich Hybrids: A Quantitative Hybridization Format

HANS SÖDERLUND and KAI KORPELA

22.1.1 Principle and Applications

In most applications of hybridization assays the target nucleic acid is immobilized onto a filter, where it is then allowed to react with a labeled probe. In the sandwich hybridization reaction, the target sequence is kept in solution while allowed to react with two probes, one for capture, and one for detection (Ranki et al., 1983). The two probes should recognize two essentially adjacent sequences on the target but must not recognize each other. The advantage of sandwich hybridization is that it removes the step in which the target is immobilized. This step is time-consuming but, more important, when crude samples are immobilized biological compounds other than nucleic acids may also be bound causing unspecific binding of detector probe. Furthermore the use of two probes increases the specificity of the reaction and removes the potential error caused by cloned probes recognizing the vector in addition to the actual target (Chou et al., 1983).

A common factor in filter hybridization is that the reactions are carried out in mixed phase and are slowed down due to accessibility problems for the reacting molecules (Flavell et al., 1974). Both the primary collision between the two DNA strands and the subsequent winding process are retarded. Consequently, to shorten the incubation time the hybridization step should preferably be performed in solution.

In the affinity-based hybrid collection method (Syvänen et al., 1986) the sandwich hybridization is performed in solution. The capture probe carries an affinity tag by which it efficiently can be bound to an immobilized counterpart which has a high affinity for the tag. The detector probe is labeled by any detectable group described, e.g., elsewhere in this book. After the hybridization the capture probe, indepently of whether it is hybridized or not, is separated from other components by affinity capture

Hybridization in solution

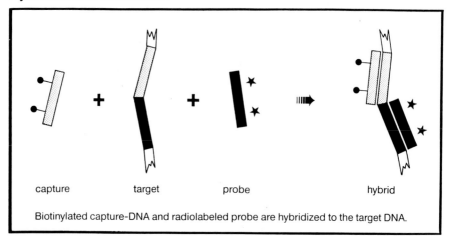

capture   target   probe   hybrid

Biotinylated capture-DNA and radiolabeled probe are hybridized to the target DNA.

Affinity-based hybrid collection

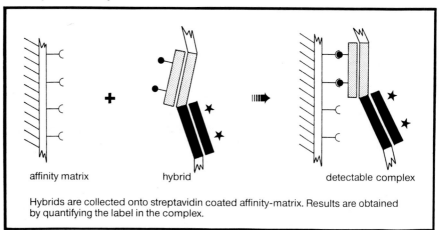

affinity matrix   hybrid   detectable complex

Hybrids are collected onto streptavidin coated affinity-matrix. Results are obtained by quantifying the label in the complex.

The affinity based hybrid collection assay

prior to quantification of the bound label. The principle of the method is shown in the figure. The hybridization and capture steps are represented as independent and subsequent steps; however, sometimes it is convenient to hybridize and capture simultaneously.

When appropriately assembled the system gives a quantitative test for the target sequence. The assay requires somewhat more preparations than a direct filter hybridization, but is, when set up, more convenient to run with long series of samples. The main application areas of the affinity-based hybrid collection system are when a given test is frequently repeated and/or when many crude samples should be analyzed. This is the case, e.g., in diagnostics (Ranki et al., 1988).

The affinity capture method is quite versatile and can be modified depending on the question and on available reagents. Some considerations are:

1. *Probes:* At least two probes are used, one for capture and one for detection. Increased sensitivity is obtained by using several probe pairs for the same target. Synthetic oligonucleotides and polynucleotides may be used. It is, however, imperative that the two probes do not hybridize to each other. Vector sequences included in the probes will not disturb the assay, but it should be noted that many otherwise unrelated vectors contain the same indicator gene (e.g., *lac* z) or polylinker sites which cross-hybridize to each other.

2. *Affinity pairs:* It should be noted that when affinity pairs are used for detection (Sect. 3) it is advantageous to introduce as many label moieties per probe molecule as possible. For capture the important factor is rather that all capture probe molecules are modified, while the label density is less critical. Three different affinity pairs are generally used.
   − Biotin/streptavidin: The capture probe is biotinylated, e.g., by nick translation or by using photobiotin. Avidin or streptavidin is immobilized onto a solid support (Syvänen et al. 1986). The pair has a very high affinity and is stable even in very harsh conditions. It is the method of choice when biotin is not used as the marker in the detector probe (see Sect. 4.1).
   − Homopolynucleotides: The capture probe is tagged with, e.g., poly A using terminal transferase and a poly U or oligo T resin is used for capture (Morrisey and Collins, 1989). Specific nucleotide sequences can also be used but homopolynucleotides give faster capturing rates and a more versatile solid phase.
   − Hapten/antibody: The capture probe is modified to contain a hapten group while the corresponding antibody is immobilized (Syvänen et al., 1986). Several haptens are available as has been described elsewhere in this volume.

3. Solid phase: The use of coated tubes or microtiter plates is often very convenient and allows the use of available equipment used in immunoassays. The drawback is that such surfaces have a limited capacity of binding and thus limit the amount of capture probe which can be used in the hybridization step. Different kinds of resins usually give much higher binding capacity but many introduce additional steps in the handling. High capacity streptavidin-coated beads are commercially available both as latex (Fluoricon avidin assay particles, Baxter Healthcare Corporation, USA.) and magnetic (Dynabead, Dynal AS, Norway) particles.

4. Detectable label: The affinity capture system is adaptable to most non-radioactive systems described elsewhere in this volume. We have good experience with time-resolved fluorescence, and with enzyme-mediated colorimetric, fluorometric, and luminescent measurements.

Even a biosensor has recently been described in this hybridization format (Olson et al. 1991).

5. Sample treatment: Samples of different origin may obviously require different pretreatment. The most straightforward way, adaptable to most situations, includes sample solubilization with ionic detergents, such as SDS, whereafter the DNA is denatured by boiling in alkaline solution. The specimen is directly introduced into the hybridization assay after neutralization (Jalava et al., 1990). Tissues, such as biopsy specimens, high concentrations of serum, or large amounts of cells are preferably pretreated with proteinase K and SDS (Korpela et al. 1991). Bacterial cells are usually surrounded by a rigid cell wall and the DNA cannot quantitatively be released without treatment with suitable enzymes, such as lysozyme (Korpela et al., 1987).

6. Hybridization conditions: The affinity capture and sandwich hybridization systems do not introduce any parameters not known in other hybridization assays. High concentrations of SDS must be avoided when using hapten/antibody pairs for capture.

22.1.2 Quantification of HPV DNA by Affinity Capture

Here a test procedure is described in which human papilloma virus DNA in a crude cell sample is quantified using a biotinylated single stranded clone as capture probe and a ^{35}S-labeled plasmid clone as detector probe. The procedure is essentially as in the AffiProbe HPV kit made by Orion Pharmaceutica, Biotechnology.

Standard reagents
- The AffiProbe kits, including the coated plates, are available from Sangtec Medical, Bromma, Sweden.
- Photobiotin is available from Life Technologies Limited, Renfrewshire, Scotland.
- The scintillation fluid used is, e.g., Aquasol (Du Pont, Stevenage Hertfordshire, U.K.) or OptiPhase „HiSafe" (Pharmacia LKB Biotechnology AB, Uppsala, Sweden).

Standard solutions
- Sample preparation solution: 0.2 M NaOH, 0.5% SDS, 0.2 mg/ml herring sperm DNA
- Hybridization solution: 2.4 M NaCl, 0.24 M sodium-citrate, 0.4 M PO$_4$-buffer (pH 6.6), 0.07% Ficoll, 0.7% polyvinylpyrrolidone, 1.8% polyethyleneglycol, 4.4 mg/ml BSA, 0.6 M acetic acid.

1. The capture probe is a M13 clone with a 1.7 kb insert specific for the target. The DNA is biotinylated using photobiotin as described by the manufacturer. In essence, 25 μl of the photoprobe and 25 μl of the DNA (1 mg/ml) is illuminated for 15 min at 0 °C after which 50 μl of 0.1 M Tris-HCl, pH 9.0, is added. Free photobiotin is removed by two extractions with 100 μl isobutanol. The biotinylated DNA is further precipitated with ethanol. The plasmid DNA (pBR322 with a 2.8 kb insert) is

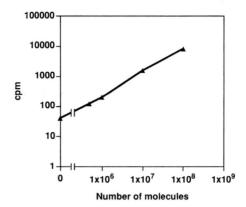

Standard curve for detection of
HPV 18 in the AffiProbe test

labeled with [^{35}S] dCTP by nick translation to a specific activity of $2-4$ $\times$ 10^8 cpm/µg of DNA. The clones have been described in detail elsewhere (Jalava et al., 1990). Prior to use, the detector probe is denatured by boiling and cooled on ice. Do not boil the capture probe! It is destroyed by boiling. The two probes are combined to give 0.5 fmol detector and 10 fmol capture per 20 µl assay.

2. The sample, containing less than approximately 5 million cells, is boiled in 60 µl of 0.1 M NaOH and 1% SDS for 5 min in an eppendorf tube. Then 30 µl of the hybridization mixture and 20 µl of the probe-capture mixture is added. The final hybridization conditions are: 0.66 M NaCl, 65 mM sodium citrate, 0.3 mM EDTA, 0.1 M phosphate buffer (pH 6.6), 0.3% SDS, 0.02% Ficoll, 0.2% polyvinylpyrrolidone, 0.5% polyethyleneglycol (4000), 1.2 mg/ml BSA. The solution is incubated at 65°C for 3h and then transferred into a streptavidin-coated well of a microtitration plate. The plate is covered by adhesive tape. The capture reaction is allowed to proceed for 2h at 37°C with shaking. Then the wells are washed six times with 0.1× SSC, 0.2% SDS preheated to 50°C. Finally the bound hybrids are eluted twice with 0.2 M NaOH and transferred into a scintillation tube, scintillation fluid is added, and the radioactivity measured.

The standard curve of the AffiProbe test for HPV 18 is shown in the figure. Different numbers of the standard plasmid molecule containing HPV 18 as insert have been analyzed in the assay. The sensitivity for detecting HPV 18 is about 5×10^5 molecules.

22.1.3 Special Hints for Application and Troubleshooting

– The system is easily adaptable to nonradioactive tracers described elsewhere in this volume. Since most laboratories still use radioactivity as a reference system that format is described here.

- The total amount of biotin in the system must not exceed the binding capacity of the immobilized streptavidin. Thus, avoid too high a concentration of capture probe, and carefully remove free biotin after the biotinylation reaction. Samples usually do not contain enough free biotin to affect the capture reaction.
- Poorly purified probe preparations may contain chromosomal DNA. If both probes are contaminated with DNA from the same host (e.g., *Escherichia coli*) direct hybridization may occur causing background.
- If the concentration of target exceeds that of the probes, signal reduction will be observed.
- NaOH-denatured samples must be neutralized prior to hybridization. Errors in pipetting may lead to an alkaline final pH in which no hybridization will occur.

References

Chou S,Merigan M (1983) Rapid detection and quantitation of human cytomegalovirus in urine through DNA hybridization. N Eng J Med 308:921–925

Flavell RA, Birfelder EJ, Sanders PM, Borst P (1974) DNA-DNA hybridization on nitrocellulose filters, 1, General considerations on non-ideal kinetics. Eur J Biochem 47:535–543

Jalava T, Kallio A, Leinonen AW, Ranki M (1990) A rapid solution hybridization method for detection of human papilloma viruses. Mol Cell Probes 4:341–352

Korpela K, Laaksonen M, Kallio A, Söderlund H, Pettersson U, Kyrönseppä H, Ranki M (1992) Detection of Plasmodium falciparum DNA in a microtitration plate based hybridization test. FEMS Microbiol Lett 90:173–178

Korpela K, Buchert P, Söderlund H (1987) Determination of plasmid copy number with nucleic acid sandwich hybridization. J Biotechnol 5:267–277

Morrisey DV, Collins ML (1989) Nucleic acid hybridization assays employing dA-tailed capture probes. Single capture methods. Mol Cell Probes 3:189–207

Olson JD, Panfifli PR, Zuk RF, Sheldon EL (1991) Quantitation of DNA hybridization in silicon-based system: application to PCR. Mol Cell Probes 5:351–358

Ranki M, Syvänen AC, Söderlund H (1988) Nucleic acids in viral diagnosis. In: Halonen P, Murphy FA, Ohasi M, Turano A (eds) Laboratory diagnosis of infectious diseases. Springer-Verlag, Berlin, pp 132–151

Ranki M, Palva A, Virtanen M, Söderlund H (1983) Sandwich hybridization as a convenient method for detection of nucleic acids in crude samples. Gene 21:77–85

22.2 Detection of DNA:RNA Target:Probe Complexes with DNA:RNA-Specific Antibodies

FRANCOIS COUTLÉE, ROBERT H. YOLKEN, and RAPHAEL P. VISCIDI

22.2.1 Principle and Applications

We have devised a novel enzyme immunoassay (EIA) for the quantitation of biotinylated DNA:RNA hybrids. The method involves a homogeneous hybridization reaction between sample nucleic acids and a complementary biotinylated nucleic acid probe. The detection of DNA extracted from samples (Coutlee et al., 1989a; Viscidi et al., 1989) or amplified with the polymerase chain reaction (PCR) (Bobo et al., 1990; Coutlee et al., 1991a) is accomplished with biotinylated RNA probes. The biotin-labeled RNA probes are generated by in vitro transcription, in the presence of biotinylated UTP (bio-11-UTP), of a DNA template (Coutlee et al., 1989b). The RNA probe can be synthesized from sequences cloned into vectors with RNA polymerase promoters. The DNA template used to generate the RNA probe can also be prepared by PCR with a nested set of primers, one of these primers containing sequences of the T7 RNA polymerase promoter at its 5' end (Schowalter et al., 1989; Stoflet et al., 1988). The amplified segment can be transcribed in vitro into biotinylated single-stranded RNA (Coutlee et al., 1989b). The detection of RNA requires the use of a biotinylated DNA probe which is labeled with biotinylated dUTP (bio-11-dUTP) in a standard nick translation reaction (Viscidi et al., 1989; Coutlee et al., 1989a; Coutlee et al., 1990a).

After completion of the hybridization reaction, the quantitation of biotinylated DNA:RNA hybrids is performed in microtiter plates (Viscidi et al., 1989; Coutlee et al., 1989a; Coutlee et al., 1989b). Biotin-labeled DNA:RNA hybrids are captured onto wells of a microtiter plate coated with an anti-biotin antibody. Following removal of unbound nucleic acids by washings, the amount of probe-target nucleic acid hybrids bound to the solid phase is measured by the addition of an enzyme-labeled monoclonal antibody (Mab) directed against DNA:RNA hybrids and the appropriate enzyme substrate. This anti-DNA:RNA Mab recognizes specifically the helix structure of DNA:RNA or RNA:RNA hybrids but exhibits little reactivity with single- or double-stranded DNA or single-stranded RNA (Bogulawski et al., 1986; Kinney et al., 1989; Viscidi et al., 1989; Yehle et al., 1987). The reactivity of the Mab is independent of the base composition of the hybrid. The enzymatic degradation of substrate is measured in a fluorometer (fluorescent substrates) or in a spectrophotometer (colorigenic substrates) (Coutlee et al., 1989c) and is proportional to the quantity of biotin-labeled hybrids immobilized on the solid phase.

This method of detection of nucleic acids is versatile and has the advantage of combining features of both hybridization reactions and the widely used EIA format. The hybridization reaction between DNA and single-stranded RNA probes is completed in 30 min. The rapidity of the reaction is attributable to the single-stranded nature of RNA probes and the favorable reassociation kinetics of homogeneous hybridization reactions (Coutlee et al., 1989b). DNA probes require longer incubation periods (Coutlee et al., 1989a). The hybridization reaction can be performed under very stringent conditions, with a reannealing temperature of 78°C for long DNA-RNA sequences (Newman et al., 1989; Viscidi et al., 1989; Coutlee et al., 1989a). Since the efficient formation of hybrids can occur over a wide latitude around the optimal temperature, small changes in hybridization temperatures have a minimal effect on assay results (Coutlee et al., 1989a).

The EIA format provides a simple, nonisotopic, and objective method to measure nucleic acids (Yolken et al., 1989). The assay achieves sensitivity endpoints identical to those of ^{32}P-based assays (Coutlee et al., 1989a; Coutlee et al., 1990a). The EIA for DNA:RNA hybrids is linear over two logs of target nucleic acid and can detect 5 pg of complementary DNA with RNA probes (Coutlee et al., 1989b) or 0.5 pg of RNA with DNA probes (Coutlee et al., 1989a; Viscidi et al., 1989). The combination of this assay with PCR allows for detection of 10 viral DNA copies per test (Bobo et al., 1990; Coutlee et al., 1991a; Coutlee et al., 1991b; Coutlee et al., 1992). The signal in the EIA depends on the length of DNA:RNA hybrids, but hybrids as small as 25 bp can be detected. The substrate incubation period can be shortened with a fluorogenic substrate for alkaline phosphatase. For those laboratories without access to a fluorometer, colorigenic substrates can replace fluorogenic substrates but require a longer incubation for substrate degradation to reach a comparable sensitivity (Coutlee et al., 1989c).

This assay has been applied mainly for the detection of viral agents in body fluids. It has been used to detect single-stranded RNA viruses such as HIV-1 (Viscidi et al., 1989; Yolken et al., 1991), HTLV I and HTLV II (Viscidi et al., 1991), influenza virus (Viscidi et al., 1987), and picornavirus (Newman et al., 1989), and for double-stranded RNA viruses such as rotavirus (Viscidi et al., 1987). Detection of the transcriptional activity of human papillomavirus has been assessed by quantitating viral mRNA in clinical specimens and in established cancer-derived cell lines (Coutlee et al., 1991c). By using subgenomic probes, the assay could determine which open reading frame was actively transcribed in a specimen. The assay has also been applied to the analysis of eukaryotic mRNA as demonstrated by a study on IL-1 and IL-2 expression in cell lines (Coutlee et al., 1990a). Combined with PCR, it has been used to detect HIV-1 proviral DNA in peripheral blood mononuclear cells of infants and adults (Coutlee et al., 1990b; Coutlee et al., 1991a) and in stools (Yolken et al., 1991), *Chlamydia trachomatis* (Bobo et al., 1990) in cervical specimens, and human papillomavirus type 16 in cell lines and cervical lavages (Coutlee et al., 1991d; Coutlee et al., 1992).

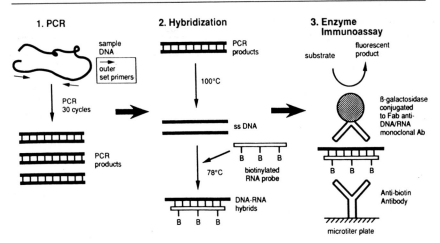

Detection of PCR-amplified DNA segment by enzyme immunoassay (EIA). Viral DNA is amplified with outer primers in clinical samples. The products of amplification are reacted with a biotinylated RNA probe. This probe is synthesized by amplification of a stock of viral DNA by a set of primers nested within the fragment amplified with the outer set of primers. One of the nested set of primers has sequences from the T7 RNA polymerase promoter appended at its 5' end. Transcription of the amplified nested fragment with biotinylated UTP generates the biotinylated RNA probe. After the hybridization reaction between DNA fragments generated with primers of the outer set and the biotinylated RNA probe, the biotin-labeled hybrids are captured by an anti-biotin antibody and detected with an enzyme-labeled monoclonal antibody against DNA:RNA hybrids. (From reference Yolken 1991)

22.2.2 Reaction Scheme

The assay for detection of amplified DNA, designated PCR-EIA, is depicted in the accompanying figure.

22.2.3 Detection of PCR-Amplified DNA by Enzyme Immunoassay

- Riboprobe transcription system for RNA synthesis (Promega)   **Labeling reagents**
- RNasin ribonuclease inhibitor (Promega)
- T7 RNA polymerase and DNase RQ (Promega)
- Dithiothreitol (DTT) (Bethesda Research Laboratories)
- Diethylpyrocarbonate (DEPC) (ICN Biochemicals)
- Bio-11-UTP and bio-11-dUTP (Enzo Inc.)
- NAP-5 Sephadex G-50 columns (Pharmacia)
- Nick translation kit (Life Technologies/BRL)
- 10× PCR buffer, *Taq* DNA polymerase and mineral oil (Perkin Elmer Cetus)
- dNTPS (dATP, dGTP, dCTP, dTTP) 100 mM solutions at pH 7.5 (Pharmacia)

Solution hybridization reagents
- Proteinase K, Tween 20, Nonidet P-40 (Sigma Chemical)
- Guanidinium thiocyanate (Fluka)
- SDS, NaCl, Na-citrate (ICN Biochemicals)
- Hepes (Sigma Chemical)

Enzyme immunoassay for biotinylated DNA-RNA hybrids
- Triton X-100, gelatin (Sigma Co)
- Monoclonal anti-biotin antibody (Boehringer Mannheim)
- 4-methylumbelliferyl β-d-galactoside (Boehringer Mannheim)
- 4-methylumbelliferyl phosphate (Boehringer Mannheim)
- Diethanolamine (ICN biochemicals)
- Fab′ fragment of an anti-DNA-RNA hybrid Mab labeled with β-d-galactosidase or alkaline phosphatase (Dr. Robert J. Carrico from Ames division, Miles laboratories Inc., Elkhardt, IN).
- Black microtiter plates (Microfluor B, 96 well U-microplate) (Dynatech Laboratories)
- Fluorometer: Dynatech Microfluor microtiter plate fluorometer (detection wave length 365 nm, emission wave length 450 nm)

Labeling solutions
- 10× Transcription buffer: 400 mM Tris-HCl pH 7.5, 60 mM $MgCl_2$, 20 mM spemidine and 100 mM NaCl
- 10× PCR buffer: 25 mM $MgCl_2$, 100 mM Tris-HCl pH 8.3, 500 mM KCl
- elution buffer: solution of 0.5% SDS in DEPC-treated H_2O
- TE buffer: 10 mM Tris-HCl pH 7.5, 1 mM EDTA
- Nick translation enzymes: 0.4 units per μl of DNA polymerase and 40 pg per μl of DNase I in 50 mM Tris-HCl pH 7.5, 5 mM Mg-acetate, 1 mM 2-mercaptoethanol, 50% (v/v) glycerol, and 100 μg/ml of bovine serum albumin
- dNTP nick translation mixture: 0.2 mM dCTP, 0.2 mM dGTP, 0.2 mM dATP in 500 mM Tris-HCl pH 7.8, 50 mM $MgCl_2$, 100 mM 2-mercaptoethanol, and 100 μg/ml nuclease free bovine serum albumin
- Stop buffer: 0.3 M EDTA pH 8.0
- DEPC-treated H_2O: autoclave deionized water with 0.1% DEPC

Hybridization solutions
- SDS 10%: 10 gm of SDS in 100 ml of deionized H_2O
- 1 M Hepes, pH 7.4
- 0.5 M EDTA, pH 8.0
- 20× SSC (3 M NaCl, 0.3 M Na-citrate, pH 7
- Hybridization solution: 4× SSC; 20 mM Hepes; 2 mM EDTA; 0.5% SDS
- PCR lysis buffer: addition to sample of Tween 20 (final concentration of 0.5% [v/v] and Nonidet P-40 (final concentration of 0.5% [v/v] and proteinase K (final concentration of 250 μg/ml), depending on the sample total volume

Enzyme immunoassay solutions
- Carbonate buffer: 0.06 M carbonate buffer, pH 9.6
- PBS: 10 mM phosphate buffer pH 7.0, 100 mM NaCl
- Washing buffer: PBS-0.05% Tween 20 (PBST)

- Conjugate solution: Fab′ fragment of a Mab against DNA:RNA hybrids conjugated with β-d-galactosidase or alkaline phosphatase diluted to 0.025 µg/ml in PBST, 0.5% gelatin, 0.5% mouse serum
- MBG substrate solution: 0.1 mM 4-methylumbelliferyl β-d-galactoside in 10 mM phosphate buffer, pH 7.0; 100 mM NaCl; 1 mM MgCl$_2$; 50 µg/ml bovine serum albumin
- Diethanolamine buffer: 50 mM diethanolamine, pH 9.6; 0.1 mM MgCl$_2$
- MBA substrate solution: 0.1 mM 4-methylumbelliferyl-phosphate in diethanolamine buffer

- DNA can be purified by a standard phenol-chloroform extraction procedure (Maniatis et al., 1982) and resuspended in TE buffer or by treatment of cell suspensions (with a cell concentration of 10^7 cells/ml 10 mM Tris-HCl, pH 8.0) with PCR lysis buffer for 2 h at 45°C followed by boiling for 10 min. Cell lysates are stored at −70°C until tested. **Sample preparation for DNA detection**

The one step acid aquanidinium thiocyanate-phenol-chloroform extraction method is suggested (Chomczynski et al., 1987). **Sample preparation for RNA analysis**

Synthesis of nested DNA fragments by PCR

1. Mix in a 500 µl microfuge tube:
   - 10 µl   (1 pg) viral DNA template in TE
   - 10 µl   each nested primer at a concentration of 5 µM
   - 10 µl   deoxynucleoside triphosphates solution (2500 µM each)
   - 0.5 µl   *Taq* DNA polymerase (5 U/µl)
   - 10 µl   10× PCR buffer (optimize MgCl$_2$ concentration for each new primer pair)
   - 59.5 µl   dH$_2$O
2. Mix well.
3. Overlay samples with two drops of mineral oil to prevent evaporation of reagents.
4. Amplify in a DNA thermal cycler heat block through 30 cycles of denaturation at 94°C for 1 min, primer reannealing at 55°C for 1 min, and primer extension at 72°C for 1 min.

Preparation of biotinylated RNA probe from nested DNA fragments

1. Mix in a 1500 µl microfuge tube:
   - 10 µl   PCR-amplified DNA (±1 µg nested DNA fragment)
   - 10 µl   10× transcription buffer (Melton et al., 1984)
   - 10 µl   100 mM DTT
   - 3 µl   RNasin (100 units)
   - 10 µl   rATP, rCTP, and rGTP mixture (2.5 mM each)
   - 5 µl   bio-11-UTP at 20 mM
   - Complete with DEPC-treated dH$_2$O to 98 µl.
2. Mix well and centrifuge briefly
3. Add 2 µl T7 RNA polymerase (40 U)
4. Incubate at 40°C for 90 min
5. Add 2 µl DNase I (2 U) for 30 min at 37°C
6. Stop the digestion reaction with 1 µl of 0.5 M EDTA, pH 8.0

22.2.3

7. Inactivate at 95 °C for 5 min. Cool to room temperature.
8. Labeled RNA is separated from unincorporated biotinylated UTP by Sephadex G-50 chromatography:
   Wash NAP-5 columns with 3 ml elution buffer.
   Dilute the RNA probe in 400 µl elution buffer.
   Filter the diluted RNA probe onto the column.
   Discard the first 500 µl collected from the column.
   Elute the RNA probe from the column with 1 ml elution buffer.
   The solution collected from the column with contain the RNA probe.
   Aliquot the probe and store at −70 °C until used.

Nick translation reaction for labeling of DNA

The procedure is adapted from Rigby et al. 1977.

1. Into a 1500 µl microfuge tube on ice add:
   5 µl   dNTP nick translation mixture
   5 µl   template DNA (1 µg)
   5 µl   bio-11-dUTP at 0.3 mM
   30 µl   H$_2$O
2. Mix briefly and add 5 µl nick translation enzyme.
3. Mix gently and centrifuge briefly.
4. Incubate at 15 °C for 60 min.
5. Add 5 µl stop buffer.
6. Purify the biotinylated DNA probe by ethanol precipitation (Maniatis et al., 1982).
7. Store at 4 °C.

Enzyme immunoassay for DNA-RNA hybrids

1. In microfuge tubes of 1500 µl mix:
   100 µl of nucleic acids
   100 µl of hybridization buffer containing
   − 0.4 µg/ml of biotinylated DNA probe for detection of RNA, or
   − 2 ng/ml of biotinylated RNA probe for detection of DNA
2. Vortex reaction mixture.
3. Denature mixture in a boling water bath for 3−5 min.
4. Incubate for 30 min (RNA probe) or 16 h (DNA probe) at 78 °C for long DNA-RNA duplexes. Optimize the hybridization temperature for shorter DNA-RNA hybrids.
5. After completion of the hybridization reaction, cool samples to room temperature.
6. Add 20 µl 10% Triton X-100.
7. Use black microtiter plates which have been previously coated overnight at 4 °C with 50 µl per well of anti-biotin Mab at a concentration of 1 µg/ml of carbonate buffer.
8. Wash each well of the microtiter plate six times with PBST.
9. Dispense 50 µl of hybridized mixture in to each well and incubate for 1 h at 37 °C.
10. Wash six times with PBST each well.
11. Add 50 µl anti-DNA-RNA Mab conjugate solution.
12. Incubate at 37 °C for 30 min.

13. Wash each well six times.
14. Add 50 μl MBG substrate solution (when using the β-galactosidase conjugate) or 50 μl MBA substrate solution (when using the alkaline phosphatase conjugate) per well.
15. Incubate for 2 h at 37 °C for the MBG substrate solution or 20 min at 37 °C for the MBA substrate solution (Coutlee et al., 1989c).
16. Measure the amount of fluorescent methylumbelliferone generated by the enzymatic degradation of substrate in a fluorometer.

22.2.4 Special Hints for Application and Troubleshooting

– The polyclonal anti-biotin antibody sold by Sigma is no longer appropriate to coat microtiter plates for this EIA.
– The use of streptavidin to capture biotinylated DNA:RNA hybrids is appropriate for biotin-labeled DNA probes (nick translation) or biotinylated oligonucleotides but has a variable efficiency with biotin-labeled RNA probes. Haptens other than biotin can work appropriately in this assay including, DIG-[11]-dUTP or DIG-[11]-UTP.
– In enzyme immunoassays for DNA:RNA hybrids, longer DNA:RNA targets produce more intense signals since the epitope for the Mab is repetitive. Sensitivity may decrease with short hybrids, especially those less than 150 bp. If biotinylated oligonucleotide probes are used, the concentration of oligonucleotide has to be reduced down to 1 ng/ml or the signal can be completely lost. Inadequate removal of unincorporated biotinylated nucleotides from the probe decreases its binding to the solid phase and reduces the intensity of specific signal.
– RNA probes with hairpin loops will react with the Mab and need to be diluted down to 1 ng/ml in order to avoid background reactivity of probe (Coutlee et al., 1989b; Kinney et al., 1989).
– The use of colorigenic substrates requires a prolonged incubation period for substrate degradation unless a enzymatic cycling assay is performed (Coutlee et al., 1989c).

Acknowledgements. This work was supported in part by the Fonds de la recherche en santé du Québec (FRSQ, Canada) and the National Institute for Health USA.

References

Bobo L, Coutlee F, Yolken RH, Quinn TC, Viscidi RP (1990) Diagnosis of Chlamydia trachomatis infection by detection of amplified DNA with an enzyme immunoassay. J Clin Microbiol 28:1968–1973
Bogulawski SJ, Smith DE, Michlak MA et al. (1986) Characterization of a monoclonal antibody to DNA-RNA and its application to immunodetection of hybrids. J Immunol Meth 89:123–130

Chomczynski P, Sacchi N (1987) Single-step method of RNA isolation by acid guanidinium thiocyanate-phenol-chloroform extraction. Anal Biochem 162:156–159

Coutlee F, Yolken RH, Viscidi RP (1989a) Nonisotopic detection of RNA in an enzyme immunoassay format using a monoclonal antibody against DNA-RNA hybrids. Anal Biochem 181:153–162

Coutlee F, Bobo L, Mayur K, Yolken RH, Viscidi RP (1989b) Immunodetection of DNA with biotinylated RNA probes: a study of reactivity of a monoclonal antibody to DNA-RNA hybrids. Anal Biochem 181:96–105

Coutlee F, Viscidi RP, Yolken RH (1989c) Comparison of colorimetric, fluorescent, and enzymatic amplification substrate systems in an enzyme immunoassay for detection of DNA-RNA hybrids. J Clin Microbiol 27:1002–1007

Coutlee F, Rubalcaba EA, Viscidi RP, Murphy P, Lederman HW (1990a) Quantitative detection of messenger RNA by solution hybridization and enzyme immunoassay. J Biol Chem 265:11601–11604

Coutlee F, Bingzhi Y, Bobo L, Mayur K, Yolken RH, Viscidi RP (1990b) Enzyme immunoassay for detection of hybrids between PCR-amplified HIV-1 DNA and a RNA probe: PCR-EIA. AIDS Res Hum Retrovir 6:775–784

Coutlee F, Shah K, Yolken RH, Viscidi R (1991c) Analysis of human papillomaviruses messenger RNA in cancer cell lines and in cervical biopsies. J Clin Microbiol 29:968–974

Coutlee F, Hawwari A, Bobo L, Shah K, Dalabeta G, Hook N, Yolken R, Viscidi R (1991d) Infection génitale à papillomavirus de type 16 chez des patientes asymptomatiques. Union Médicale du Canada 120:3A

Coutlee F, St-Antoine P, Olivier C, Kessous A, Voyer H, Berrada F, Begin P, Giroux L, Viscidi RP (1991a) Evaluation of infection with HIV-1 with a nonisotopic assay for detection of PCR amplified HIV-1 proviral DNA. J Clin Microbiol 29:2461–2467

Coutlee F, Viscidi RP, Saint-Antoine P, Kessous A, Yolken RH (1991b) The polymerase chain reaction: a new tool for the understanding of the pathogenesis and detection of HIV-1 at the molecule level. Mol Cell Probes 5:241–259

Coutlee F, St-Antoine P, Olivier C, Kessous A, Voyer H, Berrada F, Begin P, Giroux L, Viscidi RP (1991a). Inhibitors of Taq polymerase: a potential cause for discordant results for detection of HIV-1 DNA with PCR. J Infect Dis 164:817–818

Coutlee F, Bobo L, Dalabetta G, Hook N, Shah K, Viscidi RP (1992) Evaluation of the prevalence of HPV-16 infection with PCR and a nonisotopic hybridization assay. J Med Virol, in press

Kinney J, Viscidi RP, Vonderfecht SL, Eiden JJ, Yolken RH (1989) Monoclonal antibody assay for detection of double-stranded RNA: application for the detection of group A and non-group A rotavirus. J Clin Microbiol 27:6–12

Langer PR, Waldrop AA, Ward DA (1981) Enzymatic synthesis of biotin labeled polynucleotides: a novel acid affinity probes. Proc Natl Acad Sci USA 78:6633–6637

Maniatis T, Fritsch EF, Sambrook J (1982) Molecular cloning: A laboratory manual. Cold Spring Laboratories, Cold Spring Harbor, NY

Melton DA, Krieg PA, Rebagliati MR et al. (1984) Efficient in virto synthesis of biologically active RNA and RNA hybridization probes from plasmids containing a bacteriphage SP6 promoter. Nucleic Acids Res 12:7035–7056

Newman C, Modlin J, Yolken RH, Viscidi RP (1989) Solution hybridization and enzyme immunoassay for biotinylated DNA-RNA hybrids to detect enteroviral RNA in cell culture. Mol Cell Probe 3:375–382

Schowalter DB, Sommer SS (1989) The generation of radiolabeled DNA and RNA probes with polymerase chain reaction. Anal Biochem 177:90–94

Stoflet ES, Koeberl DD, Sarhar G, Sommer SS (1988) Genomic amplification with transcript sequencing. Science 239:491–494

Rigby PWJ, Dieckmann M, Rhodes C, Berg P (1977) Labeling deoxyribonucleic acid to high specific activity in vitro by nick translation with DNA polymerase. J Mol Biol 113:237–251

Viscidi RP, O'Meara C, Farzadegan H, Yolken RH (1989) Monoclonal antibody solution hybridization for detection of human immunodeficiency virus nucleic acids. J Clin Microbiol 27:120−125

Viscidi RP, Yolken RH (1987) Abstract R.30.5 in Abstracts of VII International Congress of Virology, Edmonton, Canada, National Research Council of Canada, p 211

Viscidi RP, Hill PM, Shuo G, Cerny E, Vlahlv D, Farzadegan H, Halsey N, Kelen D, Quinn TC (1991) Diagnosis and differentation of HTLVI and HTLVII infection by enzyme immunoassay and synthetic peptides. J Acquired Immune Deficiency Syndrome 4:1190−1198

Yehle CO, Patterson WL, Boguslawski SJ (1987) A solution hybridization assay for ribosomal RNA from bacteria using biotinylated DNA probes and enzyme-labeled antibody to DNA-RNA hybrids. Mol Cell Probes 1:177−193

Yolken RH, Coutlee F, Viscidi RP (1989) New Prospects for the diagnosis of viral infections. Yale J Biology Medicine 62:131−139

Yolken RH, Li S, Perman J, Viscidi RP (1991) Persistent diarrhea and fecal shedding of retroviral nucleic acids in children infected with human immunodeficiency virus. J Infect Dis 164:61−66

22.3 Capturing of Displaced DNA Strands as a Diagnostic Method

MARY COLLINS

22.3.1 Principle and Applications

The displacement of DNA strands from a partially duplex DNA substrate can be used as a highly specific and sensitive assay for the detection of nucleic acids [1]. The principle of the assay is diagrammed in the figure. A probe complex is constructed that consists of a probe strand and a signal strand. The probe strand contains two regions, t and s, that are homologous to the DNA or RNA molecule of interest. The labeled signal strand (s') is shorter than the probe strand and is complementary to the s region of the probe strand. The DNA or RNA of interest, or analyte, can hybridize to the single-stranded portion of the probe strand (t) resulting in a triple-stranded intermediate. Migration of the branch point will rapidly resolve such an intermediate, releasing the labeled signal strand. This signal strand can be separated from intact complexes and detected. Release of the signal strand will be proportional to the concentration of analyte in the sample, resulting in a quantitative measurement of the amount of analyte in the original sample.

Strand displacement can be carried out using either immobilized probe complexes or probe complexes in solution. Use of immobilized complexes

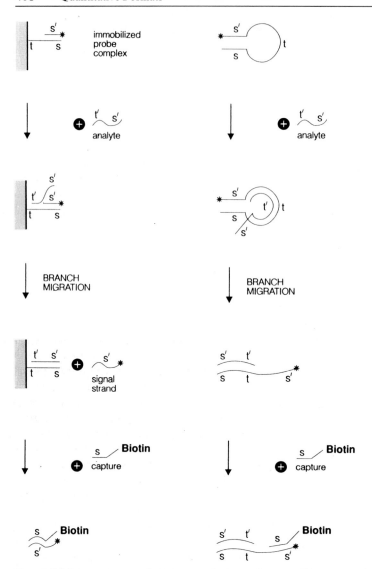

Strand displacement assay. *Left:* noncovalent complex on solid support; *right:* covalent complex in solution

results in release of labeled probe into solution, where it can be readily separated from the solid phase and measured [2]. The assay can also be carried out completely in solution, followed by separation of the intact probe complexes from the displaced signal strands [1]. For example, size exclusion columns or gel electrophoresis can be used to separate released signal strands from intact probe complexes. Solution phase reactions have the advantage of more rapid hybridization kinetics.

An alternative format for solution phase strand displacement reactions uses a third nucleic acid strand to "capture" the released signal strand (see figure). This third strand is homologous to a portion of the released signal strand. However, because it is shorter than the displaced signal strand and does not contain sequences that are able to interact with the single-stranded region of the probe complex (t), it is incapable of mediating strand displacement by itself. The "captured" signal strand is then either separated on the basis of size, or if a biotinylated capture strand is used, an avidin matrix can be used to isolate displaced signal strands. Methods and experimental results using a capturing strand in a strand displacement assay are described.

22.3.2 Strand Displacement Assay

The reactions were carried out using standard conditions for nucleic acid **Methods** hybridization (50 µl of 0.3 M NaCl; 0.1 M Tris-HCl, pH 8.0; 10 mM EDTA: 65 °C for 30−60 min).

Probe complexes were constructed in single-stranded M13 origin vectors as described by [1]. Fragments of a human albumin cDNA [3] were used as model analytes. Briefly, both the 1 kb PvuII-HincII fragment of albumin cDNA (target strand) and the 0.5 kb BglII-HincII (signal strand) were subcloned in an inverted orientation in a derivative of the M13 origin vector pSDC12 [4] and single-stranded template DNA was prepared. The presence of inverted repeats outside the cloning sites allows cleavage of the displacement complex from the single-stranded template DNA by BamHI. This releases a complex in which the signal strand and target strand are covalently linked at one end. Intramolecular hybridization of the signal to the target strand results in the formation of the displacement complex. Displacement of the signal strand with the analyte simply unfolds this "covalent" complex (see figure). A second small inverted repeat was located between the signal strand and target strand cloning sites. Cleavage at this site with EcoRI breaks the covalent linkage between the signal and target strands, which are then held together only by base-pairing. Displacement of the signal strand from this "noncovalent" complex releases the signal strand into solution (see figure). Complexes were labeled with ^{32}P either by polynucleotide kinase or by ligation of a ^{32}P-labeled oligonucleotide to the 3' end of the complex.

Capture strands were constructed by subcloning either a 300 bp TaqI-PstI fragment from the albumin cDNA into M13mp19 (Yanish-Perron et al., 1985) or a 280 bp XbaI-PstI fragment of the albumin cDNA into a deletion derivative of M13mp7 [5] that lacks sequences between the two PvuII sites at nucleotides 5960 and 6363. Single-stranded M13 template DNA was prepared and biotinylated with photobiotin as described [7]. Alternatively, a 300 nucleotide primer-extended capture strand was synthesized using a biotinylated primer and standard primer extension conditions. In each case, the biotinylated capture strand contains sequences complementary to an internal portion of the 500 nucleotide signal strand.

Model analytes were made by subcloning an 800 bp fragment from the albumin cDNA into M13mp11 [8]. Template DNA was prepared and linearized with *Hae*III.

Captured signal strands were separated either by gel electrophoresis or on a streptavidin-agarose matrix. Samples were bound to streptavidin agarose in 0.3 M NaCl, 0.1 M Tris-HCl pH 8.0, 5× Denhardt's solution, and 1 mM EDTA and the matrix was washed several times in this solution and counted to determine the amount of ^{32}P-labeled signal strand bound to the matrix.

Results Strand displacement and capturing were demonstrated using model displacement complexes prepared as described in methods. Incubation of ^{32}P-labeled probe complexes with analyte resulted in displacement of the signal strand, which wash then "captured" by hybridization to the biotinylated capture strand. Upon completion of the reaction, samples were analyzed by electrophoresis on an agarose gel and by binding to streptavidin agarose.

The first experiment (see table) was carried out using a "noncovalent" complex (see Methods and figure), composed of a separate signal strand hybridized with a target strand. Displacement of the signal strand with analyte releases the free labeled signal strand into solution. The labeled signal strand is then "captured" with a biotinylated capture strand homologous to an internal portion of the signal strand. Complex, analyte, and biotinylated capture strand were incubated in a 50 μl hybridization reaction for 60 min at 65 °C. After the reaction, an aliquot was removed for analysis by gel electrophoresis. The rest of the sample was bound to strep-

Strand displacement and capturing on streptavidin agarose

Experiment	Probe complex	Analyte	Capture strand	Amount bound to streptavidin agarose (% cpm)
1	Noncovalent (0.2 pmol)	0.2 pmol	Primer-extended (0.5 pmol)	46
	Noncovalent (0.2 pmol)	0 pmol	Primer-extended (0.5 pmol)	2
2	Covalent (0.1 pmol)	0 pmol	Photobiotin (0.16 pmol)	3
	Covalent (0.1 pmol)	0.01 pmol	Photobiotin (0.16 pmol)	11
	Covalent (0.1 pmol)	0.05 pmol	Photobiotin (0.16 pmol)	34
3	Noncovalent (0.2 pmol)	0.2 pmol	Photobiotin (0.8 pmol)	62
	Noncovalent (0.2 pmol)	0.2 pmol	Photobiotin (0 pmol)	1
	Covalent (0.2 pmol)	0.2 pmol	Photobiotin (0.8 pmol)	57
	Covalent (0.2 pmol)	0.2 pmol	Photobiotin (0 pmol)	3
4	Covalent (0.02 pmol)	0.02 pmol	Primer-extended (0.04 pmol)	26
	Covalent (0.02 pmol)	0 pmol	Primer-extended (0.04 pmol)	5

tavidin agarose, and the matrix was washed and counted. Gel analysis indicated that no probe complex was displaced in the absence of analyte and that all of the signal strand was displaced in the presence of analyte (gel data are not shown for any experiments). Similarly, binding to streptavidin agarose was dependent upon the presence of analyte. However, in this and subsequent experiments, some nonspecific binding was observed and the efficiency of binding of displaced, captured signal strands to streptavidin agarose was not 100%.

Experiment 2 (see table) was carried out similarly to experiment 1 except that a covalent complex was used. In this case, binding of the analyte to the probe complex simply unfolds the probe complex, making the signal strand sequence available for hybridization to the capture strand. In the absence of analyte, no captured signal was observed on the autoradiograph of the gel, but addition of 0.01 pm and 0.05 pm of analyte resulted in capturing of approximately 20% and 80% of the labeled complex, respectively (gel data not shown). Thus, by gel assay, capturing of labeled probe complexes was dependent upon the amount of input analyte, and nonspecific capturing of undisplaced probe complexes was below detectable levels. A similar conclusion was drawn by analysis of ^{32}P cpm detected on the streptavidin support, in that increasing levels of analyte resulted in the binding of higher levels of probe complex. Thus, both noncovalent and covalently closed probe complexes can be used for this reaction.

Experiment 3 was carried out to compare the efficiency of the strand displacement and capturing reactions using noncovalent and covalent complexes. Gel analysis indicated that, for both complexes, no displacement was observed in the absence of analyte, and all of the complex was displaced and captured by the addition of analyte and capture strand. Approximately, the same efficiency of binding to the streptavidin matrix was observed for both the covalent and noncovalent complexes, indicating that one format is not prefered.

Two types of capturing strands were assessed in these experiments. In experiments 2 and 3 the capturing strand was single-stranded M13 template DNA that had been biotinylated with photobiotin. Experiments 1 and 4 were carried out to assess the use of a capture strand made by primer extension of a 5′ biotinylated oligonucleotide. As seen in these experiments, both types of capturing strands work with both covalent and noncovalent complexes. Gel analysis of experiment 4 indicated that no displacement was observed in the absence of analyte and that addition of analyte resulted in complete displacement. Similarly, binding to the streptavidin matrix was dependent upon the presence of analyte.

Discussion

The displacement of a labeled signal strand from a probe complex can be used as an assay system for the detection of specific DNA sequences. The extent of the displacement reaction is dependent upon the concentration of analyte within the sample [1]. The reaction follows standard nucleic acid hybridization kinetics. The rate-limiting step is the initial hybridization of the analyte to the probe complex; branch migration rapidly results in displacement of the signal strand [9]. However, the rate of branch migration and displacement can be slowed significantly by using mismatched analyte-probe complex pairs [1]. This contributes to the increased specificity of the strand displacement reaction. In addition, branch migration and displacement are greatly inhibited by the presence of small deletions or insertions [1]. Thus, strand displacement reactions can be used to distinguish between related, but nonidentical, sequences.

In each of the experiments shown above, displacement and capturing were dependent upon the presence of analyte. Addition of a capturing strand to the reaction mixtures did not result in capturing of probe strands in the absence of analyte. Thus, capture strand does not form stable three-stranded intermediates with undisplaced probe complexes. This allows one to carry out the displacement and capturing reactions in a single step.

In the experiments shown in the table, although displacement reactions were essentially complete, only a portion of the displaced signal strands bound to streptavidin agarose. Several factors contribute to this result. One is the integrity of the support; washing too vigorously can result in loss of complexes from the support presumably due to loss of streptavidin agarose particles. Although degradation of the DNA is a formal possibility, analysis of unbound material by gels showed no evidence of degradation. Another possibility is that formation and resolution of displacement complex intermediates results in loss of labeled signal strands from the support. If the analyte molecule binds to the capture strand first and then binds to the single-stranded region of the probe complex (t), migration of the analyte strand through the probe complex is inhibited by the presence of the capture strand bound to the analyte. This intermediate, which contains probe complex, analyte, and capture strand, can bind to the streptavidin support. However, resolution of this intermediate by strand exhange could occur and potentially result in the release of free signal strand from the support. In an experiment in which analyte was prehybridized with capture strand and then hybridized with probe complex, resolution of this intermediate complex was observed with a half-life of about 20 min (data not shown). Thus, resolution of intermediates bound to the matrix could result in release of signal from the matrix during the washing step. However, this reaction can be minimized by using probe complexes in excess of both analyte and capture strand. Excess probe complexes would drive the initial hybridization and displacement reaction with the analyte. Capturing of the displaced signal strand would then occur as a second step.

In the experiments described, ^{32}P was used to label the signal strands. Other methods of labeling DNA are also appropriate for this assay format. Capture strands could also be immobilized on a support in an assay format in which the final detection step is carried out on a surface. Strand displacement could also be used in conjunction with the polymerase chain reaction (PCR) [10]. Analyte sequences within a test sample could be amplified by PCR and the reaction products could then be assayed by strand displacement. This would provide a highly specific method for detecting a particu-

22.3.2

lar nucleotide sequence within the amplified products. Amplified products could be incubated with a labeled displacement complex and a biotinylated capture strand for a few minutes. The reaction would then be passed over an avidin matrix and detection of the signal would follow. Amplification of the analyte would eliminate the need for a highly sensitive signal detection method. Strand displacement would enhance the specificity of the reaction, eliminating detection of spurious amplification products and provide a format for detecting a postive signal.

Acknowledgements. I thank Ed Fritsch, Ken Jacobs, Stan Wolf and Leonard Lerman for many contributions to the ideas presented here.

References

1. Collins M, Fritsch EF, Ellwood MS, Diamond SE, Williams JI, Brewen JG (1988) A novel diagnostic method based on DNA strand displacement. Mol Cell Probes 2:15−30
2. Wolf SF, Haines L, Fisch J, Kremsky J, Dougherty JP, Jacobs K (1987) Rapid hybridization kinetics of DNA attached to submicron latex particles. Nucleic Acids Res 15:2911−2926
3. Lawn RM, Adelman J, Bock SC, Franke AE, Houch CM, Najarian RC, Seeburg PH, Wion KL (1981) The sequence of human serum albumin cDNA and its expression in E. coli. Nucleic Acids Res 9:6103−6114
4. Levison A, Silver S, Seed B (1984) Minimal size plasmids containing an M13 origin for production of single-strand transducing particles. J Mol Appl Gen, 2:507−517
5. Yanisch-Perron C, Vieira J, Messing J (1985) Improved M13 phage cloning vectors and host strains: nucleotide sequences of the M13mp18 and pUC19 vectors. Gene 33:103−119
6. Messing J, Crea R, Seeburg PH (1981) A system for shotgun DNA sequencing. Nucleic Acids Res 9:309−321
7. Welcher AA, Torres AR, Ward DC (1986) Selective enrichment of specific DNA, cDNA and RNA sequences using biotinylated probes, avidin and copper-chelate agarose. Nucleic Acid Res 14:10027−10044
8. Norrander J, Kempe T, Messing J (1983) Construction of improved M13 vectors using oligonucleotide-directed mutagenesis. Gene 26:101−106
9. Green C, Tibbets C (1981) Reassociation rate limited displacement of DNA strands by branch migration. Nucleic Acids Res 9:1905−1918
10. Saiki RK, Gelfand DH, Stoffel S, Scharf S, Higuchi R, Horn GT, Mullis KB, Erlich HA (1988) Primer-directed enzymatic amplification of DNA with a thermostable DNA polymerase. Science 239:487−491

22.4 Nonradiative Fluorescence Resonance Energy Transfer

RICHARD A. CARDULLO

22.4.1 Basic Principles of Nonradiative Fluorescence Resonance Energy Transfer

The method of nonradiative fluorescence resonance energy transfer (FRET) allows separation between fluorescent molecules to be made with Ångstrom resolution over distances between 10 and 100 Å. The technique was first conceived of by Förster (see 1957 for historical review) and has been used to measure intra- and intermolecular distances between, for instance, different sites on polynucleotides (Wells and Cantor 1977; Cardullo et al., 1988), different reactive sites on a protein (Taylor et al., 1981; Chantler and Tao 1986; Herz et al., 1989; Johnson et al., 1990), proteins and lipids in a model membrane system (Valenzuela et al., 1992; Wolf et al., 1992), or proteins on the cell surface (Shaklai et al., 1977; Fung and Stryer 1978; Stryer et al., 1982; Holowka and Baird 1983; Yarden and Schlessinger 1987; Carraway et al., 1989; Azevedo and Johnson 1990; Cardullo et al., 1991). Because of its sensitivity over molecular distances the FRET technique has been dubbed as the "spectroscopic ruler" (Stryer and Haugland 1967). Although extremely useful for probing macromolecular

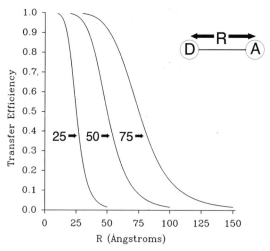

Theoretical transfer efficiency curves as a function of the separation distance (R) between donor (D) and acceptor (A). Numbers represent R_0 values of 25, 50, and 75 Å

interactions, FRET is a sophisticated technique that requires exact determinations of critical spectroscopic parameters. To emphasize these parameters it is critical to understand the mechanism of electronic energy transfer in fluorescence assays.

The principle of FRET is simple: a donor molecule absorbs light at some frequency which temporarily places that molecule into a higher energetic electronic state. Before the electron can decay down to its ground state, the close proximity of another molecule results in a transfer of energy from the donor molecule to an acceptor molecule by a dipole-induced dipole interaction. The donor molecule has therefore given up its energy in a nonradiative fashion and placed the acceptor molecule into its excited state, where it can now decay radiatively at its characteristic fluorescence emission wavelength. In using FRET, only one of the two molecules needs to be fluorescent. However, most users of this technique employ both fluorescent donor and acceptor molecules whose spectrofluorometric properties become modulated when energy transfer occurs. The conditions of energy transfer are quite stringent:

1. The donor emission spectrum must overlap the acceptor's absorbance spectrum.
2. The donor and acceptor fluorophores must be sufficiently aligned so that an acceptor dipole can be induced by the donor.
3. The donor and acceptor must be separated by some minimum distance so that the probability of energy transfer is high.

The degree of energy transfer, also known as the transfer efficiency, E_t, is related by the Förster equation:

$$E_t = \frac{R_0^6}{R^6 + R_0^6} \tag{1}$$

where R is the separation distance, in Ångstroms, between the donor and acceptor and R_0 is the separation distance for a particular donor/acceptor pair where the transfer efficiency equals 0.5. It is because the transfer efficiency falls off inversely as the sixth power of the separation distance that FRET is a useful tool for measuring molecular distances both quantitatively and qualitatively. The transfer efficiency, and therefore the molecular separation distance, can be determined by either measuring donor lifetimes or steady state methods. I shall consider these different methods in the next section.

22.4.2 Methodologies for Measuring Transfer Efficiencies

There are four major ways to measure the transfer efficiency due to FRET when the donor is excited. These include:

1. A decrease in the donor lifetime
2. A decrease in the steady state fluorescence emission intensity of the donor (donor quenching)

3. An increase in the steady state fluorescence emission intensity of the acceptor (sensitized emission)
4. A change in the fluorescence polarization.

Of these four methods, 1 and 2 are the most widely used and the most amenable for measuring transfer efficiencies, and in this chapter I shall focus on these two methods.

A molecule M, which is in an excited electronic state M^* can lose energy by a number of both radiative and nonradiative mechanisms:

$$M^* \longrightarrow M \quad k_t = k_r + k_{nr}$$

where k_t is the total rate constant for radiative (k_r) and nonradiative (k_{nr}) processes. In the absence of nonradiative processes the donor fluorophore will decay at some characteristic rate. However, when a competing non-radiative process occurs (e.g., FRET), there is a *decrease* in the lifetime of the donor. In general, the transfer efficiency, E_t, is given by:

$$E_t = \frac{\tau_{D,A}}{\tau_{D,U}} \tag{2}$$

where $\tau_{D,A}$ is the excited state lifetime of the donor in the presence of acceptor, and $\tau_{D,U}$ is the excited state lifetime in the absence of acceptor. Using a fluorescence lifetime machine, measurements of donor lifetimes in the presence and absence of acceptor can be made and the transfer efficiency of can be calculated using Eq. 2. The advantage of lifetime measurments is that they are extremely sensitive so that accurate determinations of the E_t can be made. However, few investigators have ready access to this equipment as it is expensive and difficult to maintain.

As opposed to fluorescence lifetime equipment, most investigators have access to a steady state fluorimeter and it is possible to make fairly accurate determinations of transfer efficiency using such instrumentation. The degree of energy transfer can be quantified as a decrease in the donor's fluorescence emission intensity or by an increase in the fluorescence emission intensity of the acceptor (sensitized emission). The most common method is to quantify the transfer efficiency by measuring the amount of donor quenching in the presence and absence of acceptor. In this mode the transfer efficiency is:

$$E_t = 1 - \frac{F_{D,A}}{F_{D,U}} \tag{3}$$

where $F_{D,A}$ is the fluorescence intensity of the donor in the presence of acceptor (i.e., the transfer condition) and $F_{D,U}$ is the fluorescence intensity in the absence of acceptor. In the case of ligand-receptor interactions or nucleic acid hybridization the measurement of $F_{D,U}$ (see the second figure) must be made with an unlabeled complementary molecule. By measuring $F_{D,U}$ (or $\tau_{D,U}$) in this way, other donor quenching mechanisms besides FRET are accounted for. Often it is more convenient to express the transfer efficiency using the degree of donor emission quenching rather than measuring the fluorescence intensities directly. In this case:

a   b

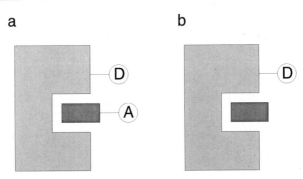

Illustration of a ligand-receptor interaction as determined by FRET. The interaction between receptor and ligand can be determined by tagging the receptor with a fluorescent donor molecule, D, and the ligand with a fluorescent acceptor molecule, A. **a** To accurately determine the transfer efficiency between donor and acceptor, a decrease in either donor lifetime or steady state donor fluorescence intensity is measured. **b** Parallel experiments must also be performed in the presence of unlabeled ligand to determine the modulation in fluorescence parameters in the presence of unlabeled ligand. The transfer efficiency and therefore the separation distance between donor and acceptor can be determined quantitatively as outlined in the text

$$E_t = \frac{q_{D,A} - q_{D,U}}{1 - q_{D,U}} \tag{4}$$

Whether one is making lifetime or steady measurements, an accurate determination of R_0 must be made. Measurements of R_0 are easily achieved if certain spectral parameters can be measured. In general, R_0 is given by:

$$R_0 = 9.765 \times 10^3 \, (x^2 \, J(\lambda) \, Q_D \, n^{-4})^{1/6} \tag{5}$$

where $J(\lambda)$ is the overlap integral (in $cm^3 \, M^{-1}$) between the donor emission spectrum and the acceptor absorbance spectrum, x^2 is the orientation factor between donor and acceptor, Q_D is the quantum yield of the donor in the absence of acceptor, and n is the refractive index of the medium between donor and acceptor. Quantitatively, the overlap can be calculated as:

$$J(\lambda) = \int_0^\infty \varepsilon(\lambda) f(\lambda) \lambda^4 \, d\lambda \tag{6}$$

where $\varepsilon(\lambda)$ is the molar extinction coefficient of the acceptor and $f(\lambda)$ is the normalized fluorescence spectrum of the donor which is defined as:

$$f(\lambda) \, d\lambda = \frac{F(\lambda) \, d\lambda}{\int_0^\infty F(\lambda) \, d\lambda} \tag{7}$$

and $F(\lambda)$ is the fluorescence emission intensity of the donor. The orientation factor, x^2 is perhaps the most troublesome factor since it can vary anywhere between 0 and 4. In reality, the orientation between a population of donor and acceptor molecules are randomly oriented both spatially and temporally. It can be also shown that under randomizing conditions, the average value of x^2 ($<x^2>$) is equal to ⅔ (Dale et al. 1979). For most macromolecular interactions in solution using a x^2 value of ⅔ is entirely appropriate and even minor polarizations of donor and acceptor molecules will not lead to major variations in the determination of R_0. However, in certain geometries, the investigator using energy transfer as a quantitative tool must be conscious of the possibility that certain macromolecular interactions may lead to extreme values in x^2. In particular, energy transfer in highly polarizing situations such as in biological membranes or in nucleic acids, may lead to situations where dipoles are either orthogonally aligned (leading to no possibility for a dipole induced-dipole) or aligned in parallel so that x^2 is equal to 4. Even in the parallel case, the difference between R_0 in the aligned and random situation is only 30%.

22.4.3 Using Energy Transfer to Monitor Nucleic Acid Hybridization

As an example of the utility of FRET in molecular biology, I will describe assays which can be easily performed with a steady state fluorimeter to monitor nucleic acid hybridization in solution. With modifications, this method can also be used to detect hybridization in either solid phase assays (i.e., Northen or Southern Blots) or within single cells using a modified fluorescence microscope (Cardullo et al., 1991; Herman, 1989) although these require specialized equipment such as lasers and sensitive detection devices. At the moment we are able to monitor hybridization in solution in the attomole (10^{-18} mole) range in 100 microliter volumes. The sensitivity of FRET rivals radioisotopic assays and because it works only over molecular distances gives one the added advantage of not having to remove unhybridized nucleic acids before detection is possible.

Fluorescent Labeling Strategies Fluorescent labeling of nucleic acids has been outlined elsewhere in this text. In general, one must first determine the position of labeling (either 5′ or 3′) and the length of complementary oligonucleotides to be used. Initial studies in our laboratory used fluorescein as the donor (excitation = 472nm, emission = 517nm) and rhodamine as the acceptor (excitation = 550nm, emission = 577nm). These fluorophores were attached onto a hydrocarbon linker at the 5′ ends of complementary oligonucleotides (ODNTs) between 6 and 20 nucleotides in length (see the last figure, scheme a). In the case of fluoresecein and rhodamine, R_0 was found to be between 49 and 51 Å, which made this donor/acceptor pair useful for quantifying distances up to about 80 Å (Cardullo et al., 1988).

For monitoring longer nucleic acids, a strategy such as that illustrated in the last figure (scheme b) can be used. Here the nucleic acid to be detected can be virtually any length and two different ODNTs are chosen, one of which is labeled at the 5′ end and the other at the 3′ end such that they are separated only by a few bases when they hybridize to the complementary nucleic acid. When using fluorescein and rhodamine attached to a hexyl

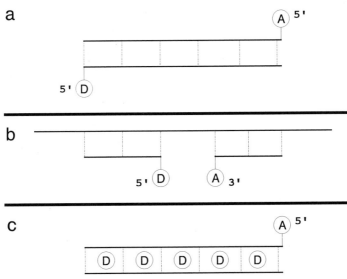

Diagram of three different ways to monitor nucleic acid hybridization using FRET. **a** To monitor the hybridization of short oligonucleotides the terminal bases at the 5′ ends can be labeled with either donor or acceptor fluorophores. Hybridization can then be monitored using FRET. **b** Longer stretches of DNA or RNA can be detected by using two complementary strands, one which is labeled at the 5′ end and the other labeled at the 3′ end. The complementary probes are chosen so that upon hybridization the donor and acceptor molecules are separated by only a few bases (4−10) so that the resulting transfer efficiency is high. **c** By using an intercalating dye like acridine orange or ethidium bromide, only one end of a single oligonucleotide needs to be labeled. The energy transfer therefore occurs by an antenna effect where many donors are used to excite a single acceptor. Figure modified from Cardullo et al. (1988)

spacer, separations of 4−8 bases work best. As with most hybridization assays, conditions must be worked out (e.g., temperature, ionic strength, and complementary ODNT length) to get hybridization to occur. In physiological salt solutions, we have found that hybridization of large nucleic acids (>30 nucleotides long) to labeled, complementary ODNTs greater than 10 nucleotides occurs only when the hybrids are first heated to 70°C for 30 min and then cooled to room temperature (20°C) before measurement. During this cooling phase, hybridization can be monitored by following either an increase in acceptor fluorescence or a decrease in donor fluorescence. For smaller oligonucleotides (<20 nucleotides) the hybridization temperature is significantly less (<50°C).

A somewhat simpler strategy is to label only one end of a nucleic acid and to use an intercalating dye such as ethidium bromide or acridine orange as the other fluorophore (scheme c in the last figure). This method has the advantage of having to covalently label only one nucleic acid and to use multiple intercalated dye molecules as either a donor or acceptor. The main disadvantage of this method is that many intercalating dyes undergo significant spectral shifts upon binding and the inability to quantify donor and acceptor distances because of the unknown location and number of participating dye molecules. Nevertheless, this method is probably the most useful qualitative assay of nucleic acid hybridization available to most investigators and represents a major advantage over existing radioisotopic assays.

Experimental Protocol for Monitoring Nucleic Acid Hybridization Perhaps the most useful aspect this technology is to accurately follow hybridization of nucleic acids in solution at relatively low concentrations. In all other assays for detecting nucleic acid hybridization, the hybrids can only be detected after removal of nonhybridized nucleic acid. Because FRET creates a distinctive signal between donor and acceptor fluorophores that occurs only when conjugated nucleic acids are hybridized, this method can be used even in the presence of nonhybridized nucleic acids.

As mentioned above, the easiest way to monitor hybridization is to use a steady-state fluorimeter. In all cases, conditions need to be established which minimize erroneous determinations of fluorescence intensity such as inner filter effects. In our studies, we routinely used a Perkin-Elmer spectrofluorimeter (model MPF-3) with the excitation and emission slit widths set at 8 nm. In addition, all quantitative FRET experiments should be performed at the so-called "magic-angle" and ideally Glan-Thompson polarizers should be used. Finally, since hybridization of nucleic acids is temperature dependent, the sample chamber should be temperature controlled with a temperature range of 0 °C to 80 °C. Quartz cuvettes should be used in all quantitative experiments and for most of our studies we use cuvettes which can accommodate volumes from 90 to 200 microliters (optical pathlength = 0.3 cm).

A typical experiment is outlined below:

1. *Determine the background fluorescence of the buffer.* In our experiments, we use a standard phosphate buffered saline (PBS) and measure the background fluorescence of 85 μl. This can either be done at a single wavelength (either the donor emission wavelength or the acceptor emission wavelength) while exciting for the donor fluorophore or alternatively an emission scan can be performed over the entire wavelength range of interest. We find it useful to perform emission scans in order to monitor any spectral shifts that may occur during hybridization.
2. *Determine the fluorescence intensity of the donor.* Determine the level of donor fluorescence that is easily measured in the spectrofluorimeter. Ideally, this should be in a range where the fluorescence emission intensity varies linearly with fluorophore concentration. This value will be the base-line value for determining the amount of quenching that occurs after the complementary strand is added.
3. *Determine the fluorescence intensity of the donor in the presence of the unlabeled complementary strand.* In an ideal experiment, the unlabeled complementary strand would not modulate the fluorescence intensity

of the donor after hybridization. However, we have found that this is never the case, and that the fluorescence emission intensity of the donor generally decreases somewhat upon hybridization. The degree of quenching ($q_{D,U}$ in Eq. 4) depends on a number of different parameters including the number of nucleotides pairs in the hybrid, the length of the hydrocarbon spacer between the terminal nucleotide and the fluorophore, the temperature, and the ionic strength of the solution. In all cases the degree of donor quenching should be determined at saturation (i.e., at a concentration where all of the complementary strands are hybridized and no further donor quenching occurs).

4. *Determine the fluorescence intensity of the donor in the presence of the acceptor labeled complementary strand.* Once the degree of quenching due to unlabeled complement has been determined, the degree of donor quenching in the presence of acceptor labeled complement (i.e., the transfer condition) can be determined. If the transfer condition is satisfied, the donor emission fluorescence will be quenched beyond the quenching seen in the presence of unlabeled complements as measured above. The amount of quenching in the presence of acceptor ($q_{D,A}$) is used along with the amount of quenching in the absence of acceptor labeled complement ($q_{D,U}$) to determine the transfer efficiency using Eq. 4.

5. *Establish the reproducibility of the hybridization by melting and annealing the hybrids.* Because the FRET signal is present only when hybridization between labeled complementary strands occurs, the signal will be rapidly abolished when the temperature is increased beyond the melting temperature. When the donor is excited, there is an increase in the donor emission fluorescence intensity and a decrease in the acceptor fluorescence intensity when the temperature is increased and the duplex is melted. Indeed, we routinely generate melting and annealing curves which approximate hyperchromicity curves using the identical probes and the T_ms (defined as the midpoint transition temperature) are virtually identical in all cases. Slight differences in shape between FRET and absorbance derived temperature curves reflect chromophore interactions with the duplex and the exquisite spatial resolution of FRET.

Energy Transfer as a Quantitative Tool

The experimental protocol outlined above for determining hybridization is extremely sensitive for dtermining if hybridization between specific nucleic acid sequences has occurred but the information gained is highly qualitative. FRET can also be used quantitatively to determine either the amount of hybridized material or the distance donor and acceptor fluorophores in the hybrid.

To determine the number of copies of a polynucleotide the best configuration is the one outlined in part b in the last figure with saturating amounts of donor and acceptor labeled oligonucleotides. A dose-response curve is generated with saturating amounts of 5′ and 3′ labeled oligonucleotides at different concentrations of a complementary hybrid. Transfer

Calculation of transfer efficiencies and corresponding separation distances for ODNT's of chain length n labeled at the 5′ ends with the donor fluorescein (F) or the acceptor rhodamine (R).

n	$q_{F,R}$	$q_{F,U}$	E_1	$R(Å)^a$
8	0.632 ± 0.046	0.265 ± 0.021	0.501 ± 0.035	48.6 ± 1.0
12	0.423 ± 0.030	0.265 ± 0.013	0.215 ± 0.052	60.9 ± 2.5
16	0.295 ± 0.017	0.262 ± 0.013	0.045 ± 0.018	81.5 ± 4.9

[a] Calculated using an R_0 of 49.1 Å and E_t in the text. (Data modified from Cardullo et al., 1988)

efficiency is then calculated as a function of unlabeled polynucleotide concentration. Once this calibrating data is established it is a simple matter to determine unknown copy numbers of specific DNA or RNA sequences in solution.

Of course the other quantitative application for FRET is to determine the separation distance between donor and acceptor fluorophores on hybrids. This is useful mostly for those researchers studying folding in structures like tRNA or in short hybrid oligonucleotides as shown in the last figure. Once transfer efficiencies are measured one can determine the separation distance using the Förster equation (Eq. 1). With the proper equipment and care, FRET can be a useful tool for measuring distances on the molecular scale. The table shows some steady state FRET data for duplex hybrids varying in length between 8 and 16 nucleotides. Using these data and a mathematical model that approximate a double helix, we showed that the fluorescent probes fluorescein and rhodamine were extended away from the helix and parallel to the helical axis (Cardullo et al., 1988).

22.4.4 Summary

In this chapter I have shown that nonradiative fluorescence resonance energy transfer (FRET) can be used both qualitatively and quantitatively to detect events at the molecular level. For qualitative studies, the investigator can use a standard steady state fluorimeter to monitor either donor quenching or sentisized emission. For more quantitative work, as in the determination of molecular distances with Ångstrom resolution, much more sophisticated equipment needs to be employed (e.g., a fluorescence lifetime machine). Although I have focused on solution studies, FRET can also be applied to solid state assays and in fluorescence microscopes. In addition, FRET can be used to quantify many other events of interest in biology besides nucleic acid structure including protein structure, membrane dynamics, and the elucidation of signal transduction pathways (Cardullo et al., 1991).

References

Azevedo JR, Johnson DA (1990) Temperature-dependent lateral and transverse distribution of the epidermal growth factor receptor in A431 plasma membranes. J Membrane Biol 118:215−224

Cardullo RA, Agrawal S, Flores C, Zamecnik PC, Wolf DE (1988) Detection of nucleic acid hybridization ny nonradiative fluorescence resonance energy transfer. Proc Natl Acad Sci USA 85:8790−8794

Cardullo RA, Mungovan RM, Wolf DE (1991) Imaging membrane organization and dynamics. In: Dewey G (ed) Biophysical and biochemical aspects of fluorescence microscopy. Plenum, New York, pp 231−260

Carraway KL 3d, Koland JG, Cerione RA (1989) Visualization of epidermal growth factor (EGF) receptor aggregation in plasma membranes by fluorescence resonance energy transfer. Correlation of receptor activation with aggregation. J Biol Chem 264:8699−8707

Chantler PD, Tao T (1986) Interhead fluorescence energy transfer between probes attached to translationally equivalent sites on the regulary light chains of scallop myosin. J Mol Biol 192:87−99

Dale RE, Eisinger J, Blumberg WE (1979) The orientation factor in intramolecular energy transfer. Biophys J 26:161−194

Förster T (1959) Transfer mechanisms of electronic excitation. Discuss Farraday Soc 27:7−17

Fung BK-K, Stryer L (1978) Surface density determination in membranes by fluorescence energy transfer. Biochemistry 17:5241−5248

Herman B (1989) Resonance energy transfer microscopy. In: Taylor DL, Wang Y-L (eds) Fluorescence microscopy of living cells in culture, part B. Meth Cell Biol 30:219−243

Herz JM, Johnson DA, Taylor P (1989) Distance between the agonist and noncompetitive inhibitor sites on the nicotinic acetylcholine receptor. J Biol Chem 264:12439−12448

Holowka D, Baird B (1983) Structural studies on the membrane-bound immunoglobulin E-receptor complex: II. Mapping of distances between sites on IgE and the membrane surface. Biochemistry 22:3475−3484

Johnson DA, Cushman R, Malekzadeh R (1990) Orientation of cobra α-toxin on the nicotinic acetylcholine receptor. J Biol Chem 265:7360−7368

Shaklai N, Yguerabide J, Ranney HM (1977) Interaction of hemoglobin with red blood cell membranes as shown by a fluorescent chromophore. Biochemistry 16:5585−5592

Stryer L, Haugland RP (1967) Energy transfer: a spectroscopic ruler. Proc Natl Acad Sci USA 58:719−726

Stryer L, Thomas DD, Carlsen WF (1982) Fluorescence energy transfer measurements of distances in rhodopsin and the purple membrane protein. Methods Enzymol 81:668−678

Taylor DL, Reidler J, Spudich JA, Stryer L (1981) Detection of actin assembly by fluorescence energy transfer. J Cell Biol 89:362−367

Valenzuela CF, Kerr JA, Johnson DA (1992) Quinacrine binds to the lipid-protein interface of the Torpedo acetylcholine receptor: a fluorescence study. J Biol Chem 267:8238−8244

Wells BD, Cantor CR (1977) A strong ethidum binding site in the acceptor stem of most or all transfer RNAs. Nucleic Acids Res 4:1667−1679

Wolf DE, Winiski AP, Ting AE, Bocian KM, Pagano RE (1992) Determination of the transbilayer distribution of fluorescent lipid analogues by nonradiative fluorescence resonance energy transfer. Biochemistry 31:2865−2873

Yarden Y, Schlessinger J (1987) Epidermal growth factor induces rapid, reversible aggregation of the purified epidermal growth factor receptor. Biochemistry 26:1443−1451

22.5 Colorimetric Assays with Streptavidin/ Avidin-Coated Surfaces

RUDOLF SEIBL and STEFANIE KOEHLER

22.5.1 Principle and Applications

Streptavidin- or avidin-coated surfaces, e.g., tubes, microtiter plates, and magnetic or nonmagnetic beads, are widely used to anchor target molecules via biotin on a solid phase (Wilchek and Bayer, 1988; Uhlen, 1989). This allows rigid washing and fast separation of specific from unspecific signals. The detection of the bound target molecule is usually performed with a second labeling and detection system independent of the biotin-streptavidin/avidin system, for example with a DIG hapten recognized by a DIG-specific antibody (see Sect. 3.2 and 3.3). It is also possible to exchange the molecules used for binding and detection and to bind target molecules to an antibody-coated solid phase with an antigen and to use the biotin label for the detection reaction via streptavidin/avidin. The detection step is based on the quantitative measurement of the systems described in Part II. The use of coated tubes or microtiter plates compatible with a robotic system allows automation of the quantitative analysis of many samples in parallel.

For the analysis of nucleic acids, further described here, there are several possibilities for incorporating the biotin and detection labels and for combining the labels with the target molecule:

1. Hybridization with two labeled probes: Both labels are fixed to specific nucleic acid sequences and the target molecule is detected in a sandwich hybridization (see Sect. 22.1) (Jalava et al., 1990; Syvänen et al., 1986). The biotin label can be incorporated into a nucleic acid complementary to a part of the target nucleic acid with chemical or enzymatic methods (see Sect. 4). Chemically synthesized oligonucleotides can be biotinylated at the 5' end through amino-linked coupling, at the 3' end by chemical synthesis or the incorporation of biotin-nucleoside triphosphates with terminal transferase, or internally by the incorporation of biotin-phosphoamidites. Plasmids or DNA fragments derived from plasmids can be labeled with the standard techniques using the incorporation of biotin-nucleoside-triphosphates by polymerases. The detection label can be incorporated identically if a hapten such as DIG is used (see Sect. 3.2). Other possibilities for detection probes are described in Sects. 5–10.
2. Hybridization with one labeled probe: The biotin capture probe is prepared and used similar to sandwich hybridization but the detection label is incorporated into newly synthesized nucleic acid sequences

complementary to the target sequence in a process which is part of the assay. Therefore the labeling reaction is performed individual for each sample. Modified nucleoside-triphosphates, e.g., digoxigenin-labeled nucleotides, are incorporated by polymerases into sequences complementary to the target sequence resulting in a labeled target-specific sequence which can be bound to the solid phase by a specific nucleic acid hybridization step with the biotin-labeled capture probe (Syvänen et al., 1990). The detection label can also be incorporated by labeled primer. If target amplification systems, such as PCR or NASBA (see Part III), are combined with the specific incorporation of modified nucleotides the sensitivity of the assays can be dramatically increased (Lanzillo, 1990; Lion and Haas, 1990; Balaguer et al., 1991).

In an alternative reaction scheme the biotin label is incorporated into newly synthesized nucleic acid, which is bound to the solid phase and detected in a hybridization step with a detection probe (Syvänen et al., 1988; Harju et al., 1990).

3. Assays without hybridization: The biotin capture label and the detection label are incorporated simultaneously into the same nucleic acid strand by a polymerase-catalyzed reaction. This assay system is very fast because the hybridization step is eliminated. However, specificity is only generated by the specific primer binding reaction. This type of assay can also be used for the measurment of components of the polymerase reaction, especially for the quantification of the enzymatic activity of polymerase or the analyses of promoter structures (Eberle and Seibl, 1992).

A DNA ligase can also be used to join two differentially labeled nucleic acids dependent on the presence of the target nucleic acid (Nickerson et al., 1990).

Also, target amplification systems can combine the biotin capture label and the detection label into one nucleic acid stretch. In a PCR or LCR assay (see Part III) the resultant double-stranded DNA fragment can be bound to the solid phase by biotin and detected via a detection label if the individual PCR primer or ligated oligonucleotides are labeled with biotin and, for example, digoxigenin or fluorescein (Wahlberg et al., 1990; Landgraf et al., 1991; Dahlen et al., 1991).

As an example the quantitative detection of digoxigenin-labeled nucleic acid in streptavidin-coated microtiter plates after sandwich hybridization is described in detail below.

22.5.2 Quantitative Detection of DIG-Labeled Nucleic Acid

− Anti-DIG:POD conjugate (Boehringer Mannheim)   **Standard**
− ABTS substrate and buffer (Boehringer Mannheim)   **reagents**
− Bovine serum albumin (Boehringer Mannheim)
− Tris-HCl (Boehringer Mannheim)
− EDTA, NaCl, Na_2HPO_4, NaH_2PO_4, NaOH, Na-citrate (Merck)

22.5.2

Standard — TE buffer: 10 mM Tris-HCl, 1 mM EDTA, pH 7.5/25 °C
solutions — Hybridization solution: 50 mM sodium phosphate buffer, pH 5.4, 0.75 M NaCl, 0.074 M Na-citrate, 0.05% bovine serum albumin
— Conjugate buffer: 100 mM Tris-HCl, pH 7.5, 0.9% NaCl, 1% bovine serum albumin
— NaCl solution (0.9%)
— ABTS solution (1.9 mM)

Sandwich The example describes the sandwich assay using biotin- and digoxigenin-
hybridi- labeled oligonucleotides. The sensitivity of this assay is in the range of
zation 100–200 pg analyte DNA. This corresponds to 0.5–1 ng/ml (0.5–1 fmol). It is also possible to use labeled DNA fragments for a sandwich hybridization. In this case it is useful to add 40% formamide to the hybridization solution.

1. Dilute the analyte DNA in TE buffer in the range of 10–500 ng/ml.
2. For denaturation mix 90 µl of the DNA dilution with 10 µl 5 N NaOH and incubate 10 min at room temperature.
3. Pipette 20 µl of the denaturation mix into a well of a streptavidin-coated microtiter plate and add immediately 180 µl of the hybridization solution containing 200 ng/ml of the digoxigenin-labeled oligonucleotide and 200 ng/ml of the biotin-labeled oligonucleotide. Incubate the hybridization mixture 3 h at 37 °C.
4. Suck off the hybridization mix and wash the well five times with 0.9% NaCl.
5. Add 200 µl/well anti-DIG:POD conjugate (200 mU/ml) in conjugate buffer and incubate 1 h at 37 °C.
6. Wash the plate again five times with 0.9% NaCl.
7. Add 200 µl ABTS substrate solution and incubate 30–60 min at 37 °C.
8. The absorbance is measured with an ELISA reader at 405 nm.

References

Balaguer P, Terouanne B, Boussioux AM, Nicolas JC (1991) Quantification of DNA sequences obtained by polymerase chain reaction using a bioluminescent adsorbent. Anal Biochem 195:105–110
Dahlen P, Iitiae A, Mukkula VM, Hurskainen P, Kwiatkowski M (1991) The use of europium (Eu3+) labeled primers in PCR amplification of specific target DNA. Mol Cell Probes 5:143–149
Harju L, Jaenne P, Kallio A, Laukkanen ML, Lautenschlager I, Mattinen S, Ranki A, Ranki M, Soares VRX, Soederlund H, Syvänen AC (1990) Affinity-based collection of amplified viral DNA: application to the detection of human immunodeficiency virus type 1, human cytomegalovirus and human papillomavirus type 16. Mol Cell Probes 4:223–235
Jalava T, Kallio A, Leinonen AW, Ranki M (1990) A rapid solution hybridization method for detection of human papillomavirus. Mol Cell Probes 4:341–352
Landgraf A, Reckmann B, Pingoud A (1991) Direct analysis of polymerase chain reaction products using enzyme-linked immunosorbent assay techniques. Anal Biochem 198:86–91

Lanzillo JL (1990) Preparation of digoxigenin-labeled probes by the polymerase chain reaction. BioTechniques 8:621−622

Lion T, Haas OA (1990) Nonradioactive labeling of probe with digoxigenin by polymerase chain reaction. Anal Biochem 188:335−337

Nickerson DA, Kaiser R, Lappin S, Stewart J, Hood L, Landegren U (1990) Automated DNA diagnostics using an ELISA-based oligonucleotide ligation assay. Proc Natl Acad Sci USA 87:8923−8927

Seibl R, Eberle J (1992) Quantification of reverse transcriptase activity by ELISA. J Virol Meth, in press

Syvänen AC, Laaksonen M, Soederlund H (1986) Fast quantification of nucleic acid hybrids by affinity-based hybrid selection. Nucleic Acids Res 14:5037−5048

Syvänen AC, Bengtstroem N, Tenhunen J, Soederlund H (1988) Quantification of polymerase chain reaction products by affinity-based hybrid collection. Nucleic Acids Res 16:11327−11338

Syvänen AC, Aalto-Setaelae K, Harju L, Kontula K, Soederlund H (1990) A primer-guided nucleotide incorporation assay in the genotyping of apolipoprotein E. Genomics 8:684−692

Uhlen M (1989) Magnetic separation of DNA. Nature 340:733−734

Wahlberg J, Lundeberg J, Hultman T, Holmberg M, Uhlen M (1990) Rapid detection and sequencing of specific in vitro amplified DNA sequences using solid phase methods. Mol Cell Probes 4:285−297

Wilchek M, Bayer EA (1988) Review: the avidin-biotin complex in bioanalytical applications. Anal Biochem 171:1−32

Appendix I Suppliers of Reagents

AB Sangtec Medical
P.O. Box 20045
S-16102 Bromma
Sweden

Aldrich Chemical Co.
1001 West Saint Paul Avenue
Milwaukee, WI 53233
USA

Aldrich-Chemie GmbH
Riedstraße 2
Postfach 1120
D-7924 Steinheim
Germany

American Type Culture Collection
12301 Parklawn Drive
Rockville, MD 20852
USA

Amersham International
White Lion Road
Amersham
Buckinghamshire HP7 9LL
UK

Amersham International
Lincoln Place
Green End
Aylesbury
Buckinghamshire HP20 2TP
UK

Amersham Buchler GmbH
Gieselweg 1
D-3300 Braunschweig
Germany

Amersham Corp.
2636 South Clearbrook Drive
Arlington Heights, IL 60005
USA

Amersham Canada Ltd.
1166 South Service Road West
Oakville, Ontario L6L 5T7
Canada

Applied Biosystems, Inc.
850 Lincoln Center Drive
Foster City, CA 94404
USA

Aurion
ImmunoGold Reagents and Accessories
Costerweg 5
NL-6702-AA Wageningen
The Netherlands

Baker
222 Red School Lane
Phillipsburg, NJ 08856
USA

Baxter Healthcare
909 Orchard Street
Mundelein, IL 60060
USA

BDH Chemicals Ltd.
Broom Road
Poole
Dorset BH12 4NN
UK

BDH Chemicals Ltd.
c/o Gallard Schlessinger Industries, Inc.
584 Minncola Avenue
Carle Place, NY 11514-1731
USA

Biocell
Cardiff Business Technology Centre
Senghenydd Road
Cardiff CF2 4AY
UK

Biorad
Chemical Division
3300 Regatta Boulevard
Richmond, CA 94804
USA

Bio-Rad Laboratories GmbH
Heidemannstraße 164
Postfach 45 01 33
D-8000 München 45
Germany

Boehringer Mannheim GmbH
Sandhofer Straße 116
Postfach 31 01 20
D-6800 Mannheim 31
Germany

Boehringer Mannheim UK Ltd.
Bell Lane
Lewes
East Sussex BN7 1LG
UK

Boehringer Mannheim Biochemicals
P.O. Box 50414
Indianapolis, IN 46250
USA

Boehringer Mannheim Canada Ltd.
200 Micro Boulevard
Laval, Quebec H7V 3Z9
Canada

Clontech Laboratories, Inc.
4030 Fabian Way
Palo Alto, CA 94303
USA

Coy Laboratory Products, Inc.
14500 Coy Drive
Grass Lake, MI 49240
USA

Cyberfluor, Inc.
179 John Street
Toronto, Ontario M5T 1X4
Canada

Daikin Industries Ltd.
Umeda Center Building
2-4-12, Nakazaki-Nishi
Kita-Ku, Osaka, 530
Japan

Dako Ltd.
16 Manor Courtyard
Hughenden Avenue
High Wycombe
Bucks HP13 5RE
UK

Digene Diagnostics
2301-B Broadbirch Drive
Silver Spring, MD 20904
USA

DuPont Co.
Biotechnology Systems
P24 Barley Mill Plaza
Wilmington, DE 19898
USA

Dynal AS
P.O. Box 158
Skoyen
N-0212 Oslo 2
Norway

Enzo Diagnostics, Inc.
325 Hudson Street
New York, NY 10013
USA

Enzo Biochem
c/o Ortho Diagnostic Systems GmbH
Karl-Landsteiner-Straße 1
D-6903 Neckargemünd
Germany

Epicentre Technologies
1202 Ann Street
Madison, WI 53713
USA

Farmitalia Carlo Erba SPA
Via C. Imbonati 24
I-20159 Milano
Italy

Fisher Scientific
2775 Pacific Drive
P.O. Box 4829
Norcross, GA 30091
USA

Fisher Scientific
52 Fadem Road
Springfield, NJ 07081
USA

Fluka Chemical Corp.
980 South Second Street
Ronkonkoma, NY 11779-7238
USA

Fluka Chemie AG
Industriestraße 25
P.O. Box 260
CH-9470 Buchs
Switzerland

Flow Laboratories Ltd.
Woodcock Hill, Harefield Road
Rickmansworth, Ontario WOZ 1PQ
Canada

Fresenius AG
Borkenberg 14
D-6370 Oberursel
Germany

Glen Research Corp.
44901 Falcon Place
Sterling, VA 22170
USA

Hendley
Oakwood Hill Industrial Estate
Loughton, Essex
UK

ICN Biomedicals
2485 Guenette Street
St. Laurent, PO H4R 2R2
Canada

Igen, Inc.
1530 E. Jefferson Street
Rockville, MD 20852
USA

Leica GmbH
Lilienthalstraße 39–45
Postfach 1651
D-6140 Bensheim 1
Germany

Life Technologies/BRL, Inc.
8717 Grovemont Circle
P.O. Box 6009
Gaithersburg, MD 20877
USA

Life Technologies/BRL, Inc.
Burlington, Ontario L7P 1A1
Canada

Life Technologies/BRL GmbH
Dieselstraße 5
Postfach 1212
D-7514 Eggenstein-Leopoldshafen
Germany

Life Technologies/BRL, Inc.
Trident House, Renfrew Road
P.O. Box 35
Paisley PA3 4EF
UK

Mallinckrodt, Inc.
Paris By-Pass
P.O. Box M
Paris, KY 40361
USA

Marabuwerke
Erwin Martz GmbH & Co.
D-7146 Tamm
Germany

Medor GmbH
Laboratorien für Biochemie und
Klinische Chemie
Arzberger Straße 5
D-8036 Herrsching
Germany

E. Merck
Frankfurter Straße 250
Postfach 4119
D-6100 Darmstadt
Germany

E. Merck (EM) Industries, Inc.
5 Skyline Drive
Nawthorne, NY 10532
USA

New England Biolabs
32 Tozer Road
Beverly, MA 01915-5599
USA

New England Biolabs
Postfach 2750
D-6231 Schwalbach
Germany

Novabiochem AG
Post Box
CH-4448 Läufelfingen
Switzerland

Perkin-Elmer Cetus
761 Main Avenue
Norwalk, CT 06859
USA

Perkin-Elmer Cetus
Bahnhofstraße 30
D-8011 Vaterstetten
Germany

Pharmacia LKB Biotechnology AB
Björngatan 30
S-75182 Uppsala
Sweden

Pharmacia Biosystems GmbH
Munzinger Straße 9
D-7800 Freiburg
Germany

Pharmacia LKB
800 Centennial Avenue
P.O. Box 1327
Piscataway, NJ 08855-1327
USA

Pharmacia LKB
9319 Gaither Road
Gaithersburg, MD 20877
USA

Pharmacia LKB Canada, Inc.
500, Boulevard Morgan
Baie d'Urfé, PQ H9X 3V1
Canada

Polysciences, Inc.
400 Valley Road
Warrington, PA 18976-2590
USA

Schleicher & Schuell GmbH
Postfach 4
D-3354 Dassel
Germany

Schleicher & Schuell, Inc.
10 Optical Avenue
Keene, NH 03431
USA

Serva, Inc.
Fine Biochemicals
50 A&S Drive
Paramus, NJ 07652
USA

Sigma Chemical Co.
P.O. Box 14508
St. Louis, MO 63178-9916
USA

Sigma Chemical Co.
2255-B Queen Street East
P.O. Box 10
Toronto, Ontario M4E 9Z9
Canada

Sigma Chemie GmbH
Grünwalder Weg 30
D-8024 Deisenhofen
Germany

Sigma Chemical Company, Ltd.
Fancy Road
Poole
Dorset BH17 7NH
UK

Southern Biotechnology Associates, Inc.
Birmingham, AL 35226
USA

Southern Biotechnology Associates, Inc.
c/o Atlanta Chemie-
und Handels-Ges. mbH
Carl-Benz-Straße 7
Postfach 10 52 60
D-6900 Heidelberg
Germany

Stratagene
11099 North Torrey Pines Road
La Jolla, CA 92037
USA

Syngene, Inc.
3252 Holiday Court, Suite 101
La Jolla, CA 92037
USA

Tropix, Inc.
47 Wiggins Avenue
Bedford, MA 01730
USA

US Biochemical
P.O. Box 22400
Cleveland, OH 44122
USA

Vector Laboratories, Inc.
30 Ingold Road
Burlingame, CA 94010
USA

Whatman Ltd.
Springfield Mill
Maidstone, Kent ME14 2LE
UK

Worthington Biochemical Corp.
Hals Mill Road
P.O. Box 650
Freehold, NJ 07728
USA

Worthington Biochemical Corp.
c/o Millipore GmbH
Siemensstraße 20
D-6078 Neu-Isenburg
Germany

Appendix II Suppliers of Equipment

American Bionetics, Inc.
21377 Cabot Boulevard
Hayward, CA 94545
USA

Anachem
Charles Street
Luton
Beds LU2 0EB
UK

Applied Biosystems, Inc.
850 Lincoln Center Drive
Foster City, CA 94404
USA

Applied Biosystems, Inc.
Kelvin Close
Birchwood Science Park
Warrington
Cheshire WA3 7PB
UK

Becton Dickinson
Diagnostic Instrument Systems
7 Loveton Circle
P.O. Box 999
Sparks, MD 21152−0999
USA

Berthold
Calmbacher Straße 22
Postfach 160
D-7547 Wildbad 1
Germany

Betagen, Inc.
100 Beaver Street
Waltham, MA 02154
USA

Bio-Rad Laboratories
220 Maple Avenue
P.O. Box 708
Rockville Center, NY 11571
USA

Dow Chemical Co.
c/o Genetic Research Instrumentation Ltd.
Gene House, Dunmow Road
Felsted, Dunmow
Essex CM6 3LD
USA

Dynatech Deutschland GmbH
Justinus-Kerner-Straße 32
D-7306 Denkendorf
Germany

Fisons/Gallenkamp
Bishop Meadow Road
Loughborough
Leicestershire LE11 0RG
UK

Flow Laboratories Ltd.
Woodcock Hill, Harefield Road
Rickmansworth, Ontario W0Z 1PQ
Canada

GATC GmbH
Fritz-Arnold-Straße 23
D-7750 Konstanz
Germany

Grant (BDH/Merck)
Broom Road
Poole
Dorset BH12 4NN
UK

Hamamatsu Photonics Deutschland GmbH
Arzberger Straße 10
D-8036 Herrsching
Germany

Hamilton Bonaduz AG
Postfach 26
CH-7402 Bonaduz
Switzerland

Heraeus Holding GmbH
Fritz-Berne-Straße 47
D-8000 München 60
Germany

Hoefer Scientific Instruments
654 Minnesota Street
P.O. Box 77387
San Francisco, CA 94107
USA

Hölzel GmbH
Bernöder Weg 7
D-8250 Dorfen
Germany

Intelligenetics, Inc.
700 E. El Camino
Mountain View, CA 94040
USA

Kontron Instruments
Postfach 13 80
D-8056 Neufahrn
Germany

Medor GmbH
Laboratorien für Biochemie
und Klinische Chemie
Arzberger Straße 5
D-8036 Herrsching
Germany

Millipore GmbH
Hauptstraße 87
D-6236 Eschborn
Germany

National Biosciences
3650 Annapolis Lane
Plymouth, MN 55447
USA

Omega Specialty Instrument Co.
4 Kidder Road, Unit 5
Chelmsford, MA 01824
USA

OWL Scientific Plastics, Inc.
P.O. Box 566
Cambridge, MA 02139
USA

Perkin-Elmer Cetus
761 Main Avenue
Norwalk, CT 06859
USA

Perkin-Elmer Cetus
Bahnhofstraße 30
D-8011 Vaterstetten
Germany

Pharmacia Biosystems GmbH
Munzinger Straße 9
D-7800 Freiburg
Germany

Pharmacia LKB Ltd.
Midsummer Boulevard
Central Milton Keynes
Buckunghamshire MK9 3HP
UK

Schleicher & Schuell GmbH
Postfach 4
D-3354 Dassel
Germany

DATE DUE

DEMCO, INC. 38-2971